GENETIC ASPECTS OF PLANT NUTRITION

DEVELOPMENTS IN PLANT AND SOIL SCIENCES

VOLUME 8

Also in this series

1. J. Monteith and C. Webb, eds.,
   Soil Water and Nitrogen in Mediterranean-type Environments.
   1981. ISBN 90-247-2406-6
2. J. C. Brogan, ed.,
   Nitrogen Losses and Surface Run-off from Landspreading of Manures.
   1981. ISBN 90-247-2471-6
3. J. D. Bewley, ed.,
   Nitrogen and Carbon Metabolism.
   1981. ISBN 90-247-2472-4
4. R. Brouwer, I. Gašparíková, J. Kolek and B. C. Loughman, eds.,
   Structure and Function of Plant Roots.
   1981. ISBN 90-247-2405-8
5. Y. R. Dommergues and H. G. Diem, eds.,
   Microbiology of Tropical Soils and Plant Productivity.
   1982. ISBN 90-247-2624-7
6. G. P. Robertson, R. Herrera and T. Rosswall, eds.,
   Nitrogen Cycling in Ecosystems of Latin America and the Caribbean.
   1982. ISBN 90-247-2719-7
7. D. Atkinson et al., eds.,
   Tree Root Systems and Their Mycorrhizas.
   1983. ISBN 90-247-2821-5
9. J. R. Freney and J. R. Simpson, eds.,
   Gaseous Loss of Nitrogen from Plant-Soil Systems.
   1983. ISBN 90-247-2820-7

Series ISBN 90-247-2405-8

GENETIC ASPECTS OF PLANT NUTRITION

Proceedings of the First International Symposium on
Genetic Aspects of Plant Nutrition
Organized by the Serbian Academy of Sciences and Arts,
Belgrade, August 30–September 4, 1982

Edited by

M. R. SARIĆ
B. C. LOUGHMAN

Chapters indicated with an asterisk in the table of contents were first published in *Plant and Soil,* Vol. 72 (1983)

1983 **MARTINUS NIJHOFF PUBLISHERS**
a member of the KLUWER ACADEMIC PUBLISHERS GROUP
THE HAGUE / BOSTON / LANCASTER

Distributors

for the United States and Canada: Kluwer Boston, Inc., 190 Old Derby Street, Hingham, MA 02043, USA

for all other countries: Kluwer Academic Publishers Group, Distribution Center, P.O.Box 322, 3300 AH Dordrecht, The Netherlands

Library of Congress Cataloging in Publication Data

```
Main entry under title:

Genetic aspects of plant nutrition.

   (Developments in plant and soil sciences ; v. 8)
   Includes index.
   1. Plants--Nutrition--Genetic aspects--Congresses.
I. Sarić, Miloje R.  II. Loughman, B. C.  III. Srpska
akademija nauka i umetnosti.  IV. Series.
QK867.G38  1983       581.1'335       83-2333
ISBN 90-247-2822-3
```

ISBN-13: 978-94-009-6838-7   e-ISBN-13: 978-94-009-6836-3
DOI: 10.1007/978-94-009-6836-3

Copyright

© 1983 by Martinus Nijhoff/Dr W. Junk Publishers, The Hague.
Softcover reprint of the hardcover 1st edition 1983

All rights reserved. No part of this publication may be reproduced, stored in a retrieval system, or transmitted in any form or by any means, mechanical, photocopying, recording, or otherwise, without the prior written permission of the publishers,
Martinus Nijhoff/Dr W. Junk Publishers, P.O. Box 566, 2501 CN The Hague, The Netherlands.

Contents

Preface IX

M. R. Sarić, Theoretical and practical approaches to the genetic specificity of mineral nutrition of plants* 1

Section I: Cytological and anatomical changes in different genotypes caused by altered nutrient supply

Ch. Hecht-Buchholz, Light and electron microscopic investigations of the reactions of various genotypes to nutritional disorders* 17

D. Kramer, Genetically determined adaptations in roots to nutritional stress: correlation of structure and function* 33

S. M. Fatalieva, E. D. Gulieva and O. F. Melikova, Ultrastructure of the root cells of two pea genotypes depending on Se in the growth medium 41

Section II: Absorption, translocation and accumulation of ions in different genotypes

R. B. Clark, Plant genotype differences in the uptake, translocation, accumulation, and use of mineral elements required for plant growth* 49

W. D. Jeschke, Cation fluxes in excised and intact roots in relation to specific and varietal differences* 71

E. Alcántara and M. D. de la Guardia, Genotypic differences in calcium and magnesium nutrition in sunflower 87

A. H. de Boer and H. B. A. Prins, A study of the electrophysiological organization in roots of two *Plantago* species: direct measurement of ion transport to the xylem using excised roots 93

R. R. Duncan, Concentration of critical nutrients in tolerant and susceptible sorghum lines for use in screening under acid soil field conditions 101

N. V. Gujova, S. M. Fatalieva and A. Sh. Kerimova, The effect of selenium on ion uptake at two nutrient levels 105

M. Holobradá, Differences in sulphate and phosphate uptake and utilization within *Zea mays* L. species 111

W. J. Horst, Factors responsible for genotypic manganese tolerance in cowpea (*Vigna unguiculata*)* 117

B. Jocić and M. R. Sarić, Efficiency of nitrogen, phosphorus, and potassium use by corn, sunflower, and sugarbeet for the synthesis of organic matter* 123

V. Mego, Differences in phosphate absorption in various barley genotypes 129

A. Maggioni and Z. Varanini, Free-space binding and uptake of ions by excised roots of grapevines ... 133
I. Michalík, The accumulation of phosphate in roots of different genotypes of maize ... 139
N. E. Nielsen and J. K. Schjørring, Efficiency and kinetics of phosphorus uptake from soil by various barley genotypes* ... 145
S. Pettersson and P. Jensén, Variation among species and varieties in uptake and utilization of potassium* ... 151
J. Wieneke, Application of root zone feeding for evaluation of ion uptake and efflux in soybean genotypes* ... 159
T. Zaharieva, Effect of genotype and iron, applied to soil, on the chemical composition and yield of corn plants ... 165

Section III: The influence of mineral nutrition on physiological, and biochemical processes of genotypes

B. C. Loughman, S. C. Roberts and C. I. Goodwin-Bailey, Varietal differences in physiological and biochemical responses to changes in the ionic environment* ... 173
M. Popp, Genotypic differences in the mineral metabolism of plants adapted to extreme habitats* ... 189
A. Bottacin, M. Saccomani, P. Spettoli and G. Cacco, NaCl-induced modifications of nitrogen absorption and assimilation in salt tolerant and salt resistant millet ecotypes (*Pennisetum typhoideum* L. Rich.) ... 203
P. Büscher and N. Koedam, Soil preference of populations of genotypes of *Asplenium trichomanes* L. and *Polypodium vulgare* L. in Belgium as related to cation exchange capacity* ... 209
F. S. Chapin, III, Adaptation of selected trees and grasses to low availability of phosphorus* ... 217
H. Eggers and W. D. Jeschke, Comparison of K^+–Na^+ selectivity in roots of Fagopyrum and Triticum ... 223
C. H. Hommels, A. A. Sterk and O. Gy. Tanczos, Genetic differentiation in Taraxacum and its relation to mineral nutrition ... 229
D. Jelenić and V. Hadži-Tašković Šukalović, The effect of nitrogen on the activity of some enzymes of nitrogen metabolism during ontogenesis of maize kernel hybrids ... 237
H. Königshofer, Changes in ion composition and hexitol content of different *Plantago* species under the influence of salt stress* ... 243
K. Konstantinov, V. Lazić, M. Denić, Č. Radenović and V. Furtula, Influence of chlorate ions on some characteristics of maize seedlings with different protein content ... 251

B. Krstić and M. R. Sarić, Efficiency of nitrogen utilization and photosynthetic rate in C_3 and C_4 plants — 255

D. Kuiper, Genetic differentiation of various physiological parameters of *Plantago major* and their role in strategies of adaptation to different levels of mineral nutrition — 261

L. Nátr, Genotypic differences in growth and photosynthesis of young barley plants — 269

J. Repka, The effects of mineral nutrition on the photosynthetic and respiratory activities of leaves of winter wheat and maize varieties — 279

J.-P. Wacquant and N. Bouab, Nutritional differentiation within the species *Dittrichia viscosa* W. Greuter, between a population from a calcareous habitat and another from an acidic habitat* — 285

Section IV: The influence of mineral nutrition on yield and quality of different genotypes

K. Mengel, Responses of various crop species and cultivars to fertilizer application* — 295

G. Schilling, Genetic specificity of nitrogen nutrition in leguminous plants* — 311

P. Andonova and T. Kudrev, Changes in contents of N, P and K in maize hybrids at different nutrient regimes — 325

W. G. Braakhekke, A triaxial ratio diagram and its use in comparative plant nutrition — 331

R. D. Graham, W. J. Davies, D. H. B. Sparrow and J. S. Ascher, Tolerance of barley and other cereals to manganese-deficient calcareous soils of South Australia — 339

B. C. Hemming and G. A. Strobel, Bacterial disease and the iron status of plants — 347

B. Lásztity, Fertilizers and nutrient relations in some genotypes of cereals — 359

Z. Sarić and A. H. Fawzia, Nitrogen fixation in soybean depending on variety and *R. japonicum* strain — 365

B. Todorčić, B. Bertić, M. Šeput, V. Vukadinović, I. Folivarski and V. Laktić, Characteristics of sugar beet varieties on chernozem semigley in Baranja — 371

Section V: Genetical investigations concerned with selection of genotypes for a more effective use of mineral elements

W. H. Gabelman and G. C. Gerloff, The search for and interpretation of genetic controls that enhance plant growth under deficiency levels of a macronutrient* — 379

P. B. Vose, Rationale of selection for specific nutritional characters in crop improvement with *Phaseolus vulgaris* L. as a case study* — 395

M. Dambroth and N. El Bassam, Low input varieties: definition, ecological requirements and selection*   409

G. Alagarswamy and N. Seetharama, Biomass and harvest index as indicators of nitrogen uptake and translocation to the grain in sorghum genotypes   423

B. Bochev, E. Neikova-Bocheva, G. Ganeva and T. Frolozhki, Genetic basis of mineral nutrition in *Triticum aestivum*. II. Effect of the cytoplasm on the absorption of nutrient elements   429

G. Cacco, G. Ferrari and M. Saccomani, Genetic variability of the efficiency of nutrient utilization by maize (*Zea mays* L.)   435

J. R. Caradus, Genetic differences in phosphorus absorption among white clover populations*   441

P. J. Goodman, Genetic variation in nitrogen nutrition of grasses and cereals: possibilities of selection   447

P. R. Furlani, R. B. Clark, W. M. Ross and J. W. Maranville, Variability and genetic control of aluminium tolerance in sorghum genotypes   453

E. Kiss, A. Bálint, K. Debreczeni and J. Sutka, Genetic basis of N-utilization in wheat   463

V. Kovačević, The ear-leaf percentage of nitrogen, phosphorus and potassium in maize (*Zea mays* L.) inbred lines and their diallel progeny   471

I. Mihaljev and R. Kastori, Genotypic variation and inheritance of mineral element content in winter wheat   477

M. G. Zaitseva and N. K. Zubkova, Genetical differences in the cation accumulation capacity of mitochondria   483

V. K. Shumny and B. I. Tokarev, Genetic variability in the systems of absorption and utilization of mineral elements. A review   487

Preface

The idea of addressing the problem of the genetic specificity of mineral nutrition at an international level arose four years ago in a proposal for this topic to be included in the program of the II Congress of the Federation of European Societies for Plant Physiology (FESPP) as a separate section. The Organising Committee of the II Congress of FESPP which was held in Santiago de Compostella in 1980 arranged a special session and it was clearly successful. A special scientific meeting where the genetic aspects of plant nutrition in their widest sense could be presented and discussed comprehensively appeared to be necessary and that is how this Symposium came to be organized by the Serbian Academy of Sciences and Arts.

Much progress has already been achieved in this field, and bearing in mind the importance of this problem, particularly at the present moment, it is necessary for us both to acquaint ourselves with what has been achieved so far, and even more to direct attention and effort to the fundamental problems for the future.

The multidisciplinary approach necessary for investigating the genetic basis for differences in capacities for absorption, translocation and utilization of ions on the one hand and the tolerance of excess nutrients and toxic ions on the other is well illustrated in the list of contents and authors in this volume. Whole plant physiologists, cell biologists, soil scientists, geneticists and plant breeders all contribute and in some individual papers the expertise of the joint authors covers a very wide field. Perhaps our discussions would have benefited by the presence of more plant breeders but it was obvious that a number of the laboratory-based contributors particularly enjoyed their contact with those whose main interest is field work and *vice versa*. In deciding to publish the volume we had the benefit of discussions with both the Serbian Academy and Mr. Plaizier, the representative of Martinus Nijhoff. Our aim has been to produce the volume with as little delay as possible and the cooperation of all concerned has been much appreciated.

We are very grateful for the excellent facilities placed at our disposal by the Academy and for the hospitable way in which they received us. We hope that this Symposium will be the first of many on this topic and we look forward to meeting again in the United States in 1985.

M. R. SARIĆ
B. C. LOUGHMAN

Theoretical and practical approaches to the genetic specificity of mineral nutrition of plants

M. R. SARIĆ
Institute of Biology, Faculty of Natural Sciences, 21000 Novi Sad, Yugoslavia

Key words   Genetic specificity   Mineral nutrition   Organic matter   Yield

Summary   Mineral nutrition of plants is one of the most important factors controlling biomass production. However, the efficiency of utilizing certain elements of mineral nutrition in biomass production is highly related to the genetic specificity of plants. The present paper deals with problems and former results regarding plant mineral nutrition presented from the genetic aspects. Particular attention has been devoted to the increased efficiency of using both the natural fertility of soils and mineral fertilizers by creating and utilizing suitable cultivars and hybrids, increased efficiency of using mineral nutrients under certain ecological conditions, plant-specific role of microorganisms in enriching soil with nitrogen and soluble forms of other elements, role of genetic specificity of mineral nutrition in plants in solving the problems of environmental pollution, principles of evaluating the genetic specificity of mineral nutrition in plants, genotype features influencing uptake of mineral nutrients, criteria for evaluating the genetic specificity of mineral nutrition of plants, and also to the methods for selecting genotypes for specific soil types, and mineral nutrition.

Introduction

The problem of genetic, *i.e.* the varietal specificity of mineral nutrition, was emphasized about sixty years ago. During recent years, we have examined this problem from both the theoretical and practical point of view. Due to progress in the experimental approach and also to other improvements, this problem has been studied from various methodological aspects in certain plant species at various levels of plant organization. The most important reason for the currently increasing interest in studying this problem is the energy crisis, *i.e.* the economic situation resulting in the permanent increase in prices of mineral fertilizers, as well as in investments in soils which are only partly arable.

The papers dealing with this problem have been presented in comprehensive reviews in which the results obtained by numerous investigators are cited. The genetic specificity of mineral nutrition has been discussed in general[5,15,17,23], as a particular problem in creating the genotypes suitable for certain nutritional conditions[3,7,27,30], from the aspect of individual plant species[2,24,25,26], and specific elements[1,8,18]. In past years the problem of the genetic specificity of mineral nutrition in plants was emphasized at scientific meetings. In addition, special publications dealing with mineral nutrition have been issued recently[10,11,14,21,31]

On the basis of the results obtained it can be concluded that individual plant

M.R. Sarić and B.C. Loughman (eds.), Genetic Aspects of Plant Nutrition.
ISBN-13: 978-94-009-6838-7
©1983 *Martinus Nijhoff/Dr W. Junk Publishers, The Hague/Boston/Lancaster.*

species show specific requirements for certain mineral nutrients while the response of higher botanical categories may be discussed in general only. When the specificity of nutrition is considered at the level of a species, however, the plant species belonging to the same higher botanical category, *i.e.* genus of a certain family, should be taken into account. Today, plant physiologists devote much of their attention to the efficiency of both C_3 and C_4 plants in utilizing mineral nutrients. Frequently it is difficult to form a general judgement of the evidence presented since various growing conditions are employed, and also, far more representatives of the C_3 plants are introduced in experiments than those of the C_4 plants. In our opinion, the investigation of the efficiency of the C_3 and C_4 plants in utilizing nitrogen should include not only a great number of plant species belonging to those two plant groups, but also a great number of their cultivars.

There is no doubt that differences occur in the effect of mineral nutrition both among species and among cultivars, *i.e.* genotypes belonging to the same species. Numerous data presented in the reviews contribute to such a statement[4, 6, 9, 12, 13, 16, 19, 20, 22, 28, 29, 31]. The range of the variance of these effects among cultivars is often similar, but it can be greater than among plant species.

We shall leave the participants at the Symposium to give their personal contributions to the subject of the genetic specificity of mineral nutrition while we are only emphasizing the fact that in previous papers, concentration and content of mineral elements in plant tissues have been most frequently used as a criterion for the occurrence of genetic specificity of mineral nutrition in plants. Cultivar specificity of mineral nutrition is manifested through differences in ion uptake, transport, distribution, and accumulation in individual organs, as well as in their re-utilization, efflux, and role in the metabolic processes.

A number of old cultivars, poorly influenced by mineral nutrients, are frequently characterized by a more intensive growth of their vegetative organs. Also, the ratio of carbohydrates utilized for their respiration to that accumulated in their generative organs is higher. Further, they utilize greater amounts of water in their process of transpiration. On the other hand, new idiotypes, in addition to small concentrations of individual ions in their organs, *i.e.* an evident response to applied fertilizers, should have a small leaf area, a high photosynthetic rate per unit of leaf area, intensive photochemical reduction of CO_2 in chloroplasts, low photorespiration, and maximum yield.

So far we have concluded that the mechanism of genetic specificity of mineral nutrition includes many morphological and anatomical features, as well as physiological and biochemical processes influencing the differences in uptake and content of certain elements among genotypes, thereby specifically affecting the structural and metabolic shifts. In general, various criteria have been developed in evaluating the results obtained for the genetic specificity of mineral nutrition in plants. There is plentiful evidence for the response of individual genotypes of different plant species to certain elements. Most of the investigations however are reduced to information about the differences between genotypes having different contents of certain elements. Considering previous results and particularly the enormous economic importance of modern achievements and the approach to this problem, the primary objective of the Symposium is to provide an opportunity for the presentation of the latest views regarding the genetic specificity of mineral nutrition in plants. The topics of the Sections, in our opinion, indicate directions leading toward solving this problem.

I shall attempt to throw light on some fundamental aspects regarding the genetic specificity of mineral nutrition in plants.

The increased efficiency of using both the natural fertility of soils and mineral fertilizers by creating and utilizing suitable cultivars and hybrids

The main reason for our interest in the genetic specificity of mineral nutrition in plants lies in the question of how to make progress in using the natural fertility of both the productive soils and those that are naturally acid, low in nutrients, drainage, *etc.*, as well as mineral fertilizers by growing suitable species, *i.e.* cultivars. According to Wright's monograph[31], 21.0% of the world's land belongs to the first, 40.8% to the second, and 38.2% to the third class. Hence, to increase the agricultural output the application of mineral fertilizers ought to be increased, particularly in underdeveloped countries where poor soils are frequently found. Also, in the same monograph there are data telling us that in the year 2000, the production of nitrogen fertilizer will be four times, and of phosphorus and potassium three times as high as in 1975. The prices of mineral fertilizers will be increasing primarily due to the shortage in energy supply, thus increasing the costs of crop production. Besides, the resources of mineral fertilizers are being depleted to the point which may drastically decrease food production. For that reason, an increase in the efficiency of using mineral fertilizers is a question of particular importance, not only because of the increase in production costs but also because of a limited supply of their raw materials. It can be foreseen that, in addition to increasing prices, their application will be decreasing in developed agricultural regions while increasing in underdeveloped ones. These trends will cause an increase in the overall consumption of fertilizers. Hence, the problem of the genetic specificity of mineral nutrition in plants is present in both developed and under-developed countries. The efficiency of mineral fertilizers may be increased by introducing new techniques in their application and by improving their composition. Our attention should be devoted to creating new cultivars characterized by high potentials for utilizing mineral nutrients.

To understand the distinctions in efficiency of plants in absorbing and utilizing mineral nutrients, clearly outlined conditions of nutrition are needed. Requirements of certain species and cultivars for individual elements of mineral nutrition may be discussed in terms of optimum concentrations of all the elements present, various concentrations of an optimum nutrient solution, various concentrations of individual elements, or deficiency or excess of a certain element in a nutrient solution. To establish the differences among the genotypes in responding to individual mineral elements under all the conditions mentioned above, different genotypes ought to be included in the investigation. Today we use the terms plant-efficient, plant-nonefficient, plant-tolerant, plant-nontolerant, salt-sensitive, salt-tolerant, *etc*. Maybe plant-efficient should refer to those plants having the highest dry matter weight at the lowest concentration of a certain element in a medium, while plant-tolerant should refer to those sustaining high concentrations of certain ions, *i.e.* surviving under either natural or artificial conditions without evidently reducing the production of organic matter.

The application of excessive amounts of mineral fertilizers is not only economically unjustified but such doses may be harmful to a great number of both the plant species and cultivars owing to their specific biological features, and may affect their nutritive values, as well as provoke a shift in the biological balance of nature.

The amounts of certain elements comprising a mineral fertilizer are also of importance. It is necessary to establish the correct fertilizer ratios taking into account the soil fertility and the influence of individual fertilizers, as well as a number of the biotic and abiotic factors affecting metabolic pathways and also the physiological-biochemical characteristics of a plant species. Why do we mention this? Because the synthesis and contents of proteins, oils, sugars, and other organic compounds, occurring specifically in individual plant species, are directly influenced. It is therefore of interest to refer to the following four variants of applying mineral fertilizers per hectare and yield when we discuss the increase in efficiency of using both the natural fertility of soils and applied mineral fertilizers:

— currently used amounts of mineral fertilizers and the higher yield,
— reduced amounts of mineral fertilizers without decrease in current production,
— reduced amounts of mineral fertilizers and the higher yield, or
— increased amounts of mineral fertilizers and the higher yield.

It seems that the assumption mentioned first is the most likely to be true taking into account the existing plant species and cultivars to be grown under certain agroecological conditions. This statement is in agreement with our ten-year investigations on the four plant species. The evidence obtained illustrates the differences between wheat, maize, sugar beet, and sunflower plants regarding mineral nutrition. The averages presented in Fig. 1 show that the most striking effect of mineral fertilizers was recorded in wheat (yield increased by 80% as compared to the control), in maize and sugar beet (by 30%), and sunflower (by only 18%). It should be emphasized that the cited values were obtained on chernozem in the continental part of Yugoslavia. In addition, the specificity of

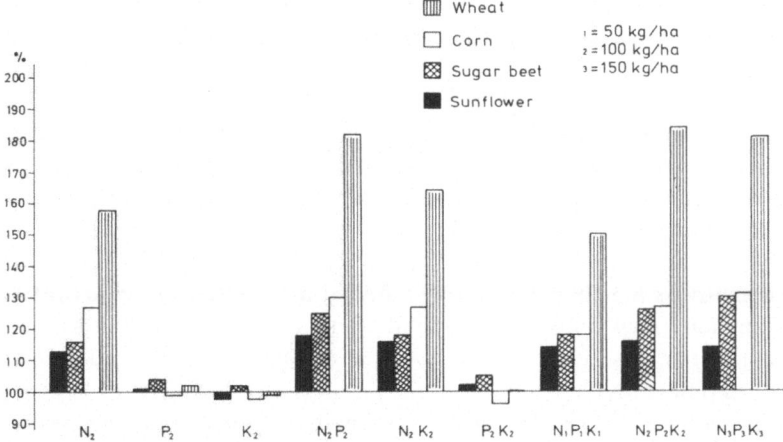

Fig. 1. Effects of N, P and K in some plant species (ten-year average).

physiological-biochemical processes and the influence of mineral fertilizers on the yield quality of the plants under investigation should be taken into account.

The increased efficiency of using mineral nutrients under certain ecological conditions

Today there is an opinion that in the production of organic material it is necessary to use those plant species which give the greatest yield in certain areas of land surfaces. By using solar energy converted by plants into organic compounds which may be further processed into various forms to be utilized by man, we can attain our goal.

Much has been said about the influence of a great number of abiotical factors upon ion uptake and content. Also, it is known that individual cultivars respond differently to environmental conditions. Therefore, it can be assumed that ion uptake is directly related to temperature, light intensity, pH, soil humidity, *etc*. It is worth mentioning that a correlation was found between winter hardiness and the uptake of phosphorus, possibly due to the increased content of phospholipids and other phosphorus compounds present which combine with water to a greater extent, affecting a decrease in the content of free water and increasing the resistance to low temperatures. No evidence, however, has been presented as to whether there is a correlation between winter hardiness and ion uptake. The same is true for light intensity, particularly the correlation between the effect of low light intensity upon individual genotypes and the specificity of uptake of certain ions. In addition, it should be taken into account that these abiotical factors influence not only ion uptake and content but also translocations of ions into individual organs. Therefore, the amounts of certain ions in individual plant organs are the indicator of the metabolic processes under certain living conditions. For this reason, it is particularly important to create cultivars which utilize mineral nutrients more efficiently under unfavourable ecological conditions.

Finally, the differences in pH either of root cell sap, root surface or nutrient solution in certain plant species and cultivars may account for the increase in ion or cation uptake, *i.e.* the increase in efficiency of utilizing mineral nutrients and particularly insoluble ones. Also, the buffering capability of the roots of individual cultivars may control ion uptake and it may be assumed that individual cultivars are characterized by different ionic efflux and influx rates. Such distinctions may influence the uptake and accumulation of ions into individual plant organs. It is not clear, however, why certain anions and cations are absorbed differently.

The importance of the genetic specificity of mineral nutrition of wild species occurring in natural phytocenosis should be emphasized. Little work has been done, however, on this aspect of the problem.

The plant-specific role of microorganisms in enriching soil with nitrogen and soluble forms of other elements

Microorganisms like nitrogen-fixing, ammonifying, denitrifying, and phosphate releasing bacteria inhabit the rhizospheres of certain plant species where favourable living conditions occur. The specificity of the rhizosphere microflora is primarily due to various characteristics of root exudates. It has been found that the more amino acids and other N-compounds are released from the root, the more proteolytic microorganisms appear. In the plants releasing less N-compounds, however, the numbers of nitrogen-fixing bacteria may be increased thus influencing the nitrogen balance in the rhizosphere. This is probably affected by denitrifying bacteria occurring in great numbers in the rhizosphere. If so, the question may arise as to how much their nitrate reductase affects nitrate utilization by various plant species and cultivars.

Due to the activity of the rhizosphere microflora, transformation of all the substances originating either from a root or a microorganism occurs. This results in the formation of plant assimilates, the release of ammonia from amino acids, nitrate formation, and the release of phosphates from phospholipids and nucleotides. The microbiological processes operating in rhizosphere "restore" it to maintain favourable living conditions for plant growth and development. The harmonious relationship between a plant species and microorganisms inhabiting its rhizosphere has developed during the process of evolution. Therefore, the specific character of such a system and differences in both the qualitative and quantitative composition of rhizosphere microflora are evident. The most striking example is the symbiosis between the leguminous plants and the nodule bacteria. As a result, definite associations of the crossing groups of the leguminous plants and the species of the nodule bacteria characterized by different nitrogen fixation rates occur. In addition, it has been found that nitrogen fixation depends upon both plant species and cultivars and the bacterium present. It is reasonable to expect such a specificity between certain cultivars and the strains of microorganisms, belonging to other physiological groups, to occur.

The role of genetic specificity of mineral nutrition in plants in solving the problems of environmental pollution

The results showing that definite plant species and cultivars absorb certain mineral nutrients can have a practical application in solving the problem of environmental pollution, particularly in decontamination and detoxification. The investigations in the field of geophytochemistry concerning the content of certain radioactive elements in different soil types resulted in the detailed description of their location while their content in plants points to the specific nature of their utilization by plants. On the basis of the evidence on the content of radioactive elements in plants, certain plant organs of specific species or entire plants serving as decontaminants might be separated. Thus, for instance, sugar beet accumulates enormous amounts of uranium and radium into its leaves, whereas oats and maize accumulate them into roots to a far greater extent than into their leaves. In addition, in vegetative organs of cereals or leguminous plants, the content of radioactive strontium was found to be 5–6 times as high as in their generative organs. The contents of heavy metals such as Cd, Hg, and Pb, which are highly toxic, differed among individual plant species and among their products. Consequently, certain plant species may be used to decontaminate soils, waters, *etc*. Today particular attention is devoted not only to certain plant species but also to certain cultivars, namely genotypes differing in the uptake of these harmful elements. At this point it should be mentioned that the content of cadmium in certain lettuce cultivars was 300 times as high as in other cultivars. There is also evidence that both the resistant and susceptible cultivars absorb toxic elements, but in the former transfer into leaves occurred. Some authors stated, however, that the resistant cultivars absorbed the toxic substances to a far lesser extent than the susceptible ones. The toxicity of individual elements at relatively low concentrations is relieved by the presence of certain organic acids from the Krebs cycle having the role of chelates. Therefore, some evidence showed that there is a relationship between the occurrence of deleterious elements and evident increase in the contents of citric and malic acids in roots of the resistant cultivars, while aluminium affected the decrease (by 2–2.5 times) in the organic acids in the susceptible cultivars.

The principles of evaluating the genetic specificity of mineral nutrition in plants

The investigation of the genetic specificity of mineral nutrition in plants brings to view the fact that numerous methodological problems frequently make the proper explanation of the results obtained impossible. For that reason, particular attention must be devoted to the agreement among those principles listed in Table 1.

To compare the results obtained concerning the genetic specificity of mineral nutrition in certain plant species, the above mentioned principles ought to be fulfilled.

We shall not deal with the cited principles in detail, but we shall emphasize that, frequently, the results of experiments with genotypes characterized by various durations of the growing season are comparable. Thus for instance, such differences may amount to 20–30 days or more. The differences in the duration of certain phases of development (*e.g.* from flowering to maturity) also occur.

Table 1. Principles in studying the genetic specificity of mineral nutrition in plants

1. Genetically uniform material
2. Identical nutrient solutions or soil type
3. Identical ecological conditions
4. Genotypes with the same growing season
5. The same specific phases in ontogenesis of a certain plant species
6. To use the same most suitable plant organs for evaluating the genetic specificity of mineral nutrition in the species under the investigation
7. In addition to total content, to investigate the forms of individual elements
8. To discuss the results obtained on the basis of concentration and content

The genotype features influencing uptake of mineral nutrients

It is stated by many authors that roots have a prevailing effect in the genetic specificity of mineral nutrition in plants due to the absorption of various nitrogen forms (nitrates and ammonium ions).

Nitrates are reduced prior to taking part in metabolic processes. In the case of the reductive breakdown, a genotype exhibits nitrogen deficiency despite its high absorption rate and content in a plant. Such a disturbance is due to the inadequate influence of an enzyme catalyzing the reduction (nitrate and nitrite reductase). It should be emphasized that the breakdown occurs not only in roots but also in other organs.

A number of genotype features are summarized in Table 2, without considering in detail their significance in the genetic specificity of mineral nutrition in plants.

Criteria for evaluating the genetic specificity of mineral nutrition in plants

Various criteria are employed in evaluating the genetic specificity of mineral nutrition in plants. In Table 3, some of those criteria are presented.

Table 2. Some characteristics of genotypes influencing uptake of mineral nutrients

1. MORPHOLOGICAL FEATURES OF ROOTS: root type (primary, secondary), weight, length, topography, absorption area, cortex diameter.
2. MORPHOLOGICAL FEATURES OF LEAF: shape, size, thickness, position.
3. MORPHOLOGICAL FEATURES OF STEM: diameter, length, number of elements of transporting vessels and their material.
4. RATIO OF ABOVE-GROUND PART TO ROOT
5. PHYSIOLOGICAL PROCESSES: photosynthesis, transpiration, respiration, distribution, reutilization of inorganic and organic compounds
6. BIOCHEMICAL PROCESSES: enzyme activity, synthesis pathway of organic compounds (sugars, proteins, lipids), contents of phytohormones, amino and organic acids.
7. POLYPLOIDY AND HYBRIDITY LEVELS

It has come out clearly, however, taking into consideration the above mentioned, that the concentration and content of individual mineral nutrients in plant tissues are the most striking criteria in evaluating the genetic specificity of mineral nutrition, primarily in young plants. A far smaller number of results discuss the concentration and content of individual ions during the ontogenesis of a plant species. The evaluation of the genetic specificity of mineral nutrition of different genotypes is also important from the practical point of view enabling the balance between their mineral nutrition and both yield and quality to be achieved. Unfortunately, this criterion is less employed since it is necessary to grow plants during the whole growing season, under either natural or artifical conditions.

Methods for selecting genotypes for certain soil types and mineral nutrition

It is confirmed today that great differences exist between genotypes regarding the contents of certain mineral nutrients. Such differences influence a great

Table 3. Criteria for evaluating the genetic specificity of mineral nutrition in plants

1. Cytological and anatomical features of cells and subcellular units
2. Morphological features of plants
3. Uptake, exudation, translocation, distribution, and re-utilization of ions
4. Concentration and content of mineral nutrients
5. Total content and forms of ions
6. Physiological and biochemical processes
7. Total dry weight
8. Yield and quality

number of structural-morphological characteristics and metabolic processes. Consequently, genetic polymorphism in uptake of mineral nutrients occurs in plants. The origin of such a polymorphism is uncertain. Many authors state that these differences develop during evolution, either by the natural adaptation of species to certain ecological conditions or by weak cumulative mutations. The methods of selection, *i.e.* of the creation of new cultivars characterized by specific requirements for mineral elements, have resulted from the statement mentioned above. In selecting such genotypes other characteristics such as yield and quality should be taken into account. In other words, the balance between the physiological and biochemical processes ought to be maintained in the genotypes responding specifically to mineral nutrition. Due to the complexity of this problem in creating such cultivars uncommon methods in the process of the evolution will be employed also. In addition to creating new genotypes, these investigations will contribute to the explanation of the genetic control of mineral nutrition in plants.

In our opinion, screening should be adjusted to the growing conditions (natural and artificial), plant age (from germination to the end of growing season), and to the criteria employed in evaluating and selecting the genotypes characterized by a particular response to mineral nutrition. The greatest problem lies in the fact that some analyses to be introduced cause fatal injuries to a plant under investigation, while, on the other hand, offsprings for further analyses are needed. Finally, a question arises as to whether the requirements for certain mineral elements are the same in a cultivar, namely a genotype characterized by identical morphological features. Genetic investigations will be directed toward selecting the genotypes characterized by high efficiency in utilizing both the natural fertility of soils and mineral fertilizers. Consequently, high overall production of organic matter, as well as agricultural and economical yields will be attained.

The biological potential of individual plant species for the synthesis of organic matter is conditioned by their genetic characteristics. In past years, and particularly recently science has made great progress in improving the ratio of agricultural to biological yield of new cultivars, whereas the overall synthesis of organic matter has remained almost at the same level. Thus, for instance, the ratio of the biological to agricultural yield of wheat was 2:1, whereas today this ratio is changing to the benefit of kernel prodution. On the other hand, the overall production of organic matter of new cultivars remains the same as it was in old cultivars. Of course, the ratio is influenced by a great number of factors, and particularly by mineral nutrition. Further increase in overall synthesis of organic compounds is rather limited. Therefore, it seemes reasonable that future breeding programs will be directed toward creating genotypes characterized not only by higher production of organic matter but also by a high utilization of mineral nutrients.

In attempting to summarize the possible directions in creating new genotypes, the following four combinations of the requirements of genotypes for certain mineral nutrients and the synthesis of the organic matter may occur:

— genotypes with small concentration of certain mineral nutrients and a high rate of organic matter synthesis,
— genotypes with small concentration of certain nutrients and low rate of organic matter synthesis,
— genotypes with high concentration of certain nutrients and a high rate of organic matter synthesis,
— genotypes with high concentration of certain nutrients and low rate of organic matter synthesis.

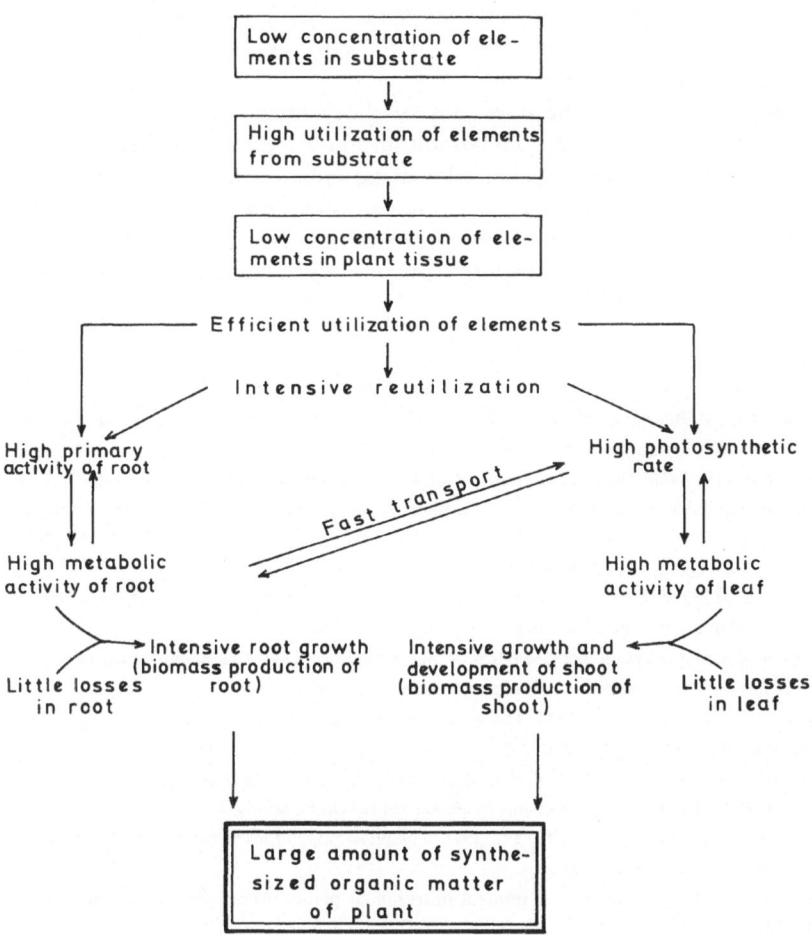

Fig. 2. Characteristics of primary physiologic and biochemical processes in the genotypes with low nutrient concentrations and high organic matter synthesis.

Economically, the first combination mentioned is the one which produces the best results since such genotypes will require smaller amounts of mineral fertilizers accompanied by a high rate of organic matter synthesis that will result in high yield. The characteristics of fundamental physiological and biochemical processes operating in such genotypes are summarized in Fig. 2. In the fourth combination cited, the processes would have the reverse character. It should be emphasized that the concentration and content of any of the mineral nutrients represent no satisfactory criterion since the synthesis of the organic matter is dependent not only upon their concentration but also upon their utilization in metabolic processes. Also, greater or smaller variations of individual nutrients in the genotypes belonging to a single plant species may be expected to occur. In other words, the differences in the nitrogen concentration among genotypes may be smaller than for calcium or some other element. Besides, the genetic variability of concentrations and contents of mineral nutrients possibly is also due to their concentrations in nutrient solution.

Complex as they are, the questions of the genetic specificity of mineral nutrition of plants need close collaboration among plant physiologists, biochemists, geneticists, breeders, microbiologists, and agrochemists. Many details are not included in this paper since it is not possible to treat all aspects of this rather broad subject. However, other papers to be presented at the Symposium and posters will contribute to better understanding of this particular problem.

References

1. Brown J C, Ambler I E, Chaney R L and Foy C D 1972 Differential responses of plant genotypes to micronutrients in agriculture. Soil Sci. Soc. Am. Proc. 389–418.
2. Chapin F S III 1980 The mineral nutrition of wild plants. Annu. Rev. Ecol. Syst. 11, 233–260.
3. Clark R B and Brown J C 1980 Role of the plant in mineral nutrition as related to breeding and genetics. pp. 45–70. *In* Moving up the Yield Curve Advances and Obstacles. Ed. L S Murphy *et al*. Am. Soc. Agron. Madison, WI.
4. Epstein E 1963 Selective ion transport in plants and its genetic control. *In* Desalination Research Conference. Nat. Acad. Sci. Washington. pp 284–298.
5. Epstein E 1972 Physiological genetics of plant nutrition. *In* Mineral Nutrition of Plants: Principles and Perspectives. Ed. E. Epstein. New York. pp 325–344.
6. Epstein E and Jefferies R L 1964 The genetic basis of selective ion transport in plants. Annu. Rev. Plant Physiol. 15, 169–184.
7. Epstein E, Norlyn J D, Rush D W, Kingsbury R W, Kelley D B, Cunningham G A and Anne F Wrona 1980 Saline culture of crops: a genetic approach. Science 210, 399–404.
8. Foy C D, Chaney R L and White M C 1978 The physiology of metal toxicity in plants. Annu. Rev. Plant Physiol. 29, 511–566.
9. Gerloff G C 1963 Comparative mineral nutrition of plants. Annu. Rev. Plant Physiol. 14, 107–124.
10. IAEA/FAO 1967 Isotopes in plant nutrition and physiology. Proc. symp. Vienna, September 1966. IAEA, Vienna.

11 Jung A G (Ed.) 1978 Crop tolerance to suboptimal land conditions. Am. Soc. Agric. Special Pub. N° 2, 343.
12 Klimaševski E L 1969 Fiziologia sortovoj specifiki kulturnih rasteni v svjazi s kornevim pitaniem (Physiology varietal specificity of cultivated plants in conection with nutrition of root). In Fiziologija i biohimija sorta, AN SSSR, Irkutsk. pp 19–32.
13 Klimaševski E L 1974 Problema genotipičeskoj specifiki kornevogo potanija rastenii (The problem of gentoypic specificity of plant root nutrition). In Sort i udobrenie. AN SSSR, Irkutsk. pp 11–54.
14 Klimaševski E L (Ed.) 1974 Sort i udobrenie (Variety and nutrition). AN SSSR, Irkutsk. 283.
15 Klimaševski E L and Cerniševa N F 1980 Aktuelnie voprosi genetičeskoj variabilnosti mineralnogo pitania kulturnih rastenii (Urgent problems of genetic variability of cultivated plants mineral nutrition). Fiziolog. i biohim. kultur. rastenii. 4, 375–388.
16 Kruckeberg A R 1959 Ecological and genetic aspects of metallic ion uptake by plants and their possible relation to wood preservation. In Marine Boring and Fouling Organisms. Ed. D L Ray. Univ. of Washington Press, Seattle. pp 526–536.
17 Läuchli A 1976 Genotypic variation in transport. In Encyclopedia of Plant Physiology. Ed. U Lüttge and M G Pitman. Vol. 2. Part B. Springer-Verlag, Berlin. pp 372–393.
18 Lessman G M 1972 Differential response of plant genotypes to magnesium fertilization. In Magnesium in the Environment. Soils, Crops, Animals and Man. Fort Valley State College. pp 177–192.
19 Milikan C R 1961 Plant varieties and species in relation to the occurrence of deficiencies and excesses of certain nutrient elements. J. Aust. Inst. Agric. Sci. 27, 220–233.
20 Myers W M 1960 Genetic control of physiological processes: consideration of differential ion uptake by plants. In A Symposium on Radioisotopes in the Biosphere. Ed. R S Caldecott and L A Snyder. Univ. of Minnesota, Minneapolis. pp 201–226.
21 Rejmers R E and Klimaševski E L (Eds.) 1969 Fiziologija i biohimija sorta (Physiology and biochemistry of variety). AN SSSR, Irkutsk. pp 173.
22 Sarić M R 1974 Značenie problemi sorotovoj specifiki mineralnogo pitania (The importance of the problem of variety specificity of mineral nutrition). In Sort i udobrenie. AN SSSR, Irkutsk. pp 54–60.
23 Sarić M R 1981 Genetic specificity in relation to plant mineral nutrition. J. Plant Nutr. 3, 743–766.
24 Sarić M R and Kovačević V 1980 Genetska specifičnost mineralne ishrane kukuruza (The genotype specificity related to mineral nutrition in maize). In Fiziologija kukuruza. Ed. J Belić. Srpska akademija nauka i umetnosti, Beograd. pp 127–144.
25 Sarić M R and Kovačević V 1981 Sortna specifičnost mineralne ishrane pšenice (Varietal specificity of wheat mineral nutrition). In Fiziologija pšenice Ed. J. Belić. Srpska akademija nauka i umetnosti, Beograd. pp 61–77.
26 Sarić M R and Kovačevic V 1981 Sortna specifičnost mineralne ishrane šećerne repe (Varietal specificity in sugar beet nutrition). In Fiziologija šećerne reče. Ed. J Belić. Srpska akademija nauka i umetnosti, Beograd. pp 57–72.
27 Šumni V K, Tokarev B I and Dedov V M 1981 Genetiko-selekcionie aspekti mineralnogo pitanija rastenii (Genetical-breeding aspects of plant mineral nutrition). Seljskohozjajstvenaja biologija 2, 185–192.
28 Vose P B 1963 Varietal differences in plant nutrition. Herb. Abstr. 33, 1–13.
29 Vose P B 1974 Ocenka i ispoljzovanie odzivčivosti sortov seljskohozjajstvenih rastenii na

uslovija mineralnogo pitania (Utilization and recognition of nutritional variation in crop plants). *In* Sort i udobrenie. AN SSSR, Irkutsk pp 61–71.
30 Vose P B 1982 Effects of genetic factors on nutrition requirements of plants. *In* Contemporary Bases for Crop Breeding. Ed. P B Vose and S Blixt (*In press*).
31 Wright J M (Ed.) 1976 Plant adaption to mineral stress in problem soils. Cornell Univ. Ithaca, New York. 420 p.

SECTION I
CYTOLOGICAL AND ANATOMICAL CHANGES IN DIFFERENT GENOTYPES CAUSED BY ALTERED NUTRIENT SUPPLY

Light and electron microscopic investigations of the reactions of various genotypes to nutritional disorders

CH. HECHT-BUCHHOLZ
Institut für Nutzpflanzenforschung, Pflanzenernährung, Technische Universität Berlin, D-1000 Berlin

Key words Adaptation of genotypes Cell structure Efficiency Mineral deficiency Mineral toxicity Nutritional disorder Tolerance

Summary A review is given on light and electron microscopic investigations about the reactions of various genotypes to nutritional disorders such as mineral deficiency and mineral toxicity. Microscopic investigations have been carried out to find initial symptoms of nutritional disorders in plant tissue in order to improve diagnosis and to gain information about disturbed metabolism. Recent investigations have been focussed on changes of cell structure which indicate adaptive mechanisms towards mineral stress in order to explain tolerance and efficiency mechanisms. The influence of mineral deficiency or excess of minerals on the cell structure of different genotypes will be described. Special attention will be drawn to cytological changes in connection with the adaptation of plant genotypes towards mineral stress.

Introduction

Nutritional disorders in plants, mineral deficiency or excess of minerals, cause retardation of growth, impairment of metabolism, and damage to root and shoot tissue. The reaction of different genotypes to nutritional disorders varies and depends on their past genetic adaptation to nutrient supply conditions. Genotypes vary in their requirements for mineral nutrients and their ability to tolerate mineral deficiency or mineral excess conditions. A good example of the fact that different crops require certain nutrients in different amounts is that of boron. Dicotyledonous crops require a considerably higher amount of boron than monocotyledons of the Gramineae family such as barley, wheat, rye and rice. Taxonomically, the requirements for mineral nutrients and the ability to tolerate mineral stress conditions differ between orders, families and species but also between varieties and cultivars within a species. Plants differ in their tolerance to mineral stress and the efficiency of genotypes in the use of phosphate, nitrogen, potassium, and trace elements varies widely. Agricultural research has put the emphasis on investigations of tolerance to excess of salt, aluminium and manganese. Because of the immense importance of plant adaptation to mineral stress in so-called problem soils[54], screening programmes for tolerance and efficiency are being investigated.

Current plant nutrition research includes, besides chemical and analytical procedures, methods using the light and electron microscope to find incipient symptoms of nutritional disorders in order to improve diagnosis[13, 14] and in order to gain more information about impairment of cell metabolism[19]. Recent research work has been focussed on changes in cell structure which can indicate adaptive mechanisms under conditions of mineral stress. In addition investigations have been carried out with the light and electron microscope to find differences in cell structure and anatomy which could explain the difference in tolerance of genotypes towards excess and deficiency of minerals.

Review of results

Cytological and anatomical changes caused by nutritional disorders
Cytological and anatomical changes occurring in nutrient-deficient plants have been investigated using both the light and electron microscope. Characteristic symptoms have been established for different disorders such as potassium, calcium, iron and manganese deficiency. One of the interesting observations in this work is that for a given nutrient deficiency the symptoms are the same for different plant species.

Potassium deficiency The influence of potassium deficiency on cell structure was studied in 37 plant species from 18 families[9]. No differences in microscopical symptoms of potassium deficiency were found in these genotypes. Characteristic microscopical symptoms in potassium-deficient tissue are the collapsing of cells, swelling of cells in the adjacent tissue and aggregation of dead cells forming dark layers.

Calcium deficiency For calcium deficiency too the changes of cell structure are similar in different plant species[10]. These cytological changes include impairment of the cell membranes and as a result of this a rapid cell autolysis. This has been described in barley[39, 40], apple[3] and potatoes[22].

Iron deficiency Nutritional disorders which cause chlorosis and necrosis lead to disorganization of the chloroplast structure. Vesk *et al.*[51] found that chloroplast aberrations in mineral-deficient plants greatly varied among species, within leaves, and even within single cells. Examination of initial and progressive stages of deficiency suggests that the process of chloroplast disorganization in iron-deficient leaf tissue is similar in different genotypes. Characteristic of iron deficiency is an inhibition of grana stack formation and a disintegration of existing grana stacks. In contrast to a chloroplast with sufficient iron supply (Fig. 1), an iron-deficient chloroplast lacks grana[48] (Hecht-Buchholz, unpublished). Only long parallel orientated thylacoids are present (Fig. 2). Such iron-deficient chloroplasts resemble young poorly differentiated plastids. In severe stages of iron deficiency the disorganization of the chloroplasts leads to irregular or parallel aggregations of lamellae which are often accompanied by an increase in

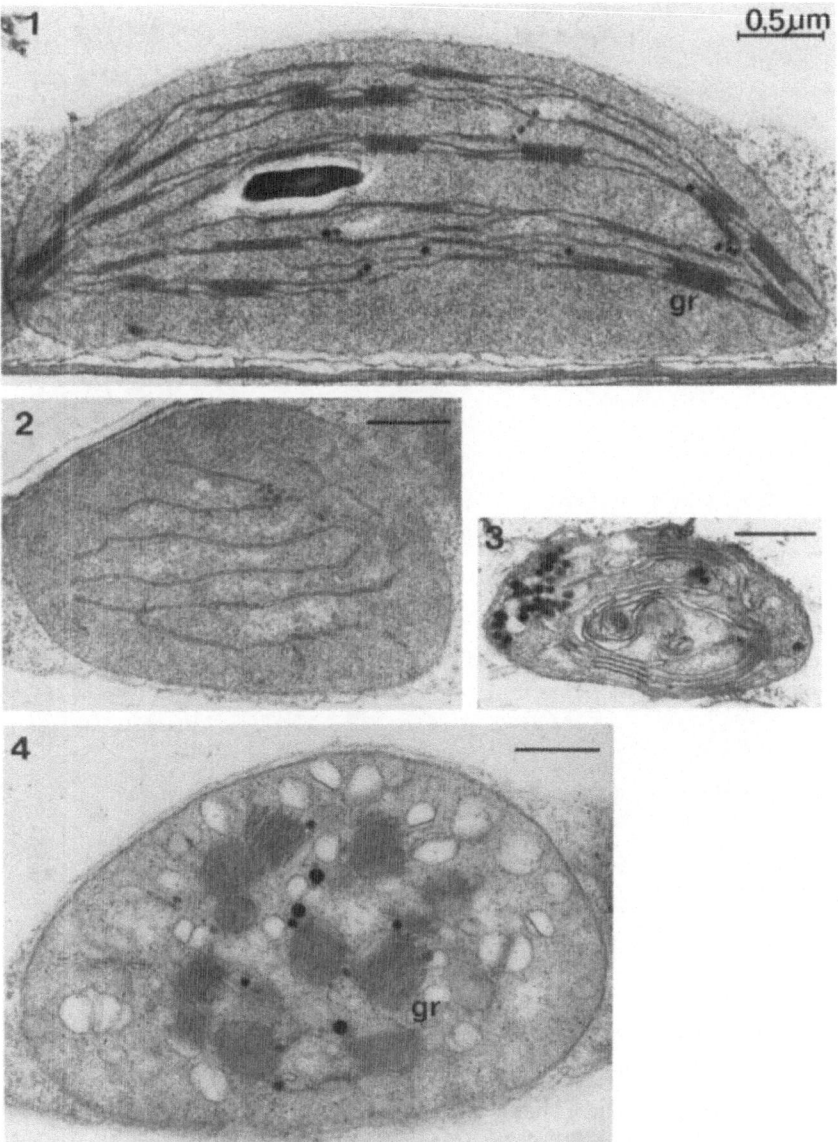

Fig. 1. Chloroplast of sunflower with sufficient iron supply. Grana stacks (gr) are visible.

Fig. 2. Chloroplast of Fe-deficient sunflower lacking grana. Only long parallel orientated thylacoids are present.

Fig. 3. Chloroplast of maize with severe iron deficiency. There are irregular or parallel aggregations of lamellae accompanied by an accumulation of lipid globuli.

Fig. 4. Mn-deficient chloroplast of soybean. The intergranal lamellae are destroyed. Grana stacks (gr) are still visible. Some of the grana compartments are swollen.

Figs. 1–4 Final magnification: 23,700 × .

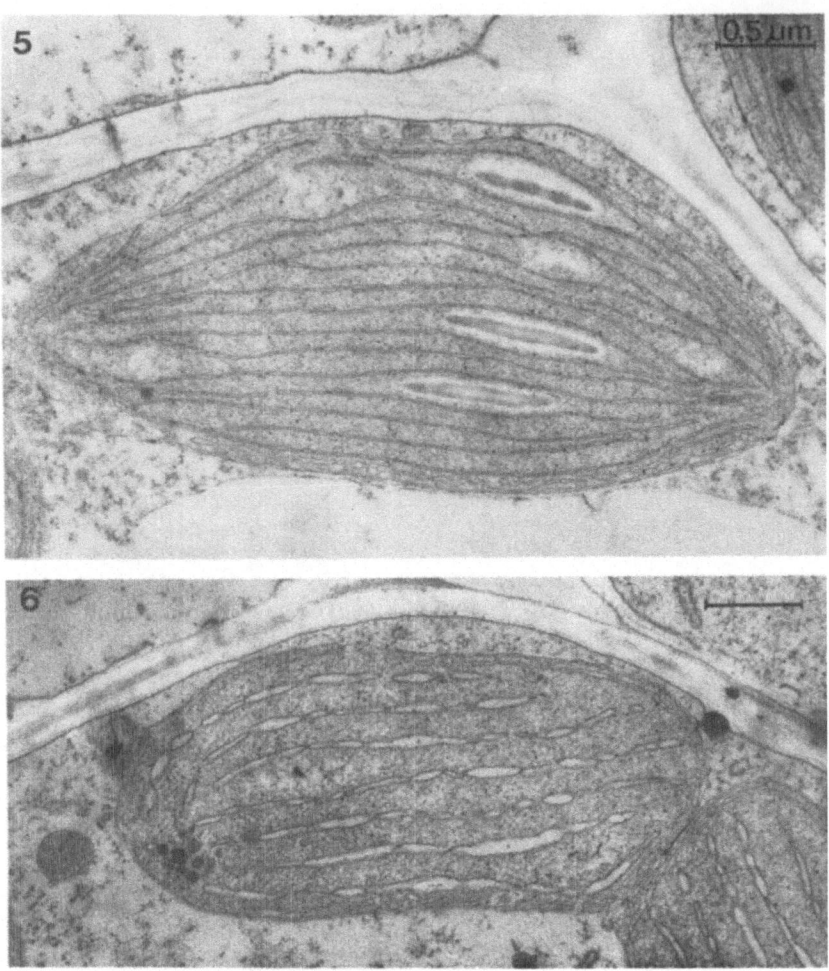

Fig. 5. Bundle sheath chloroplast of maize with sufficient iron supply.

Fig. 6. Bundle sheath chloroplast of iron-deficient maize. The thylacoids are swollen.

Figs. 5 and 6. Final magnification: 27,360 ×.

lipid globuli (Fig. 3). These changes are not characteristic of iron deficiency, but occur when senescent chloroplasts are converted to chromoplasts[50].

The bundle sheath chloroplasts of C_4 plants do not contain grana even under normal environmental conditions (Fig. 5). Iron deficiency in the bundle sheath cell chloroplasts (Fig. 6) simply results in a swelling of the intergranal thylacoids[49].

Manganese deficiency Differences in chloroplast structure caused by nutritional disorders in different genotypes can be due to differing experimental conditions; age and insertion of the investigated leaves, time of sampling, stage of deficiency. Mercer et al.[41] described a general disorganization of manganese-deficient chloroplasts in spinach. This disorganization appeared to be a rather severe stage of manganese deficiency. Manganese deficiency apparently also causes similar cytological changes in different plant species. In contrast to iron deficiency the initial changes resulting from manganese deficiency appear to be a distintegration of the intergranal lamellae with the grana stacks remaining relatively resistant (Fig. 4). This has been observed in spinach[46], in soybean (Hecht-Buchholz, unpublished), and in *Vicia faba*[43]. Homann[26] suggested that there is "no such thing as a normal pattern of deficiency symptoms in higher plants cultivated in absence of manganese". The plants which were investigated by Homann responded to manganese deficiency in different ways. Chloroplasts of spinach, tobacco and pokeweed (*Phytolacca americana*) became severely disorganized as a result of manganese deficiency, but chloroplasts of manganese-deficient coffeeweed (*Cassia obtrusifolia*) looked healthy in spite of the fact that the leaves were chlorotic. In the manganese-deficient cassia chloroplasts there were higher stacked grana in comparison with the normal chloroplasts. Also far fewer chloroplasts per cell were found in comparison with the plants with sufficient manganese. According to Homann, it appears as if cassia plants try to adapt themselves to the lack of manganese by packing whatever traces of manganese are available into the higher stacked grana of a few chloroplasts. This view has been supported by analysing the chloroplasts for their manganese content. The forming of higher stacked grana as a response to manganese deficiency seems to be an interesting example of an adaptation of a plant genotype to a nutritional disorder.

Boron deficiency Boron deficient meristematic tissues divide continuously without differentiation. This has been observed in dicotyledonous plants in the cambium, in abscission layers and in growing storage tissue[11,14]. Disturbed differentiation and irregular cell division occur also in monocotyledonous plants, but irregular cell division is restricted to the apical meristems[31].

Phosphorus deficiency The occurrence of purple anthocyanin cell sap coloration in phosphate-deficient plants depends on the plant species. Anthocyanin as a result of phosphorus deficiency has been observed in various plant species such as tomatoes and cauliflower[7] but does not occur in various other species.

Zinc deficiency A microsymptom of zinc deficiency is the development of a red coloration of unknown origin. This coloration does not appear in all genotypes. It has been observed in the cell sap of single cells in cotton and in the walls of lignified cells in Panicum but not in onions[47].

Cytological and anatomical changes in relation to tolerance

Molybdenum excess For molybdenum excess, different species differ widely in their tolerance. Some species accumulate high amounts of molybdenum without displaying any damage or retardation of growth, whereas others are very sensitive. This difference could be explained simply by different tissue tolerance to molybdenum accumulation, but might also result from a difference in molybdenum distribution in the plant. In an investigation in which high amounts of molybdenum were supplied to three plant species, bean, sunflower and tomato[20], it was found that the growth of only tomato was severely affected by molybdenum excess. Molybdenum excess often causes a yellow coloration of plants[12] which is probably due to a complex of molybdenum with phenolic compounds. In our experiment with molybdenum excess in bean this was not observed, but in sunflower, the roots, shoot hairs and the leaf veins did display a yellow coloration. In tomatoes the yellow colour was found in the growing zones of the youngest leaves, in the shoot hairs and in the root tips. Using the light microscope orange-yellow globules and yellow vacuoles were found in the meristematic tissue of the apical parts of the shoot and the roots of this species. In sunflower yellow coloration was observed only in the cells in vicinity of the conducting elements in the shoot, leaves and roots. We suggest that in the molybdenum-sensitive tomato molybdenum impairs the metabolism of the meristematic cells causing irregular growth of the youngest leaves. In molybdenum-tolerant plant species molybdenum can be withdrawn from the cytoplasm of the meristematic zones and is accumulated in cell vacuoles located in cells near the conducting elements. This difference in molybdenum distribution could be an explanation for the different tolerance of different genotypes to molybdenum excess.

Manganese excess Manganese excess causes manganese-precipitation in plant cells. Bussler[8] described them in hair cells of *Phaseolus vulgaris* and in the epidermis of the stem of *Pisum*. According to Horst and Marschner[27] Mn-precipitation in *Phaseolus vulgaris* was primarily localized in the cell walls in the vicinity of leaf vessels. Autoradiographic studies[28] with ^{54}Mn indicated a causal

---

Figs. 7–9. Parts of cortex cells of root tips of *Zea mays* and *Phaseolus vulgaris*. P = proplastids, M = mitochondria, Mb = microbodies, V = vacuole, St = starch, N = nucleus. Final magnification: 13,050 ×.

Fig. 7. The proplastids in *Zea mays* (untreated) display a dense matrix. The proplastids contain starch.

Fig. 8. The proplastids in *Zea mays* after treatment with NaCl are swollen and display a light matrix. There are accumulations of mitochondria and microbodies visible. The distribution of ribosomes is changed in comparison with the untreated tissue.

Fig. 9. The proplastids in *Phaseolus vulgaris* are much more swollen than in *Zea mays* after treatment with NaCl.

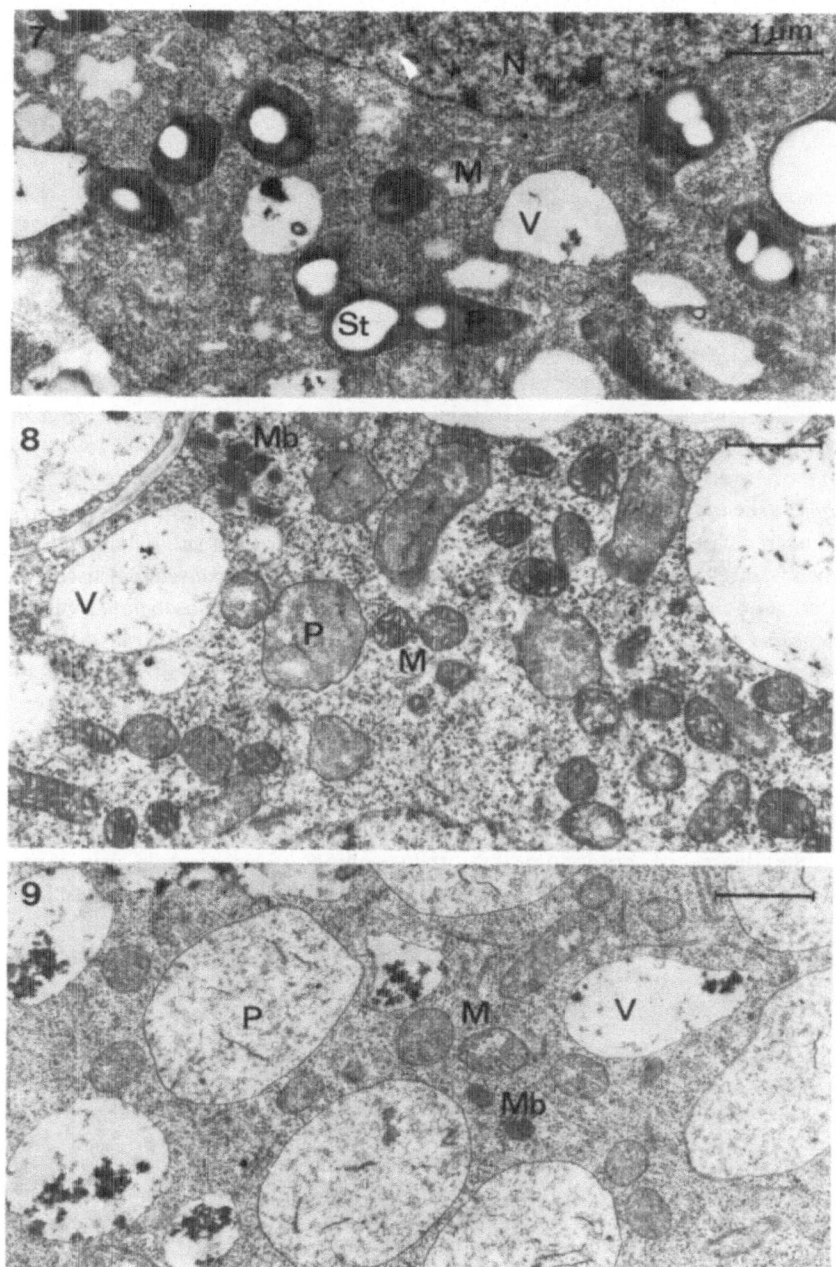

Table 1. Effect of treatment with NaCl (25 mM) on K content (mM g^{-1} dry matter) of leaf tissue and fine structure of chloroplasts

Plant species	K content mM g^{-1} dry matter		Fine structure of chloroplasts after Na treatment
	Untreated	NaCl-treated	
Bean	0.62	0.09	+++ swollen
Barley	1.02	0.32	+ swollen
Sugarbeet	1.55	1.51	unchanged

relationship between manganese tolerance and the more homogenous distribution of manganese in the tissue. In tolerant genotypes local accumulation and distribution of manganese is inhibited or retarded.

Sodium toxicity Electron microscopic investigations of different plant species have been carried out in order to gain more information on salt tolerance[17, 21]. These investigations have shown that sodium treatment of excised leaf tissue or excised root tips resulted in a swelling of plastids, which was associated with a decrease of potassium concentration in the tissue. The swelling of plastids and the extent of potassium loss was related to the degree of known salt sensitivity of the plant species investigated (Tables 1 and 2). Chloroplasts of bean tissue were much more swollen than those of barley and their potassium concentration was also much more severely decreased than in barley. In sugarbeet which is known to be sodium-tolerant, sodium treatment did not change potassium concentration nor was the chloroplast ultrastructure changed. In root tips also associated with the degree of potassium loss, the swelling of proplastids was more pronounced in bean (Fig. 9) than in the less salt-sensitive maize (Table 2 and Fig. 8).

Table 2. Effect of treatment with NaCl (25 mM) on K content (mM g^{-1} dry matter) and proplastids of root tips

Plant species	K content mM g^{-1} dry matter		Fine structure of proplastids after NaCl treatment
	Untreated	NaCl-treated	
Maize	0.65	0.17	++ swollen
Bean	0.66	0.04	+++ swollen

In maize root tips, besides swelling of proplastids, additional changes of cell ultrastructure occurred after treatment with sodium (Fig. 8). These changes included accumulation of numerous mitochondria, rough endoplasmic reticulum, microbodies and an alteration in the ribosome distribution pattern[19]. Biochemical investigations have been carried out to determine whether these changes were indicative for tolerance mechanisms[18,44]. But there was no evidence from respiration measurements that the increase of mitochondria could be indicative of a so called sodium efflux pump. The protein content of the sodium-treated maize root tips was not increased in spite of the fact that the ribosome distribution in the form of polysomes indicated the presence of an active protein synthesis. Microbodies and endoplasmic reticulum probably play roles in detoxification processes as they are increased in senescent and stressed tissue[4,5]. The accumulation of microbodies in sodium-stressed maize is indicative of detoxification processes. It is an interesting question as to whether the higher salt tolerance of maize in comparison to bean is not only caused by a better ability of maize to retain potassium but also by more active detoxification reactions in this species.

Iron toxicity A further example of how electron microscopic investigations together with biochemical work are used to characterize different tolerance in genotypes is evidenced from the finding of Lai and Hou[35] in rice types differing in tolerance towards excess iron. In the rice type Japonica, which is well adapted to submerged paddy soil, more peroxysome-like microbodies and a higher O_2 quotient and higher activity of oxidizing enzymes such as catalase, glycolic acid oxidase and cytochrome oxidase were found than in Indica which is poorly adapted to submerged paddy soils. The increased number of microbodies and the increased activity of oxidizing enzymes in Japonica was attributed to a higher oxidizing capacity of the roots which is important for tolerance towards iron toxicity.

Aluminium toxicity There has been much speculation as to the mechanisms of tolerance towards excess aluminium. Aluminium excess causes stunted root tips, destruction of root cap cells, swelling and destruction of root epidermis and cortex cells resulting in a disintegrated outer root shape (Figs. 11, 12). This was found in investigations with wheat[15,25], sugarbeet[29], maize[1] and barley[23]. As far as different cultivars are concerned they differ enormously in their susceptibility towards aluminium excess. Some cultivars of the same species are extremely damaged whereas others are scarcely affected. Flemming and Foy[15] suggested in their publication on root structure of aluminium-affected wheat varieties that the varietal aluminium tolerance of difference genotypes can be attributed to variable resistance of the root meristem to irreversible destruction. Evidence now exists to support this view.

Two cultivars of barley, Dayton (Al-tolerant) and Kearney (Al-sensitive) were

Figs. 10–12. Root tips of barley. Longitudinal sections. Magnification: 94 ×.

Fig. 10. Cultivar Dayton without Al exposure.

Fig. 11. Cultivar Dayton after 24 h 9 ppm Al exposure. The cells of the root are damaged but the proximal meristem (arrow) is relatively intact.

Fig. 12. Cultivar Kearney after 24 h 9 ppm exposure. The cells of the root are damaged, the proximal meristem including the so called quiescent centre is severely affected.

investigated recently using light and electron microscopy[24]. Initial changes of cell structure after only a few hours of exposure to aluminium were an increase in vacuolation of the root cap and the meristematic cells of the root tip and also the presence of destroyed cells at the boundary between the root cap and the proximal meristem (Fig. 13). These changes occurred in both cultivars. An important difference between the two cultivars was the different resistance of the proximal meristem to damage. The proximal meristem of the aluminium-sensitive cultivar Kearney was severely damaged after aluminium treatment (Fig. 12), whereas the proximal meristem of the aluminium-tolerant cultivar Dayton remained relatively intact (Fig. 11). The resistance of the proximal meristem of the aluminium-tolerant cultivar Dayton led to a regeneration of the root tip after the old root cap had been destroyed (Fig. 14). Barlow[2] showed that root tips from which the root cap has been artificially removed can form a new root cap due to the regenerative activity of the proximal meristem. This is also true for Al-stressed root tips, provided the proximal meristem is still alive. Genotypes are tolerant to a certain amount of Al when they are able to regenerate the root tips and thus continue apical root growth. Based on resistance against irreversible damage to the proximal meristem a screening procedure has been developed[42]. As it was shown in our investigation with barley[24], the protection of the root structure and meristematic activity during Al exposure was enhanced by increasing the calcium or magnesium supply. Increase in the calcium and magnesium supply protected the roots from aluminium toxicity more in the aluminium-tolerant cultivar than in the aluminium-sensitive one.

Changes of cell structure and anatomy which indicate adaptive mechanisms

Recent research work has been focussed on cytological changes indicative of adaptive response towards conditions of mineral stress. Morphological changes which indicate adaptive reactions are summarized in literature for salt-stressed plants by authors like Poljakoff-Mayber and Mayer[45] or Waisel[52]. Formative effects of salinity as Waisel has called them in his book about halophytes include an increase in leaf thickness, an increase in cell size and vacuoles resulting in water storing succulent cells, increase in the conductive system and induction of salt glands. Such special anatomical structures, which indicate a response to salinity, occur in halophytes but they can also be induced in glycophytes. Salt-induced anatomical changes are known also in cultivated plants such as cotton and sugarbeet. Some genotypes react to salinity with xeromorphic features such as decrease in epidermal cell size, increase in number of stomata per unit leaf area and increase in waxy layers and cuticle thickness. The differing response to salt stress depends on the type of salinity and greatly on genetical differences.

Cytological changes which indicate an adaptive response to nutritional disorders and which have been found recently using the electron microscope include the induction of transfer cells. Transfer cells are characterized by

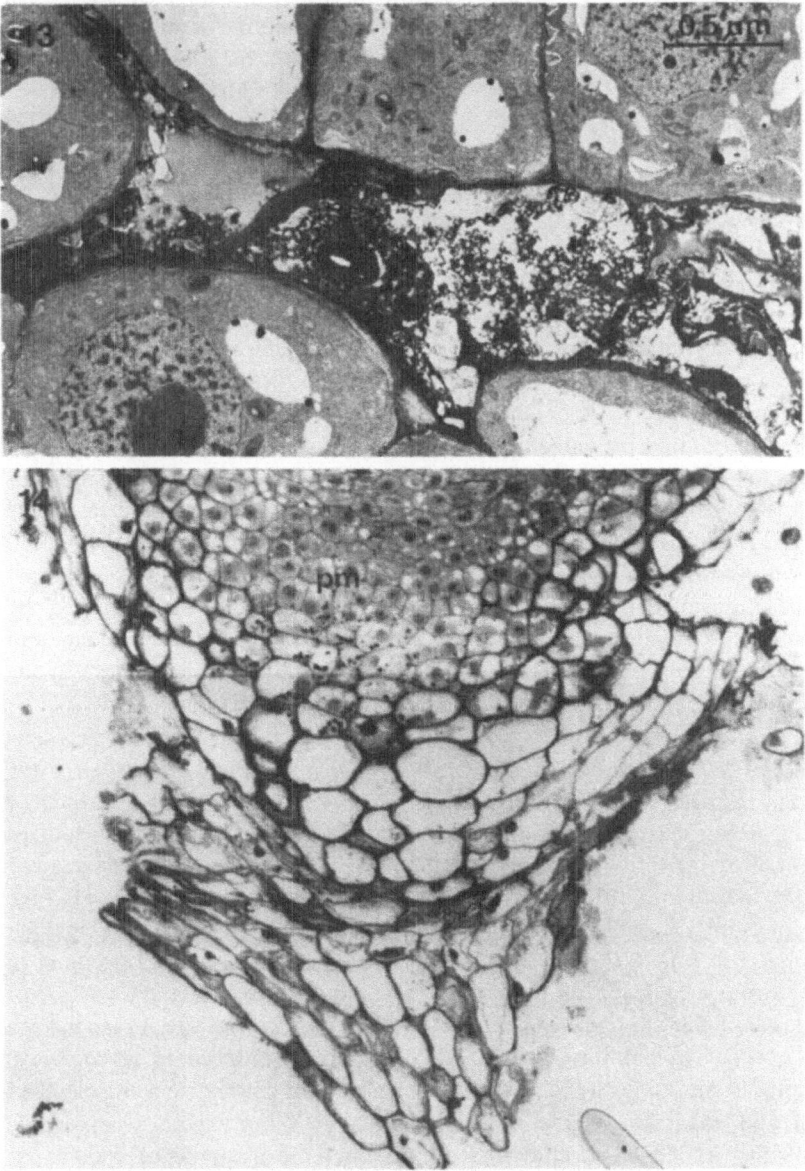

Fig. 13. Autolysed cells in the meristem region of a root tip of Kearney after 24 h exposure to 9 ppm Al. The adjacent cells seem to be intact. Final magnification 3240 ×

Fig. 14. Regenerating root tip of barley after 4 days' Al exposure. The proximal meristem (pm) is forming new root cap cells after the old root cap has been destroyed.

protuberances of cell wall and plasmalemma thus providing vigorous transport of solutes[16]. They have normally been found in conducting tissue between xylem and phloem in minor veins, in salt glands and in nitrogen-fixing root nodules. Transfer cells have been described as a response to iron deficiency[33, 37, 38] and as a response to salt excess. Their induction is dependent on the genotype. Salt-induced xylem transfer cells have been found in connection with sodium exclusion in a salt-tolerant variety of soybean[34] in *Phaseolus vulgaris*[33] and in *Trifolium alexandrinum*[53]. In a recent publication by Winter[53] not only salt-induced xylem transfer cells, but also phloem transfer cells in leaves are described as being involved in export and intraveinal recycling of sodium in the moderately salt-tolerant clover *Trifolium alexandrinum*.

References

1 Baier R, Münnich H, Heinke F and Göring H 1976 Zytologische Untersuchungen zur Wirkung von Aluminium auf Maiswurzeln. Wiss. Z.d. Humboldt. Univ. zu Berlin. Math. Nat. R. XXV, 840–844.
2 Barlow P 1974 Regeneration of the cap of primary roots of *Zea mays*. New Phytol. 73, 937–954.
3 Bangerth F 1970 Die Stippigkeit der Äpfel, ein noch immer ungelöstes Problem der Fruchtphysiologie. Gartenbau Wissenschaft 35, 91–120.
4 Berger C and Schnepf E 1970 Entwicklung und Altern der Spadix-Appendices von *Sauromatum guttatum* Schott und *Arum maculatum*. L. I. Veränderung der Feinstruktur. Protoplasma 69, 237–251.
5 Bergfeld R and Falk H 1968 Geordnete Aggregationen von endoplasmatischem Retikulum in den weißen Hochblättern von *Davidia involucrata*. Z. Pflanzenphysiol. 59, 297–300.
6 Bogorad L, Pires C, Swift H and McIrath W J 1958 The structure of chloroplasts in leaf tissue of iron deficient Xanthium. Brookhaven, Symp. Biol. 11, 132–136.
7 Buchholz Ch 1964 Phosphormangelerscheinungen bei Gemüsepflanzen. Die Phosphorsäure 24, 175–179.
8 Bussler W 1958 Manganvergiftung bei höheren Pflanzen. Z. Pflanzenernaehr. 81, 256–265.
9 Bussler W 1962 Comparative examinations of plants suffering from potash deficiency. Verlag Chemie, Weinheim/Bergstraße.
10 Bussler W 1963 Die Entwicklung von Calcium-Mangelsymptomen. Z. Pflanzenernaehr. Dueng. Bodenkd. 100, 53–58.
11 Bussler W 1964 Die Bormangelsyptome und ihre Entwicklung. Z. Pflanzenernaehr. Bodenkd. 105, 113–126.
12 Bussler W 1970 Die Molybdänmangelsymptome und ihre Entwicklung. Z. Pflanzenernaehr. Bodenkd. 125, 50–64.
13 Bussler W 1974 Microscopical and microchemical characteristics of deficiency diseases. Plant Analysis and Fertilizer Problems, Proc. of the 7th Internat. Coll. Hannover, pp 83–92.
14 Bussler W 1981 Microscopical possibilities for the diagnosis of trace element stress in plants. J. Plant Nutr. 3, 115–128.
15 Fleming A L and Foy C D 1968 Root structure reflects differential aluminium tolerance in wheat varieties. Agron. J. 60, 172–176.

16 Gunning B E 1977 Transfer cells and their roles in transport of solutes in plants. Sci. Prog. Oxf. 64, 539–568.
17 Hecht-Buchholz Ch and Marschner H 1970 Veränderungen der Feinstruktur von Zellen der Maiswurzelspitze bei Entzug von Kalium. Z. Pflanzenphysiol. 63, 416–427.
18 Hecht-Buchholz Ch, Marschner H and Pflüger R. 1971 Einfluß von Natriumchlorid auf Mitochondrienzahl und Atmung von Maiswurzelspitzen. Z. Pflanzenphysiol. 65, 410–417.
19 Hecht-Buchholz Ch 1972 Wirkung der Mineralstoffernährung auf die Feinstruktur der Pflanzenzelle. Z. Pflanzenernaehr. Bodenkd. 132, 45–68.
20 Hecht-Buchholz Ch 1973 Molybdänverteilung und -verträglichkeit bei Tomate, Sonnenblume und Bohne. Z. Pflanzenernaehr. Bodenkd. 136, 110–119.
21 Hecht-Buchholz Ch, Mix G and Marschner H 1974 Effect of NaCl on mineral content and fine structure of cells in plant species with different salt tolerance. Proc. of the 7th Internat. Coll. on Plant Analysis, Hannover. pp 147–156.
22 Hecht-Buchholz Ch 1979 Calcium deficiency and plant ultrastructure. Comm. Soil Sci. Plant Anal. 10, 67–81.
23 Hecht-Buchholz Ch and Foy C D 1981 Effect of aluminium toxicity on root morphology of barley. Plant and Soil 63, 93–95.
24 Hecht-Buchholz Ch and Schuster 1983 Effect of Al-toxicity on barley varieties. *In preparation*.
25 Henning S J 1975 Aluminium toxicity in the primary meristem of wheat roots. Ph.D. Thesis. Oregon State Univ. Corvallis.
26 Homann P H 1967 Studies on the manganese of the chlorplasts. Plant Physiol. 42, 997–1007.
27 Horst W J and Marschner H 1978 Symptome von Mangan-Überschuß bei Bohnen (*Phaseolus vulgaris*). Z. Pflanzernernaehr. Bodenkd. 141, 129–142.
28 Horst W J 1980 Genotypische Unterschiede in der Mangan-Toleranz von Cowpea (*Vigna unguiculata*). Angew. Botanik 54, 377–392.
29 Kesser M, Neubauer, B F, Hutschinson F E and Verrill D B 1977 Differential aluminium tolerance of sugarbeet cultivars, as evidenced by anatomical structure. Agron. J. 69, 347–350.
30 Kluge M and Ting J P 1978 Crassulacean acid metabolism. Analysis of an ecological adaption. Springer-Verlag, Berlin, Heidelberg, New York.
31 Koronowski P 1961 Anatomische Veränderungen an Mais und anderen Getreidearten bei Bormangel. Z. Pflanzenernaehr. Dueng., Bodenkd. 94, 53–67.
32 Kramer D, Homheld V, Landsberg E and Marschner H 1980 Induction of transfer cell formation by iron deficiency in the root epidermis of *Helianthus annuus* L. Planta, Berlin 147, 335–339.
33 Kramer D, Läuchli A, Yeo A R and Gullasch J 1977 Transfer cells in roots of *Phaseolus coccineus*: Ultrastructure and possible function in exclusion of sodium from the shoot. Ann. Bot. 41, 1031–1040.
34 Läuchli A, Kramer D and Stelzer R 1974 Ultrastructure and ion localization in xylem parenchyma cells of roots. *In* Membrane Transport in Plants. Eds U Zimmermann and J Dainty. pp 363–371. Springer-Verlag, Berlin-Heidelberg-New York.
35 Lai K L and Hou C R 1976 Studies on the differentiation of physiological characteristics of the roots of different type rice plants (*Oryza sativa* L.). Taiwan Agricultural Center, Bulletin of the Phytotron, Oct. 10, 17–21.
36 Lamprecht J 1961 Die Feinstruktur der Plastiden von *Tradescantia albiflora* (Kth.) bei Eisenmangelchlorose. II. Elektronenmikroskopische Untersuchungen. Protoplasma 53, 162–199.

37 Landsberg, E-Chr 1980 Fe-deficiency stress-induced development of transfer cells in the epidermis of red pepper roots. Plant Physiol. 65 (Suppl.), 83.
38 Landsberg E Chr 1982 Transfer cell formation in the root epidermis: a prerequisite for Fe efficiency? J. Plant Nutrition 5, 415–429.
39 Marschner H and Günther J 1964 Ionenautnahme und Zellstruktur bei Gerstenwurzein in Abhängigkeit von der Calcium-Versorgung. Z. Pflanzenernaehr. Dueng. Bodenkd. 107, 118–136.
40 Marinos N G 1962 Studies on submicroscopic aspects of mineral deficiency. I. Calcium deficiency in the shoot apex of barley. Am. J. Bot. 49, 834–841.
41 Mercer F V, Nittim I and Possingham J V 1962 The effect of manganese deficiency on the structure of spinach chloroplasts. J. Cell Biol. 15, 379–381.
42 Moore D P, Kronsted W E and Metzger R J 1978 Screening wheat for aluminium tolerance. *In* Plant Adaptation to Mineral Stress in Problem Soils. Eds. M J Wright and S A Ferrari. pp 287–295. Cornell Univ. Ithaca, N.Y.
43 Neigengerd E and Hecht-Buchholz Ch 1982 Elektronenmikroskopische Untersuchungen einer Virusinfektion (BYMV) von *Vicia faba* bei gleichzeitigern Mineralstoffmangel. Z. Pflanzenernaehr. Dueng. Bodenkd. (*In press*).
44 Önal M and Hecht-Buchholz Ch 1978 Einfluß von Natriumchlorid auf den Protein- und RNS-Gehalt von Maiswurzelspitzen in Beziehung zu Veränderungen der Ribosomenanordnung. Z. Pflanzenphysiol. 90, 193–200.
45 Poljakoff-Mayber and Mayer A M (Eds.) 1964 Salt tolerance of plants. Israel Programme for Scientific Translation Ltd. IPST Cat. No. 2066.
46 Possingham J V, Vesk M and Mercer F V 1964 The fine structure of leaf cells of manganese-deficient spinach. J. Ultra-structure Res. 11, 68–83.
47 Rahimi A and Bussler W 1978 Makro- und Mikrosymptome des Zinkmangels bei höheren Pflanzen. Z. Pflanzenernaehr. Bodenkd. 141, 567–581.
48 Spiller S and Terry N 1980 Limiting factor in photosynthesis. II. Iron stress diminishes photochemical capacity by reducing the number of photosynthetic units. Plant Physiol. 65, 121–125.
49 Stocking C R 1975 Iron deficiency and the structure and physiology of maize chloroplasts. Plant Physiol. 55, 626–631.
50 Thomson W W 1966 Ultrastructural development of chromoplasts in Valencia oranges. Bot. Gaz. 127, 133–139.
51 Vesk M, Possingham J V and Mercer F V 1966 The effect of mineral nutrient deficiencies on the structure of leaf cells of tomato, spinach and maize. Austr. J. Bot. 14, 1–18.
52 Waisel Y 1972 Biology of Halophytes. Acad. Press. New York, London.
53 Winter E 1982 Salt tolerance of *Trifolium alexandrinum* L. III. Effects of salt on ultrastructure of phloem and xylem transfer cells in petioles and leaves. Aust. J. Plant Physiol. 9, 221–226.
54 Wright M J and Ferrari S A (Eds.) 1976 Plant Adaptation to Mineral Stress in Problem Soils. Proc. of a workshop at Beltsville, Md. Cornell Univ. Ithaca, N.Y.

Genetically determined adaptations in roots to nutritional stress: correlation of structure and function

D. KRAMER
Botanisches Institut, Fachbereich Biologie, Technische Hochschule Darmstadt, Schnittspahnstr. 3, D-6100 Darmstadt, FRG

Key words Iron deficiency   Nutritional stress   Soil salinity

Abstract   Many plant species are able to adapt to problem soil conditions such as salt toxicity and nutrient deficiencies by modifying transport systems in their roots. This is frequently correlated with anatomical and ultrastructural developments. The role of such developments is discussed in two specific cases: soil salinity and iron deficiency.

Introduction

The vast majority of higher plants absorb water and nutrients through the root system. Only aquatic species and some epiphytic plants such as the Bromeliaceae rely exclusively on absorption at the leaf surface. The evolution of a root system was one of the preconditions for plants being able to conquer environments that are not permanently irrigated. The roots enable plants to grow during dry periods by making water and salts available from the soil, either over a fairly large distance from the surface by growing to greater depth, or from a large soil volume by branching and growing laterally.

All plants have a requirement for distinct concentrations of ions within cells and tissues, and it is principally the root that is determined to regulate uptake and transport of nutrients. This allows control of composition of the xylem sap, which supplies the nutrients and water to the overground organs.

Generally the composition of the soil solution and that of the xylem sap are quite different. One of the most important differences is the concentration of potassium, which is about 100 times higher in the xylem sap than in a normal soil. Other elements like Na^+ or Cl^- may be present in the soil at a concentration that could never be tolerated in the cell sap of the shoot. The concentrations of ions in natural soils are dependent on the chemical composition of the soil material, the content and direction of flow of water and other parameters. This requires that the plant is able to adapt quickly by modifying the uptake rates of the different elements in order to maintain the concentration of elements within a certain range. This may illustrate the key role of the root system in the supply with water and inorganic solutes of the shoot.

What physiological adaptations then are found that enable plants to survive under "non-ideal" conditions, and how are these associated with specializations at the anatomical level?

The root can be regarded as a highly specialized organ, composed of different zones and various tissues that perform distinct functions. Only the first few millimeters of a growing root have apparently no other "function" but to grow and to develop into the first functional stage, which is characterized by the formation of a suberin lamella in the tangential walls of the endodermis, the death of the xylem vessels and the differentiation of sieve elements. The most interesting feature in this part of the root, if it is growing in a well-balanced nutrient solution, is the xylem parenchyma. These cells surround the proto- and metaxylem vessels in the form of a closed ring. The cells are relatively rich in cytoplasm and have prominent mitochondria and rough endoplasmic reticulum. Half-bordered pits are also found in the cell walls between the vessels and xylem parenchyma cells, in which areas the cell wall is very thin[14]. Many authors have suggested that water and most nutrients are absorbed in this zone of the root[15,18]. According to this hypothesis, nutrients would enter the symplast somewhere in the cortex, by either active transport or diffusion through the plasmalemma; water is thought to follow passively down the water-potential gradient between the external medium (rhizosphere + cortical apoplast) and the symplast. This first passage of ions through the plasmalemma is thought to be selective, so that ions may be accumulated or excluded. The ions would then be transported via the symplastic pathway (*i.e.* the plasmodesmata) into the stele. Transport through the plasmalemma of the xylem-parenchyma cells into the apoplast of the vessel would consequently be a second possible step at which the composition of the xylem sap could be controlled. Whereas the significance of the uptake mechanisms in selection of nutrients is not questioned, the significance of this second membrane-transport step at the symplast/vessel border is still not clear. Läuchli *et al.*[13] applied X-ray microanalysis to freeze-substituted barley roots and found that the concentrations of potassium in the xylem parenchyma and (mature) vessels were significantly different. However, some authors still maintain that membrane transport across the xylem-parenchyma plasmalemma is purely diffusive[3]. It has even been proposed that ions may reach the vessels entirely symplastically (*i.e.* before maturation of the vessels) and that they are released during degradation of the cytoplasm[2].

At the primary stage of development, the casparian strip forms a barrier within the apoplast, separating the apoplastic pathways of stele and cortex. By using lanthanum as tracer Tanton and Crowdy[24] and Robards and Robb[21] have convincingly shown that the casparian strip prevents the influx of ions from the rhizosphere to the stele through the cell walls. Although the significance of this observation is widely accepted, it should be recognized that the casparian strip may also serve to prevent *leakage* of water and solutes from the stele back into other parts of the apoplast thus losing water and a part of the nutrients that have been concentrated during the energy-consuming uptake and lateral transport.

This part of the root is short in length and is followed by a zone of secondary thickening in the dicotyledons, or by a tertiary thickened endodermis in the monocotyledons. The function of this part of the root appears to be mainly to develop lateral roots and to maintain longuitudinal transport to the shoot. However, as will be discussed later, there is some evidence that xylem-sap composition may be secondarily modified at the proximal parts of the root.

These are the essential structures in roots that seem to be important in the provision of the shoot with water and nutrients. Their significance may be emphasized by considering some examples of the structural modifications shown at the response to conditions of stress. Stress in this connection means that the root has, temporarily or permanently, to work under conditions that require increased efficiency in order to maintain an adequate supply to the shoot. The aim of the following sections is not intended to cover the nature of the damage suffered by plants or the inhibitory effects of stress conditions that are only rarely found in their natural environment. The aim here is rather to draw attention to examples in which particular reaction patterns enable the plant to cope with the prevailing stress in the edophic environment.

Adaptations to salinity

Salinity is a problem of increasing importance for crop production in numerous parts of the world. Many areas that need to be taken into plant cultivation lie in arid or semiarid parts of the world. They either already contain high soil concentrations of Na^+ and other potentially toxic anions like Cl^-, SO_4^{2-}, CO_3^{2-} and HCO_3^-, or are going to accumulate these ions due to the combined effects of irrigation practices and high evaporative demand.

Plants that are adapted to grow on saline soils have one common feature: their rate of K^+-uptake is not inhibited by Na^+, and may even be stimulated[6]. Although not many salt-tolerant plants are typical "salt excluders", in which Na^+ is prevented from reaching the leaves, roots from all plants that are able to adapt to saline conditions show a significant selectivity of K^+ over Na^+ (ref.[4]). Some plants, however, like many mangroves and salt-tolerant grasses, are typical "salt excluders" that transport no Na^+ to the shoot. The following examples deal with structural variations that have been found in the absorbing part of the root as response to salt stress.

Kramer et al.[11] found that in *Atriplex hastata* the epidermal cells of roots develop into transfer cells when plants are grown under saline conditions. These transfer cells are restricted to the absorbing part of the root. X-ray microanalysis of deep-frozen fracture planes of this area showed that the major site of K^+/Na^+ discrimination is situated in the epidermis. However, the conclusion that the transfer cells were responsable for this[11] turned out to be incorrect[8]. As will be shown later, root-epidermal transfer cells are developed by many plants as reaction to iron deficiency in the root medium. In fact it was found that transfer cells did not occur in *Atriplex* roots when the Fe^{3+}-concentration in the nutrient solution was increased in relation to the concentration of NaCl. The conclusion is thus that NaCl somehow reduces the ability of the plant to absorb Fe, by a mechanism that is not understood so far. Thus as yet no ultrastructural modifications have been observed that can be directly related to K^+-uptake under salt-stress conditions.

The situation is different if the subsequent part of the lateral pathway, that is the symplastic transport of ions, is considered. Stelzer and Lauchli[23] have observed that in a grass species that is highly salt-tolerant and belongs to the extreme Na-excluders, *Puccinellia peisonis*, small but plasma-rich cortical cells are found adjacent to the passage-cells. Their physiological significance is unclear, but it could be that these specialized cortical cells improve the efficiency of the symplast in controlling lateral transport from the cortex to the stele. This, however, would imply that the symplastic transport itself can be controlled – an assumption for which experimental evidence is lacking. One significant finding in this context is that it has been observed by X-ray microanalysis that neighbouring cells that are symplastically connected may have very different ion ratios[10,26].

Poljakoff-Mayber has calculated from data of Ginsburg[19] that the width of the Casparian strip depends greatly on the environmental conditions. If expressed as the ratio of the width of the casparian strip to the width of the whole tangential wall, this ratio is highest in halophytes and smallest in mesophytes.

Within the stele the most significant changes are observed in the xylem-parenchyma cells. Yeo *et al.*[27] observed a dramatic increase in the number of cisternae of rough endoplasmic reticulum in the xylem-parenchyma cells of salt-treated *Zea mays*. Further, in the xylem-parenchyma cells of salt-stressed *Atriplex hastata* roots, the ratio of cytoplasm to vacuole is higher than in control plants[11], the main components of the cytoplasm being mitochondria and rough endoplasmic reticulum.

The increased number of mitochondria can be explained by the assumption that at the symplast/apoplast border, the plasmalemma of the xylem-parenchyma cells, transport processes are intensified that are energy-consuming: relative accumulation of potassium and exclusion of Na^+ has also been measured by X-ray microanalysis[11,27].

It is more difficult to find an explanation for the increase of rough endoplasmic reticulum (RER). High proportions of RER have been observed in other systems, such as the developing leaf bladder cells of the halophyte *Mesembryanthemum crystallinum* and the xylem-parenchyma transfer cells of salt-stressed bean plants, *Phaseolus coccineus*[9]. Stelzer *et al.*[22] and van Steveninck *et al.*[25] have demonstrated that Cl^- is localized in the lumen of the RER. This provides support for the hypothesis that the ER represents an intracytoplasmic compartment that, particularly under saline conditions, can serve in the sequestration of ions within the cytoplasm, thus protecting enzymes that are located in the cytosol. Intracellular symplastic transport of salt might thus proceed via the desmotubule, the small inner tubule of the plasmodesmata, which is considered by many authors to be connected with the ER.

The structures that have been described so far are involved in the absorption and the lateral, symplastic transport of ions from the external medium into the xylem vessels. From some plants it is known, however, that the ionic content of the xylem sap can be controlled secondarily during the upward transport to the shoot. In *Phaseolus* it has been found that Na$^+$ accumulates in the proximal part of the root and the lower parts of the stem when the plants are treated with NaCl[5,20]. Cytological investigations revealed that here xylem-parenchyma cells are differentiated as transfer cells[9], with the wall ingrowths restricted to the area of the half-bordered pits. The cytoplasm of these cells is, again, characterized by numerous mitochondria and cisternae of RER. By X-ray microanalysis it has been demonstrated that at these sites Na$^+$ is reabsorbed from the xylem sap in exchange for K$^+$, which is secreted into the vessels[9]. Sodium might either be stored in surrounding cells or excreted into the soil[17].

Reactions to iron deficiency

Iron is essential for plants as a component of many enzymes. Plants suffering from iron deficiency become chlorotic because iron is essential for chlorophyll synthesis. Iron deficiency occurs naturally wherever pH or other factors keep iron in a form that is insoluble in the soil water. However, species are found in all plant taxa that are able to overcome this problem. In the so-called "iron-efficient plants", to which most dicotyledons belong, a definite reaction pattern is observed, namely release of hydrogen ions and chelating compounds and an increase of the reducing capacity in the rhizosphere. Many monocotyledons show these reactions too, except for the acidification of the rhizosphere, although there are some iron-inefficient monocotyledons such as the *Poaceae*, which show no obvious reaction at all (*cf.* ref.[1]). It is interesting that distinct morphological changes are observed when the reaction patterns as described above are initiated: many species develop short thickened secondary roots, around which zones of acidification and Fe^{3+} reduction can be localized[16]. However, it was only recently shown that in these root segments epidermal cells may show features characteristic of transfer cells. The wall labyrinth of these cells is built up on the outer peripheral walls, so the increased plasmalemma surface faces the rhizosphere[12]. The cytoplasm of these cells contains numerous large mitochondria with well-developed internal membranes. At the same time xylem-parenchyma and companion cells in the stele become significantly richer in cytoplasm. Since this discovery of transfer cells induced by iron deficiency, a wide range of plants has been investigated recently. So far, eleven dicotyledonous species have been examined (*Helianthus annuus, Capsicum annuum, Pisum sativum, Cicer arietinum, Cucumis sativus, Atriplex hastata, A. hortensis, Arachis hypogaea, Solanum tuberosum, Coleus* sp., *Mesembryanthemum crystallinum*) and all of them developed epidermal transfer cells in response to iron deficiency (Kramer, unpublished results). These transfer cells differ only in some minor cytological details.

No transfer cells, however, have been found in monocotyledonous species so far in response to iron deficiency, irrespective whether they show an increased reducing capacity and increased capacity for iron uptake, like *Allium porrum* and *Maranta* sp., or no reaction at all like *Zea mays* and *Setaria* sp. One, still speculative, conclusion might be drawn from these observations. In terms of reducing capacity and Fe-uptake capacity, some of the monocotyledons are at least as efficient as many dicotyledons. But compared with all the dicotyledons examined, the monocotyledons acidify the root environment to a much lesser extent. Thus, in looking for the functional significance of iron-deficiency-induced transfer cells, the question should be considered whether the wall labyrinth is a feature of cells with high rates of proton extrusion.

Although plants may be able to adapt to diverse stress conditions in the soil, such as deficiency of phosphorous or potassium, nothing is known about changes in the ultrastructure of the root in response to such situations. In the author's opinion such reactions to nutritional stress provide promising tools for further research in root cytology.

References

1 Brown J C 1978 Mechanism of iron uptake by plants. Plant, Cell Environ. 1, 249–257.
2 Davis R F and Higinbotham N 1976 Electrochemical gradient and K^+ and Cl^- fluxes in excised corn roots. Plant Physiol. 57, 129–136.
3 Dunlop J and Bowling D J F 1971 The movement of ions into the xylem exudate of maize roots. I. Profiles of membrane potential and vacuolar potassium activity across the root. J. Exp. Bot. 22, 434–444.
4 Epstein E 1972 Mineral Nutrition of Plants: Principles and Perspectives. New York; Wiley and Sons.
5 Jacoby B 1965 Sodium retention in excised bean stems. Physiol. Plant. 18, 730–739.
6 Jeschke, W D 1980 Roots: cation selectivity and compartmentation, involvment of protons and regulation. *In* Plant Membrane Transport: Current Conceptional Issues. Eds. R M Spanswick, W J Lucas and J Dainty. Elsevier/North Holland Biomedical Press.
7 Kramer D 1979 Ultrastructural observations on developing leaf bladder cells of *Mesembryanthemum crystallinum* L. Flora 168, 193–204.
8 Kramer D 1981 Structure and function in absorption and transport of nutrients. *In* Structure and Function of Roots. pp 303–307. Eds. R Brouwer, O Gasparikova, J Kolek and B C Loughman. Martinus Nijhoff/Dr W. Junk Publishers, The Hague/Boston/London.
9 Kramer D, Läuchli A, Yeo A R and Gullasch J 1977 Transfer cells in roots of *Phaseolus coccineus*: Ultrastructure and possible function in exclusion of sodium from the shoot. Ann. Bot. 41, 1031–1040.
10 Kramer D and Preston J 1978 A modified method for X-ray microanalysis of bulk-frozen plant tissue and its application to the problem of salt exclusion in mangrove roots. *In* Microsc. Acta Suppl. 2. Microprobe Analysis in Biology and Medicine. Eds. P Echlin and R Kaufmann. Hirzel, Stuttgart.

11 Kramer D, Anderson W P and Preston J 1978 Transfer cells in the root epidermis of *Atriplex hastata* L. as response to salinity: a comparative cytological and X-ray microprobe investigation. Aust. J. Plant Physiol. 5, 739–747.
12 Kramer D, Römheld V, Landsberg E and Marschner H 1980 Induction of transfer cell formation by iron deficiency in the root epidermis of *Helianthus annuus* L. Planta, Berlin 147, 325–339.
13 Läuchli, A, Spurr A R and Epstein E 1971 Lateral transport of ions into the xylem of corn roots. II. Evaluation of a stelar pump. Plant Physiol. 48, 118–124.
14 Läuchli A, Kramer D, Pitman M G and Lüttge U. 1974 Ultrastructure of xylem parenchyma cells of barley roots in relation to ion transport to the xylem. Planta, Berlin 119, 85–99.
15 Läuchli A, Pitman M G, Lüttge U, Kramer D and Ball E 1978 Are developing xylem vessels the site of exudation from root to shoot? Plant, Cell Environ. 1, 217–223.
16 Marschner H, Kalisch A and Römheld V 1974 Mechanism of iron uptake in different plant species. *In* Plant Analysis and Ferilizer Problems. Ed. J Wehrmann. pp 273–281. German Society of Plant Nutrition.
17 Marschner H and Ossenberg-Neuhaus H 1976 Langstreckentransport von Natrium in Bohnenpflanzen. Z. Pflanzenernaehr. Dueng. Bodenkd. 139, 129–142.
18 Pitman M G 1972 Uptake and transport of ions in barley seedlings. II. Evidence for two active stages in transport to the shoot. Aust. J. Biol. Sci. 25, 243–257.
19 Poljakoff-Mayber A 1975 Morphological and anatomical changes in plants as a response to salinity stress. *In* Plants in Saline Environments. Eds. A Poljakoff-Mayber and J Gale. Springer, Berlin, Heidelberg, New York.
20 Rains D W 1969 Sodium and potassium absorption by stem tissue of bean and cotton. Plant Physiol. 44, 547–554.
21 Robards A W and Robb M E 1974 The entry of ions and molecules into roots: an investigation using electron-opaque tracers. Planta, Berlin 120, 1–12.
22 Stelzer R, Läuchli A and Kramer D 1975 Interzelluläre Transportwege des Chlorids in Wurzeln intakter Gerstepflanzen. Cytobiologie 10, 449–457.
23 Stelzer R and Läuchli A 1977 Salz- und Überflutungstoleranz von *Puccinellia peisonis*. II. Strukturelle Differenzierung der Wurzel in Beziehung zur Funktion. Z. Pflanzenphysiol. 84, 95–108.
24 Tanton T W and Crowdy S H 1972 Water pathways in higher plants. II. Water pathways in roots. J. Exp. Bot. 23, 600–618.
25 Stevenick R F M van, Stevenick M E van, Hall T A and Peters P D 1974 X-ray microanalysis and distribution of halides in *Nitella translucens*. *In* Electron Microscopy 1974. Eds. J V Sandars and D J Goodchild. Canberra A. C. T.: The Australian Academy of Sciences.
26 Stevenick R F M van, Stevenick M E van, Stelzer R and Läuchli A. 1980 Electron probe X-ray microanalysis of ion distribution in *Lupinus luteus* L. seedlings exposed to salinity stress. pp 489–490. *In* Plant Membrane Transport: Current Conceptional Issues. Eds. R M Spanswick, W J Lucas and J Dainty. Elsevier/North Holland Biomedical Press.
27 Yeo A R, Kramer D, Läuchli A and Gullasch J 1977 Ion distribution in salt stressed mature *Zea mays* roots in relation to ultrastructure and retention of sodium. J. Exp. Bot. 28, 17–29.

Ultrastructure of the root cells of two pea genotypes depending on Se in the growth medium

S. M. FATALIEVA, E. D. GULIEVA and O. F. MELIKOVA
Institute of Botany, Azerbaijan Academy of Sciences, Baku, USSR

Introduction

Selenium influences the morphological and functional peculiarities of two pea genotypes but morphometric indicators do not help to explain the cause of these changes. Taking into account different plant reactions to the presence of Se in the cultivation medium the supposition was made that the functional changes can be related to ultrastructural changes.

Materials and methods

Pea (*Pisum sativum* L.) var. "Kurdamirsky uluchenny", var. "Aznihi" and maize var. "Zakatalskaya uluchenny" were grown in water culture on 1/5 Knop's nutrient mixture and on the third day transferred to fresh nutrient solution for 7 days:

I. —Se (control);
II. +0.5 mg Se/l (promoting dose);
III. +5 and 10 mg Se/l for pea and maize, respectively (inhibitory doses).

The tissues were fixed in 3% glutaraldehyde on phosphate buffer (pH 7.4). After O_sO_4 fixation the material was embedded into a mixture of EPON+araldite. Ultrathin sections were obtained by means of the LKB—8802A ultratome. The sections on the grids were contrasted with 2% uranyl acetate solution and lead hydroxide according to Reynolds and examined in the electron microscope JEM-100B.

Results

The fragments of meristematic cells of two genotypes of pea and maize have typical properties for the cells of this zone (Fig. 1).

The fragments of maize root cells (III variant) are presented in microphotographs. The number of free ribosomes in the cytoplasm did not decrease (Figs. 2, 3). Not all mitochondria are subjected to destructive changes. In a row with normal, large, oval ortodoxal mitochondrias changed ones were seen where a decrease in the number of crystae and the formation of optically empty regions occurred. No changes in the cell envelope, endoplasmic reticulum and Golgi apparatus were found. The cell fragments from the meristematic zone of pea roots (var. Aznihi), which is relatively tolerant to the presence of selenium in the growth medium are presented (III variant). As can be seen from the photographs

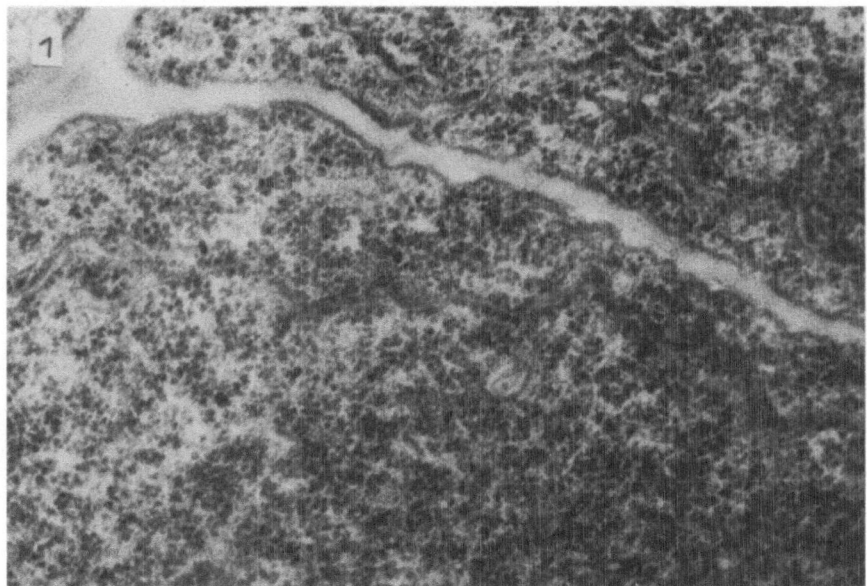

Fig. 1. Fragments of mesistematic cells (× 40,000).

not all the cells are changed. There are cells with a great number of ribosomes, while in some other cells they decrease. The cell walls of some cells are thickened, and contain plasmadesmata. Often the Golgi apparatus is near the cell wall and the vesicles are separated from the wall, move to the plasmalemma (Figs. 4, 5) and are in contact with it.

In a row with normal mitochondria reduced numbers of crystae occur and we observe contact between mitochondria and the tonoplast. The leucoplasts with big starch grains do not differ from the control variants (I).

The cell fragments from the meristematic zone of pea var. Kurdamirsky uluchenny, which is sensitive to Se, are shown. In separate cells progressive vacuolization and sedimentation of solid material on the tonoplasts is marked and significant thickening of the cell wall is observed. Considerable changes take place in the Golgi apparatus (Figs. 6, 7).

Discussion

Repka et al.[3,4] investigated the effect of macronutrient deficiency on the chloroplast structure of maize plants and concluded that the deficiency of macronutrients and nitrogen in particular, resulted in destructive changes in chloroplasts. The effect of calcium, phosphorus and nitrogen deficiency caused

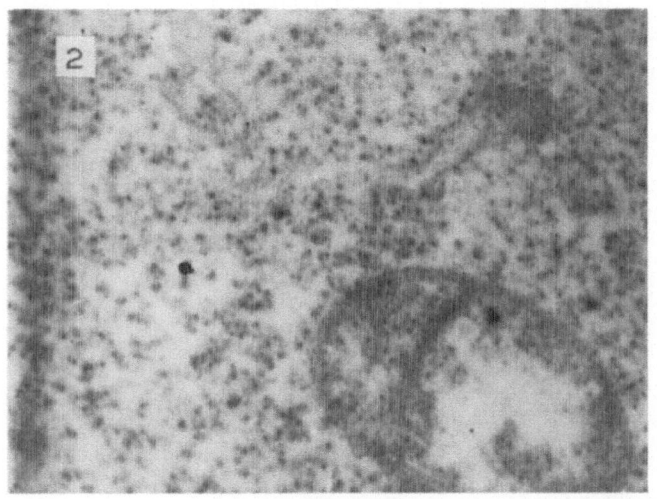

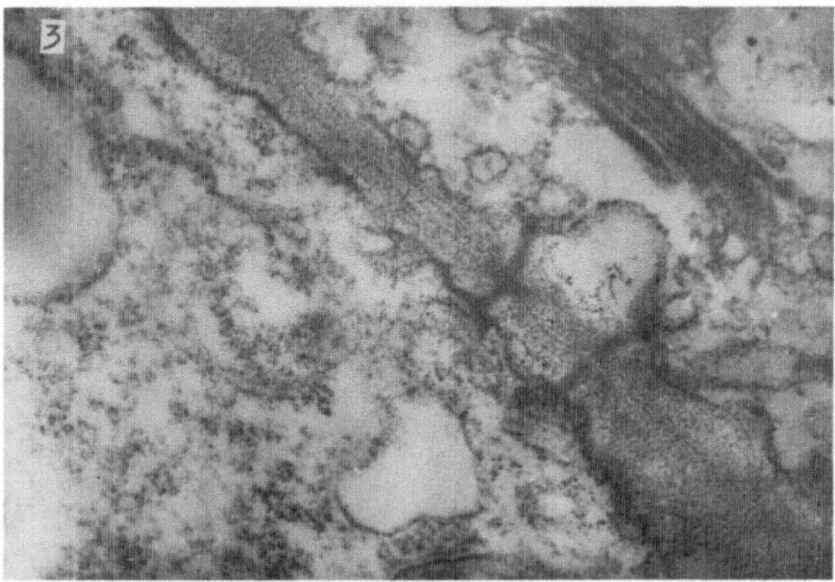

Figs. 2 and 3. Maize root cells from mesistematic zone (×40,000).

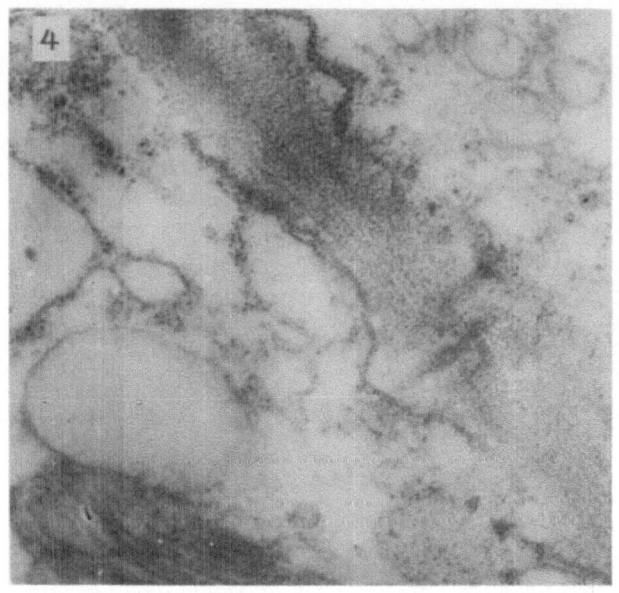

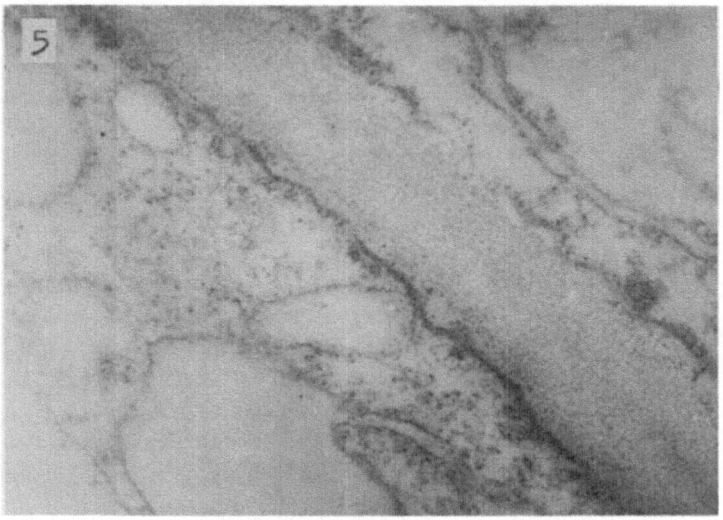

Figs. 4 and 5. Cells of roots of relatively steady pea (× 40,000).

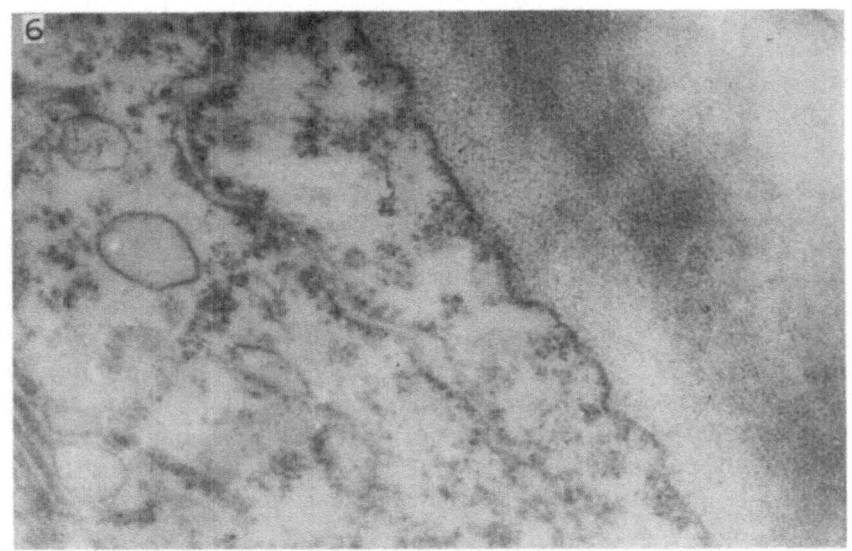

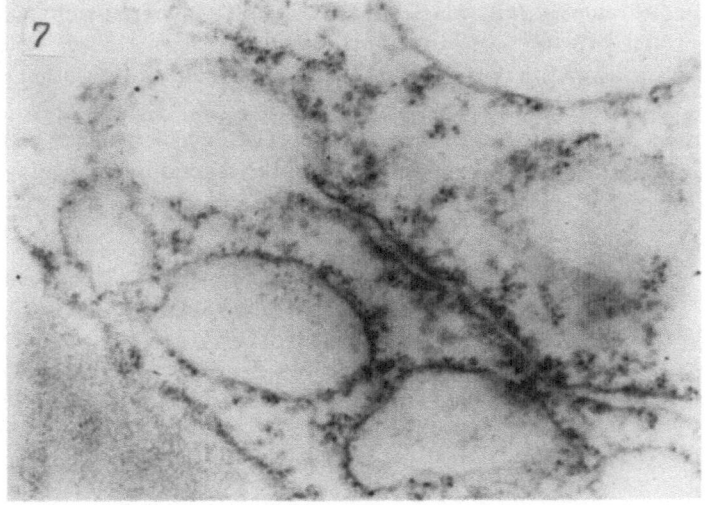

Figs. 6 and 7. Cells of roots of pea sensitive to Se.

destructive changes in the mitochondria and endoplasmatic reticulum of the elongation zone and of the cortex parenchyma cells[1,2].

The concentration of Se used suppresses the growth of all 3 types. The concentration for maize (10 mg/l) is twice as high as for pea (5 mg/l), but the effect on biomass accumulation is equal. Both pea genotypes responded similarly, but in order to emphasize differences the same concentrations were used (5 mg/l).

It follows, from the microphotographs presented, that selenium in the growth medium influences the ultrastructure of meristematic cells differently, depending on the type. Maize is relatively tolerant and shows insignificant changes in the ultrastructure of cell organelles.

More destructive changes are manifest in the sensitive pea genotype. First of all, we have registered concentrated thickening of the cell wall and significant changes occur in the Golgi apparatus. This phenomenon is often met and in comparison with the cells of the control variant and tolerant genotype it is well developed; polarization is increased and vesicles of different sizes are formed at the edge. It is known that in plant cells the Golgi apparatus is a more variable organelle, all components of which change during the process of differentiation. In the cells of pea meristems the zone of the Golgi apparatus has a structure characteristic of differentiated cells, whose functional activity increases with the formation of many vesicles.

The channels of endoplasmic reticulum become wider at the ends and vacuoles are formed. In separate cells progressive vacuolization and sedimentation of the solid material on the tonoplasts take place, which is often met in the parenchymal cells. The mielino-like structures are formed in the vacuoles. Some changes take place in a relatively tolerant pea type, so that here the preservation of structure is marked.

Thus, in genotype sensitive to Se, destructive changes of the cell organelles take place, which leads to premature ageing of the cell and to functional weakening.

References

1. Molikova O F 1975 Vlianie nedostatka kal'cia, fosfora i azota na ul'trastrukturu kletok kornya. Tezisy dokl. XII Mezhdunarodnogo botanicheskogo kongressa, Leningrad.
2. Melikova O F and Gulieva E D 1979 The changes of the cortex parenchyma cells of pumpkin and pea roots on the nitrogen deficiency. Proc. of the 1st Intern. Symp. on Plant Nutrition. Varna, Bulgaria.
3. Repka J, Sarich M, Marek I and Zima M 1972 Vlianie nedostatka makroelementov na strukturu i produktivnost fotosinteza u rasteniy. Fiziologia rasteniy 18.
4. Repka J, Sarich M, Marek I and Zima M 1973 Vlianie nedostatka makroelementov na strukturu hloroplastov parenhimnoi obkladki provodiashchih puchkov kukurzy. Fiziologia rasteniy 20, 4.

SECTION II
ABSORPTION, TRANSLOCATION AND ACCUMULATION OF IONS IN DIFFERENT GENOTYPES

Plant genotype differences in the uptake, translocation, accumulation, and use of mineral elements required for plant growth*

RALPH B. CLARK
Department of Agronomy and U.S. Department of Agriculture, Agricultural Research Service, University of Nebraska, Lincoln, NE 68583, USA

Key words Boron Calcium Copper Genotypic differences Iron Magnesium Manganese Mineral nutrition Molybdenum Nitrogen Phosphorus Plant adaptation Plant breeding Potassium Zinc

Abstract Plant genotypes differ in their uptake, translocation, accumulation, and use of mineral elements. Examples of genotype differences to iron, nitrogen, phosphorus, potassium, calcium, magnesium, manganese, boron, copper, zinc, and molybdenum are discussed. Current knowledge is sufficient to indicate that many crop plants can be improved for the efficient use of mineral elements and better adaptation to mineral stress conditions.

Introduction

Plant species differ extensively in the uptake, translocation, accumulation, and use of mineral elements. If this were not so, the great diversity of plants now known would not exist. Diversity among genotypes, cultivars, varieties, lines, inbreds, etc., within a plant species for mineral element uptake, translocation, distribution, and use have been recognized for many years. Numerous reviews have been written on the subject[5,17,22,24,29,52,61,69,70,73,79,103,110,123,125,127,156,159,180,189]. Large differences among genotypes for mineral elements should be recognized. Many differences are under genetic control, but their expression may be altered dramatically when plants are grown under different environments. These genotype differences help explain plant adaptations to many mineral stress conditions noted throughout the world[61,103,153,189]. These genotype differences provide the basis for better adaptation and survival under unique mineral stress conditions.

The mechanisms by which specific plant genotypes are able to grow and produce under mineral stress conditions while others do not, or by which one genotype is able to accumulate or concentrate higher amounts of an element compared to another, differ with the mineral element and the plant species. This article discusses genotype differences within plant species affecting uptake of mineral elements, their translocation and distribution, accumulation, and utilization. Genotype differences to mineral element toxicities or excesses are not considered.

* Published as Paper No. 6875, Journal Series, Nebraska Agricultural Experiment Station.

M.R. Sarić and B.C. Loughman (eds.), *Genetic Aspects of Plant Nutrition.*
ISBN-13: 978-94-009-6838-7
©1983 Martinus Nijhoff/Dr W. Junk Publishers, The Hague/Boston/Lancaster.

Since uptake, translocation, and accumulation processes are interrelated, these factors will be discussed together for each element.

Iron

Except for possibly N and P, more information is known about plant genotype differences to Fe than for the other mineral element essential to plant growth. Iron is often used as a model for the study of other mineral elements. Iron deficiency is such a problem for many plants grown on many neutral to alkaline, especially calcareous, soils of the world that widespread attention has been given to this element. Another reason why plant genotype differences to Fe have been studied so extensively is because actively growing plant tissues require a continuous supply of Fe from external sources to remain green and healthy.

Plant genotype differences to Fe have been recognized in many plant species for many years and several review articles have been written[15,16,20,21,24,52,53,67,185]. In addition, a recent international symposium and other scientific meetings on the subject have been held in which Fe nutrition and genotype differences to Fe were discussed[67,183,184]. Genotype characteristics affecting Fe deficiency symptoms have focused on ability of plants to solubilize, absorb, and more efficiently use Fe. Mechanisms and processes whereby some plant genotypes are more efficient or effective in absorbing and using Fe than other genotypes have been discussed[11b,20,21,24,53,134a]. Briefly, those genotypes that have the ability to make the rhizosphere Fe more available for root absorption and are able to enhance the availability of Fe inside the plant are considered to be "Fe-efficient". Properties that more "Fe-efficient" plant genotypes usually exhibit are: (a) their roots have a greater ability to reduce Fe^{3+} to Fe^{2+} by producing H^+ and/or Fe^{3+} reductants when needed, (b) the plants show fewer complications from interfering elements like P, Ca, Cu, Zn, Mo, Al, and heavy metals, and (c) the plants have chelating and storage compounds or exhibit chemical/photochemical processes inside the plant that regulate Fe availability and utilization[11a]. Although not as yet fully understood, Fe efficiency is associated with physiological activities and dynamics of storage and metabolism reactions associated with energy transfer processes, chloroplast development, nitrogen metabolism, and Fe distribution and storage. Numerous enzymes and redox agents that are vital to plant metabolic processes including photosynthesis, respiration, cellular protection, nitrogen fixation, and many other general metabolic functions require Fe as an indispensable cofactor[11b].

Much of the Fe measured in plant foliage or roots may be found stored in phytoferritin or precipitated on root surfaces; thus the active Fe fraction in plant tissues must be known before interpreting plant total concentrations of Fe. Higher total Fe concentrations have often been reported in Fe deficient tissues than in green tissues[15,111,175,185]. Iron concentrations in normal green tissue may vary extensively with the plant species and the plant part. In general, leaf tissues that contain less than about 50 µg/g dry weight border on Fe deficiency. Tissues containing more than 300 µg/g dry weight border on Fe toxicity. Iron distribution studies show that roots of green plants may contain 5 to 20 times more Fe than is found in the foliage[18,20,21,63]. Roots of Fe deficient plants may contain as much as 100 times more Fe than the foliar parts[20,21]. Iron absorption into the root vascular tissues for transport in the stem requires prior reduction to Fe^{2+} (ref.[41]).

Iron is translocated through the xylem as Fe^{3+}-citrate[176,177,178]. Citrate concentrations in the xylem increased in response to increased Fe in the xylem[20,21]. Citrate concentrations and Fe translocation in the xylem paralleled each other regardless of the methods used to impose Fe deficiency stresses[36]. Iron trapped outside the roots decreased the Fe supply in the xylem[25]. Organic acids other than citric acid (*e.g.* malic, acetic, transaconitic) were ineffective, and isocitric was partially effective, in chelating and transporting Fe in stem exudates[56]. Citrate concentrations are often as high or higher in Fe deficient plants as in green plants, thus total organic acid concentrations in plants may not be related to the amount of citrate in the xylem for Fe transport[57,99,109,151].

Iron in plant tissues is not readily mobilized and retranslocated to other tissues[25]. Thus, green plants deprived of Fe soon become Fe deficient in the newly developing foliar tissues, while older leaves remain green. The elements P, Ca, Cu, Mn, Mo, Al, and heavy metals interact so extensively with Fe that these elements can inhibit the reduction of Fe^{3+} to Fe^{2+} and override the effect of higher Fe concentrations in plant tissues[53]. Thus, Fe concentrations *per se* may not represent active Fe in plant tissues, and the effective Fe status inside plants.

Nitrogen

Nitrogen is used in large amounts as a fertilizer element and is assimilated into integral parts of cell composition and metabolism making studies on plant genotype differences to N (especially as it relates to protein) numerous. A limited number of examples are considered here. Nitrogen absorption and translocation processes are somewhat more complex than many of the other elements because N taken into the plant as NO_3^- and NH_4^+ is usually converted into organic compounds. Other elements like K and Cl are not. Many of the complexities and factors affecting N uptake and transport processes are described elsewhere[98,100,163].

Intraplant species differences to N were recognized many years ago[93,97,168]. Maize (*Zea mays* L.) inbred lines and hybrids showed differences to N when the plants were grown at low levels of N[168]. The hybrids exhibited N uptake and assimilation patterns similar to those of the inbred parents. Harvey[93] noted similar responses among maize and tomato [*Lycopersicum esculentum* (L.) Mill.] genotypes for NO_3^- and NH_4^+ uptake from nutrient solutions. Differences were noted in N uptake and assimilations in high and low protein lines of maize which had been selected 40 generations for high and low protein[97]. The widest divergence among the hybrids occurred at NO_3^- levels resembling those of fertile soil conditions.

Differences in plant protein must be considered when considering differential N uptake and accumulation by plant genotypes. Selecting for high protein is an integral part of many plant breeding programmes[39,40,58,94,97,102,122,138,165,179].

Two wheat (*Triticum aestivum* L.) genotypes identified as possessing different leaf concentrations of NO_3^- showed that NO_3^- absorption per g root weight was about 30% lower in one genotype than in the other after two hours absorption time[98]. Proportionately lower differences in NO_3^- concentrations continued as the genotypes absorbed NO_3^- up to six hours. The genotype that accumulated higher amounts of NO_3^- in the leaves ('Anza') however, had lower NO_3^- absorption rates. 'Anza' also had lower nitrate reductase activity than the other genotype ('UC 44-111'). The 'UC 44-111' genotype produced more protein than 'Anza'. 'Anza' showed lower straw protein levels, but more of the vegetative protein was remobilized to the grain. For each comparable unit of N fertilizer added, protein concentration in the grain increased similarly in both genotypes, but protein in the straw increased more for 'UC 44-111' than for 'Anza'. Even though 'UC 44-111' had a greater capacity to assimilate reduced N, it was less able to remobilize the reduced N to the grain as protein.

Variations in the translocation of N from leaves and stems to the grain were reported in other studies with wheat[39,59,64,91,102,108,122]. Johnson *et al.*[102] found that high protein genotypes generally had lower stover N concentrations than low protein genotypes. A low grain protein genotype had the ability to absorb more soil N, but was less able to redistribute the N to the grain in comparison to the other genotypes. A high protein genotype had lower N in the leaves, but higher N in the grain. On the other hand, McNeal *et al.*[122] found that even though a high protein wheat genotype absorbed about 20% more N than a low protein genotype, it translocated about 6% less N to the grain.

Recently, Lal *et al.*[108] found genotype differences in the translocation of N from vegetative plant parts (loss in dry weight) to the grain in both wheat and triticale (*Triticale hexaploide* Lar.). Distribution patterns for N taken into wheat plants differed with genotype and with plant part[91]. Differences ranging from 48 to 75% were noted in the partitioning of total N in the grain compared to other tissues for plants grown under similar conditions. Allocation of N to the grain appeared to be related to the plant sink size (dry matter production, grain size, and grain number). It was proposed that sink size differences among plants could be altered through plant breeding and greater N efficiency within plants could be improved. Greater N accumulation rates in wheat grains were associated with increased N per seed[39]. Rates and duration of N deposition varied among the genotypes. Dubois and Fossati[64] noted that grain protein concentration and N partitioning depended on the translocation of carbon compounds. Greater redistribution of N from the leaves and stems to the grain were noted when wheat plants were deprived of N or had less than adequate amounts of N during rapid growth phases compared to plants supplied with sufficient N[130].

Genotype differences among maize inbreds and hybrids were noted for NO_3^- absorption, translocation, accumulation, and partitioning[11,43,95,150,156,163] Nitrogen absorption differences of inbreds and hybrids made from the same inbreds were similar in 5- to 8-week-old plants as they were in 12-week-old plants[43]. This indicated that young plants may be used to predict or evaluate differences in older plants. Although N absorption rates differed, plant N concentrations did not differ. Differences were noted among the maize genotypes for total N, NO_3^-, and reduced N in most plant parts as well as in N absorption, reduction, and partitioning. Relationships between the parents and F_1 progeny indicated that N partitioning did not always carry over into the F_1 parent. Unreduced NO_3^- accumulated in the stems of some genotypes of maize[163]. Reed and Hageman[150] also noted maize genotype differences for $_3^-$ accumulation in the stalks and leaf midribs. Even though higher NO_3^- uptake was associated with more extensive root systems, genotype differences for partitioning of NO_3^- were not apparent.

Hay et al.[95] noted that the proportion of N translocated into maize grains before and after the silk stage was genotype dependent. For example, one maize hybrid translocated 56% of its N to the grain after silk compared to 43% for another hybrid. Beauchamp et al.[11] suggested that the potential exists for screening maize genotypes for differences in N translocation, efficiency, and accumulation.

In other studies, N uptake and translocation appeared to be independent of each other[154]. Maize genotypes with high N uptake showed both high and low N harvest indicies.

Sorghum [*Sorghum bicolor* (L.) Moench] genotypes that differed in dry matter yield, N content, and dry matter yield per unit N when grown with limited N varied over 2-fold in N uptake rates per unit root weight[75]. Genotypes with higher dry matter yields and higher N contents exhibited lower NO_3^- uptake rates compared to genotypes with lower dry matter yields and lower N contents. Nitrate uptake rates were more than three times higher for young (24- to 34-day-old) plants than for 44-day-old plants. Forty-four-day-old plants absorbed N more than 2-fold faster than 54-day-old plants. Total N accumulated by the plants increased as plants aged even though uptake rates decreased. Genotype differences in NO_3^- uptake rates varied by 50% or more at all plant ages. Genotypes showing the highest uptake rates as young plants usually had higher uptake rates as older plants. Partitioning of the N in grain compared to the leaves, stalks, and roots did not necessarily follow N uptake rate patterns. One sorghum genotype (SC150-6 × SC150-9) partitioned nearly 35% of the total plant N into the grain compared to 10 to 15% for two other genotypes (TX3934 × GH-8-17 and SC110 × SC120) studied. A much lower proportion of the N remained in the roots of SC150-6 × SC150 than in the other genotypes.

Differences in N concentrations among the genotypes as a function of age were relatively small, thus, the genotype responses were manifested through the partitioning of the N from the various plant parts to the grain.

Sorghum genotypes show marked differences in N efficiency (dry matter produced per unit N, grain produced per unit N, and grain produced per unit N × grain N/stover N ratio), but environmental factors strongly affected genotype responses from year to year[121]. Both male and female parents influenced the hybrid responses to N for plants grown in the field. Large differences were noted among genotypes for N partitioning between grain and stover.

Dry matter yield varied by 45% among 146 tomato strains grown with limited N[136]. Nitrogen efficiency ratios (dry matter produced per unit N) varied among the genotypes. Nitrogen uptake and translocation from the roots to the tops could not explain the genotype differences to N. At equivalent N concentrations, N-efficient plants produced larger lower leaves and maintained more normal tissues. Higher plant N efficiency ratios were associated with higher stem weights and lower stem N concentrations.

Phosphorus

Examples of plant genotype differences in P uptake, accumulation, and use are numerous[12,42,52,70,79,103,114,189]. These reports, as well as others, describe genotype responses to P and give explanations as to why certain genotypes are able to adapt to various levels of P, especially to low P.

Uptake and accumulation of P by maize has been studied extensively by scientists at the Pennsylvania State University[6,7,8,9,10,14,82,83,84,85,173,174]. Differences among maize genotypes with high and low P accumulation properties were attributed to root size, the ability to compete effectively for P, the ability to tolerate detrimental effects of competing or harmful elements like Al, and to root phosphatase activities[51]. Phosphorus uptake kinetics by roots were not different between two maize hybrids known to accumulate different amounts of P^{140}. Leaf uptake rates, however, were different for the two hybrids. The differences were attributed to the relative strengths of the P carrier bond[140] and to the differences in the formation of organic P compounds[141]. Other studies on P uptake showed that maize inbreds and hybrids varied in the types and sizes (dry weights and lengths) of their roots, Imax (maximum influx) rates, Km (Michaelis-Menten) constants, and Cm (P concentration remaining in solution when net influx was zero) values[131]. Diversity in these root traits among the genotypes were reflected in their P uptake when grown in the field. Differences among maize genotypes for P uptake were 2- to 3-fold[161]. Early maturing maize genotypes appeared to absorb and accumulate more P than late maturing genotypes[37]. Variation in the growth of maize genotypes was partially explained by factors affecting vigor during early growth and not by P absorption characteristics[74].

Marked differences in the growth and P uptake were noted among sorghum genotypes grown in low P soils[55] and in nutrient solutions containing limited P[54]. Differences were noted among sorghum genotypes for distribution of P between the roots and the tops between the upper leaves and the lower leaves[54]. Some genotypes contained more P in the roots than in the tops; others contained more P in the tops than in the roots including more P in the upper leaves than in the lower leaves. Phosphorus concentrations and contents and dry matter produced per unit P varied extensively among the genotypes.

Phosphorus uptake rates varied among four sorghum genotypes[77]. Plant age and P treatment level were important factors affecting the process. The genotypes selected for the study showed high and low responses to growth, P concentrations and contents, dry matter produced per unit P, and P distribution. The genotype that showed the highest uptake rates when young (24 days old) exhibited the highest uptake rates as older plants (38 and 52 days old), and had the highest uptake rate as P levels increased. Likewise, the genotype with the lowest uptake rate as young plants maintained the lowest uptake rate when older and also had the lowest uptake rate as P levels increased. Differences among genotypes diminished as the plants aged, but their relative rankings among the genotypes continued. The genotype with the highest dry matter yield at each age had the lowest P uptake rate, and the genotype with the lowest dry matter yield at each age had the highest P uptake rate. In the same study, partitioning of P among the plant parts varied; the largest differences were noted in the older plants. The genotype considered to be the most "P-efficient" had lower top/root ratios of P, but higher upper leaf/lower leaf P ratios, and higher dry matter produced per unit P than "P-inefficient" genotypes. The "P-efficient" genotype appeared to have greater ability to retranslocate P from inactive to active tissues, lower P uptake rates, and larger root systems. Growth and P absorption rates differed when grown with organic and inorganic sources of P[76]. It should be pointed out that favourable P efficiency traits in plants may be detrimental to the availability and utilization of certain other elements like Fe and Cu[28,31,32,33,34].

Differences in P absorption and translocation by bean (*Phaseolus vulgaris* L.) genotypes were also noted[112,152,188]. Excised roots from 59 snap bean lines showed large variations in their P absorption rates[112]. However, the absorption rate was negatively correlated with root dry weight. Using excised roots of the same weight, large differences in P absorption appeared among the genotypes. Higher pretreatment levels of P resulted in lower P absorption rates. The relative differences between "P-efficient" (higher dry matter produced per unit P) and "P-inefficient" genotypes were similar at each level of P. Growth in bean lines grown with limited P (2 mg/plant) differed by 72% and dry matter produced per unit P differed by 77%. Differences were also noted among bean lines grown at high P[152].

Potassium

Differences in the accumulation of K in plants[13,65,89] have led scientists to investigate variations in K uptake and transport. Potassium uptake rates among tobacco (*Nicotiana tabacum* L.) genotypes grown at relatively low levels of K were genotype specific, but the differences disappeared with increased levels of K[96].

Potassium absorption potentials were similar in soybean [*Glycine max* (L.) Merr.] genotypes grown with low K, even though the root morphology was markedly different[146]. At relatively high K levels, the K absorption potential for one soybean genotype was nearly double that of another genotype. Uptake efficiencies for K and Rb (and SO_4) were different for barley (*Hordeum vulgare* L.) genotypes[101,139,162]. Pettersson[139] reported that Rb uptake by barley genotypes was controlled by the plasmalemma of root cells. Influx differences of about 75% were noted among the genotypes. In later studies, K contents inside plants did not correlate well with influx rates[101]. Genotypic differences were attributed to net K (Rb) transport to the shoot. Differences in both influx and efflux were noted.

In other studies, Glass and Perley[80] studied the differences among barley genotypes for K influx and transport rates. No efflux of K was noted for these genotypes. A high correlation was obtained between growth rates and K influx rates. Close relationships were also noted between K (Rb) absorption and H^+ release by barley roots which could possibly be used as a screening technique for the determinaton of K differences in plant genotypes[81]. Differences of 200 to 250% in the translocation rates of K were found among ryegrass (*Lolium multiflorum* Lam. and *L. perenne* L.) genotypes[66]. These differences were attributed to the relative rates of K uptake by the plants.

In extensive studies on genotype differences to K, 66 bean[166] and 156 tomato[120] genotypes were screened for variations to K efficiency (dry matter produced per unit K absorbed) in plants grown with limited K (5 mg/plant). Plant dry matter yields varied by 300% among the bean genotypes and by 79% among the tomato genotypes. Genotypes grown at high K showed fewer differences than genotypes grown at low K. The ability of some bean genotypes to produce normal growth under low K conditions was not due to seed size or greater competition among root systems. Variations in K efficiency among the bean genotypes were not associated with higher K absorption or with higher K concentrations in the "K-efficient" plants, nor with Na substitution for K, but with K utilization inside the plants. The "K-efficient" tomato genotypes had 39% less K and 29% more Na in their tissues when grown at low levels of K. Differences in the uptake and translocation of K from the roots to the shoots could not explain the genotype responses. Partial substitution of Na for K was important in K efficiency, but much of the K efficiency was associated with functions of K for which Na could not substitute.

Calcium and magnesium

Certain maize genotypes grew better than others at low levels[60]. One genotype required 50 mg Ca/g for normal growth compared to only 10 mg Ca/g for another genotype. Calcium and Mg accumulations in leaf samples of 31 maize inbreds ranged from 0.28 to 1.10% Ca and from 0.3 to 1.5% Mg when grown in gravel culture[158]. Thirteen inbred lines of maize grown in the field showed ranges in the Ca concentration between 0.69 and 1.78% and Mg concentration between 0.06 and 0.34%. Thirty-one inbred and hybrid entries ranged from 0.50 to 1.90% Ca and from 0.45 to 1.17% Mg[159]. Thomas et al.[174] noted ranges of 0.74 to 1.26% Ca and 0.11 to 0.22% Mg and Gorsline et al.[83] noted ranges of 0.84 to 1.14% Ca and 0.11 to 0.21% Mg in field-grown maize inbreds and single-cross hybrids. Even though the accumulations of Ca and Mg were heritable, Ca and Mg concentrations had no obvious relationships to the accumulation of other elements[83,174]. Differences in Ca and Mg accumulation by mature field-grown maize plants could be predicted from young plants grown under greenhouse conditions[8]. The genotypes maintained their relative differences when grown at various locations[173]. High or low accumulations of a particular element by one genotype did not mean that other elements accumulated in a similar manner[8]. Calcium and Mg accumulations in maize were associated with soil pH[115], maturity class of plants[38], temperature[145], and with deficiencies of other elements like Cu[19].

Studies similar to those discussed for maize were conducted for Ca and Sr accumulation in barley, wheat, and soybean plants[71,106,107,147,148,149,190]. The element with the greatest differences for accumulation was Ca[147]. Differential responses of other plant genotypes to Ca and Mg have been noted[143,169,170].

Maize inbreds grown on low Ca and Mg soils differed in their Ca and Mg concentrations and in dry matter yields[50]. Some of these inbreds were grown in nutrient solutions to better understand genotype responses to low and high Ca and Mg[44,47]. A more "Ca-efficient" genotype ('Oh43'), produced more dry matter, developed fewer Ca deficiency or toxicity symptoms, and produced greater dry matter per unit Ca when grown at both low and high Ca than did a more "Ca-inefficient" genotype ('A251')[47]. 'Oh43' produced equivalent dry matter at about one-fourth the level of Ca in solution as did 'A251'. High levels of K and Mg, known to interact with Ca, were more detrimental to 'A251' than to 'Oh43'. Concentrations of Ca in 'A251' leaves and roots were slightly higher than in 'Oh43' which indicated that Ca uptake and translocation could not explain the differences between the genotypes. 'A251' appeared to have a higher solution Ca requirement and a higher critical Ca concentration than 'Oh43'. Differences between the inbreds for functional Ca requirement appeared to be relatively small.

A more "Mg-efficient" genotype ('B57'), produced more dry matter, developed fewer Mg deficiency symptoms, and had higher dry matter production per unit Mg than did a more 'Mg-inefficient" genotype ('Oh40B') when grown at low levels of Mg[44]. On the other hand, when these genotypes were grown at relatively high levels of Mg, 'Oh40B' responded more favorably to Mg than did 'B57'. To produce comparable dry matter at low Mg, "Oh40B' required ten times the solution level of Mg than did 'B57'. Regardless of the Mg level, 'B57' leaves contained about two times higher Mg concentrations, and the roots contained about one-half the Mg concentrations than did 'Oh40B', 'Oh40B' absorbed about as much Mg as 'B57', but 'Oh40B' was less able to translocate it as effectively to the leaves. Similar results have been observed for other maize inbreds showing differential responses to Mg[72,157,160].

Manganese

Differential uptake and accumulation of Mn has been reported for plants grown at low levels of Mn[30,78,124,126,133,134]. Oat (*Avena sativa* L.) genotype rankings for tolerance to low Mn were related to the Mn concentration, Mn uptake, and Mn contents[124]. Environmental factors (Ca, Fe, and Mn levels, pH, source of N, and temperature) influenced the distribution of Mn between the roots and shoots, but did not generally change genotype rankings for low Mn tolerance. Wheat, barley, and oat genotype differences in Mn deficiency were attributed to the inability of some genotypes to absorb Mn rather than to the higher requirement of Mn inside the plant[133]. Some genotypes with higher tissue Mn concentrations showed greater tolerance to low Mn in solution than plants with lower tissue Mn concentrations. Oat genotypes that grew best exhibited fewer Mn deficiency symptoms at low levels of Mn and had higher Ca in their tissues[30]. It was suggested that Ca substituted for Mn in certain non-specific reactions. Sorghum genotypes showed differences in dry matter production at low levels of Mn, but did not necessarily absorb more Mn (Clark, unpublished). No difference in the translocation of Mn from roots to tops were noted for the sorghum genotypes.

Boron

Plant genotype differences in response to low B have been documented[4,24,105,144,164,172,181,182]. Boron uptake appears to be controlled by the plant roots[24]. Eaton and Blair[68] found that sunflower (*Helianthus annuus* L.) leaves contained more B when grown on artichoke (*Helianthus tuberosus* L.) rootstock than when grown on its own rootstock. Conversely, artichoke on sunflower rootstock had less B than when grown on its own rootstock. Brown and Jones[29] found lower B in the leaves of tomato genotypes grown on a low B susceptible rootstock, but more B when grown on a B tolerant rootstock even though the genotypes showed differences in tolerance or susceptibility to low B. The physiological requirements for B in the leaves of both tomato genotypes appeared to be similar[23]. Differential accumulations of B have also been reported for various maize genotypes[44,82,90].

Copper

Genotype differences to Cu deficiency have been reported for many plant species[24,35,86,118,128,129]. Differences among genotypes have been attributed to variations in Cu uptake patterns or availability due to larger or smaller root systems, especially to increased root lengths with more branching[86,88,167]. Copper efficiency was associated with higher Cu in the shoots or the greater capacity to transfer Cu from roots to shoots[86,88,167]. Copper deficiency was accentuated by high P. High P also suppressed the translocation of Ca[24,27,28,35]. As such, Cu deficiency imitated Ca deficiency, and Cu deficiency may be Ca deficiency. Changes in metabolism and reductive capacity have also been noted in Cu deficient plants[24,27]. Tolerance to Cu deficiency has been linked to one or two genes[86,118], and Cu efficient traits have been transferred from Cu efficient plants like rye (*Secale cereale* L.) to triticale (*Triticum* × *Secale*) to make triticale more Cu efficient[86,88,92]. The Cu efficient trait in triticale has also been transferred to wheat, a Cu inefficient plant species[87].

Zinc

Differences among plant genotypes for the absorption, translocation, accumulation, and use of Zn have been reported for many plant species[24,32,33,34,35,48,50,113,119,137,142]. Even though the physiological activities of the roots influence Zn absorption, the interactions with many elements like P, Fe, Mn, and Cu effect Zn uptake[1,2,3,35,48,104,135,155,171,186]. Zinc deficient plants have higher concentrations of these elements as well as abnormal ratios of these elements to Zn. These elements inactivate or interact with Zn inside the plant rather than in the growth media. Roots usually show more abnormal element concentrations and ratios than tops.

Greater susceptibility to Zn deficiency in bean plants was associated with high P and Fe concentrations in the tissues[2,3]. A Zn susceptible oat genotype absorbed and accumulated more Fe than a genotype that was more tolerant to Zn deficiency[35].

Maize genotypes that showed differences in their susceptibility and tolerance to Zn deficiency in soils[50] were grown in nutrient solutions to characterize their physiological properties[48]. The susceptible genotype ('A635') had higher concentrations of P, Ca, Mg, Mn, Fe, and Cu, showed greater changes in these elements as the Zn level in the growth media varied, and had more abnormal ratios of these elements to Zn than a genotype tolerant to Zn deficiency ('H84'). 'H84' produced more dry matter, had higher shoot/root ratios of dry matter, developed fewer Zn deficiency symptoms, had higher Zn concentrations and contents, produced more dry matter per unit Zn, and translocated more Zn to the shoots compared to the roots than 'A635' when grown at low levels of Zn. 'H84' appeared to have a lower Zn requirement than 'A635'. When the genotypes were grown with adequate Zn, 'A635' responsed more to Zn than did 'H84'. In other studies, 'H84' absorbed Zn faster than the Zn susceptible genotype ('A632') in short-term studies, but Zn absorption rates could not explain differences in susceptibility or tolerance to Zn deficiency in single-cross hybrids made from these parents (Kimball, J. G., 1978, M. S. Thesis, Univ. Kentucky, Lexington, KY). In other investigations using maize genotypes with wide differences in their susceptibilities and tolerances to Zn deficiency, Peaslee et al.[137] found no differences in root absorption of Zn or translocation of absorbed Zn to shoots for the genotypes. The genotypes that produced the most dry matter at low Zn levels ('Conico' composite) also produced more dry matter per unit Zn absorbed. Like 'H84', 'Conico' also produced more dry matter with adequate Zn than did the genotype less tolerant to Zn deficiency (Pioneer '3369A' hybrid). Absorption and translocation rates of ^{65}Zn applied to leaves were greater in younger leaves than in older leaves. Translocation of ^{65}Zn was greater in '3369A' than in 'Conico'.

Molybdenum

Genotype differences to Mo deficiency have been described for brassicas, many legumes, and a wide range of dicotylendonous plants[45]. Responses of monocotyledonous plants to Mo are somewhat unusual, but maize genotypes differences to Mo have been reported[26,62,116,117,132,187]. Maize hybrids of American origin were found to be more susceptible to Mo deficiency than hybrids of Spanish origin[62]. Broad ranges of Mo concentrations were found in the leaves of various hybrids grown with and without Mo. Seedborne Mo could not account for the differences between the American and Spanish maize genotypes[62], but seedborne Mo could account for differences reported by Weir and Hudson[187]. In the former study, observed differences among the maize genotypes were attributed to differences in Mo absorption from the soil. The maize genotype 'C103' repeatedly had lower Mo than many other maize genotypes grown in acid soils at various pH values[116,117].

Brown and Clark[26] identified the maize genotype 'Pa36' as a plant that grew poorly on acid soils. 'Pa36' and 'WH' (a more tolerant genotype to low Mo) were compared when grown on an acid soil. They both showed differences in P uptake and Al tolerance; 'Pa36' was more effective in taking up P and was more Al tolerant than 'WH'[46,51]. From mineral element data, 'Pa36' continually had lower Mo than 'WH' regardless of the amendments, other than Mo, added to the soil in an attempt to alleviate the problem. When Mo was added to the soil at increased increments, 'Pa36' growth improved. At the higher Mo additions, 'Pa36' growth was normal on the acid soil.

Conclusion

Plant genotypes of many crop species show important differences in their uptake, translocation, accumulation, and use of the essential mineral elements. Advantage should be taken of these differences to improve plants for growth under defined or constrained fertilizer conditions and for the adaptation of plants to mineral stress conditions. This approach should be suited particularly to soils where fertilizer constraints exist and where soil mineral stress conditions are difficult to correct or amend. Sufficient information is available relative to plant responses to essential mineral elements to suggest that plant selection and adaptation to mineral stress conditions is feasible and practical.

References

1 Adriano D C, Paulsen G M and Murphy L S 1971 Phosphorus-iron and phosphorus-zinc relatonships in corn (*Zea mays* L.) seedlings as affected by mineral nutrition. Agron. J. 63, 36–39.
2 Ambler J E and Brown J C 1969 Cause of differential susceptibility to zinc deficiency in two varieties of navy beans (*Phaseolus vulgaris* L.). Agron. J. 61, 41–43.
3 Ambler J E, Brown J C and Gauch H G 1970 Effect of zinc on translocation of iron in soybean plants. Plant Physiol. 46, 320–323.
4 Andrus C 1955 Brittle stem, an apparently new sublethal gene in tomato. Tomato Genetics Coop. Rep. 5, 5.
5 Antonovics J, Bradshaw A D and Turner R G 1971 Heavy metal tolerance in plants. Adv. Ecol. Res. 7, 1–85.
6 Baker D E, Bradford R R and Thomas W I 1967 Accumulation of Ca, Sr, Mg, P, and Zn by genotypes of corn (*Zea mays* L.) under different soil fertility levels. pp 465–477. *In* Isotypes in Plant Nutrition and Physiology. Int. Atom. Eng. Ag., Vienna, Austria.
7 Baker D E, Jarrell A E, Marshall L E and Thomas W I 1970 Phosphorus uptake from soils by corn hybrids selected for high and low phosphorus accumulation. Agron. J. 62, 103–106.
8 Baker D E, Thomas W S and Gorsline G W 1964 Differential accumulations of strontium, calcium and other elements by corn (*Zea mays* L.) under greenhouse and field conditions. Agron. J. **56**, 352–355.
9 Baker D E, Wooding F J and Johnson M W 1971 Chemical element accumulation by populations of corn (*Zea mays* L.) selected for high and low accumulations of P. Agron. J. 63, 404–406.
10 Barber W D, Thomas W I and Baker D E 1967 Inheritance of relative phosphorus accumulation in corn (*Zea mays* L.). Crop Sci. 7, 104–107.
11 Beauchamp E G, Kannenberg L W and Hunter R B 1976 Nitrogen accumulation and translocation in corn genotypes following silking. Agron. J. 68, 418–422.
11a Bennett J H, Lee E H, Kirzek D T, Olsen R A and Brown J C 1982 Photochemical reduction of iron. II. Plant related factors. J. Plant Nutr. 5, 335–344.
11b Bennett J H, Olsen R A and Clark R B 1982 Modification of soil fertility by plant roots: Iron stress-response mechanism. What's New in Plant Physiology 13(1), 1–4.
12 Bielelski R L 1973 Phosphate pools, phosphate transport and phosphate availability. Annu. Rev. Plant Physiol. 24, 225–252.

13   Bortner C E, Wallace A M and Hamilton J L 1960 Differences in potassium, nitrogen and total alkaloid concentration in ten burley tobacco varieties. Tobacco Sci. 4, 151–155.
14   Bradford R R, Baker D E and Thomas W I 1966 Effect of soil treatments on chemical element accumulation of four corn hybrids. Agron. J. 58, 614–617.
15   Brown J C 1956 Iron chlorosis. Annu. Rev. Plant Physiol. 7, 171–190.
16   Brown J C 1961 Iron chlorosis in plants. Adv. Agron. 13, 329–369.
17   Brown J C 1963 Interactions involving nutrient elements. Annu. Rev. Plant Physiol. 14, 93–106.
18   Brown J C 1966 Fe and Ca uptake as related to root-sap and stem-exudate citrate in soybeans. Physiol. Plant. 19, 968–976.
19   Brown J C 1967 Differential uptake of Fe and Ca by two corn genotypes. Soil Sci. 103, 331–338.
20   Brown J C 1977 Genetically controlled chemical factors involved in absorption and transport of iron in plants. pp. 93–103. *In* Advances in Chemistry Series, No. 162. Ed. K N Raymond. Bioinorganic Chemistry-II. Am. Chem. Soc., Washington, D C.
21   Brown J C 1978 Mechanism of iron uptake by plants. Plant Cell Environ. 1, 249–257.
22   Brown J C 1978 Physiology of plant tolerance to alkaline soils. pp. 257–276. *In* Crop Tolerance to Suboptimal Land Conditions. Ed G A Jung. V. Am. Soc. Agron., Madison, WI.
23   Brown J C and Ambler J E 1973 Genetic control of uptake and a role of boron in tomato. Soil Sci. Soc. Am. Proc. 37, 63–66.
24   Brown J C, Ambler J E, Chaney R L and Foy C D 1972 Differential responses of plant genotype to micronutrients. pp 389–418. *In* Micronutrients in Agriculture. Eds. J J Mortvedt, P M Giordano and W C Lindsay. Soil Sci. Soc. Am., Madison, WI.
25   Brown J C and Chaney R L 1971 Effect of iron on the transport of citrate into the xylem of soybeans and tomatoes. Plant Physiol. 47, 836–840.
26   Brown J C and Clark R B 1974 Differential response of two maize inbreds to molybdenum stress. Soil Sci. Soc. Am. Proc. 38, 331–333.
27   Brown J C and Clark R B 1977 Copper as essential to wheat production. Plant and Soil 48, 509–523.
28   Brown J C, Clark R B and Jones W E 1977 Efficient and inefficient use of phosphorus by sorghum. Soil. Sci. Soc. Am. J. 41, 747–750.
29   Brown J C and Jones W E 1971 Differential transport of boron in tomato (*Lycopersicon esculentum* Mill.). Physiol. Plant. 25, 279–282.
30   Brown J C and Jones W E 1974 Differential response of oats to manganese stress. Agron. J. 66, 624–626.
31   Brown J C and Jones W E 1975 Phosphorus efficiency as related to iron inefficiency in sorghum. Agron. J. 68, 468–472.
32   Brown J C and Jones W E 1977 Fitting plants nutritionally to soils. I. Soybeans. Agron. J. 69, 399–404.
33   Brown J C and Jones W E 1977 Fitting plants nutritionally to soils. II. Cotton. Agron. J. 69, 405–409.
34   Brown J C and Jones W E 1977 Fitting plants nutritionally to soils. III. Sorghum. Agron. J. 69, 410–414.
35   Brown J C and McDaniel M E 1978 Factors associated with differential response to two oat cultivars to zinc and copper stress. Crop Sci. 18, 817–820.
36   Brown J C and Tiffin I O 1965 Iron stress as related to the iron and citrate occurring in stem exudate. Plant Physiol. 40, 395–400.

37 Bruetsch T F 1976 Physiological factors affecting the differential uptake and accumulation of phosphorus by long and short season genotypes of maize. Diss. Abstr. 37, 5472–B (1977).
38 Bruetsch T F and Estes G O 1976 Genotype variation in nutrient uptake deficiency in corn. Agron. J. 68, 521–523.
39 Brunori A, Axmann H, Figueroa A and Micke A 1980 Kinetics of nitrogen and dry matter accumulation in the developing seed of some varieties and mutant lines of *Triticum aestivum*. Z. Pflanzenzuecht. 84, 201–218.
40 Butler G W, Barclay P C and Glenday A C 1962 Genetic and environmental differences in mineral composition of ryegrass herbage. Plant and Soil 16, 214–228.
41 Chaney R L, Brown J C and Tiffin L O 1972 Obligatory reduction of ferric chelates in iron uptake by soybeans. Plant Physiol. 50, 208–213.
42 Chapin F C III 1980 The mineral nutrition of wild plants. Annu. Rev. Ecol. Syst. 11, 233–260.
43 Chavalier P and Schrader L E 1977 Genotypic differences in nitrate absorption and partitioning of N among plant parts in maize. Crop Sci. 17, 897–901.
44 Clark R B 1975 Differential magnesium efficiency in corn inbreds. I. Dry matter yields and mineral element composition. Soil Sci. Soc. Am. Proc. 29, 488–491.
45 Clark R B 1976 Plant efficiencies in the use of calcium, magnesium, and molybdenum. pp 175–191. *In* Plant Adaptation to Mineral Stress in Problem Soils. Ed. M J Wright. Cornell Univ. Agric. Exp. Stn., Ithaca, NY.
46 Clark R B 1977 Effect of aluminium on growth and mineral elements of Al-tolerant and Al-intolerant corn. Plant and Soil 45, 653–662.
47 Clark R B 1978 Differential response of corn inbreds to calcium. Commun. Soil Sci. Plant Anal. 9, 729–744.
48 Clark R B 1978 Differential response of maize inbreds to Zn. Agron. J. 70, 1057–1060.
49 Clark R B 1982 Plant response to mineral element toxicity and deficiency. pp 71–142. *In* Breeding Plants for Less Favorable Environments. Eds. M N Christiansen and C F Lewis. John Wiley & Sons, New York, NY.
50 Clark R B and Brown J C 1974 Differential mineral uptake by maize inbreds. Commun. Soil Sci. Plant Anal. 5, 213–227.
51 Clark R B and Brown J C 1974 Differential phosphorus uptake by phosphorus-stressed corn inbreds. Crop Sci. 14, 505–508.
52 Clark R B and Brown J C 1980 Role of the plant in mineral nutrition as related to breeding and genetics. pp 45–70. *In* Moving up the Yield Curve. Advances and Obstacles. Eds. L S Murphy, E C Doll and L F Welch. Am. Soc. Agron., Madison, WI.
53 Clark R B, Brown J C, Olsen R A and Bennett J H 1981 Biological effects on plants. pp 272–338. *In* Multimedia Criteria for Iron and Compounds. Envir. Prot. Ag., Cincinnati, OH.
54 Clark R B, Maranville J W and Gorz H J 1978 Phosphorus efficiency of sorghum grown with limited phosphorus. pp 93–99. *In* Eds. A R Ferguson, R L Bieleski and I B Ferguson. Proc. 8th Int. Colloq. Plant Anal. Fert. Probl., Auckland, New Zealand.
55 Clark R B, Maranville, J W and Ross W M 1977 Differential phosphorus efficiency in sorghum. pp 1–2. *In* Proc 10th Grain Sorghum Research Utilization Conference, Wichita, KS.
56 Clark R B, Tiffin L O and Brown J C 1973 Organic acids and iron translocation in maize genotypes. Plant Physiol. 52, 147–150.
57 DeKock P C and Morrison R I 1958 The metabolism of chlorotic leaves. 2. Organic acids. Biochem. J. 70, 272–277.
58 Della A and Hadjichristodoulou A 1975 Genetic variability in uptake of nitrogen at various

growth stages of barley and wheat under dryland conditions. pp 85–96. Proc. of the 3rd Research Coordination Meeting, Hahnenklee, FRG.
59 Desai R M and Bhatia C R 1978 Nitrogen uptake and nitrogen harvest index in durum wheat cultivars varying in their grain protein concentration. Euphytica 27, 561–566.
60 DeTurk E E 1941 Plant nutrient deficiency symptoms. Physiological basis. Ind. Eng. Chem. 33, 648–653.
61 Devine T E 1982 Genetic fitting of crops to problem soils. pp 143–173. *In* Breeding Plants for Less Favorable Environments. Eds. M N Christiansen and C F Lewis. John Wiley & Sons, New York, NY.
62 Dios Vidal R and Broyer T C 1965 Deficiency symptoms and essentiality of molybdenum in corn hybrids. Agrochimica 9, 273–284.
63 D'Souza T J and Mistry K B 1979 Uptake and distribution of gamma-emitting activation products ^{59}Fe, ^{58}Co, ^{54}Mn and ^{65}Zn in plants. Environ. Exp. Bot. 19, 193–200.
64 DuBois J B and Fossati A 1981 Influence of nitrogen uptake and nitrogen partitioning efficiency on grain yield and grain protein concentration of twelve winter wheat genotypes (*Triticum aestivum* L.). Z. Pflanzenzuecht. 86, 41–49.
65 Dunlop J 1975 Differences in xylem exudation by seminal roots of *Lolium* varieties. New Phytol. 74, 19–23.
66 Dunlop J and Tomkins B 1976 Genotypic differences in potassium translocation in rye-grass. pp 145–152. *In* Transport and Transfer Processes in Plants. Eds. I F Wardlaw and J B Passioura. Academic Press, New York, NY.
67 Duvick D N, Kleese R A and Frey N M 1981 Breeding for tolerance of nutrient imbalances and constraints to growth in acid, alkaline and saline soils. J. Plant Nutr. 4, 111–129.
68 Eaton F M and Blair G Y 1935 Accumulation of boron by reciprocally grafted plants. Plant Physiol. 10, 411–422.
69 Epstein E 1972 Mineral Nutrition of Plants: Principles and Perspectives. John Wiley & Sons, New York, NY.
70 Epstein E and Jefferies R L 1964 The genetic basis of selective ion transport in plants. Annu. Rev. Plant Physiol. 15, 169–184.
71 Fick G N and Rasmusson D C 1967 Heritability of Sr-89 and Can-45 accumulation in barley seedlings. Crop Sci. 7, 315–317.
72 Foy C D and Barber S A 1958 Magnesium absorption and utilization by two inbred lines of corn. Soil. Sci. Soc. Am. Proc. 22, 57–62.
73 Foy C D, Chaney R L and White M C 1978 The physiology of metal toxicity in plants. Annu. Rev. Plant Physiol. 29, 511–566.
74 Fox R H 1978 Selection for phosphorus efficiency in corn. Commun. Soil Sci. Plant Anal. 9, 13–37.
75 deFranca G E 1981 Differences in dry-matter yield and the uptake, distribution, and use of nitrogen by sorghum genotypes. Ph.D. Thesis, Univ. of Nebraska, Lincoln, NE. Diss. Abstr. 41:4018B.
76 Furlani A M C, Clark R B, Maranville J W and Ross W M 1982 Sorghum genotype differences to organic and inorganic phosphorus compounds. *In* Sorghum in the Eighties, Ed. L R House. Int. Crops Res. Inst. Semi-Arid Trop., Hyderabad, India (in press).
77 Furlani A M C, Clark R B, Maranville J W and Ross W M (in preparation). Sorghum genotype differences in phosphorus uptake rate, distribution, and root phosphatase activity.
78 Gallagher P H and Walsh T 1943 The susceptibility of cereal varieties to manganese deficiency. J. Agric. Sci. 33, 197–203.

79 Gerloff G C 1963 Comparative mineral nutrition of plants. Annu. Rev. Plant Physiol. 14, 107–124.
80 Glass A D M and Perley J E 1980 Varietal differences in potassium uptake by barley. Plant Physiol. 65, 160–164.
81 Glass A D M, Siddiqi M Y and Giles K I 1981 Correlations between potassium uptake and hydrogen efflux in barley varieties. Plant Physiol. 68, 457–459.
82 Gorsline G W, Baker D E and Thomas W I 1965 Accumulation of eleven elements by field corn (*Zea mays* L.). Penn. Agric. Exp. Stn. Bull. 725.
83 Gorsline G W, Ragland J L and Thomas W I 1961 Evidence for inheritance of differential accumulation of calcium, magnesium and potassium by maize. Crop Sci. 1, 155–156.
84 Gorsline G W, Thomas W I and Baker D E 1964 Inheritance of P, K, Mg, Ca, B, Mn, Al, and Fe concentrations by corn (*Zea mays* L.) leaves and grain. Crop Sci. 4, 207–210.
85 Gorsline G W, Thomas W I and Baker D E 1968 Major gene inheritance of Sr-Ca, Mg, K, P, Zn, Cu, B, Al-Fe, and Mn concentration in corn (*Zea mays* L.). Penn. Agric. Exp. Stn. Bull. 746.
86 Graham R D 1978 Tolerance of triticale, wheat and rye to copper deficiency. Nature, London 281, 542–543.
87 Graham R D 1978 Nutrient efficiency objectives in cereal breeding. pp 165–170. *In* Eds. A R Ferguson, R L Bieleski and I B Ferguson. Plant Nutr. 1978., Proc. 8th Int. Colloq. Plant Anal. Fert. Prob., Auckland, New Zealand.
88 Graham R D, Anderson G D and Ascher J S 1981 Absorption of copper by wheat, rye and some hybrid genotypes. J. Plant Nutr. 3, 679–686.
89 apGriffiths G and Walters R J K 1966 The sodium and potassium content of some grass genera, species and varieties. J. Agric. Sci., Camb. 67, 81–89.
90 Gupta U C 1979 Boron nutrition of crops. Adv. Agron. 31, 273–307.
91 Halloran G M and Lee J W 1979 Plant nitrogen distribution in wheat cultivars. Aust. J. Agric. Res. 30, 779–789.
92 Harry S P and Graham R D 1981 Tolerance of triticale, wheat and rye to copper deficiency and low and high soil pH. J. Plant Nutr. 3, 721–730.
93 Harvey P H 1939 Hereditary variation in plant nutrition. Genetics. 24, 437–461.
94 Haunold A, Johnson V A and Schmidt J W 1962 Genetic measurements of protein in the grain of *Triticum aestivum* L. Agron. J. 54, 203–206.
95 Hay R E, Earley E B and DeTurk E E 1953 Concentration and translocation of nitrogen compounds in the corn plant (*Zea mays* L.) during grain development. Plant Physiol. 28, 606–621.
96 Hiatt A J 1963 Varietal differences in potassium uptake by excised roots of *Nicotiana tabacum*. Plant and Soil 18, 273–276.
97 Hoener I R and DeTurk E E 1938 The absorption and utilization of nitrate nitrogen during vegetative growth by Illinois high protein and Illinois low protein corn. J. Am. Soc. Agron. 30, 232–243.
98 Huffaker R C and Rains D W 1978 Factors influencing nitrate acquisition by plants: Assimilation and fate of reduced nitrogen. pp 1–43. *In* Nitrogen in the Environment: Soil-Plant-Nitrogen Relationships, Vol. 2. Eds. D R Nielsen and J G MacDonald Academic Press, New York, NY.
99 Iljin W S 1951 Metabolism of plants affected with lime-induced chlorosis (calciose). I. Nitrogen metabolism. Plant and Soil 3, 239–256.
100 Jackson W A 1978 Nitrate acquisition and assimilation by higher plants: Processes in the root

system. pp 45–88. *In* Nitrogen in the Environment: Soil-Plant-Nitrogen Relationships, Vol. 2. Eds. D R Nielsen and J G MacDonald. Academic Press, New York, NY.
101 Jensen P and Pettersson S 1980 Varietal variation in uptake and utilization of potassium (rubidium) in high-salt seedlings of barley. Physiol. Plant. 48, 411–415.
102 Johnson V A, Mattern P J and Schmidt J W 1967 Nitrogen relations during spring growth in varieties of *Triticum aestivum* L. differing in grain protein content. Crop Sci. 7, 664–667.
103 Jung G A (Ed.) 1978 Crop tolerance to suboptimal land conditions. Am. Soc. Agron., Madison, WI.
104 Kandala J C, Sharma D and Rathore V S 1974 Iron-manganese and zinc-manganese interactions in maize seedlings. pp 379–389. *In* Proc. Use. Rad. Radiois. Stud. Plant Prod. Symp., Bhabha Atom. Res. Center, Bombay, India.
105 Kelly J F and Gabelman W H 1960 Variability in the tolerance of varieties and strains of red beet (*Beta vulgaris* L.) to boron deficiency. Am. Soc. Hort. Sci. Proc. 76, 409–415.
106 Kleese R A 1967 Relative importance of stem and root in determining genotypic differences in Sr-89 and Ca-45 accumulation by soybeans (*Glycine max* L.). Crop Sci. 7, 53–55.
107 Kleese R A, Rasmusson D C and Smith L H 1968 Genetic and environmental variation in mineral element accumulation in barley, wheat, and soybeans. Crop Sci. 8, 591–593.
108 Lal P, Reddy G G and Modi M S 1978 Accumulation and redistribution of dry matter and N in triticale and wheat varieties under water stress condition. Agron J. 70, 623–626.
109 Landsberg E Ch 1981 Organic acid synthesis and release of hydrogen ions in response to Fe deficiency stress of mono- and dicotyledonous plant species. J. Plant Nutr. 3, 579–591.
110 Lauchli A 1976 Genotypic variation in transport. pp 372–393. *In* Transport in Plants II. Tissues and organs. Vol. 2, Part B. Eds. U Lüttge and M G Pitman. Encyclopedia of Plant Physiology, New series. Springer-Verlag, Berlin.
111 Leeper G W 1952 Factors affecting availability of inorganic nutrients in soils with special reference to micronutrient metals. Annu. Rev. Plant Physiol. 3, 1–16.
112 Lindgren D T, Gabelman W H and Gerloff G C 1977 Variability of phosphorus uptake and translocation in *Phaseolus vulgaris* L. under phosphorus stress. J. Am. Soc. Hortic. Sci. 102, 674–677.
113 Loneragan J F 1976 Plant efficiencies in the use of B, Co, Cu, Mn, and Zn. pp 193–203. *In* Plant Adaptation to Mineral Stress in Problem Soils. Ed. M J Wright. Cornell Univ. Agric. Exp. Stn., Ithaca, NY.
114 Loneragan J F 1978 The physiology of plant tolerance to low phosphorus availability. pp 329–343. *In* Crop Tolerance to Suboptimal Land Conditions. Ed. G A Jung. Am. Soc. Agron., Madison, WI.
115 Lutz J A Jr, Genter C F and Hawkins G W 1972 Effect of soil pH on element concentration and uptake by maize. I. P, K, Ca, Mg, and Na. Agron. J. 64, 581–583.
116 Lutz J A Jr, Genter C F and Hawkins G W 1972 Effect of soil pH on element concentration and uptake by maize. II. Cu, B, Zn, Mn, Mo, Al, and Fe. Agron. J. 64, 583–585.
117 Lutz J A Jr, Hawkins G W and Genter C F 1971 Differential response of corn inbreds and single crosses to certain properties of an acid soil. Agron. J. 63, 803–805.
118 MacNair M R 1981 The uptake of copper by plants of *Mimulus guttatus* differing in genotype primarily at a single major copper tolerance locus. New Phytol. 88, 723–730.
119 Mahadevappa M, Ikehashi H and Aurin P 1981 Screening rice genotypes for tolerance to alkalinity and zinc deficiency. Euphytica 30, 253–258.
120 Makmur A, Gerloff G C and Gabelman W H 1978 Physiology and inheritance of efficiency in

potassium utilization in tomatoes grown under potassium stress. J. Am. Soc. Hortic. Sci. 103, 545–549.
121 Maranville J W, Clark R and Ross W M 1980 Nitrogen efficiency in grain sorghum. J. Plant Nutr. 2, 577–589.
122 McNeal F H, Boatwright G O, Berg M A and Watson C A 1968 Nitrogen in plant parts of seven spring wheat varieties at successive stages of development. Crop Sci. 8, 535–537.
123 Millikan C R 1961 Plant varieties and species in relation to the occurrence of deficiencies and excesses of certain nutrient elements. J. Aust. Inst. Agric. Sci. 27, 220–233.
124 Munns D N, Johnson C M and Jackson L 1963 Uptake and distribution of manganese in oat plants. I. Varietal variation. Plant and Soil 19, 115–126.
125 Munson R D 1970 Plant analysis: Varietal and other considerations. pp 84–104. *In* Proc. Symp. Plant Analysis. Ed. F Greer. Int. Min. Chem. Corp., Skokie, IL.
126 Murray G A and Benson J A 1976 Oat response to manganese and zinc. Agron. J. 68, 615–616.
127 Myers W M 1960 Genetic control of physiological processes: Consideration of differential ion uptake by plants. pp. 201–212. *In* Radioisotopes in the Biosphere. Eds. R S Caldecott and L A Snyder. Univ. Minnesota, Minneapolis, MN.
128 Nambiar E K S 1976 Genetic differences in the copper nutrition of cereals. I. Differential responses of genotypes to copper. Aust. J. Agric. Res. 27, 453–463.
129 Nambiar E K S 1976 Genetic differences in the copper nutrition of cereals. II. Genotypic differences in response to copper in relation to copper, nitrogen and other mineral contents of plants. Aust. J. Agric. Res. 27, 465–477.
130 Neales T F, Anderson M J and Wardlaw I F 1963 The role of the leaves in the accumulation of nitrogen by wheat during ear development. Aust. J. Agric. Res. 14, 725–736.
131 Nielsen N E and Barber S A 1978 Differences among genotypes of corn in the kinetics of P uptake. Agron. J. 70, 695–698.
132 Noonan J B 1953 Molybdenum deficiency in maize and other crops in Taree District. Agric. Gaz. (New South Wales) 64, 422–424.
133 Nyborg M 1970 Sensitivity to manganese deficiency of different cultivars of wheat, oats and barley. Can. J. Plant Sci. 50, 198–200.
134 Ohki K, Wilson D O and Anderson O E 1980 Manganese deficiency and toxicity sensitivities of soybean cultivars. Agron. J. 72, 713–716.
134a Olsen R A, Clark R B and Bennett J H 1981 The enhancement of soil fertility by plant roots. Am. Scientist 69, 378–384.
135 Olsen S R 1972 Micronutrient interactions. pp. 243–264. *In* Micronutrients in Agriculture. Eds. J J Mortvedt, P M Giordano and W L Lindsay. Soil Sci. Soc. Am., Madison, WI.
136 O'Sullivan J, Gabelman W H and Gerloff G C 1974 Variation in efficiency of nitrogen utilization in tomatoes (*Lycopersicon esculentum* Mill.) grown under nitrogen stress. J. Am. Soc. Hortic. Sci. 99, 543–547.
137 Peaslee P E, Isarangkura R and Leggett J E 1981 Accumulation and translocation of zinc by two corn cultivars. Agron. J. 73, 729–732.
138 Perez C M, Cagamapang B V, Esmama R V, Monserrate R U and Juliano B O 1973 Protein metabolism in leaves and developing grains of rices differing in grain protein content. Plant Physiol. 51, 537–542.
139 Pettersson S 1978 Varietal differences in rubidium uptake efficiency of barley roots. Physiol. Plant. 44, 1–6.
140 Phillips J W, Baker D E and Clagett C O 1971 Kinetics of P absorption by excised roots and leaves in corn hybrids. Agron. J. 63, 517–520.

141 Phillips J W, Baker D E and Clagett C O 1971 Identification of compounds which account for variation in P concentration in corn hybrids. Agron. J. 63, 541–543.
142 Polson, D E and Adams M W 1970 Differential response of navy beans (*Phaseolus vulgaris* L.) to zinc. I. Differential growth and elemental composition at excessive Zn levels. Agron. J. 62, 557–560.
143 Pope D R and Munger H M 1953 Hereditary and nutrition in relation to magnesium deficiency chlorosis in celery. Proc. Am. Soc. Hortic. Sci. 61, 472–480.
144 Pope D R and Munger H M 1953 The inheritance of susceptibility to boron deficiency in celery. Proc. Am. Soc. Hortic. Sci. 61, 481–486.
145 Porter O A and Moraghan J T 1975 Differential response of two corn inbreds to varying root temperature. Agron. J. 67, 515–518.
146 Raper C D Jr and Barber S A 1970 Rooting system of soybeans. II. Physiological effectiveness as nutrient absorption surfaces. Agron. J. 62, 585–588.
147 Rasmusson D C, Hester A J, Fick G N and Byrne I 1971 Breeding for mineral content in wheat and barley. Crop Sci. 11, 623–626.
148 Rasmusson D C, Smith L H and Kleese R A 1964 Inheritance of Sr-89 accumulation in wheat and barley. Crop Sci. 4, 586–589.
149 Rasmusson D C, Smith L H and Myers W M 1963 Effect of genotype on accumulation of strontium-89 in barley and wheat. Crop Sci. 3, 34–37.
150 Reed A J and Hageman R H 1980 Relationship between nitrate uptake, flux, and reduction and the accumulation of reduced nitrogen in maize (*Zea mays* L.). I. Genotypic variation. Plant Physiol. 66, 1179–1183.
151 Rhoads W A and Wallace A 1960 Possible involvement of dark fixation of CO_2 in lime-induced chlorosis. Soil Sci. 89, 248–256.
152 Rice R M 1974 Physiology and inheritance of differential growth response under high phosphorus levels among different lines of beans (*Phaseolus vulgaris* L.). Diss. Abstr. 35, 5220-B (1975).
153 Rorison I H (Ed.) 1969 Ecological aspects of the mineral nutrition of plants. Blackwell Sci. Publ., Oxford, England.
154 Rodriquez-P M S 1977 Varietal differences in maize in the uptake of nitrogen and its translocation to the grain. Diss. Abstr. 38, 5690-B (1978).
155 Safaya N M 1976 Phosphorus-zinc interaction in relation to absorption rates of phosphorus, zinc, copper, manganese, and iron in corn. Soil Sci. Soc. Am. J. 40, 719–722.
156 Saric M R 1981 Genetic specificity in relation to plant mineral nutrition. J. Plant Nutr. 3, 743–766.
157 Sayre J D 1952 Magnesium . . . important element in corn nutrition. Crops Soils 5, 16–17.
158 Sayre J D 1952 Mineral accumulation in corn leaves. pp 16–26. *In* Proc. 28th Ann. Mtg. Natl. Joint. Comm. Fert. Appln., Natl. Fert. Assn., Washington, D C.
159 Sayre J D 1955 Mineral nutrition of corn. pp. 293–314. *In* Corn and Corn Improvement. Ed. C F Sprague. Academic Press, New York, NY.
160 Schauble, C E and Barber S A 1958 Magnesium immobility in the nodes of certain corn inbreds. Agron. J. 50, 651–653.
161 Schenk M K and Barber S A 1979 Phosphate uptake by corn as affected by soil characteristics and root morphology. Soil Sci. Soc. Am. J. 43, 880–883.
162 Schimansky Ch and Marschner H 1971 Suitability of Rb-86 as a tracer for potassium in studies relating to potassium uptake by maize, sugar beet, and four varieties of barley. Z. Pflanzenernaehr. Bodenkd. 129, 144–147.

163 Schrader L E 1978 Uptake, accumulation, assimilation, and transport of nitrogen in higher plants. pp 101–141. *In* Nitrogen in the Environment: Soil-Plant-Nitrogen Relationships, Vol. 2. Eds. D R Nielsen and J G MacDonald. Academic Press, New York, NY.
164 Scott L E 1941 An instance of boron deficiency in the grape under field conditions. Proc. Am. Soc. Hortic Sci. 38, 375–378.
165 Seth J, Hebert T T and Middleton G K 1960 Nitrogen utilization in high and low protein wheat varieties. Agron. J. 52, 207–209.
166 Shea P F, Gerloff G C and Gabelman W H 1968 Differing efficiencies of potassium utilization in strains of snapbeans (*Phaseolus vulgaris* L.). Plant and Soil 28, 337–346.
167 Smilde K W and Henkens Ch. H. 1967 Sensitivity to copper deficiency of different cereals and strains of cereals. Neth. J. Agric. Sci. 15, 249–258.
168 Smith S N 1934 Response of inbred lines and crosses in maize to variations of nitrogen and phosphorus supplied as nutrients. J. Am. Soc. Agron. 26, 785–804.
169 Snaydon R W 1962 The response of certain *Trifolium* spp. to calcium in sand culture. Plant and Soil 16, 381–388.
170 Snaydon R W and Bradshaw A D 1961 Differential response to calcium within the species *Festuca ovina* L. New Phytol. 60, 219–234.
171 Stuckenholtz D D, Olsen R J, Gogen G and Olsen R A 1966 On the mechanism of phosphorus-zinc interaction in corn nutrition. Soil Sci. Soc. Am. Proc. 30, 759–763.
172 Tehrani G, Munger H M, Robinson R W and Shannon S 1971 Inheritance and physiology of response to low boron in red beet (*Beta vulgaris* L.). J. Am. Soc. Hortic. Sci. 96, 226–230.
173 Thomas W I and Baker D E 1971 Genetic control of nutrient uptake by corn. *In* Mineral elements in the food chain. Am. Assn. Adv. Sci., Philadelphia, PA.
174 Thomas W I, Gorsline G W, Korman C W, Ragland J L and Wernham C C 1960 Corn performance studies 1959. Penn. Agric. Exp. Stn. Prog. Rep. 220.
175 Thorne D W, Wann F B and Robinson W 1951 Hypothesis concerning lime-induced chlorosis. Soil Sci. Soc. Am. Proc. 15, 254–258.
176 Tiffin L O 1966 Iron translocation. I. Plant culture, exudate sampling, iron-citrate analysis. Plant Physiol. 41, 510–514.
177 Tiffin L O 1966 Iron translocation. II. Citrate/iron ratios in plant stem exudates. Plant Physiol. 41, 515–518.
178 Tiffin L O 1972 Translocation of micronutrients in plants. pp 199–229. *In* Micronutrients in Agriculture. Eds. J J Mortvedt, P M Giordano and L Lindsay. Soil Sci. Soc. Am., Madison, WI.
179 Vogel K P, Johnson V A and Mattern P J 1975 Re-evaluation of common wheats from the USDA World Collecton for protein and lysine content. Nebraska Agric. Exp. Stn. res. Bull. 272.
180 Vose P B 1963 Varietal differences in plant nutrition. Herbage Abstr. 33, 1–13.
181 Walker J C, Jolivette J P and Hare W W 1945 Varietal susceptibility in garden beet to boron deficiency. Soil Sci. 59, 461–464.
182 Wall J R and Andrus C F 1962 The inheritance and physiology of boron response in the tomato. Am. J. Bot. 49, 758–762.
183 Wallace A and Berry W L (Ed.) 1981 Trace element stress in plants. Effects and Methodology. J. Plant Nutr. 4, 1–741.
184 Wallace A, Brown J C, Nelson S, Lindsay W L, Miller G W and Jolley V D (Eds.) 1982 Iron nutrition and interactions in plants. J. Plant Nutr. 5, 229–1001.

185 Wallace A and Lunt O R 1960 Iron chlorosis in horticultural plants, a review. Proc. Am. Soc. Hortic. Sci. 75, 819–841.
186 Warnock R E 1970 Micronutrient uptake and mobility within corn plants (*Zea mays* L.) in relation to phosphorus-induced zinc deficiency. Soil Sci. Soc. Am. Proc. 34, 765–769.
187 Weir R G and Hudson A 1966 Molybdenum deficiency in maize in relation to seed reserves. Aust. J. Exp. Agric. Anim. Husb. 6, 35–41.
188 Whiteaker G, Gerloff G C, Gabelman W H and Lindgren D 1976 Intraspecific differences in growth of beans at stress levels of phosphorus. J. Am. Sci. Hortic. Sci. 101, 472–475.
189 Wright M J (Ed.) 1976. Plant adaptation to mineral stress in problem soils. Cornell Univ. Agric. Exp. Stn., Ithaca, NY.
190 Young W I and Rasmusson D C 1966 Variety differences in strontium and calcium accumulation in seedlings of barley. Agron. J. 58, 481–483.

Cation fluxes in excised and intact roots in relation to specific and varietal differences

W. D. JESCHKE
Lehrstuhl für Botanik I, Universität Würzburg, D-8700 Würzburg, FRG

Key words Allium Atriplex Cation selectivity Fagopyrum Halophytes Helianthus Hordeum Plasmalemma fluxes Potassium Potassium-sodium selectivity Salt tolerance Secale Sodium Suaeda Tonoplast fluxes Triticum Xylem parenchyma

Summary In this short survey differences between species and varieties in the four major mechanisms that affect selective uptake of potassium and sodium to the plant within the root are considered. These include influx selectivity, K^+/Na^+ exchange at the plasmalemma, and selectivity at the tonoplast as well as at the symplasm-xylem boundary. The affinity of various plants for potassium influx in system 1 is rather uniform although varietal differences in barley have been observed. Differences are much more pronounced for sodium influx, for which Helianthus showed rather high and Fagopyrum rather low affinity. There is substantial variation between species in the efficiency of K^+/Na^+ exchange at the plasmalemma of cortical root cells; the three cereals Hordeum, Triticum, and Secale were highly efficient while K^+/Na^+ exchange in Atriplex, Helianthus and Allium was poor, even if the cytoplasmic sodium content was accounted for. Apparently there was no direct relation between salt tolerance and K^+/Na^+ exchange. The observed differences in the efficiency of K^+-dependent sodium extrusion or K^+/Na^+ exchange were not due to the use of excised roots, they were observed also when roots of whole seedlings were investigated. At the tonoplast a 1:1 exchange of vacuolar potassium for sodium has been observed in roots of Hordeum. By this exchange sodium ions are removed from the symplasm and potassium ions are recovered from vacuoles and thus made available for transport to the shoot. Indications for specific differences in this exchange have been observed; the exchange appears to be more efficient in Helianthus than in Hordeum roots. More comparative studies are needed here. At the boundary between symplasm and xylem vessels selectivity can be set up during xylem release of cations and there are reports that suggest a preference for sodium (*Lycopersicum cheesemanii*, *Solanum pennellii*, and Suaeda) and for varietal differences amongst tomatoes. Selectivity at this boundary, the plasmalemma of the xylem parenchyma cells was described in this paper by the selectivity ratio of transport that relates the rates of xylem transport to the cytoplasmic sodium and potassium concentrations. Based on this ratio *Atriplex hortensis* was shown to discriminate for sodium during xylem release while there was little selectivity in Hordeum and possibly some discrimination in favour of K^+ in Allium roots. The data are shortly discussed in relation to salt tolerance and to the breeding of salt-tolerant crop varieties.

Introduction

In his well-known comparative study on the K^+-Na^+ selectivity in different plant species Collander has shown that whole plants behaved rather uniformly with respect to K^+ uptake but varied widely in their capability to take up or exclude Na^+ ions. These differences in Na^+ ionic relations are reflected in the wide spectrum of the ability of plants to cope with salinity which in most cases

M.R. Sarić and B.C. Loughman (eds.), *Genetic Aspects of Plant Nutrition.*
ISBN-13: 978-94-009-6838-7

relates to excessive concentrations of sodium salts. Genotypic variations in salt tolerance of crop species have been studied[3,11,37,38] and reviewed recently[12,27].

Another important aspect of Na^+ salt relations and of the ability of plants to take up sodium ions is the need for an adequate sodium content in the shoots of plants when they are to be used as cattle feed. Whereas Na^+ ions are added in sufficient or overoptimal quantities to the human diet, this is not always true for fodder plants and pastural grasses[44]. Only a few extensive studies on the genotypic variation in Na^+ uptake by pastural grasses are available[44].

Apart from the basic scientific interest there are, therefore, two main reasons to study in a comparative way the properties of plants with respect to their Na^+ salt relations. Such studies may contribute to an understanding and improvement of agricultural plants for growth on saline soils and to an improvement or better selection of plants for cattle feed having sufficient Na^+ content for the demands of cattle.

Another practical aspect could justify such investigations. When studying uptake, xylem transport, and fluxes of Na^+ rather large differences – compared to fluxes of K^+ – between species or genotypes can be expected from an extrapolation of Collander's data. This is confirmed by the work of Rush and Epstein[38], although this is not to say that genotypic differences in K^+ ionic relations need to be small, as is seen from the prominent variations in K^+ influx rates in 10 genotypes of barley[9].

Whereas the prospects of investigating Na^+ ionic relations at first sight thus appear to be good, such studies are complicated by the fact that mere uptake data will provide little information when working with Na^+ ions. With K^+ ions Glass and Perley[9] obtained significant differences already from the kinetics of ion uptake and Glass *et al.*[10] showed that genotypic differences in K^+ nutrition of barley may be observed already by measuring pH changes.

In the case of Na^+ the ionic relations of a plant can be controlled decisively at least at three sites: a) at the plasmalemma and b) the tonoplast of the cortical cells, and c) at the plasmalemma of the xylem parenchyma cells in the root. At all three sites, therefore, genotypic differences may occur and for their understanding Na^+ fluxes at these three sites ought to be measured. In addition, such studies should include the effect of K^+ ions since the three transport sites may discriminate between K^+ and Na^+ ions.

In the present paper data on Na^+ relations in different plant species will be discussed. Emphasis shall be laid on the different strategies by which glycophytic plants maintain a high $K^+ - Na^+$ selectivity in particular in their shoots or by which alternatively salt-tolerant species include Na^+ and have low selectivity for K^+.

Sites and mechanisms of $K^+ - Na^+$ selectivity

Until recently $K^+ - Na^+$ selectivity has been studied mainly using a few species and the following characteristic features have been found for barley. As the primary selective barrier within the root the plasmalemma of the cortical cells controls selectivity in two ways. Firstly it favours influx and net uptake of K^+ over that of Na^+ at low external concentrations, in the range of system 1^{35}. Na^+ influx is specifically inhibited by low concentrations of K^+ and a substantial uptake of Na^+ is possible only when K^+ ions are missing or very low in concentration[16,41] although Na^+ may be accumulated in vacuoles even in the presence of low concentrations of K^+, see below and Fig. 1. Secondly, the plasmalemma is the site of an efficient $K^+ - Na^+$ exchange system by the operation of which Na^+ ions are extruded from the cytoplasm to the external solution in exchange for K^+ (or Rb^+)[16,17,20]. This exchange of Na^+ for K^+ is mediated by proton fluxes[20,21,36] and appears to be powered by active proton extrusion. Influx and exchange selectivity were suggested to be manifestations of the same membrane transport system[20]. The minimal components of this system appear to be a) a proton pump, b) a transport site mediating influx having high affinity for K^+ and a low one for Na^+, and c) a transport site permitting net Na^+ efflux and having only negligible affinity for K^+. Genotypic differences can be expected to occur in the secificity of the uptake site and in the presence or the properties of an efflux site.

The second membrane barrier that controls $K^+ - Na^+$ selectivity is the tonoplast of cortical and possibly also the stelar parenchyma cells in the root. In some respect the tonoplast appears to be a mirror image of the plasmalemma. In barley it has been shown to favour uptake of Na^+ ions from the cytoplasm to the vacuoles[18,34]. In the opposite direction, net efflux of K^+ from vacuoles is possible[18], while Na^+ ions appear to be almost irreversibly sequestered in the vacuole[29]. Due to these properties of the tonoplast an exchange of vacuolar K^+ for cytoplasmic Na^+ is possible and for barley roots net exchange of these ions across the tonoplast has been observed[18].

Again, genotypic differences in the ability of plants to transfer cytoplasmic Na^+ to vacuoles can be expected. It should be noted, that by an influx across the tonoplast not only Na^+ ions are removed and K^+ is made available as a cytoplasmic constituent but that Na^+ together with anions like Cl^- may function as vacuolar osmotica and thereby provide a means for salt tolerance[12,20,42]. Additionally, lowering of the cytoplasmic Na^+ concentration will facilitate Na^+ uptake. The tonoplast in this way to some extent can counteract the selective properties of the plasmalemma (see the uptake of Na^+ in the presence of K^+ in barley roots, compare Fig. 1, below).

Before ions enter the apoplastic phase within the xylem vessels and are then transported to the shoot together with the transpiration stream or the volume flow generated by root pressure they have to cross another biomembrane, the plasmalemma of the xylem parenchyma cells where they are secreted or released to the xylem vessels. At this site again cation selectivity can be affected, see below.

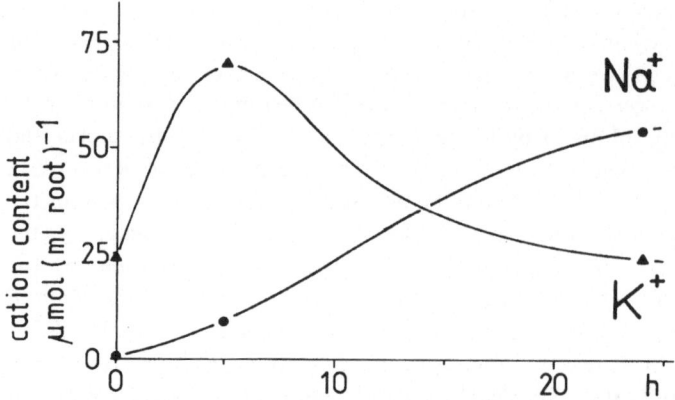

Fig. 1. Net changes in the volume content of K^+ (▲) and Na^+ (●) in excised barley (*Hordeum vulgare*, convar. distichon) roots during incubation in a solution containing 1 mM Na^+ and 0.2 mM K^+. The content at t = 0 is that of low-salt roots. Analysis was made by means of the flameless atomic absorption technique in 0.5 mm long sections taken 2.5 mm from the root tip. Composition of the incubation medium: 1 meq/l sodium phosphate buffer (pH 5.8); 0.2 mM KCl; 3 mM Ca SO$_4$; 0.5 mM Mg SO$_4$. For details of the ion profiles see Jeschke[18]; from Jeschke[20].

Comparative studies of sodium and potassium ion fluxes

a) Influx selectivity

Epstein, who pioneered the investigations on the kinetics of ion influx, has compiled[4,5] the kinetic constants for K^+ and Na^+ influx in several species and some of these data together with some more recent ones are shown in Table 1. These data show remarkably uniform values for the Michaelis constants K_m of K^+ influx in excised low-salt roots, all being around 0.2 mM (exception Agropyrum, 0.008 mM) and this is confirmed by the data obtained for Triticum, Helianthus and Fagopyrum (Table 1). At first sight this uniformity does not suggest a possibility for high genotypic variation in the affinity of plant roots towards K^+. It is most interesting, therefore, that Glass and Perley[9] found the K_m values of K^+ influx to vary in a wide range in a series of 10 varieties of barley. The K_m data, some of which are contained in Table 1, ranged between 0.011 and 0.024 mM and there were corresponding differences in the V_{max} values. High affinities (low K_m) were correlated with high maximal rates of influx, V_{max}.

Kinetic constants for influx by low-salt roots might appear to be of little significance for the performance of plants under field conditions under which they are normally in a high-salt state. As is seen from Table 1, K_m values for high-K^+ roots were considerably higher, as was observed earlier and was attributed to a regulation of influx by the internal K^+ concentration in the root[8]. Also these K_m values varied over a wide range, but the ranking between the varieties differed from that obtained for low-salt roots (Table 1). Nevertheless, the observed variation in the kinetic data of low-K^+ roots were shown to relate

Table 1. Kinetics of K^+ and Na^+ influx from solutions of low ion concentrations (System 1)*.

Species	Roots	K^+		Na^+	
		K_m	V_{max}	K_m	V_{max}
Hordeum vulgare	Low salt	0.021[a]		0.32[b]	
Helianthus annuus[c]	Low salt	0.023	1.0	0.16	7.3
Helianthus annuus	High salt	0.17	0.94	0.86	1.7
Triticum aestivum[c] cv. Carstacht	Low salt	0.023	6.7	0.4	1.28
Fagopyrum esculentum[d]	Low salt	0.02	3.2	1.56	0.63
Hordeum vulgare[e]					
cv. Fergus	Low salt	0.0109	12.06	–	–
cv. Hector	Low salt	0.0128	10.91	–	–
cv. Conquest	Low salt	0.0241	9.76	–	–
cv. Fergus	High salt	0.065	0.77	–	–
cv. Conquest	High salt	0.111	0.118	–	–
cv. Hector	High salt	0.188	2.12	–	–

[a] Epstein[4], [b] Rains and Epstein[35], [c] S. Kolibius, unpublished, [d] Eggers, this volume, [e] Glass and Perley[9].
* K_m apparent Michaelis constant in mM; V_{max} maximal rate of influx in μmol g^{-1} FW.

also to differences in the performance of the barley varieties. This was indicated by growth measurements under limiting K^+ supply or under conditions of competition between the varieties[9]. This study clearly shows the potential of kinetic measurements for finding varieties that can produce good yield also with decreased supply of fertilizers.

Although fewer data are available, the variation between species in the kinetic constants for Na^+ influx (low-salt roots) appears to be higher than those for K^+ (Table 1) and differences in K_m and V_{max} between varieties of tomato have been observed[30]. Clearly the affinity of roots for Na^+ is much lower than for K^+, but apart from Fagopyrum the K_m values ranged between 0.16 and 0.4 mM and are thus within the limits of system 1 of ion uptake (Table 1). As found for K^+ the apparent Michaelis constant was higher for high-salt roots (see the data for Helianthus, Table 1) indicating that Na^+ influx in the absence of K^+ is regulated in a similar way as was suggested for K^+ influx[8].

The kinetic data obtained for Fagopyrum roots, the species that was found to discriminate most strongly between K^+ and Na^+ amongst the plants studied by Collander[1] appear noteworthy. The apparent K_m for Na^+ influx (1.56 mM) is beyond the normal range of system 1 concentrations and as shown by the low V_{max}, buckwheat has virtually no capability to take up Na^+ from solutions of concentrations below 1 mM even when K$^+$ ions are absent. Whereas barley roots accumulate considerable amounts of Na$^+$ (final content about 75 μmol g^{-1}FW at 1 mM Na$^+$) buckwheat roots took up as little as 0.5 μmol g^{-1}FW from solutions containing 1 mM Na$^+$ (Eggers, unpublished). Apparently this species excludes Na$^+$ almost fully from its cells and appears to achieve selectivity mainly by means of a highly efficient influx selectivity (see also the contribution by H. Eggers to this volume).

By contrast to Fagopyrum, Helianthus roots showed a relatively high affinity (low K$_m$) and capacity (V$_{max}$) for Na$^+$ influx (Table 1) although this species was shown to be almost as highly K$^+$ − Na$^+$ selective as Fagopyrum[1]. Furthermore, Na$^+$ influx was much less inhibited by the presence of K$^+$ ions than it was in barley (refs.[17,35,41] and Table 3) or in Triticum roots (Kolibius, unpublished). In agreement with this observation K$^+$ influx by Helianthus roots was inhibited by the presence of Na$^+$ ions. In Helianthus roots K$^+$ − Na$^+$ discrimination, therefore, cannot be attributed to influx selectivity.

As follows from this comparison between the results obtained for sunflower and buckwheat roots, kinetic studies of Na$^+$ can be expected to reveal high genotypic variations but they do not by themselves allow to predict the performance of a plant in the discrimination between K$^+$ and Na$^+$ ions.

b) K$^+$ − Na$^+$ exchange at the plasmalemma

In barley roots a major source of K$^+$ − Na$^+$ selectivity has been shown to be K$^+$-dependent Na$^+$ efflux across the plasmalemma of their cortical cells[17]. In the presence of external K$^+$ net extrusion of Na$^+$ can be observed with sodium-loaded, excised roots[25]; the K$^+$ − Na$^+$ exchange system that is responsible for Na$^+$ extrusion operates in differentiated as well as in meristematic tissues of barley roots[20]. It must be asked whether this exchange system operates in other species, too.

Table 2 shows the sodium efflux ϕ_{co}, the cytoplasmic sodium content Q$_c$, and the potassium-dependent sodium extrusion ϕ_{co}(K$^+$-dep) for a number of species. There are marked differences between species and a relatively high Q$_c$ was found for barley, wheat and Atriplex roots. However, a prominent K$^+$-dependent sodium extrusion ϕ_{co}(K$^+$-dep) was found only for the three cereal grasses wheat, barley and rye, indicating these species to possess an efficient K$^+$ − Na$^+$ exchange system. When the rate of sodium extrusion induced by 0.2 mM K$^+$ is related to the cytoplasmic sodium content Q$_c$ and when the ratio ϕ_{co}(K$^+$-dep)/Q$_c$ is taken as a measure of exchange performance, wheat roots appear to be even more efficient in potassium- sodium exchange than barley roots (Table 2, last column).

In the other species tested so far, in *Fagopyrum esculentum, Helianthus annuus, Atriplex hortensis,* and *Allium cepa* only a small or no K$^+$-dependent Na$^+$ extrusion could be observed (Table 2). The ratio ϕ_{co} (K$^+$-dep)/Q$_c$ in all these cases was low, indicating that it was not cytoplasmic Na$^+$ content Q$_c$ that limited

Table 2. Effect of K$^+$ on the plasmalemma Na$^+$ efflux from cortical cells of roots in different species: Roots were equilibrated with 1 mM Na$^+$ solution and then the steady state fluxes were measured by means of compartmental analysis – only the plasmalemma efflux ϕ_{co} and the cytoplasmic Na$^+$ content Q$_c$ are shown –; in the steady state 0.2 mM K$^+$ was then added and the transient, K$^+$-dependent Na$^+$ efflux ϕ_{co} (K$^+$-dep) was measured*

Conditions	Na$^+$-loaded roots, (1M Na$^+$), steady state		Transient Na$^+$ efflux after addition of 0.2 mM K$^+$	
Parameter	ϕ_{co}	Q$_c$	ϕ_{co} (K$^+$-dep)	ϕ_{co}(K$^+$-dep)/Q$_c$
Allium cepa[a]	0.21	0.34	0.03	0.08
Helianthus anuus[a]	1.1	1.08	0.3	0.28
Triticum aestivum[a]	2.0	2.2	7.0	3.2
Hordeum vulgare[b]	2.1	3.6	8.2	2.3
Secale careale[c]	1.2	1.3	1.9	1.5
Atriplex hortensis[d]	2.5	4.5	0.9	0.2
Fagopyrum esculentum[e]	0.07	0.13	n.d.**	–

* Fluxes in μmol g^{-1} FW h^{-1}; cytoplasmic content in μmol g^{-1} FW.
** n.d. = not detectable.
[a] Jeschke and Nassery[2]; [b] Jeschke[19]; [c] E. Moreth, unpublished; [d] Stelter[40]; (K$^+$ = 1mM); [e] H. Eggers, unpublished.

K$^+$ – Na$^+$ exchange. Only for roots of Fagopyrum, in which no K$^+$-dependent Na$^+$ efflux could be detected, it cannot be excluded that this was due to the very low cytoplasmic Na$^+$ content (Table 2). However, since these roots almost totally exclude Na$^+$, see above, the apparent absence of K$^+$ – Na$^+$ exchange is understandable. For Helianthus, on the other hand, which is also highly K$^+$ – Na$^+$ selective in its shoots[1], the apparent absence of K$^+$ – Na$^+$ exchange appears remarkable. Possibly this low level of K$^+$ – Na$^+$ exchange is related to the low degree of influx selectivity in this species (see above) since influx selectivity and K$^+$ – Na$^+$ exchange have been suggested to be achievements of the same membrane transport system[20]. By contrast to Fagopyrum and Helianthus, *Atriplex hortensis* contains more Na$^+$ than K$^+$ in its leaves[1] and the low degree of K$^+$ – Na$^+$ selectivity at the plasmalemma (Table 2) was to be expected, therefore. This species apparently includes Na$^+$ rather than to exclude it and this undoubtedly is related to the salt tolerance of Atriplex. It should be noted however, that Na$^+$ inclusion does by no means pertain to the cytoplasm. Atriplex roots have been shown to keep the cytoplasmic Na$^+$ level similarly low as do barley roots[25].

From the prevailing results it can be concluded that efficient K$^+$ – Na$^+$ exchange is a property of certain species and so far has been found only for cereal grasses. In addition, some indications have been obtained that barley varieties differing in their salt tolerance also differ in K$^+$ – Na$^+$ exchange at the plasmalemma. Experiments are in progress to substantiate these preliminary results. It appears promising to investigate the capability of $K^+ - Na^+$ exchange when searching for genotypic differences in $K^+ - Na^+$ selectivity, in particular in Gramineae species.

c) Flux measurements with whole plants

Most experimental results on individual ion fluxes at the cellular membranes of root cells so far have been obtained with excised roots. The results, also those pertaining to $K^+ - Na^+$ selectivity, therefore, could be restricted to these roots, which lack the continuous supply of photosynthesis products from the shoots.

Some authors have studied the efflux behaviour of or have applied compartmental analysis to roots of whole plants[14,15]. During such measurements with labelled roots of whole plants tracer is not only removed from the roots by exchange with the external, unlabelled solution but also by transport to the shoot, together with the continuing flow of unlabelled ions through the xylem vessels. In previous studies xylem transport either was ignored as being small[14] or no attempt was made to evaluate the efflux data for obtaining ion fluxes at the cellular membranes[6,15].

Recently a method for studying ion fluxes in roots of whole plants has been developed, which makes allowance for the continuous transport of tracer to the shoot[23]. So far this method has been applied to Na^+ fluxes in Helianthus[23] and to K^+ and Na^+ fluxes in Hordeum seedlings[22] and some of the results are given in Table 3. As can be seen, fluxes obtained with whole plants and with excised roots compare well, only the tonoplast fluxes showed some differences (Table 3). Thus the tonoplast fluxes of Na^+ in excised roots of sunflower were significantly lower than those obtained with whole plants. As has been discussed[23] this appears to be related to the observation that intact sunflower roots – in contrast to excised ones – had not reached a level of saturation of sodium accumulation. In these roots a sizeable net Na^+ accumulation continued throughout the duration of efflux experiments. Accordingly the tonoplast influx ϕ_{cv} by far exceeded the efflux ϕ_{vc}, see Table 3.

Good agreement was found for $K^+ - Na^+$ exchange: with barley high rates of K^+-dependent net Na^+ extrusion could be induced by the addition of 0.2 mM K^+ to Na^+-loaded excised or intact roots[22]; with sunflower, on the other hand $K^+ - Na^+$ exchange was low, no matter whether excised or intact roots were used[23,24]. This shows that the low efficiency of $K^+ - Na^+$ exchange in sunflower roots (ref.[24] and Table 2) was not due to the use of excised roots and to their limited energy reserves.

A few details of Table 3 deserve to be noticed. Sodium influx ϕ_{oc} in sunflower roots was much less depressed by the continuous presence of 0.2 mM K^+ than was true for barley roots. Similarly K^+ inhibited the xylem transport of sodium much less in sunflower than in barley roots (Table 3); however, even in the absence of K^+ sodium transport across the xylem vessels ϕ_{cx} was low in

Table 3. Unidirectional steady state sodium and potassium fluxes and cytoplasmic or vacuolar contents in intact and excised roots of *Hordeum vulgare* and of *Helianthus annuus* in the absence and in the presence of K^+.[1]

Species	Hordeum[a]					Helianthus[b]		
Ion Conditions	Na^+ fluxes			K^+ fluxes		Na^+ fluxes		
Roots	1 mM Na^+ intact	1 mM Na^+ excised	1mM Na^+ 0.2 mM K^+ intact	1 mM Na^+ 0.2 mM K^+ intact	1 mM Na^+ 0.2 mM K^+ excised	1 mM Na^+ intact	1 mM Na^+ excised	1 mM Na^+ 0.2 mM K^+ intact
ϕ_{oc}	5.5	4.4	1.05	13.6	15	2.4	3.5	1.1
ϕ_{co}	2.0	2.1	0.81	7.3	5	1.5	1.1	1.0
ϕ_{cv}	1.04	0.99	0.28	2.1	7.6	1.9	0.4	0.4
ϕ_{vc}	0.86	0.99	0.19	2.0	7.6	1.3	0.4	0.4
ϕ_{cx}	3.2	2.4	0.14	6.2	10	0.29	0.46	0.12
Q_c	3.3	3.6	0.59	11.5	21.7	1.5	1.1	0.46
Q_v	76	71	48	68	69.3	35	46	44
Q_v/Q_c	23	19.6	81	5.9	3.2	24	43	96

[1] ϕ_{oc}, ϕ_{co} = plasmalemma influx and efflux; ϕ_{cv}, ϕ_{vc} = tonoplast influx and efflux; ϕ_{cx} = xylem transport; all fluxes in μmol g^{-1} FW h^{-1}. Q_c, Q_v = cytoplasmic and vacuolar ion content in μmol g^{-1} FW.
[a] *Hordeum vulgare*, convar. distichon; data from Jeschke[22]; [b] *Helianthus annuus*; data from Jeschke and Jambor[23].

sunflower compared to barley roots. The relatively small effect of K^+ on sodium transport found for sunflower, therefore, is not at variance with the finding of a high $K^+ - Na^+$ selectivity in the shoots of sunflower plants. Furthermore, the small response of sodium influx ϕ_{oc} to the presence of K^+ as found with compartmental analysis (Table 3) is consistent with the low influx selectivity observed with tracer influx studies for this species, see above. It must be asked, therefore, by which mechanism(s) the high K^+/Na^+ ratio in sunflower shoots[1] is established since neither influx nor exchange selectivity at the plasmalemma appear to be responsible. As will be shown below, selectivity at the tonoplast could be responsible.

Although as yet no data on genotypic variations in individual ion fluxes are available, it appears that the method using roots of whole plants is promising for obtaining such data. Not only does this method allow an investigation of ion flux properties of roots *in situ* on the whole plant. But it is applicable also to such species which have small roots and are, therefore, not suited for experiments with excised roots. Moreover the use of roots of whole plants allows to study effects of the shoot and of transpiration on ion fluxes in the root as well as to investigate a shoot-dependent regulation of ion fluxes in the root.

d) Selectivity at the tonoplast

Selective properties of the tonoplast generally have to be derived from indirect measurements since the tonoplast as an internal membrane is accessible to experimentation only via the plasmalemma and the cytoplasm. Compartmental analysis yields data about the steady-state distribution of K^+ and Na^+ between cytoplasm and vacuoles and by this method sodium was shown to be preferentially localised in vacuoles and potassium more so in the cytoplasm in several species, barley[34], maize[31], onion[28], (see also Table 3). This pattern of distribution was found also by measurements of the longitudinal distribution of K^+ and Na^+ along the root axis in Hordeum and Atriplex and by calculating the cytoplasmic and vacuolar ion concentrations on the basis of the high cytoplasmic volume fraction in meristematic and the high vacuolar fraction in differentiated tissues[25]. Recently electron microprobe measurements have been applied to the distribution of K^+ and Na^+ between cytoplasm and vacuoles in barley roots[33]. The resulting data confirm those obtained with the other methods, but in addition the ratio K^+/Na^+ was shown to vary radially within the root and to be higher in the stele than in the cortex, a result that cannot be obtained by the other methods.

While thus the stationary distribution between cytoplasm and vacuole seems well documented, little direct information on time-dependent changes in the distribution is available. Indications for an exchange with time of vacuolar K^+ for Na^+ were obtained from the decrease in K^+ and increase in Na^+ content in barley leaves with increasing age[11]. With the method of longitudinal ion profiles net exchange of vacuolar K^+ for Na^+ was shown to occur under appropriate experimental conditions[18]. Fig. 1 shows as an example the changes in K^+ and Na^+ contents in barley roots during incubation in a solution containing 1 mM Na^+ and 0.2 mM K^+. An initial rapid increase in K^+ content was followed by a massive decrease between 5 and 24 hrs. that was mirrored by a continuous increase in sodium content. From the magnitude of K^+ loss and Na^+ uptake it follows that K^+ ions had been accumulated in vacuoles and were then exchanged for Na^+ at a stoichiometry close to 1. The K^+ ions originating from vacuoles were then available within the cytoplasm whence they were transported through the xylem vessels; Na^+ ions were taken up from the external solution and apparently were transported via cytoplasm and tonoplast to the vacuole.

The predominant vacuolar localization of Na^+ can be seen also from the data of Table 3. Here the vacuolar contents of K^+ and Na^+ were related to their contents in the cytoplasm. The ratio Q_v/Q_c for Na^+ was much higher than for K^+ and significantly higher when both K^+ and Na^+ were present externally than in solutions containing only Na^+ besides Ca^{++} and Mg^{++}. Furthermore, the ratio Q_v/Q_c (Na^+) was somewhat higher for sunflower than for barley roots (Table 3). The data suggest than Na^+ ions are expelled from the cytoplasm to vacuoles when K^+ ions are present and that the $Na^+ - K^+$ selective system at the tonoplast that is responsible for this distribution may be more efficient in sunflower than in barley roots. Sequestration of Na^+ ions in vacuoles, therefore, could be one of the mechanisms by which sunflower roots achieve the high $K^+ - Na^+$ selectivity in the shoots. Na^+ ions taken up by the roots thereby are included in vacuoles within the root and prevented from reaching the shoot.

Besides providing a mechanism of $K^+ - Na^+$ selectivity, sequestration of Na^+ in vacuoles is thought to be a means by which salt-tolerant species take up Na^+ and use this ion as a 'cheap' vacuolar osmoticum[12,20,42]. It appears that much more comparative information about the capability of vacuoles to store Na^+ and of the tonoplast to discriminate between K^+ and Na^+ should be gathered.

e) Selectivity at the symplasm—xylem boundary

As a third site the plasmalemma of the xylem parenchyma cells can determine the overall selectivity of the plant in two ways. Firstly when released or secreted to the xylem vessels K^+ or Na^+ could be favoured or retarded and secondly Na^+ ions can be reabsorbed from the xylem fluid. This latter process has been found in some plant species like beans, maize and soybeans, which are highly efficient in excluding sodium from their shoots[13,26,39]. This reabsorption of sodium is mediated by xylem parenchyma cells in the basal parts of roots or shoots and in some cases these cells have specialized as transfer cells (Kramer, this volume). Indications have been obtained (Kolibius, unpublished) that reabsorption of Na^+ occurs also in sunflower roots.

In glycophytic species having high K^+/Na^+ ratios in their shoots a preference for K^+ during xylem release could improve the overall selectivity, indications for such a preference are limited, however (ref.[28] and below). The situation is different in halophytes[7] and salt-tolerant nonhalophytes[12] which include Na^+ in their shoots[1] rather than to prevent its uptake or to retain it in their roots. Sodium content in the leaves of halophytes is high[7] and salt-tolerant barley varieties[11] or tomato species[37] accumulate Na^+ in their leaves. In these plants the release of ions to the xylem vessels could prefer Na^+ over K^+.

In the salt-tolerant tomato *Lycopersicum cheesmanii* grown on 50 mM KCl or NaCl ($+0.5$ mM CaSO$_4$) more Na^+ than K^+ was translocated to the shoot while the salt-sensitive species *L. esculentum* translocated similar amounts of K^+ but very little Na^+ (ref.[37]). Although selectivity cannot be deduced directly from such data obtained with one-salt experiments, they suggest a preference for Na^+ transport in the salt-tolerant species. When growing tomatoes on Hoagland solution salinized up to 200 mM NaCl, Dehan and Tal[2] found much lower K^+/Na^+ ratios in leaves compared to roots in the salt-tolerant species *Solanum pennellii* but higher ones in the leaves of the salt-sensitive species *L. esculentum*. This result strongly suggests a preferred release of Na^+ to the xylem vessels in the salt-tolerant species and contrasts to those obtained for barley[32,34] according to which the K^+/Na^+ ratio in the leaves were higher than in the root but equalled that in the root cytoplasm suggesting little if any selectivity during alkali cation release to the xylem vessels, see below.

Evidence for a preference for Na^+ in xylem transport has been obtained also for the halophyte *Suaeda maritima*[7,43]. In this case K^+ increased the transport of Na^+ (5 mM Na^+) in the xylem in particular at high external K^+ concentrations.

However, K^+/Na^+ selectivity at the symplasm-xylem boundary can be described unequivocally only when the concentration or at least the content of K^+ and Na^+ in the cytoplasm and their rates of xylem transport are known. These data are available for roots of Allium[28], Hordeum (Table 3) and *Atriplex hortensis*[40] and in Table 4 the selectivity of the release of K^+ and Na^+ to the xylem vessels was calculated from the ratios of xylem transport ϕ_{cx} (or J_{cx}) and of cytoplasmic content Q_c:

$$S(\text{transport}) = \frac{\phi_{cx}(K)/\phi_{cx}(Na)}{Q_c(K)/Q_c(Na)}$$

For Allium roots this selectivity ratio was 2.8, possibly suggesting some preference for K^+, similarly for Hordeum a selectivity of 2.3 for K^+ in xylem release was calculated from the flux data obtained for whole plants (Tables 3, 4). Keeping in mind the experimental errors of determining the cytoplasmic content by means of compartmental analysis I would suggest that S(transport) is close enough to 1 and that there is little discrimination between K^+ and Na^+ during release to the xylem vessels in barley. Selectivity is set up at the plasmalemma and the tonoplast in this species.

Quite different results were obtained from measurements of xylem transport and longitudinal ion profiles in roots of the salt-tolerant *Atriplex hortensis*[40] (Table 4). Under all conditions tested S(transport) was clearly below 1 indicating a preference for Na^+. In the Table also the data obtained with low-salt roots and with 1 mM Na^+ and 1 mM K^+ (and Ca^{++}) are included, since S(transport) could be calculated on the basis of the cytoplasmic ion concentrations and the endogenous transport of the ions that were not present externally.

The low values of S(transport) found for *Atriplex hortensis* clearly show a discrimination in favour of Na^+ during xylem release in this halotolerant species. Preferential xylem transport and in addition vacuolar accumulation of Na^+ appear to be the mechanisms by which Atriplex removes K^+ from the root cytoplasm ($K^+ - Na^+$ exchange was low, Table 2). In addition, Na^+ transport in the xylem aids in the osmotic adjustment of the shoot to a saline environment. As a special adaptation and similar to other Atriplex species, *A. hortensis* can excrete excessive quantities of Na^+ into bladder hairs on the epidermis of its leaves.

Taken together, there is clear evidence for cation selectivity at the symplasm-xylem boundary only for preferred release (Suaeda, Atriplex) or reabsorption (Phaseolus, Zea) of sodium while a possible discrimination in favour of K^+ in other species (Hordeum, Allium) is still ambiguous. As follows from the results of Dehan and Tal[2] and Rush and Epstein[38] genotypic differences with respect to selectivity at this site can be expected to be found also for other species.

Table 4. Cytoplasmic Na$^+$ and K$^+$ contents or concentrations, xylem transport ϕ_{cx} of K$^+$ and Na$^+$, and selectivity ratio of xylem transport S(transport)^3 in different species.

Species	External[4] solution	Xylem transport[1] K$^+$	Na$^+$	Cytopl. content[2] or concentration K$^+$	Na$^+$	S(transport)[3]
Allium cepa[a]	1 mM Na, 1 mM K	21.6	0.65	8.5	0.76	2.8
Hordeum vulgare[b]	1 mM Na, 0.2 mM K	6.2	0.14	11.5	0.59	2.3
Atriplex hortensis[c]	0Na, 0K, (low salt)	0.8*	0.9*	90	2	0.02
Atriplex hortensis[c]	1 mM Na	3.7*	8.7	46	15	0.14
Atriplex hortensis[c]	1 mM K	8.3	0.7*	91	0.5	0.06
Atriplex hortensis[c]	1 mM Na, 1 mM K	5.8	5.4	81	2.5	0.033

[1] In μmol g^{-1} FW h^{-1}; [2] cytoplasmic content Q_c in μmol g^{-1} FW or cytoplasmic K$^+$ or Na$^+$ concentration in mM; [3] S(transport) = ϕ_{cx}(K) Q_c(Na)/ϕ_{cx}(Na) Q_c(K) or S(transport) = ϕ_{cx}(K) [Na]$_c$/ϕ_{cx}(Na) [K]$_c$;
[4] concentrations of alkali cations are given only, for full composition, see refs;
[a] Macklon, 1975, Q_c and xylem transport ϕ_{cx} from flux measurements with excised roots;
[b] Jeschke, 1982, Q_c and xylem transport ϕ_{cx} from flux measurements with intact roots;
[c] Stelter, 1979, cytoplasmic K$^+$ and Na$^+$ concentrations in mM from longitudinal ion profiles of *Atriplex* roots, see Jeschke and Stelter, 1976, ϕ_{cx} from transport measurements; the transport data designated by a star* refer to the transport of endogenous reserves.

Concluding remarks

What emerges from this short comparative survey of the mechanisms and strategies of K$^+$ − Na$^+$ selectivity is that the quantitative variation in selectivity observed by Collander[1] in fact is due to prominent qualitative differences between the species. Even plants like Fagopyrum and Helianthus that appear quantitatively similar according to Collanders data achieve the same degree of selectivity in quite different ways.

High K$^+$/Na$^+$ ratios in shoots, apparently of evolutionary advantage in glycophytes, relate to different mechanisms or strategies: to K$^+$ selectivity during influx in Fagopyrum, to influx selectivity, efficient K$^+$ − Na$^+$ exchange and to preferred vacuolar accumulation of Na$^+$ in cereal grasses (Hordeum, Triticum and Secale), to low influx and exchange selectivity, but Na$^+$ accumulation in vacuoles and possibly reabsorption of Na$^+$ from the xylem sap in Helianthus, and to low Na$^+$ influx (influx selectivity) and Na$^+$ reabsorption from the xylem vessels in Phaseolus and Zea.

Low K$^+$/Na$^+$ ratios in shoots combined with the accumulation of Na$^+$ in leaves of some halophytes being of evolutionary and ecological advantage in the saline environment, on the other hand, appear to be related to low influx selectivity, to a preference for Na$^+$ during xylem release (Suaeda. Atriplex), and to a high Na$^+$ − K$^+$ selectivity at the tonoplast (Atriplex)[40].

Only one feature appears to be common to glycophytes and to halophytes and salt-tolerant nonhalophytes. This is the ability to exclude Na^+ from the cytoplasm[20,42]. This again is achieved in different ways, by low influx of Na^+ (Fagopyrum), by the combination of influx selectivity, $K^+ - Na^+$ exchange and vacuolar sequestration of Na^+ (Hordeum) or by vacuolar sequestration of Na^+ in combination with preferred release of Na^+ to the xylem sap (roots of Atriplex).

When searching for species or varieties having higher salt-tolerance, the selective properties of the tonoplast seem most important and a low selectivity at the plasmalemma appears to be mandatory for species having a high degree of salt tolerance. Special interest should be paid also to the selectivity at the symplasm-xylem boundary. High preference for K^+ in the release to the xylem vessels at this site, no matter whether it is achieved by a retention of Na^+ in the root cytoplasm or by reabsorption of Na^+ from the xylem sap., appears to be counterproductive for salt tolerance.

It appears worth noticing that similar arguments apply to species that tend to include Na^+ and are, therefore, suitable as species which contain sufficient amounts of Na^+ for the needs of cattle even when the soil is rich in K^+ due to fertilization. In this connection it would be interesting to investigate the $K^+ - Na^+$ selective properties of those grass species like *Holcus lanatus*, *Lolium perenne* and *Anthoxanthum odoratum* that have a high Na^+ content in their leaves[44].

Acknowledgements   The experimental work from this laboratory was supported by grants of the Deutsche Forschungsgemeinschaft. Thanks are extended to Miss Andrea Schniering for technical assistance.

References

1 Collander R 1941 Selective absorption of cations by higher plants. Plant Physiol. 16, 691–721.
2 Dehan K and Tal M 1978 Salt tolerance in the wild relatives of the cultivated tomato: Responses of *Solanum pennellii* to high salinity. Irrig. Sci. 1, 71–76.
3 Dewey D R 1962 Breeding crested wheatgrass for salt tolerance. Crop Sci. 2, 403–407.
4 Epstein E 1972 Mineral Nutrition of Plants: Principles and Perspectives. Wiley, New York.
5 Epstein E 1976 *In* Encyclopedia of Plant Physiology, New Series Eds. U. Lüttge and M. G. Pitman. Springer-Verlag, Berlin—Heidelberg—New York Vol 2B pp 70–94.
6 Erlandson G 1979 Flux of potassium from wheat roots induced by changes in the water potential of the root medium. Physiol. Plant. 47, 1–6.
7 Flowers T J, Troke P F and Yeo A R 1977 The mechanism of salt tolerance in halophytes. Annu. Rev. Plant Physiol. 28, 89–121.
8 Glass A D M 1978 The regulation of potassium influx into intact roots of barley by internal potassium levels. Can. J. Bot. 56, 1759–1764.
9 Glass A D M and Perley J E 1980 Varietal differences in potassium uptake by barley. Plant Physiol. 65, 160–164.

10 Glass A D M, Siddiqi M Y and Giles K I 1981 Correlations between potassium uptake and hydrogen efflux in barley varieties. 68, 457–459.
11 Greenway H 1962 Plant response to saline substrates. II. Chloride, sodium, and potassium uptake and translocation in young plants of *Hordeum vulgare* during and after a short sodium chloride treatment. Aust. J. Biol. Sci. 15, 39–57.
12 Greenway H and Munns R 1980 Mechanisms of salt tolerance in nonhalophytes. Annu. Rev. Plant Physiol. 31, 149–190.
13 Jacoby B 1965 Sodium retention in excised bean stems. Physiol. Plant. 18, 730–739.
14 Jefferies R L 1973 The ionic relations of seedlings of the halophyte *Triglochin maritima* L. *In* Ion transport in Plants. Ed. W P Anderson. Academic Press, London-New York. pp 297–321.
15 Jensen P and Kylin A 1980 Effects of ionic strength and relative humidity on the efflux of K^+ (^{86}Rb) and Ca^{++} (^{45}Ca) from roots of intact seedlings of cucumber, oat, and wheat. Physiol. Plant. 50, 199–207.
16 Jeschke W D 1972 Wirkung von K^+ auf die Fluxe und den Transport von Na^+ in Gerstenwurzeln, K^+-stimulierter Na^+-Efflux in der Wurzelrinde. Planta, Berlin 106, 73–90.
17 Jeschke, W D 1973 K^+-stimulated Na^+ efflux and selective transport in barley roots. Ref. Jefferies pp 285–296.
18 Jeschke, W D 1977 $K^+ - Na^+$ selectivity in roots. Localization of the selective fluxes and their regulation. *In* Regulation of Cell Membrane Activities in Plants. Eds. E Marrè and O Cifferi. Elsevier Amsterdam pp 63–78.
19 Jeschke W D 1977 $K^+ - Na^+$ Exchange and selectivity in barley root cells. Effect of Na^+ on the Na^+ fluxes. J. Exp. Bot. 28, 1289–1305.
20 Jeschke W D 1979 Univalent cation selectivity and compartmentation in cereals. *In* Recent Advances in the Biochemistry of Cereals. Eds. D L Laidman and R G Wyn Jones. Academic Press, London-New York. pp 37–61.
21 Jeschke W D 1980 Involvement of proton fluxes in the $K^+ - Na^+$ selectivity at the plasmalemma; K^+-dependent net extrusion of sodium in barley roots and the effect of anions and pH on sodium fluxes. Z. Pflanzenphysiol. 98, 155–75.
22 Jeschke W D 1982 Shoot-dependent regulation of sodium and potassium fluxes in roots of whole barley seedlings. J Exp. Bot. 33, 601–619.
23 Jeschke W D and Jambor W 1981 Determination of unidirectional sodium fluxes in roots of intact sunflower seedlings. J. Exp. Bot. 32, 1257–1272.
24 Jeschke W D and Nassery H 1981 $K^+ - Na^+$ selectivity in roots of Triticum, Helianthus and Allium. Physiol. Plant. 52, 217–224.
25 Jeschke W D and Stelter W 1976 Measurement of longitudinal ion profiles in single roots of Hordeum and Atriplex by use of flameless atomic absorption spectroscopy. Planta, Berlin 128, 107–112.
26 Läuchli A and Wieneke J 1979 Studies on growth and distribution of Na^+, K^+ and Cl^- in soybean varieties differing in salt tolerance. Z. Pflanzenernaehr. Bodenkd. 142, 3–13.
27 Maas E V and Nieman R H 1978 Physiology of plant tolerance to salinity. *In* Crop Tolerance to Suboptimal Land Conditions. Ed. G A Jung. Am. Soc. Agron. Spec. Publ. 32, 277–299.
28 Macklon A E S 1975 Cortical cell fluxes and transport to the stele in excised root segments of *Allium cepa* L. I. Potassium, sodium and chloride. Planta, Berlin 122, 109–130.
29 Neirinckx L J A and Bange G G J 1971 Irreversible equilibration of barley roots with Na^+ ions at different external Na^+ concentrations. Acta Bot. Neerl. 20, 481–488.

30 Picciuro G and Brunetti N 1969 Assorbimento del sodio (^{22}Na) in radici escisse di alcune varietà di *Lycopersicum esculentum*. Agrochimica 13, 347–57.
31 Pierce W S and Higinbotham N 1970 Compartmentation and fluxes of K^+, Na^+ and Cl^- in Avena coleoptile cells. Plant Physiol. 46, 666–673.
32 Pitman M G 1965 Sodium and potassium uptake by seedlings of *Hordeum vulgare*. Aust. J. Biol. Sci. 18, 10–24.
33 Pitman M G, Läuchli A and Stelzer R 1981 Ion distribution in roots of barley seedlings measured by electron probe X-ray microanalysis. Plant Physiol. 68, 673–679.
34 Pitman M G and Saddler H D W 1967 Active sodium and potassium transport in cells of barley roots. Proc. Nat. Acd. Sci. (Wash.) 57, 44–49.
35 Rains H D W and Epstein E 1967 Sodium absorption by barley roots: Its mediation by mechanism 2 of alkali cation transport. Plant Physiol. 42, 319–323.
36 Ratner A and Jacoby B 1976 Effect of K^+, its counter anion, and pH on sodium efflux from barley root tips. J. Exp. Bot. 27, 843–852.
37 Rush D W and Epstein E 1976 Genotypic responses to salinity: differences between salt-sensitive and salt-tolerant genotypes of the tomato. Plant Physiol. 57, 162–166.
38 Rush D W and Epstein E 1981 Comparative studies on the sodium, potassium, and chloride relations of a wild halophytic and a domestic salt-sensitive tomato species. Plant Physiol. 68, 1308–1313.
39 Shone M G T, Clarkson D T and Sanderson J 1969 The absorption and translocation of sodium by maize seedlings. Planta 86, 301–314.
40 Stelter W 1979 Untersuchungen zur Kalium – Natrium – Selektivität von *Atriplex hortensis* L. Doctoral Thesis. University of Würzburg, Germany.
41 Welsh R M and Epstein E 1968 The dual mechanisms of alkali cation absorption by plant cells: their parallel operation across the plasmalemma. Proc. Nat. Acad. Sci. (Wash.) 61, 447–453.
42 Wyn Jones R G, Brady C J and Speirs J 1979 Ionic and osmotic relations in plant cells. Ref. Jeschke 1979. pp 64–103.
43 Yeo A R 1974 Salt tolerance in the halophyte *Suaeda maritima* L. Dum. Ph.D. Thesis, University of Sussex, England.
44 Zehler E 1981 Die Natriumversorgung von Mensch, Tier und Pflanze. Kalibriefe (Büntehof). 15, 773–792.

Genotypic differences in calcium and magnesium nutrition in sunflower

E. ALCÁNTARA and M. D. De La GUARDIA
Cátedra de Fisiología Vegetal, Escuela Técnica Superior de Ingenieros Agrónomos, Universidad de Córdoba, Córdoba, Spain

Key words Ca *Helianthus* spp. Mg Nutrition Transport Uptake

Introduction

There is genetic variability between species or genotypes in their adaptation to soil with high calcium or magnesium content[3,6] but little is known about the physiological basis of these adaptations. Clark[1] found differential Ca and Mg efficiencies between corn inbreds, and Kawaski and Moritsugu[5] have shown differences in the absorption of Ca by excised root of different corn and sorghum cultivars.

In sunflower, Madhok and Walker[7] studied the Mg nutrition of two *Helianthus* species, *H. annuus* L., the common cultivated sunflower and *H. bolanderi* Gray, subspecies exillis Heiser, a form endemic to serpentine soils. *H. bolanderi* grew better with high Mg and tolerated up to 50 mM Mg, while *H. annuus* reached the highest yield between 0.25 and 2 mM Mg. A high level of external Mg was apparently needed in *H. bolanderi* to increase the Mg content of the plant shoot. In another work with sunflower, Foy *et al.*[4] found differential responses in a study of 13 sunflower genotypes in an acid soil with and without Ca. Sarić and Skoric[8] studied 20 sunflower inbreds and their results indicate the existence of high genotypic specificity for individual ion content in the inbreds. Earlier genetic differences in sunflowers between inbred lines, hybrids and cultivars in their efficiency to grow in low Ca level were found[2], although later unpublished work has shown that the differences were greater when the plants were grown at several Mg levels.

To study the physiological basis of these differences two inbred lines were selected in our laboratory over several generations. These inbred were near isogenic in all characters except in their response to the Mg/Ca ratio in the nutrient solution.

Material and methods

Seeds from the two selected genotypes were treated with sodium-hypochlorite (0.5% v/v) and germinated in perlite for four days at 25°C in dark. Seedlings were transplanted to nutrient solution, keeping the roots in the dark and grown in a growth chamber equipped with 40w fluorescent tubes. Light irradiance was 40w m^{-2}, temperature $20\pm2°C$ and day length 13 h.

In two experiments plants of both inbred lines (line 1 and line 3) were grown in two types of nutrient solution: a) with constant Mg concentration (2 mM) and increasing Ca concentration (0.15; 0.5; 1.5; 5; 15 and 50 mM), b) with constant Ca concentration (5 mM) and increasing Mg concentration (0.2; 0.63; 2; 6.3; 20 and 63 mM). Both types of nutrient solutions had all the other elements in the concentration of a Hoagland solution and iron as a chelate. In both experiments the solutions were aerated and changed periodically.

A third experiment was run with 12 plants per treatment using 1/50 strength Hoagland nutrient solution, without Mg, for 10 days. The Mg was added to the 1/50 strength Hoagland solution in increasing concentrations (2; 6.3; 20; 63 and 200 μM) for 2 days. The K, Mg and Ca contents were determined in roots and leaves by atomic absorption after acid digestion.

Results and discussion

The dry matter accumulation of both lines, when grown at increasing Ca concentration (Table 1) or at increasing Mg concentration (Fig. 1), show significant differences in their response to Mg and Ca concentration. Line 1 has a maximum yield with a lower level of Ca in the nutrient solution than line 3, while the opposite occurred as the Mg level increased. From both studies it can be concluded that the maximum yield was reached when the ratio Mg/Ca in the solution was 1.3 for line 1, and between 0.13 and 0.4 for line 3. It appeared that the ratio Mg/Ca in the solution was more important than the absolute value of each of the two elements.

The calcium content in leaves of line 1 was higher than that of line 3 for all the Ca levels in the nutrient solution (Table 1). We found the opposite with the Mg content of leaves and roots when Mg concentration in nutrient solution is over 2 mM (Fig. 2). However, in the range 0.2–2 mM Mg no significant differences were

Table 1. Total dry weight accumulated and Ca content in leaves of plants of two inbred lines of sunflower, grown for 37 days in Hoagland nutrient solution modified with Ca concentration ranging from 0.15 to 50 mM. Data show the mean of three plants $\pm$ SE

	Total dry weight g		Ca content in leaves meq/g dr. wt.	
mM Ca	Line 1	Line 3	Line 1	Line 3
0.15	1.01±0.07	0.52±0.07	0.40±0.05	0.30±0.01
0.50	2.06±0.01	1.95±0.03	0.66±0.05	0.58±0.03
1.50	2.65±0.08	2.57±0.05	1.23±0.02	1.16±0.03
5	2.56±0.20	2.88±0.12	1.56±0.04	–
15	2.00±0.18	2.11±0.08	1.87±0.10	1.60±0.01
50	1.49±0.07	1.79±0.04	3.33±0.11	2.84±0.01

Ca AND Mg NUTRITION OF SUNFLOWER SPECIES

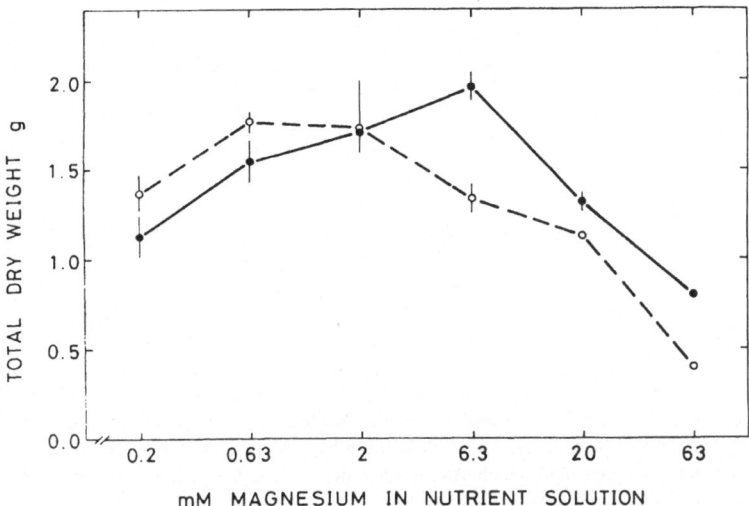

Fig. 1. Total dry weight accumulated in plants of two inbred lines of sunflower, grown for 29 days in Hoagland nutrient solution modified with Mg concentration ranging from 0.2 to 63 mM (●——●, line 1; ○– – –○, line 3). Each point is the mean of three plants. Vertical bars indicate standard errors.

found in the Mg content of the leaves, although in this range line 1 has more Mg in the roots than line 3.

This fact suggested that the root system of both lines may be controlling differently the Mg accumulated in the root or transported to the shoot. To test this hypothesis we grew plants in dilute nutrient solution without Mg for a shorter period and then Mg in increasing concentration was added to the nutrient

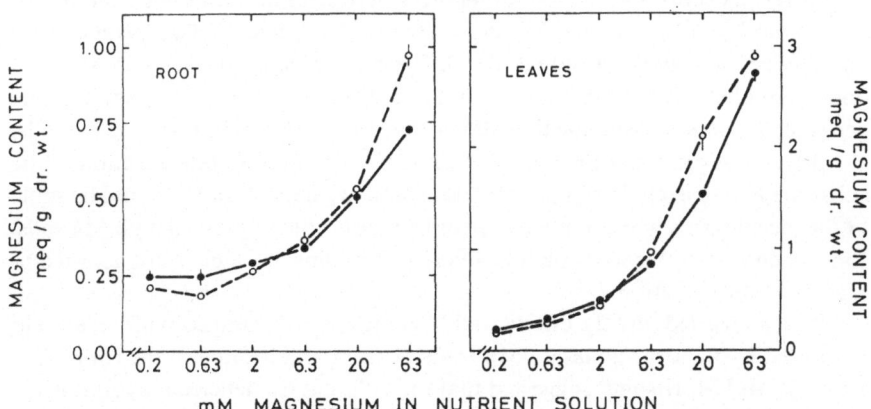

Fig. 2. Magnesium content in root and leaves of plants of two inbred lines of sunflower, grown for 29 days in Hoagland nutrient solution modified with Mg concentration ranging from 0.2 to 63 mM (●——●, line 1; ○– – –○, line 3). Each point is the mean of three plants. Vertical bars indicate standard errors.

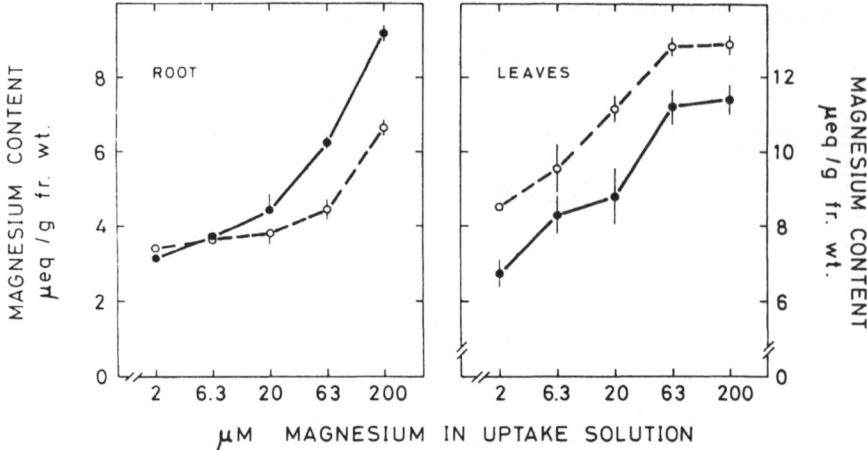

Fig. 3. Magnesium content in root and leaves of plants of two inbred lines of sunflower, grown for 10 days in a 1/50 strength Hoagland nutrient solution without Mg and then transferred for 2 days to a 1/50 strength Hoagland nutrient solution with Mg concentration ranging from 2 to 200 μM (●———●, line 1; O———O, line 3). Each point is the mean of three determinations. Vertical bars indicate standard errors.

solution. The results of the root uptake of Mg or the Mg transported to the leaves confirmed the hypothesis (Fig. 3).

In agreement with this, line 3 showed earlier Ca deficiency symptoms in the upper leaves than did line 1, when both were grown in solutions with low Ca or with high Mg. However, line 1 showed earlier Mg deficiency symptoms than line 3 in the lower leaves when both were grown at low Mg. The K content in leaves showed little differences between the two lines, however the K content in roots was in an opposite way to Mg content, line 3 had more K than line 1 at high Mg level (data not presented).

Our results show a pattern of differences between line 1 and line 3 similar to those found by Madhok and Walker[7] between the species *Helianthus bolanderi* and *Helianthus annuus*, although the differences are smaller. The behaviour of line 1 resembles *H. bolanderi* and it could be that this line possessed some of the genes of *H. bolanderi* genome that allows this species to adapt to serpentine soils.

Epstein[3] suggests that the ability to absorb Ca from low concentration in the substrate and to exclude Mg or to tolerate high Mg concentration could be some of the mechanisms which render serpentine plants able to grow in high Mg soils. Line 1 behaves in this way and it is different from line 3, which is not adapted to high Mg concentration.

Clark[1] suggested that the differential Mg efficiency between corn inbreds could be caused by factors regulating Mg translocation and located in the roots, and Kawaski and Moritsugu[5] suggested that the different Ca deficiency symptoms in leaves of corn and sorghum inbreds were partly due to the differences in the Ca absorbing capacity of the roots.

Ca AND Mg NUTRITION OF SUNFLOWER SPECIES

Our results with 2 lines of sunflower indicate that the root system of one line (line 1) is capable of absorbing more Mg, at low Mg level in the nutrient solution. (Figs. 2 and 3). However the Mg transported from root to shoot is carefully regulated, probably at the level of transport into the xylem vessels, and the Mg content in leaves is lower than in the other line (line 3). At high Mg level (over 2 mM) line 1 regulates the Mg content in the root better than line 3, probably regulating the process of uptake to cytoplasm and vacuole of cortical cells. In the high Mg range, line 3 has a higher Mg content in root and leaves than line 1 and the higher Mg contents in leaves of line 3 may induce a lower Ca level and the appearance of Ca deficiency symptoms.

References

1 Clark R B 1976 Plant efficiencies in the use of calcium, magnesium and molybdenum. *In* Plant adaptation to mineral stress (M J Wright Ed.) pp 175–192. Special Publication of Cornell University, Ithaca, N.Y.
2 De la Guardia M D, Sáiz de Omeñaca J A, Pérez Torres E B and Montes Agusti F 1980 Diferentes respuestas de las lineas, hibridos y variedades de girasol a la nutrición con bajo nivel de calcio. IX Conferencia Internacional del Girasol, vol 1, pp 239–248. Servicio de Publicaciones Agrarias, Madrid. ISBN 84–7479–12090.
3 Epstein E 1972 Mineral Nutrition of Plants: Principles and Perspectives, pp 325–392. John Wiley and Sons Inc. New York, London, Sydney, Toronto, ISBN 0–471–24340–X.
4 Foy C D, Orellana R G, Schwartz J W and Fleming A L 1974 Responses of sunflower genotypes to aluminium in acid soil and nutrient solution. Agron. J. 66, 293–296.
5 Kawaski T and Moritsugu M 1979 A characteristic symptom of calcium deficiency in maize and sorghum. Commun. *In* Soil Science and Plant Analysis, 10 (1 and 2), 41–56.
6 Läuchli A 1976 Genotypic variation in transport. *In* Encyclop. Plant Physiol. New Series. Vol. 2, part B. Eds. U Lüttge and M G Pitman. pp 372–393. Springer-Verlag, Berlin, Heidelberg, New York. ISBN 3–540–07453–8.
7 Madhok O P and Walker R B 1969 Magnesium nutrition of two species of sunflower. Plant Physiol. 44, 1016–1022.
8 Sarić M R and Skorić D 1981 Relationship between the root and above ground parts of different sunflower inbreds regarding the content of some mineral elements. *In* Structure and Function of Plant Roots. Eds. R Brouwer, O Gaspariková, J Kolek and B C Loughman. pp 399–401. Nijhoff and Junk, The Hague, Boston, London. ISBN 90–247–2510–0.

A study of the electrophysiological organization in roots of two *Plantago* species: direct measurement of ion transport to the xylem using excised roots*

A. H. de BOER and H. B. A. PRINS
Department of Plant Physiology, University of Groningen, The Netherlands

Introduction

Regulation of sodium transport across the plasmalemma of root cortical cells differs greatly between halophytes and glycophytes. Halophytes are known to accumulate salts[6] whereas glycophytes respond to salinity basically by ion exclusion[8]. The mechanism which enables glycophytes to exclude Na^+ ions is probably a coupling between K^+ uptake and Na^+ extrusion; this is indirect and a proton motive force (p. m. f.) generated by an electrogenic proton pump is thought to power this exchange[10].

Less attention has been paid to regulation of sodiumtransport to the shoot, although a selective barrier for ion transport at the symplast/xylem interface as a mechanism of salt-tolerance has been proposed by several authors[1,4]. However, there is still uncertainty as to the exact mechanisms of ion transport to the xylem[13]. One major hypothesis states that ion fluxes from outside to the symplast and from the symplast to the xylem are basically independent processes, both requiring energy. Hanson[9] transposed this hypothesis into a chemi-osmotic model for ion transport across the root. The symplast is represented as an osmotic unit bridging two apoplastic solutions, the bathing solution and the xylem, separated by the endodermis. A p.m.f. provides energy for transport of ions between the bathing solution and the symplast, whereas another p.m.f. drives the transport of ions between the symplast and the xylem. It follows from the uniformity of membrane characteristics, as proposed by the model, that the two proton pumps, one located at the plasmalemma of cortical cells, the other at the plasmalemma of xylem-parenchyma cells work in opposite directions: *i.e.* back-to-back.

Consequently the electrical potential difference between xylem sap and bathing solution or trans-root potential (TRP) will have a biphasic composition: the outer pump will hyperpolarize TRP (E_{HYP1}) while the inner pump depolarizes TRP (E_{HYP2}). Earlier studies of the structure of TRP did not reveal electrogenic pump activity at the symplast/xylem interface; seemingly the TRP of an exuding root behaved like the transmembrane electrical potential difference (PD) of a single cortical cell[1,3,14]. On the contrary Okamoto *et al.*[12] demonstrated the

* Grassland species research group, publication no. 56.
M.R. Sarić and B.C. Loughman (eds.), *Genetic Aspects of Plant Nutrition*.
ISBN-13: 978-94-009-6838-7
©1983 *Martinus Nijhoff/Dr W. Junk Publishers, The Hague/Boston/Lancaster.*

existence of back-to-back electrogenic pump activity in bean hypocotyles. They showed the existence of an electrogenic proton pump at the symplast/xylem interface (E_{HYP2}) and they ascribed to E_{HYP2} a function in ion absorption from the ascending xylem sap since E_{HYP2} was maximal (48 mV) in the elongating zone of the hypocotyl.

In the present study, two *Plantago* species differing in salt-tolerance were used; *Plantago media* L. a glycophyte and *Plantago maritima* L. a halophyte. These species differ markedly in their Na^+ transport to the shoot. No Na^+ could be detected in the shoot of *Plantago media* when cultured in 25% Hoagland + 1 mM NaCl solution, whereas the shoot of *Plantago maritima* contained 400 mmol/kg dry matter Na^+. Notwithstanding this the electrophysiological organization of roots of both the glycophyte and the halophyte is identical when plants are cultured without Na^+. In *Plantago media* as well as *Plantago maritima* two electrogenic-proton pumps contribute to TRP. The back-to-back orientation of these pumps suggests that reabsorption from the xylem sap by surrounding parenchyma cells plays an important role for the selective barrier function of the symplast/xylem interface and thus is important for the understanding of the mechanism of salt-tolerance.

Materials and methods

Plants were grown as described by Erdei and Kuiper[4], except that no NaCl was added. The experiments were carried out on individual plants, grown for 30 to 40 days in the nutrient solution. Fig. 1 depicts the set-up used to study the structure of TRP and to measure simultaneously the H^+ activity in the xylem sap and the xylem sap flowrate. The experimental set-up will be described in full detail elsewhere (De Boer, in preparation).

Results

A sufficient supply of metabolic substrates is important for the maintenance of the transmembrane PD of root cortical cells[7]. Therefore we added D-glucose, final concentration 25 mM, to the bathing solution shortly after excision of the shoot.

At the beginning of an experiment the level of TRP was in all roots of the same order -25 to -45 mV. During the experiment TRP reached a high level of 0 to -20 mV in some roots (type A), in others TRP became very negative, -80 to -100 mV (type B) and a third category roots (type C) showed intermediate TRP levels. The reaction of TRP of the three types of roots after changes in the oxygen concentration of the bathing solution (pO_2) differed markedly.

The reaction of type B roots to anoxia is discussed first (Fig. 2). When pO_2 is decreased to 8% there is a slow depolarization, but lowering pO_2 to zero results in a fast depolarization of around 40 mV. Aeration restores the original TRP and a further increase of pO_2 to 30% has no effect on TRP. As depolarization and

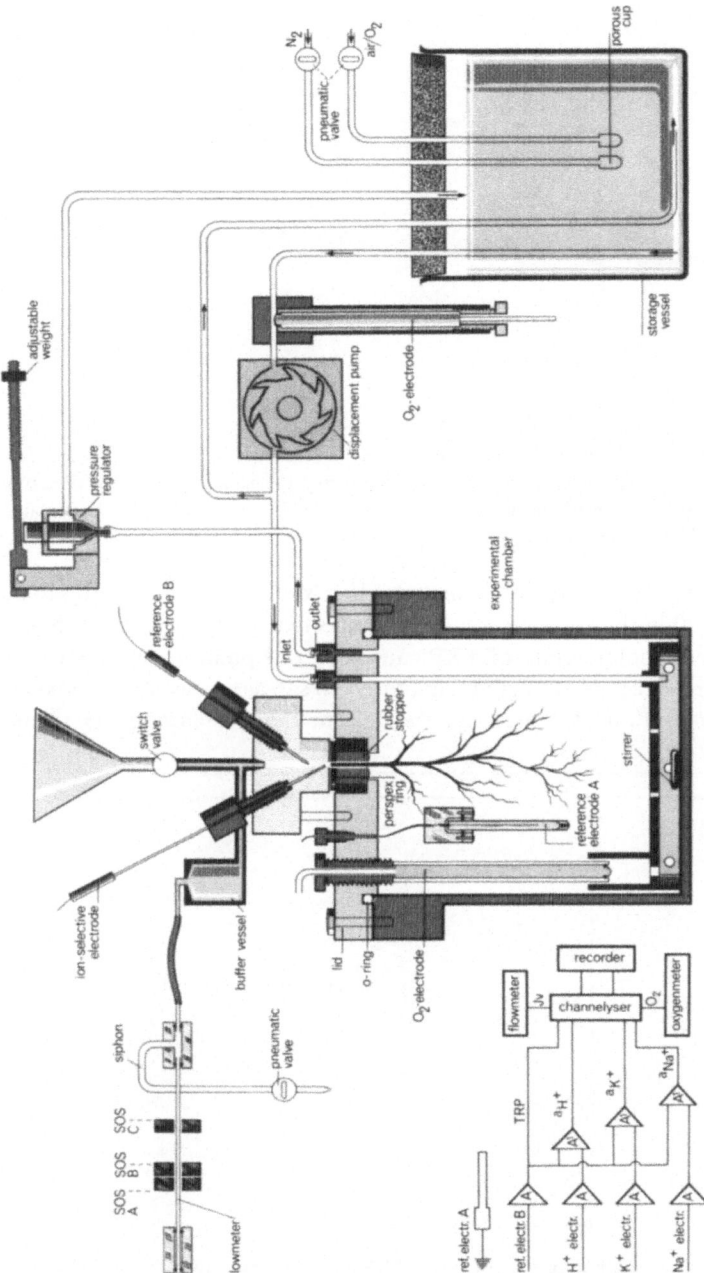

Fig. 1. The experimental set-up for simultaneous measurement of TRP, ion activities in the xylem sap and the xylem sap flowrate (J_v). A: electrometer amplifier. A^I: precision instrumentation amplifier. S.O.S.: Slotted Optical Switch.

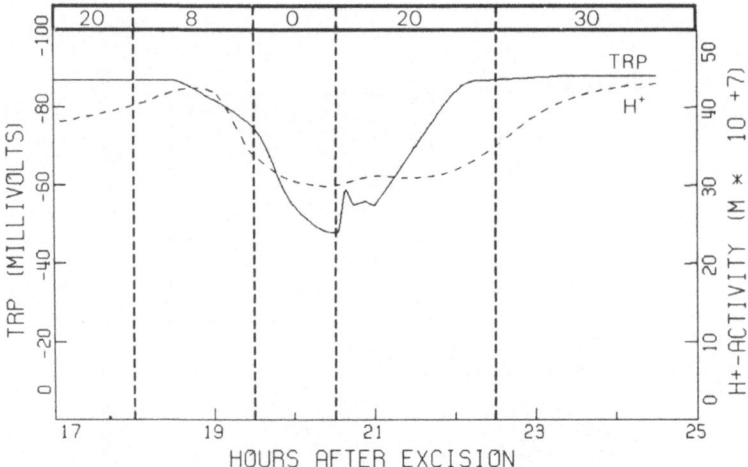

Fig. 2. Simultaneous recording of the reactions of TRP and proton-activity of the xylem sap, induced by changes in ambient pO_2, of type B roots. Numbers in the bar along the top of the figure refer to pO_2. Vertical dotted lines indicate the onset of change of pO_2. To complete the change takes about 15 minutes. This applies to Fig. 3 and Fig. 4 as well.

repolarizaton were often very fast (sometimes $10\,mV \cdot s^{-1}$), we consider this to be caused by an inhibition and re-activation respectively of electrogenic pump activity. The hyperpolarization of TRP caused by this pump is represented as E_{HYP1}. The decrease of the H^+ activity in the xylem sap as a reaction to anoxia is indicative of inhibition of proton transport from the symplast to the xylem.

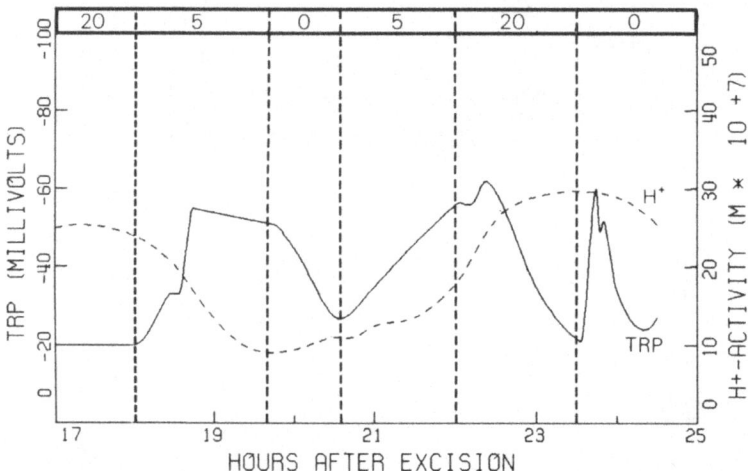

Fig. 3. Simultaneous recording of the reactions of TRP and proton-activity of the xylem sap, induced by changes in ambient pO_2, of type A roots. Numbers in the bar along the top of the figure refer to pO_2.

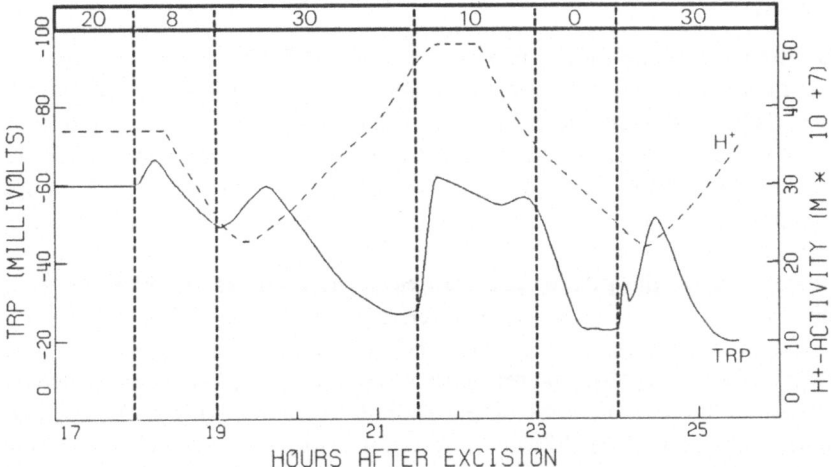

Fig. 4. Simultaneous recording of the reactions of TRP and proton-activity of the xylem sap, induced by changes in ambient pO_2, of type C roots. Numbers in the bar along the top of the figure refer to pO_2.

There was no such marked reaction of K^+ and Na^+ activities in the xylem sap (not shown here). The reaction of TRP of type A roots was bi-phasic (Fig. 3). A rapid hyperpolarization of 35 to 50 mV was observed when pO_2 was lowered to 5 or 10% and this hyperpolarized state could be maintained. Subsequent further oxygen depletion caused a depolarization of TRP similar to the reaction of type B roots. Very strikingly, the reaction of TRP after re-aeration was similar to the TRP reaction during anoxia. We consider this to be evidence for the existence of two electrogenic pump sites, operating back-to-back, both contributing to TRP. Therefore we will call the depolarization induced by anoxia E_{HYP1}, as in type B roots, and the hyperpolarization E_{HYP2}. Fig. 3 shows that the H^+ activity in the xylem sap starts to decrease when E_{HYP2} is inhibited. The H^+ activity starts to increase when, after the anoxia period, E_{HYP1} is activated, but E_{HYP2} remains inhibited.

The difference in sensitivity of the pumps for pO_2 in the bathing solution is also shown by the reaction of TRP of type C roots (Fig. 4). A decrease in pO_2 to 8% caused a small hyperpolarization of TRP, showing the activity of E_{HYP2}. A subsequent increase of pO_2 resulted in a repolarization followed by a large depolarization. From that moment on TRP of type C roots behaved like TRP of type A roots. Apparently the higher pO_2 level induced a higher activity of E_{HYP2}, which at a pO_2 of 30% results in a large increase of the H^+ activity of the xylem sap above the initial level. The H^+ activity decreases again when E_{HYP2} is inhibited by lowering pO_2 again ($t = 21.5$ h).

Discussion

On the basis of the widely different reactions of TRP in type A and type B roots, a concise model of the electrophysiological organization analogous to the model of Hanson[9] is proposed; the model is depicted in Fig. 5. In the symplasmic continuum in the cortical and stelar cells of the root, there is no gradient of transmembrane PD[11]. Essential in this model are the two electrogenic pumps working in opposite directions, which provides the basis for selective ion transport across the root.

In this context the present data can easily be understood. When E_{HYP2} is inhibited (Fig. 5), this results in a lower, more negative, TRP value. Such a hyperpolarization was observed when pO_2 was reduced to 5% (Fig. 3). The differential sensitivity to an oxygen deficit of E_{HYP2} and E_{HYP1} is to be expected on the basis of the compactness of the endodermis and pericycle. Fiscus and Kramer[5] showed that O_2 diffusion into the interior of excised roots was only enough to maintain a respiration rate there that is 65% lower than the over-all rate of the root tissue. When we subsequently made the bathing solution completely anaerobic (Fig. 3), this led to a depolarization of TRP as was to be expected, considering the potential profile in Fig 5.

The sensitivity of E_{HYP2} for oxygen deprivation also explains why an increase of pO_2 beyond 20% leads to a depolarization of TRP in type C roots (Fig. 4). Apparently in these roots E_{HYP2} is inhibited by lack of oxygen in the central part of the root even at 20% ambient pO_2.

Type B roots are similar to type A roots as far as the action and location of

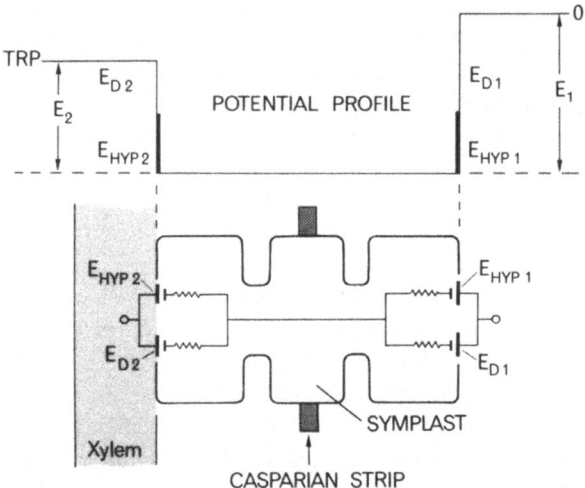

Fig. 5. Schematic illustration of the electrophysiological organization of the root. For explanation of the symbols see "Abbreviations" in text.

E_{HYP1} is concerned. Preliminary results with the slowly penetrating uncoupler CCCP ($10^{-5}M$) showed a depolarization of TRP both in type A and in type B roots; in type A roots TRP depolarized to positive values ($+10$ to $+20$ mV) whereas the reaction of TRP of type B roots to CCCP was similar to the anoxia reaction (Fig. 2), though quantitatively smaller. To fit type B roots in the model we have to assume that the electrogenic pump which builds up E_{HYP2} in type A roots, is non-electrogenic in type B roots. The marked effect of pO_2 on xylem sap proton activity suggests proton extrusion even in these types of roots. There is evidence that an electro-neutral extrusion (e.g. $K^+ - H^+$ antiport) is possible under some conditions[2].

The discovery of (electrogenic)-pump activity at the symplast/xylem interface is an important step in understanding selective ion transport to the shoot and thus for the explanation of the observed differences in sodium transport to the shoot of *Plantago media* and *Plantago maritima*. Experiments are now in progress in which sodium transport to the shoots of *Plantago media* and *Plantago maritima* is studied on the basis of simultaneous measurements of TRP and ion and waterfluxes.

References

1 Bowling D J F and Ansari A Q 1972 Control of sodium transport in sunflower roots. J. Exp. Bot. 23, 241–246.
2 Cleland R E 1976 Rapid stimulaton of $K^+ - H^+$ exchange by a plant growth hormone. Biochem. Biophys. Res. Commun. 69, 333–383.
3 Davis R F and Higinbotham N 1969 Effects of external cations and respiratory inhibitors on electrical potential of the xylem exudate of excised corn roots. Plant Physiol. 44, 1383–1392.
4 Erdei L and Kuiper P J C 1979 The effect of salinity on growth, cation content, Na^+-uptake and translocation in salt-sensitive and salt-tolerant *Plantago* species. Physiol. Plant. 47, 95–99.
5 Fiscus E L and Kramer P J 1970 Radial movement of oxygen in plant roots. Plant Physiol. 45, 667–669.
6 Flowers T J, Troke P F and Yeo A R 1977 The mechanism of salt-tolerance in halophytes. Annu. Rev. Plant Physiol 28, 89–121.
7 Graham R D and Bowling D J F 1977 Effect of the shoot on the transmembrane potentials of root cortical cells of sunflower. J. Exp. Bot. 28, 886–893.
8 Greenway H and Munns R 1980 Mechanisms of salt-tolerance in non-halophytes. Annu. Rev. Plant Physiol. 31, 149–190.
9 Hanson J B 1978 Application of the chemiosmotic hypothesis to ion transport across the root. Plant Physiol. 62, 402–405.
10 Jeschke W D 1980 Roots: cation selectivity and compartmentation, involvement of protons and regulation. *In* Plant Membrane Transport: Current Conceptual Issues. Eds. R M Spanswick, W J Lucas and J Dainty. pp 17–29. Amsterdam, Elsevier/North Holland Biomed Press, 670 pp.
11 Kelday L S and Bowling D J F 1980 Profiles of chloride concentration and PD in the root of *Commelian communis* L. J. Exp. Bot. 31, 1347–1355.

12 Okamoto H, Katou K and Ichino K 1979 Distribution of electric potential and ion transport in the hypocotyl of *Vigna sesquipedialis*. VI. The dual structure of radial electrogenic activity. Plant Cell Physiol. 20, 103–114.
13 Pitman M G 1977 Ion transport into the xylem. Annu. Rev. Plant Physiol. 28, 71–88.
14 Shone P G T 1969 Origins of the electrical potential difference between the xylem sap of maize roots and the external solution. J. Exp. Bot. 20, 698–716.

Concentration of critical nutrients in tolerant and susceptible sorghum lines for use in screening under acid soil field conditions

R. R. DUNCAN
University of Georgia, Georgia Experiment Station, Griffin, GA 30212, USA

Key words Sorghum bicolor (L.) Moench Nutritional profile Aluminium toxicity Soil pH

Introduction

Acid soil infertility is directly related to nutritional imbalances in plants grown on these problem soils. Sorghum genotypes vary tremendously in their ability to take up elements under stress conditions[1,5]. No single element concentration can be used when evaluating genotypes for tolerance/susceptibility to acid soils. The complete nutritional profile of the plant plus dry weight production and a visual rating scheme[2] can be combined to provide reasonable and reliable comparisons of genotypes grown under field conditions of acid soil stress. This information can then be used in a breeding program to develop improved acid soil tolerant germplasm[3]. This paper reports relative leaf element concentration values from tolerant and susceptible sorghum genotypes grown under five acid soil field environments from 1978 to 1980 in the southeastern United States. The leaf element uptake values are subsequently used to derive an acid soil field tolerance rating in conjunction with dry matter production values and visual rating scores to categorize sorghum genotypes for tolerance or susceptibility.

Methods and materials

Twenty-one sorghum (*Sorghum bicolor* (L.) Moench) genotypes (Table 2) were grown in a randomized complete block design with three replications in the field on three different soil types (Table 1). Data were combined on leaf element uptake values for each genotype grown for three years (1978 to 1980) on a Dyke clay loam and for one year (1980) each on a Cedar bluff silt loam and a Cecil sandy loam. Plant concentrations of Ca, Mg, K, P, Cu, Mn, Fe, Zn, Al were determined by analysis of leaf tissue (third leaf from the top) sampled after physiological grain maturity. Total above-ground dry weight (ten plants/plot) at grain harvest was measured. Visual stress ratings were taken according to Duncan[2]. An acid soil field tolerance rating scheme (ASFTR = Dry Plant Weight ÷ Σ (Elemental Interaction Ratios) × (Visual Stress Rating) was developed to evaluate the relative acid soil tolerance/susceptibility of the different genotypes for a plant breeding program.

Results and discussion

The twenty-one genotypes studied under five acid soil environments (three years on a Dyke soil and one year each on a Cedar Bluff and a Cecil soil) and their

Table 1. Soil description and range of relative elemental concentrations for three acid soils in Georgia (United States)

Location	Soil description	Element ppm				
		Al	Mn	Ca	Mg	pH
Blairsville*	Dyke clay loam Typic Rhodudult	5.2–6.0	76–108	133–165	13–26	4.7–4.8
Calhoun**	Cedar bluff silt loam Fragiaquic Paleudults	1.1–2.0	41–88	344–402	23–29	4.9–5.0
Griffin**	Cecil sandy loam Typic Hapludult	1.5–2.5	65–81	250–261	14–15	4.9–5.0

* Mean range values from 1978 to 1980. More complete breakdown of concentrations can be found in Duncan[1].
** Mean range values from 1980. Data supplied from six 15-cm subsamples per sample per sampling period. A sampling period at planting and a sampling period at harvest were combined to reflect the soil conditions throughout plant growth.

leaf elemental concentrations are listed in Table 2. NB 9040 had the highest concentrations of Al and Fe; SC0175, the highest Mn; TAM 428, the highest Zn; SC0056, the highest Cu; BSD106, the highest P, Ca, and K; and SC0723, the highest Mg. Conversely, the lowest concentrations for these elements were: SC0408-Al, Mn, Fe: SA6399-Zn, Cu, and K; NB9040-P; BT × 378 and SC0056-Ca; SC0112-Mg.

In comparing the elemental concentration values on a per genotype basis, NB9040 had high values for Al, Fe, Cu and medium-to-low values, for Mn, Zn, P, Ca, Mg, and K when compared with the other genotypes. BSD106 had high values for Zn, Cu, P, Ca, Mg, K and low values for Al, Mn, Fe. SC0283 had high values for Ca and Mg, medium values for Cu, and low values for Al, Mn, Fe, Zn, P, K.

Deciphering exactly what is happening to a specific genotype under acid soil stress conditions is difficult when evaluating strictly the element uptake values. However, by incorporating production data and visual stress ratings with the element uptake values, an overall acid soil performance rating can be devised to categorize different genotypes as to tolerance or susceptibility. Computation of the data over environments (locations, soil types, and years) will help in development of patterns for reliability and consistency of performance for specific genotypes under acid soil field stress conditions.

The relative acid soil field tolerance performance ratings for the twenty-one genotypes evaluated over five environments are included in Table 2. SC0283 is the most acid soil tolerant genotype while NB9040 is the most susceptible

Table 2. Pooled leaf elemental concentration** of selected sorghum genotypes grown under five acid soil field environments (three years on a Dyke soil, one year each on a Cedar bluff and a Cecil soil)

Genotype	Al	Mn	Fe	Zn	Cu	P	Ca	Mg	K	ASFTR*	Ranking***
	ppm					%					
NB9040	1760a	154def	720a	16.6e-h	8.3ab	0.12d	0.53de	0.16fah	1.12d-h	0.18i	21
SC723	1397ab	220abc	613abc	22.2a-d	6.3ef	0.16bc	0.66bcd	0.26a	1.14c-h	0.57ghi	14
BTx398	1261ab	223ab	614ab	19.2c-f	7.4b-e	0.13cd	0.62bcd	0.21a-e	1.23c-g	0.71fgh	11
SC0048	1243b	230ab	557a-d	23.2ab	8.1abc	0.16bc	0.67b	0.22a-d	1.37bc	0.56ghi	15
SC0418	831bc	217abc	448b-e	18.0d-g	8.0abc	0.13cd	0.56b-e	0.15fgh	1.46b	1.51cd	4
TAM428	815bc	164def	435b-f	24.8a	7.0c-f	0.14cd	0.65bcd	0.17efg	1.35bcd	0.53ghi	18
SC0112	797bc	160def	429c-f	19.0c-f	6.7def	0.14cd	0.54de	0.12h	1.31b-e	1.25de	7
SA6399-3	724bc	130ef	317ef	11.7i	4.2g	0.13cd	0.58b-e	0.22a-e	0.87h	0.64fgh	12
SC0175	688c	249a	369ef	24.0a	6.4ef	0.13cd	0.57b-e	0.20b-f	1.78a	1.02ef	9
P721	674c	133ef	380def	18.6c-g	7.0c-f	0.14cd	0.54de	0.23a-d	1.04fgh	0.56ghi	16
RTx430	629c	135ef	288ef	15.4f-i	6.7def	0.13cd	0.55cde	0.22a-e	1.17c-g	0.47ghi	19
CS3541	607c	124f	250ef	12.7hi	5.5fg	0.12d	0.62bcd	0.25ab	1.27b-f	1.27de	6
SC110	606c	161def	339ef	26.7a	7.0c-f	0.13cd	0.55cde	0.14gh	1.30b-e	0.58gh	13
SC0056	596c	196bcd	363ef	16.7e-h	8.9a	0.15bc	0.46e	0.16fgh	1.26b-f	1.95b	2
SC0322	591c	168def	336ef	20.1b-e	7.8a-d	0.17b	0.67bc	0.18d-g	1.12e-h	1.38d	5
BTx378	550c	142ef	317ef	14.9ghi	7.1b-e	0.13cd	0.46e	0.14gh	1.38bc	0.42hi	20
SC0599	492c	167def	337ef	15.4f-i	6.3ef	0.13cd	0.56b-e	0.19c-f	1.31b-e	1.20de	8
BSD106	468c	157def	247ef	22.5abc	8.6ab	0.24a	0.89a	0.24abc	1.97a	0.90efg	10
SC0283	426c	178cde	308ef	15.8f-i	7.2b-e	0.14cd	0.65bcd	0.22a-e	0.99gh	2.69a	1
SC0170	409c	128f	285ef	18.4c-g	6.4ef	0.15cd	0.54de	0.18d-g	1.47b	0.54ghi	17
SC0408	385c	126f	197f	14.0ghi	6.2ef	0.15cd	0.64bcd	0.18d-g	1.24b-g	1.76bc	3

* Acid Soil Field Tolerance Rating = Dry plant weight (above ground, 10 plants/plot, g) ÷ Σ (interaction ratios: Ca/Mg, K/Ca, K/Mg, Ca/P, Mg/P, K/P, Al/Fe, Al/Mn, Fe/Mn, Al/Zn, Zn/Cu, Fe/Zn, Mn/Zn, Al/Cu, Fe/Cu, Mn/Cu) × (Visual Stress Rating after Duncan[2].
** Third leaf from top of plant.
*** Highest acid soil tolerance = 1; moderate acid soil tolerance = 2 to 8; low acid soil tolerance = 9; susceptible = 10 to 21.
Mean separations are within columns according to Duncan's Multiple Range Test at the 5% level of significance.

genotype evaluated in this study. Based on field research during the past three years, the performance of the genotypes listed and their progeny derivations has closely paralleled the tolerance/susceptibility ratings and rankings. In general, the genotypes ranked 1 to 9 (ASFTR > 1.00) have provided some level of tolerance to acid soil field conditions. SC0283 is the superior and consistently the most acid tolerant genotype evaluated in the southeastern United States. Genotypes ranked 10 to 21 (ASFTR < 1.00) have been consistently susceptible to acid soil stress conditions. The acid soil rating scheme is proving to be a reliable indicator of relative tolerance/susceptibility for different sorghum genotypes.

The identification of tolerant sorghum genotypes has enhanced the breeding effort in Georgia for development of reliable acid soil tolerant germplasm which performs consistently under pH conditions ranging from 4.5 to 5.0[3].

Conclusions

The acid soil field tolerance rating scheme was developed for use in a breeding program to aid in the identification of relative levels of tolerance/susceptibility for different sorghum genotypes. The ASFTR values are not necessarily fixed and should vary with environmental extremes. Computation over environments (locations, soil types, years) will help in development of patterns of relative overall performance under acid soil field stress conditions. Inclusion of as much data as possible should theoretically add to the reliability of the ratings. Exclusion of certain elements or addition of different elements than reported in this paper would change the actual ASFTR values. Nevertheless, genotypic comparisons based on the same inputs should provide ratings which will aid plant breeders in determining the performance capabilities of various genotypes when subjected to acid soil field stress conditions. The rating scheme is simply another "tool" to assist the plant breeder, but should not be the sole criterion used in a soil stress breeding program. Continued refinement of the rating scheme will improve its reliability as a plant breeding strategem.

References

1 Duncan R R 1981 Variability among sorghum genotypes for uptake of elements under acid soil field conditions. J. Plant Nutr. 4, 21–32.
2 Duncan R R 1981 Visual field rating scheme for plants grown on acid soils. Sorghum Newsl. 24, 69.
3 Duncan R R 1981 Registration of GP1R acid soil tolerant sorghum germplasm population (Reg. No. GP73). Crop Sci. 21, 637.
4 Duncan R R 1982 Categorization of sorghum genotypes for tolerance or susceptibility to acid soil stress environments. Sorghum Newsl. 25, 24–25.
5 Duncan R R, Dobson Jr J W and Fisher C D 1980 Leaf elemental concentrations and grain yield of sorghum grown on an acid soil. Commun. Soil. Sci. Plant Anal. 11(7), 699–707.

The effect of selenium on ion uptake at two nutrient levels

N. V. GUJOVA, S. M. FATALIEVA and A. Sh. KERIMOVA
Moscow State University, Moscow; Institute of Botany Azebaijan Academy of Sciences, Baku, USSR

Key words Maize Nutrition Pea Potassium Toxicity

Introduction

The importance of Se for plants has not been fully investigated. However, in experiments with soil and water cultures the established influence of Se on the growth of soy beans, mustard, millet[1] and some other species[2,6] has been shown. Plant reaction on the presence of Se in growth media depends on the species character[4], that changes the metabolic activity. The processes of absorption and accumulation of Se and other elements are determined by complex biological and biochemical factors[3,4]. Many factors influence Se uptake by plants: the ratio of different forms of Se, interaction of macro and microelements, the content of Ca and Ba, P and S[6]. The aim of the present investigation is to study effects of nutrition level and Se on ion intake.

Materials and methods

The objects of this study were maize (var. Zakatalskaya uluchsennaya) and pea (var. Aznihi) which were stable and sensitive to the presence of Na_2SeO_3 in the growth medium respectively. The plants were grown in water culture on 1/5-strength or full nutrient solution with addition of Se according to the variants: I. – Se(control), II. +0.5 mg Se/l (promoting dose) III. +5.0 and 10.0 mgSe/l for pea and maize, respectively (inhibitory doses). After 7 days growth on nutrient solution with Se potassium and phosphorus intake was studied. Absorption was studied during 6 h according to diminution of elements from 1 or 1/5-strength Hoagland-Arnon I or Knop nutrient solution. Ca and K were determined on the flame photometer and in a separate experiment the kinetics of potassium intake was studied. The plants were grown for 3 weeks on the Hoagland nutrient solution with Se. Absorption was measured with the frequency of one counting during 5–10 min. The registration of potassium ion concentration in the medium was carried out by K^+-sensitive electrode. The rate of potassium uptake was expressed as microequivalents per minute.

Results

During the first 2 weeks the maize seedlings grew better on 1/5-strength, whereas pea seedlings on full nutrient solution yielded better than those on 1/5 nutrient.

Both species grew better in the 2nd variant than the controls and in the 3rd

Table 1. Effect of Se on the ion uptake of seedling maize

Variant	% of control			
	P	K	Ca	NO_3
II	232	252	190	101
III	162	258	140	114
I→III	283	174	153	62.5

variant the growth was suppressed. The data on the ion uptake by seedlings, growing on 1/5-strength nutrition solution, are presented in Table 1. The results are expressed relative to the controls, which are taken for 100. The ion uptake by maize seedlings increased with the exception of NO_3. In all variants with pea, except III variant, the uptake is decreased. The plants were grown on full nutrient solution in the experiment and were taken from I and III variants with absorption periods of 2 and 4 h.

From the results in the Table it follows that Se stimulates the uptake of K more than the uptake of P and Ca (Table 2). The change in the potassium uptake in the experiments was also shown with the use of a K^+ specific electrode. It was established that Se stimulates the uptake of potassium by maize seedlings for one plant, especially in the 2nd variant. The potential absorptive activity of roots with respect to potassium rises with the increase of Se dose in the growth medium (Fig. 1).

The seedlings of the 2nd variant absorbed K more intensively. The highest Se dose (III var.) during the first hour of the experiment decreases and then increases the intake of K, though the weight of the roots of this variant lags behind the 1 and 2 variants by 5-fold.

The calculation per 1 g of fresh weight showed that the potential absorbing activity increased with the rise of Se in the growth medium.

Table 2. Effect of Se on the ion uptake of seedlings maize (% of the control)

Growing	Absorption	P		K		Ca	
		2 h	4 h	2 h	4 h	2 h	4 h
I	III	147	178	271	268	102	128
I	III	–	159	–	242	–	115
III	I→III	98	140	283	107	77	77

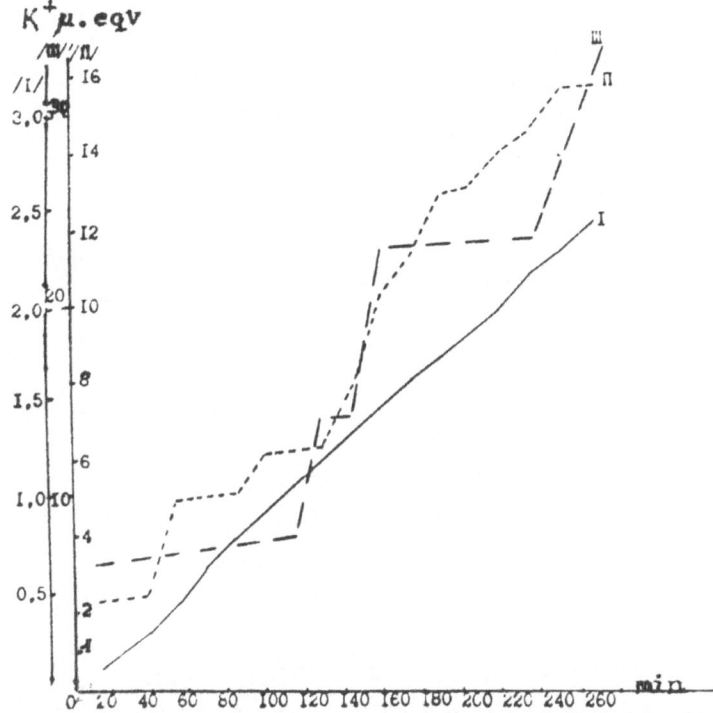

Fig. 1. The kinetics of K absorption by maize roots. I, II, III – variants. Measures were held on the medium without Se.

Discussion

We proceeded from the supposition that maize is relatively stable and that pea is sensitive to the presence of Se in the growth medium. This supposition is based on the same effect of growth inhibition with the concentration for maize 10, and for pea 5 mg/1. Certainly, maize does not possess special stability comparable with Astragalus, the accumulator of Se. The plants we used react differently to various nutrient and Se levels. Selenium can be toxic for plants, but small concentrations stimulate the growth of many plants[1,2].

Se stimulated the uptake of K, P, and Ca by maize seedlings in the 2nd variant (Table 1), when growth has also been stimulated. This especially occurs during short-period exposures (Table 2). The removal of plants from the I variant to the III causes an increase in the intake of ions studied, with the exception of NO_3. In all our experiments with Se the intensification of P uptake is marked. The increase of P content in the plants under the Se treatment has been shown by Singh and Singh[7]. Having used high doses of P they discarded the toxicity of Se. Perhaps, the intensification of P uptake is an adaptive reaction to Se.

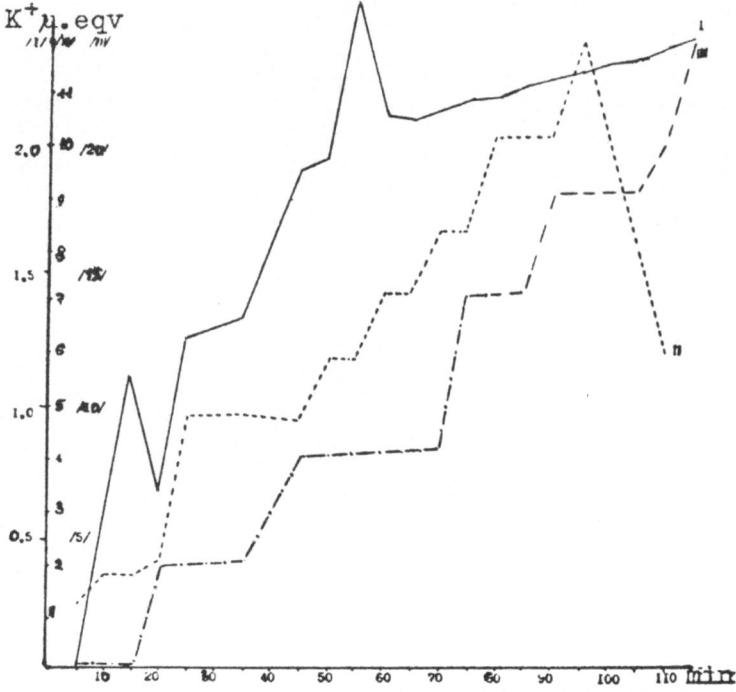

Fig. 2. The kinetics of K absorption by pea roots. I, II, III – variant. Measures were held on the medium without Se.

Plant nutrition is regulated by mechanisms which provide selectivity of ion transfer selectivity from the environment. In long and short-term experiments, and also on the removal of plants without Se to the medium with Se, the intensification of K uptake is observed. In the III variant it may be related to the inhibition of growth, which is reduced 5–6 times. Perhaps changes of membrane permeability takes place and potassium passively enters the roots. In the 2nd variant potassium entry can not be explained by the inhibition of maize seedlings growth. Perhaps another mechanism acts here. It may be supposed that the enhancement of a proton pump serves as the mechanism of Se detoxication: intense intake of potassium indicates this possibility.

References

1 Ermakov V V and Koval'skiy V V 1974 Biologicheskoe znachenie selena. M. Nauka.
2 Fatalieva S M, Guzhova N V 1978 Deistvie selenita natria na rost i energoobmen korney goroha i kukuruzy. Zh. Sel'sko hozyaistvennaya biologia 12.
3 Giessel-Nielsen 1979 Isotopes and radiation in research on soil-plant relationship. Vienna, p 427. Colombo, 1978. Jointly organized by IAEA and FAO.

4 Kramer W and Ziechman W 1972 Zur pflanzenphysiologischen Bedeutung des Selens. Z. Pflanzenernähr und Bodenkd. 132 (3), 191.
5 Middliton L J and Russel R 1958 The interaction of cations in absorption by plant tissues. J. Exp. Bot. 9, 115.
6 Ravikovich S and Margolin M 1959 The effect of barium chloride and calcium sulphate in hindering selenium absorption by lucerne. Empire J. Exptl. Agric. 27, 107, 235.
7 Singh M and Singh N 1979 The effect of forms of selenium on the accumulation of selenium, sulfur, and forms of nitrogen and phosphorus in forage cowpea (*Vigna sinensis*). Soil Sci. 127, 264.
8 Shrift Alex 1969 Aspects of selenium metabolism in higher plants. Annu. Rev. Pl. Physiol. 20, 475.

Differences in sulphate and phosphate uptake and utilization within *Zea mays* L. species

MARGITA HOLOBRADÁ
Institute of Experimental Biology and Ecology, C.B.E.V., Slovak Academy of Science, 814 34 Bratislava, Czechoslovakia

Introduction

Differences in anion uptake and their assimilation between species are due to different regulation mechanisms. Various rates of sulphate uptake and metabolism in *Zea mays*, *Pisum sativum* and *Gossipium hirsutum* demonstrate genetic differences among plant species in quantitative, and in some measure also in qualitative aspects of sulphur metabolism, indicating various requirements for sulphur supply[1,3,4,6,9]. The genetic control of the differences of some processes of uptake and metabolism of sulphate for instance is more apparent when growing the species under conditions of nutrient stress as it was found for cotton and pea roots when studying the activity of ATP-sulphurylase and APS-kinase[9], or comparing the uptake and reduction of sulphate by maize and cotton plants at deficiency and excess of sulphate[6].

Within the species genotype differences are more marked under stress conditions than under normal growing conditions. Since uptake and utilization of anions are closely related to various kinds of stress conditions, one part of the paper is concerned with comparing three genotypes with different cold resistance in their sulphate and phosphate uptake in response to chilling temperature. Some aspects of anion incorporation into organic compounds are also discussed.

Materials and methods
Two- to 3-day-old seedlings of cultivars of *Zea mays* L. (Český bíly Zajíčkův konský zub, VIR 17, STA 1871, CE 380, CE 250 and LG 9) with 4 cm long roots were treated with ^{35}S-sulphate and ^{32}P-phosphate using the labelling with carrier-free ^{35}S and ^{32}P (37 kBq Na$_2$^{35}SO$_4$.ml^{-1} of 1/5 strength Knops nutrient solution and 3.7 kBq KH$_2$^{32}PO$_4$.ml^{-1} of the same nutrient medium). For ^{35}S-infiltration the labelling was 75 kBq ^{35}S.ml^{-1}.

The uptake and assimilation of sulphur in excised roots using the method of infiltration of ^{35}S-sulphate was followed in cv. Český bíly Zajíčkův konský zub and in VIR 17. The methods of ^{35}S-infiltration and the determination of ^{35}S-compounds were described earlier[2,3].

The zonality of ^{35}S-sulphate uptake and accumulation along the root axis was studied with seedlings of cv. VIR 17, STA 1971 and CE 380 when incubating the seedlings in ^{35}S-labelled nutrient solution for 5 and 15 min uptake interval. The experimental procedures are given in the paper of Holobradá[5].

Anion uptake and utilization at various temperatures was studied with the seedlings of cultivars having different sensitivity to chilling temperature at 5°C (cv. CE 380, CE 250 and LG 9). The seedlings were cultivated at 25°C and 5°C in nutrient solution labelled with ^{35}S-sulphate and ^{32}P-phosphate. Cooling lasted 1 hour and 1–7 days respectively. The incorporation of ^{32}P into organic compounds was followed after 24 and 168 hours (1–7 days) cooling. Experimental procedures and methods are described by Holobradá et al.[7] and Mistrík and Kolek[8].

Results and discussion

In the primary seminal excised roots of VIR 17 and Český bíly Zajíčkův konský zub labelled organic sulphur was found after 30 min of exposure to infiltrated ^{35}S-sulphate. We found no differences in quality of reduction of ^{35}S-sulphate into cysteine and methionine in either cultivar[3,4]. But, in the roots of VIR 17 the rate of reduction to free S-amino acids was 0.32–4.5% of total infiltrated ^{35}S through 0.5–8 hours exposure (Table 1), while in the roots of Český bíly Zajíčkův konský zub it was only 0.1–0.5%[3].

The zonality of ^{35}S-uptake and accumulation along the axis of primary roots was compared between VIR 17, STA 1971, CE 380 and Český bíly Zajíčkův konský zub. The zonality is illustrated by Fig. 1 and Fig. 2, where ^{35}S-uptake after 5 and 15 min is demonstrated, using the results with VIR 17, STA and CE380. The 5-min uptake interval (Fig. 1) was sufficient to confirm the zonality of the root in sulphate uptake, the higher uptake capacity of the apical part of the root and the important role of root hairs in the uptake of various cultivars. There were differences in the rate of the uptake between VIR 17, STA 1971 and that of CE 380. Also the maximum uptake of the last cultivar was a little further away from the apex than those of VIR 17 and STA 1971.

Prolonging the uptake to 15 min reduces the differences between cultivars. The zonality of uptake after 15 min ^{35}S-uptake, demonstrated on Fig. 2 by results with

Table 1. The ^{35}S-content of excised primary seminal maize roots after incubation for 0.5 to 8 h after labelling and the percentage of the activity found in the soluble organic ^{35}S compounds fraction [cv. "VIR 17"). Redrawn from Holobradá[4]

^{35}S-incubation time (h)	Total ^{35}S-activity $\mu Ci/g$	Soluble organic ^{35}S-compounds %
0.5	12.22	1.0
1	11.87	1.76
4	12.30	2.43
8	11.83	4.05

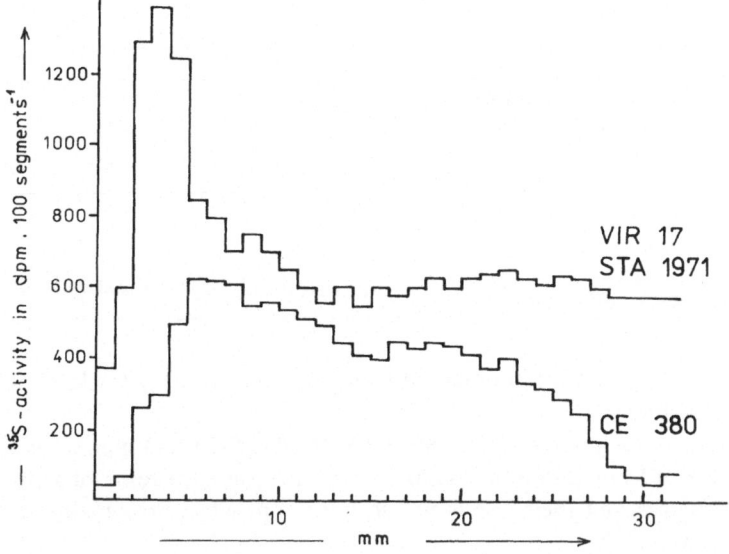

Fig. 1. The zonality of ^{35}S-sulphate uptake by roots of various cultivars of *Zea mays* after 5 min uptake interval.

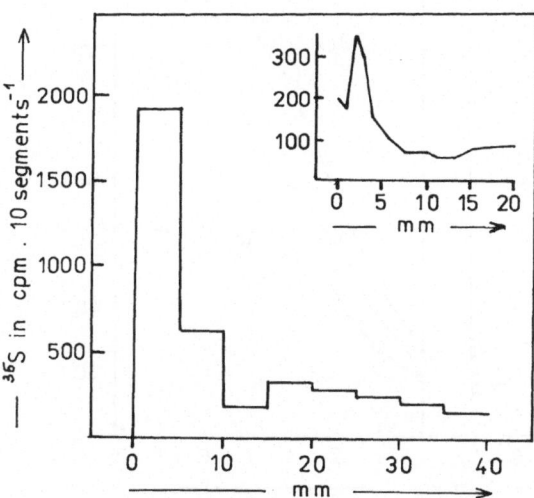

Fig. 2. S-sulphate uptake and distribution after 15 min uptake. In the upper part of the fig. the zonality of uptake during tissue differentiation and by root hairs is given.

Table 2. Changes in ^{32}P-phosphate uptake after 1 h exposure at 5°C by various cultivars of *Zea mays* L. roots compared with ^{32}P-uptake at 25°C (%)

Cultivar	Test temperature (°C)	
	25°	5°
CE 380	100	14.8
CE 250	100	18.7
LG 9	100	20.4

VIR 17 and STA 1971, differs in other cultivars very slightly; only the level of sulphate is a little higher.

The differences in the uptake of CE 380 and VIR 17 (STA 1971) suggest that the initial uptake of 5 min duration may be, in some measure, the result of some genotypic differences, but their role is not so important when prolonging the uptake to 15 min.

More evident genetic differentiation was observed when we compared cultivars CE 380, CE 250 and LG 9 from the aspect of their resistance to chilling temperatures. A marked inhibition of ^{32}P uptake occurred already after 1 h at 5°C (Table 2). The most intensive decrease of phosphate uptake was found in cv. CE 380, very sensitive to chilling during its initial phases of ontogenesis; the lowest decrease of uptake was in cv. LG 9, having a rapid initial development and

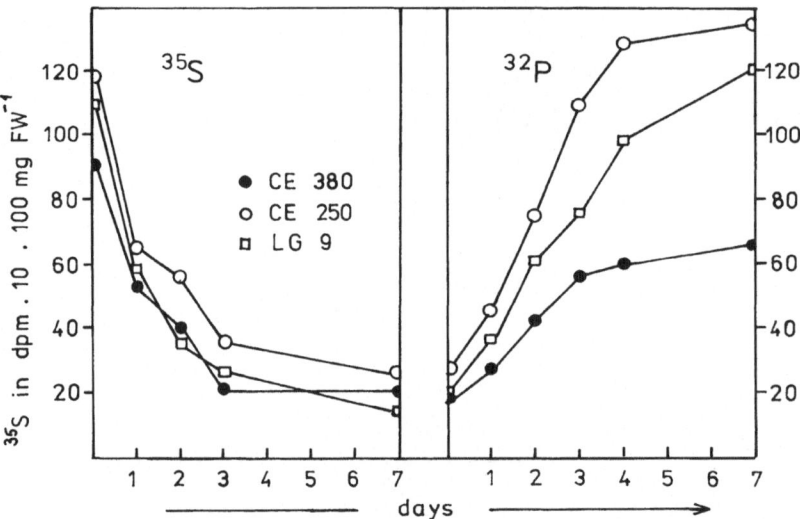

Fig. 3. The uptake of ^{35}S-sulphate and ^{32}P-phosphate by roots of various cultivars of *Zea mays* after 1–7 days exposure to 5°C. (Redrawn from Holobradá *et al.*[6])

good tolerance to low temperature. In spite of good tolerance, even LG 9 after 30 min of ^{32}P-uptake decreased its uptake to 20.4% compared to 25°C. The rate of uptake by two other cultivars CE 380 and CE 250, more sensitive to low temperature, was only 14.8% and 18.7%.

The reaction of the three cultivars to 1–7 days chilling at 5°C was totally opposite when ^{35}S-uptake is compared with the uptake of ^{32}P (Fig. 3). While ^{35}S-uptake decreased for all cultivars independently of the degree of cold-resistance, the uptake of ^{32}P, after an initial fall increased during the experiment, indicating some degree of adaptation to cooling by cultivars with higher cold resistance: LG 9 and CE 250.

Differences between the cultivars in their incorporation of ^{32}P into organic compounds, after 24 and 168 hrs 5°C (Fig. 4), and the incorporation of ^{35}S into proteins were also found. Sulphate reduction and incorporation into soluble proteins was inhibited after 7 days chilling to 4.5% compared to control plants at 25°C (Mistrík, unpublished).

In spite of some adaptability in relation to phosphate uptake (Fig. 3), the uptake rate becomes lower than the uptake by control plants. The chilling temperature affects not only the uptake, but also the fractionation of ^{32}P-compounds (Fig. 4). In roots of CE 380 and CE 250 too after cooling, the fraction of

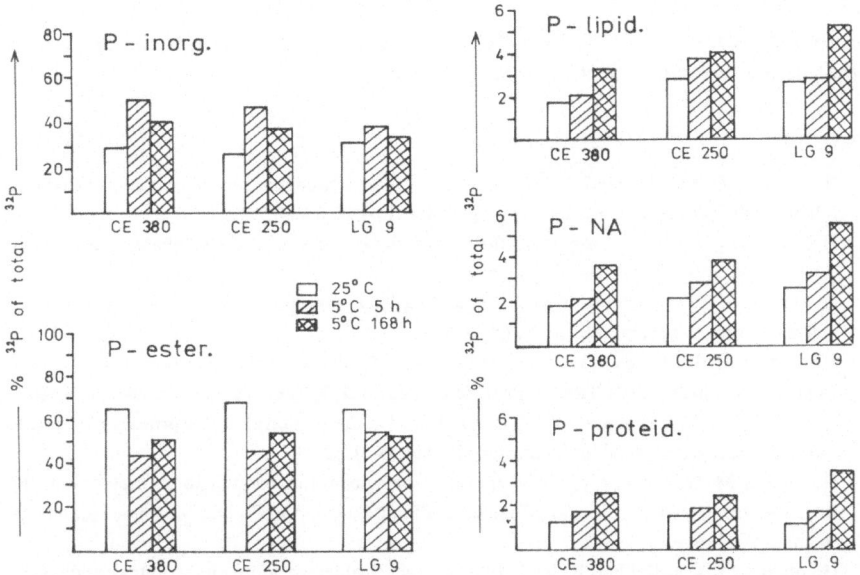

Fig. 4. The percentage of incorporation of ^{32}P-phosphate into organic compounds related to the total uptaken ^{32}P amounts in cv. CE 380, CE 250 and LG 9. ○ P-inorg. = inorganic ^{32}P-phosphate; P-ester. = low molecular weight ^{32}P-compounds + free nucleotides; P-lipid. = ^{32}P-phospholipids; P-NA = ^{32}P-nucleic acids; P-protein = ^{32}P-phosphoproteins. (Redrawn from Mistrík and Kolek, 1982.)

inorganic ^{32}P increased to higher values than in LG 9. Phosphate metabolism was well-preserved, or even stimulated in roots of cold-resistant cv. LG 9 compared to the other two cultivars. The genetic specificity in phosphorus metabolism of LG 9 manifested itself more after 7 days cooling, when the effect of adaptation processes was more evident. At 25°C most of the absorbed ^{32}P-phosphate was found in the fraction of low molecular compounds and free nucleotides. In response to 24 hours cooling, the metabolic assimilation of phosphate into low-molecular compounds, especially in CE 380 and CE 250 was inhibited, but inorganic phosphate was markedly increased. Although macromolecular P-compounds (Fig. 4) increased slightly after 24 h cooling, the increase was more obvious after 168 hours (7 days), particularly in LG 9.

On the basis of these results we suppose that differences in this phase of ontogenesis in relation to phosphate uptake and utilization have indicated higher cold-resistance of LG 9. Cultivars CE 250 and CE 380, characterized by low cold resistance, have appeared in accordance with this characteristic in the present experiments.

Nevertheless we have to emphasize that when measuring uptake and utilization (Fig. 3 and unpublished data) cold resistance did not appear, as it has been found for phosphorus. We have pointed out that differences found in experiments with seedlings, are characteristic only for this early stage of ontogenesis and can not be predicted in later stages. We cannot say, even to a limited degree, whether they will be found in later phases of ontogenesis and to what degree we can expect to observe them.

References

1 Holobradá M 1969 Dynamics of the dry matter, S- and N-contents in pea and maize grown in full and deficient nutrient medium. Biológia (Brat.) 24 (47), 524–534.
2 Holobradá M 1970 Some aspects of the uptake and conversion of ^{35}S-sulphate to the organic form in plants. Biológia (Brat) 25(10), 667–671.
3 Holobradá M 1971 Incorporation of ^{35}S-sulphate in organic compounds of excised roots of pea and maize plants. Biológia (Brat) 26(1), 27–32.
4 Holobradá M 1974 The uptake and assimilation of sulphur in primary roots. Structure and Function of Primary Root Tissues. Proc. of Symp.(Ed. J. Kolek), Veda, Bratislava, 455–460.
5 Holobradá M 1977 Changes in sulphate uptake and accumulation along the primary root during tissue differentiation. Biológia Plantarum (Praha) 19(5), 331–337.
6 Holobradá M 1981 Absorption and transport of sulphate in relation to its metabolism. R. Brouwer et al. (Eds.) Structure and Function of Plant Roots, pp 249–252, Martinus Nijhoff/Dr. W. Junk Publishers.
7 Holobradá M, Mistrík I and Kolek J 1981 The effect of temperature on the uptake and loss of anions by seedling roots of Zea mays L. Biológia Plantarum (Praha) 23(4), 241–248.
8 Mistrík I and Kolek J 1982 Effect of low temperature on ^{32}P incorporation into organic phosphorus compounds in maize (Zea mays L.) roots. Biológia (Brat) 37, 725–730.
9 Shevyakova N I and Holobradá M 1974 Aktivacya sulfata u rastenij pri izbytke i nedostatke sery. Fiziol. Rast. 21(5), 988–994.

Factors responsible for genotypic manganese tolerance in cowpea (*Vigna unguiculata*)

W. J. HORST
Institut für Pflanzenernährung (330), Universität Hohenheim, D-7000 Stuttgart 70, FRG

Key words Cowpea Genotypical differences Manganese Manganese tolerance Manganese toxicity *Vigna unguiculata*

Abstract In experiments with 29 cowpea genotypes considerable variation in Mn tolerance could be found. Ranking according to Mn tolerance was almost the same in sand and water culture. Mn tolerance is not related to greater vigour or exclusion of Mn from uptake and translocation, but depends mainly on the internal tolerance to excess Mn especially in the leaf tissue.

Growth depression by Mn excess is characterized by local accumulation of Mn, deposition of Mn oxides, and typical macro-symptoms on the older leaves (brown spots→clorosis→shedding of the leaves). Autoradiographic studies with ^{54}Mn and extraction of the leaves with methanol and H_2O indicate a causal relationship between Mn tolerance and the more homogenous distribution of Mn in the tissue. In tolerant genotypes local accumulation and deposition of Mn is inhibited or retarded.

Mn applied to the petioles of fully expanded leaves induces the same toxicity symptoms on the leaf blades as Mn absorbed by the roots. There is a good agreement between the rankings of the different genotypes for Mn tolerance according to the depression of shoot dry matter production by Mn excess in long term pot experiments and the appearance of toxicity symptoms after application of Mn to the petioles.

The regulation of Mn tolerance at the leaf tissue level allows a quick and non-destructive screening of large numbers of genotypes for Mn tolerance.

Introduction

Considerable genotypic differences in Mn tolerance of cowpea have been reported by Kang and Fox[10]. This genetic potential may be exploited for the breeding of high yielding varieties adapted to acid and waterlogged soils, in which Mn frequently limits plant growth. Breeding progress, however, is very dependent on the number of genotypes which can be tested. Therefore quick methods of screening for Mn tolerance are required based on a better understanding of the physiological and morphological reasons for genotypical differences in Mn tolerance. Mn tolerance has been attributed to low Mn uptake, limited translocation of Mn from the roots to the shoots and/or greater tolerance of high Mn levels within the plant tissue[4]. The present paper reports on studies on the factors which are responsible for Mn tolerance in cowpea and prospects for the development of quick screening techniques based on these results.

M.R. Sarić and B.C. Loughman (eds.), Genetic Aspects of Plant Nutrition.
ISBN-13: 978-94-009-6838-7
©1983 Martinus Nijhoff/Dr W. Junk Publishers, The Hague/Boston/Lancaster.

Materials and methods

Cowpea (*Vigna unguiculata*) seeds were supplied by the International Institute of Tropical Agriculture (IITA), Ibadan, Nigeria. The genotypes were selected from the Germplasm Collection (Tvu numbers according to the IITA Germplasm Catalogue) and the breeding program. The sand culture experiments (quartz sand) were conducted in a heated greenhouse with 25°C day and 20°C night temperatures. The pots were percolated twice daily with fresh nutrient solution containing Mn concentrations between 10^{-3} and 10^{-1} mM. The age of the plants at the end of the experiment was 45 days.

Plants in solution culture experiments were grown in 5 l plastic buckets under controlled environmental conditions in a growth chamber with 30/25°C day/night temperatures, 60% rel. humidity and 25,000 lux light intensity. Nutrient solutions were comparable to those of the sand culture experiment. Mn concentrations were between 5×10^{-4} and 10^{-2} mM. Solutions were changed every 2 days. Plants were harvested after 13 to 28 days.

Mineral elements in the plant tissue were determined after drying at 105°C and ashing overnight at 450°C. The ash was dissolved in 6 N HCl containing hydroxylammoniumchloride. After appropriate dilution, Mn, Fe, and Ca were determined by atomic absorption spectrometry. For studies on Mn distribution within the leaves 9 day old plants were supplied for 4 days with ^{54}Mn labelled MnSO$_4$(5×10^{-3} mM), rapidly frozen at -30°C, freeze dried and exposed to X-ray films.

The chemical forms of Mn in the leaf were estimated by sequential homogenization of the fresh tissue for 1 min, first in 80% methanol, then in water in a cold room at 4°C. After centrifugation the extracted Mn was determined in the supernatant and in the residue.

In some experiments Mn was applied directly to individual leaves of intact plants grown in water culture: tissue paper was saturated with MnSO$_4$ solution (50 mM), tied around the petiole and covered by Al-foil.

Results and discussion

Considerable differences in Mn tolerance of 29 cowpea genotypes were found in a sand culture experiment (Fig. 1) when shoot dry matter production at different Mn concentrations is compared. Unfortunately, in the field experiments of Kang and Fox[10] different genotypes were used. It is therefore not possible to make comparisons between genotypical Mn tolerance in sand culture and under field conditions. In the same manner as for soybean[3], reduction in shoot dry matter and induction of Mn toxicity symptoms by Mn excess are closely related. In cowpea there was a typical sequence of Mn toxicity symptoms: small dark-brown spots on leaf blades→yellowing→dessication and finally shedding of the leaves. Ranking of the genotypes according to Mn tolerance was almost the same in sand and in water culture[6]. In agreement with the results with soybean[1,5], higher Mn tolerance in cowpea is not related to greater vigor ("dilution effect") or exclusion of Mn from uptake and translocation, but depends mainly on the tissue tolerance to excess Mn especially in the leaves (Fig. 2). There is no relation between leaf Mn concentration at a given Mn supply and depression of dry matter production by Mn excess. In Fig. 2 the Mn concentration of old leaves has been chosen because they accumulate most of the Mn and show Mn toxicity symptoms first.

Ca and Fe nutrition[9,11] have been suggested to be involved in Mn tolerance of

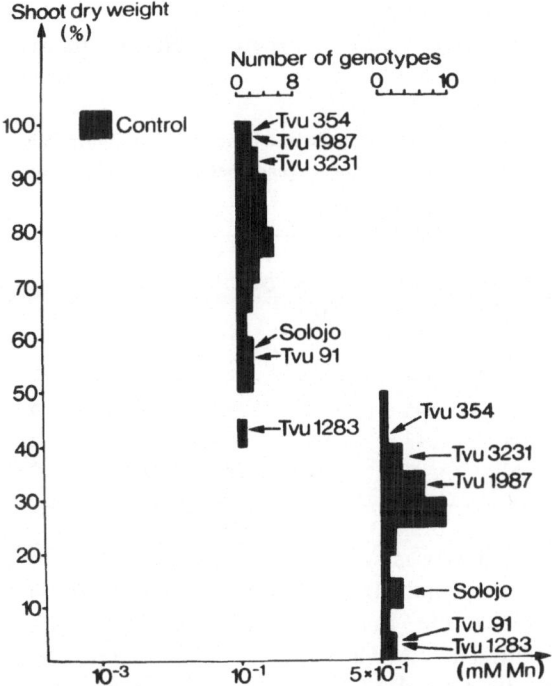

Fig. 1. Effect of Mn supply on shoot dry matter production (percent of controls) of 29 cowpea genotypes. Sand culture. Duration of the experiment 45 days.

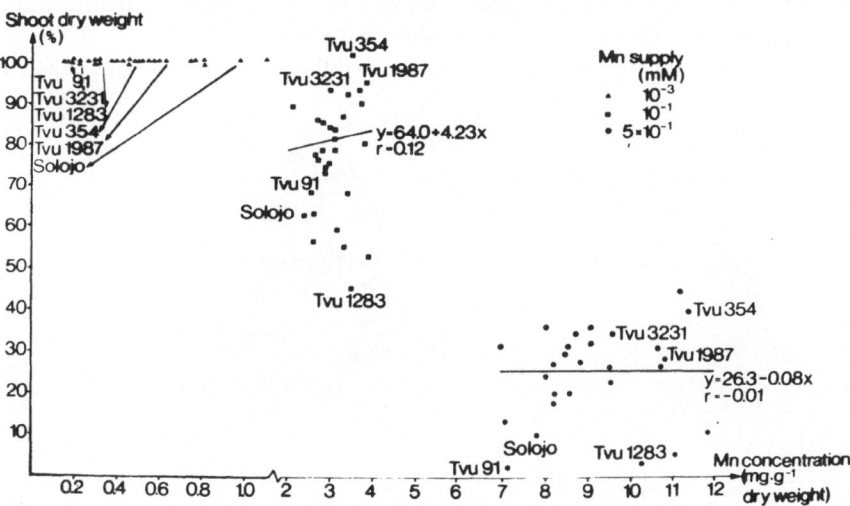

Fig. 2. Relation between Mn concentrations of older leaves and shoot dry matter production (percent of control) of 29 cowpea genotypes with increasing Mn supply. Sand culture. Duration of the experiment 45 days.

other plant species. Comparison of the mineral ion concentrations of leaves of tolerant and sensitive genotypes (Table 1) do not suggest correlations between Mn/Ca and Mn/Fe interactions and differences in Mn tissue tolerance in cowpea.

Table 1. Effect of Mn excess on dry matter production (shoots) and mineral ion concentrations in leaves of 3 extremely Mn tolerant (t.) and Mn sensitive (s.) cowpea genotypes. Sand culture. Duration of the experiment 45 days.

Mn supply (μM)	Dry weight (g)		Concentration Ca^{++}		(mval 100 g^{-1} Mn^{++}		dry weight) Fe^{++}	
	t.	s.	t.	s.	t.	s.	t.	s.
10^{-3}	7.7	7.1	120	145	1.1	1.2	0.9	0.9
10^{-1}	7.4	4.0	104	113	9.4	9.4	0.9	1.4
5×10^{-1}	2.6	0.4	81	94	30.0	29.0	1.0	2.2

Distribution of Mn within the leaf was studied by macroautoradiography (Fig. 3). Despite similar amounts of total Mn in the leaves Mn distribution is genotypically completely different. In the Mn sensitive genotype Tvu 91 distinct local accumulations of Mn occur. These Mn accumulations correspond to the dark brown spots shown to be typical Mn toxicity symptoms[8] and are precipitates of Mn oxides[2]. In contrast, in the Mn tolerant genotype Tvu 1987, the distribution of Mn is quite uniform. Higher Mn tolerance therefore is not related to precipitation and inactivation of Mn as Mn oxides but to a more homogenous distribution of Mn within the leaf tissue.

Extraction of the fresh leaf tissue also reveals different chemical forms of Mn in

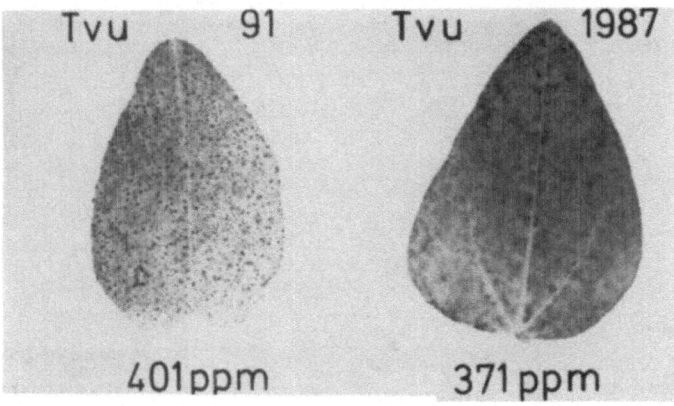

Fig. 3. Content and distribution of Mn (^{54}Mn) in primary leaves of cowpea. Autoradiographs. Age of the plants at the beginning of the Mn treatment 9 days. Then Mn supply (5×10^{-3}mM, toxic) for 4 days. Tvu 91 = Mn sensitive, Tvu 1987 = Mn tolerant.

the leaves of Mn-tolerant and Mn-sensitive genotypes (Fig. 4). Only a small percentage of the tissue Mn can be extracted by 80% methanol compared to the subsequent H_2O extract. However, the non-extractable Mn fraction is somewhat higher in the non-tolerant genotype Tvu 91. With $CaCl_2$ in the initial methanol extract, much more Mn can be removed from the leaf tissue, particularly in the tolerant genotype Tvu 1987, indicating a higher proportion of exchangeable Mn. Local accumulation and precipitation of Mn as Mn oxides in sensitive genotypes therefore is related to a higher proportion of "non-extractable" Mn. A higher fraction of extractable Mn and a more homogenous distribution of Mn in the leaves is typical for tolerant genotypes.

Further evidence of the importance of the leaf tissue itself for genotypical differences in Mn tolerance is shown by experiments in which Mn is applied directly to individual leaves of intact plants via the petioles. Petiole-applied Mn induces the same Mn toxicity symptoms as root-applied Mn. The reduction in dry matter production by Mn excess in the sand culture experiment (compare Fig. 1) is in agreement with the induction of toxicity symptoms after application of Mn to the petioles: Tvu 1987, Tvu 3231, and Tvu 354 being less sensitive than Tvu 91, Tvu 1283, and Solojo (Fig. 5). The significance of this correlation is confirmed by linear regression. Also with petiole application of Mn there was no relationship between Mn concentrations in the leaf tissue of tolerant and sensitive genotypes and the severity of Mn injury as indicated by Mn toxicity symptoms. The results clearly show that genotypical Mn tolerance in cowpea is controlled at the leaf tissue level. This allows a quick and non destructive screening of large numbers of genotypes for Mn tolerance by application of Mn to the petioles of individual leaves of intact plants and rating of the induced Mn toxicity symptoms. An even further simplification of the technique by treatment with Mn of isolated leaf segments looks promising.

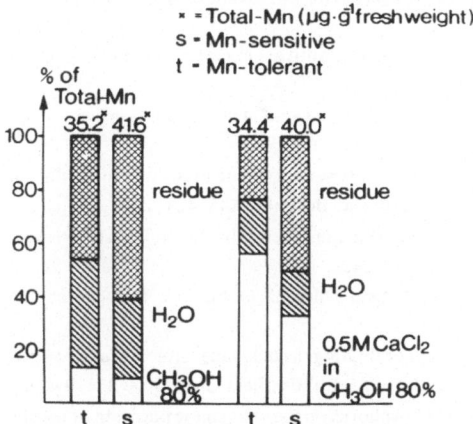

Fig. 4. Mn in extracts from fresh primary leaves of Mn-tolerant (Tvu 1987) and Mn-sensitive (Tvu 91) cowpea genotypes. 1 min extraction at $+4°C$. Preculture of plants see legend to Fig. 3.

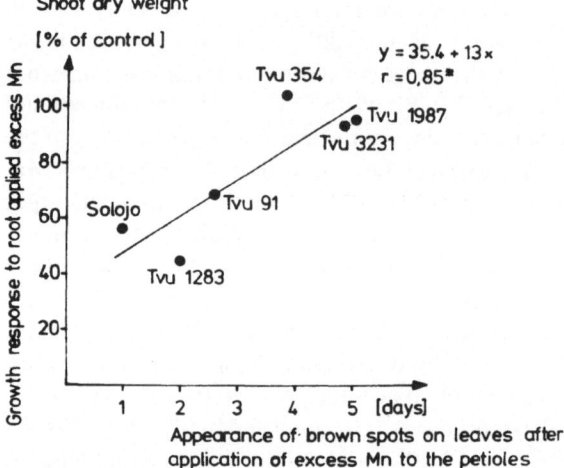

Fig. 5. Relation between growth response of 6 cowpea genotypes to root applied excess Mn in a sand culture experiment (0.1 mM Mn, 45 days) and induction of first Mn toxicity symptoms on leaf blades by application of excess Mn to the petioles of the first trifoliate leaves (50 mM Mn).

References

1 Brown J C and Jones W E 1977 Manganese and iron toxicities dependent on soybean variety. Comm. Soil. Sci. Plant Anal. 8, 1–15.
2 Bussler W 1958 Manganvergiftung bei höheren Pflanzen. Z. Pflanzenernaehr. Bodenkd. 85, 256–265.
3 Carter O G, Rose I A and Reading P F 1975 Variation in susceptibility to manganese toxicity in 30 soybean genotypes. Crop Sci. 15, 730–732.
4 Foy D C, Chaney R L and White M C 1978 The physiology of metal toxicity in plants. Ann. Rev. Plant Physiol. 29, 511–566.
5 Heenan D P and Carter O G 1976 Tolerance of soybean cultivars to manganese toxicity. Crop Sci. 16, 389–391.
6 Horst W J 1980 Genotypische Unterschiede in der Mangan-Toleranz von Cowpea (*Vigna unguiculata*). Angew. Botanik 54, 377–392.
7 Horst W J 1982 Quick screening of cowpea genotypes for manganese tolerance during vegetative and reproductive growth. Z. Pflanzenernaehr. Bodenkd. 145, 423–435.
8 Horst W J and Marschner H 1978 Symptome von Mangan-Überschuss bei Bohnen (*Phaseolus vulgaris* L.). Z. Pflanzenernaehr Bodenkd. 141, 129–142.
9 Isermann K 1975 Mögliche Ursachen der Mangan-Toleranz bestimmter Reis-Sorten. Z. Pflanzenernaehr. Bodenkd. 138, 235–247.
10 Kang B T and Fox R L 1980 A methodology for evaluating the manganese tolerance of cowpea (*Vigna unguiculata*) and some preliminary results of field trials. Field Crops Res. 3, 199–210.
11 Le Mare, P H 1977 Experiments on effects of phosphorus on the manganese nutrition of plants. II. Interactions of phosphorous, calcium and manganese in cotton grown in nutrient solutions. Plant and Soil 47, 607–620.

Efficiency of nitrogen, phosphorus, and potassium use by corn, sunflower, and sugarbeet for the synthesis of organic matter

B. JOCIĆ and M. R. SARIĆ
Faculty of Agriculture, Institute of Field and Vegetable Crops, Novi Sad, Yugoslavia Faculty of Natural Sciences, Institute of Biology, Novi Sad, Yugoslavia

Key words Corn Nitrogen Phosphorus Potassium Sugarbeet Sunflower

Summary A three-year experiment was conducted in natural conditions on chernozem soil to examine the efficiency of nitrogen, phosphorus, and potassium use by corn (C_4 type), sunflower and sugarbeet (C_3 type) grown in optimum conditions of mineral nutrition ($N_{100}P_{100}K_{100}$ kg/ha). Plant materials were analysed for the concentration of nitrogen, phosphorus, and potassium and dry matter mass per individual plant parts and the whole plant.

Leaves of different age, of all three plant species, were analysed to find eventual differences in the efficiency of use of nitrogen, phosphorus, and potassium in the synthesis of organic matter depending on leaf age.

It was found that corn had the lowest concentration of the elements studied but the highest dry matter mass. In other words, corn was more efficient than sunflower or sugarbeet in the use of these elements for the synthesis of an organic matter unit. Such results were arrived at in both sets of analyses, i.e., the analyses of leaves performed in the course of ontogenetic plant development as well as the analyses of leaves of different age.

Introduction

Recent reports have indicated that C_3 and C_4 type plants differ considerably in their efficiency of nitrogen use in the synthesis of organic matter[1,3]. This standpoint is related to the inclusion of nitrogen in the enzymes peculiar to the processes of carboxylation and metabolism in the two plant types. The use of nitrogen for the synthesis of enzymes is much lower with C_4 than with C_3 plants. Unfortunately, the above studies included a limited number of representatives of the plant species examined, analyses did not cover entire ontogenetic plant development, and no data were given on the efficiency of other elements.

Our objective was to assay the efficiency of nitrogen, phosphorus, and potassium use in the synthesis of organic matter by three plant species during their vegetative season.

Materials and methods

Experiments were conducted in field conditions on chernozem soil. Corn (C_4), sunflower (C_3), and sugarbeet (C_3) plants were grown at optimum level of mineral nutrition ($N_{100}P_{100}K_{100}$kg/ha). The experiments were repeated for several years in stationary plots, rotating the crops in the following

M.R. Sarić and B.C. Loughman (eds.), Genetic Aspects of Plant Nutrition.
ISBN-13: 978-94-009-6838-7
©1983 *Martinus Nijhoff/Dr W. Junk Publishers, The Hague/Boston/Lancaster.*

sequence: sugarbeet – corn – sunflower – winter wheat. Plant materials were sampled five times during the vegetative season (see Table).

	Corn	Sunflower	Sugarbeet
1st date	5–7 leaves, May 27	5–7 leaves, May 20	8–12 leaves, May 24
2nd date	11–14 leaves, June 22	12–14 leaves, June 10	intensive growth of leaves, June 15
3rd date	tasselling, July 15	flowering, July 7	intensive growth of roots, July 16
4th date	milk maturity, August 21	milk maturity, August 3	intensive sugar accumulation, August 20
5th date	full maturity, October 5	full maturity, September 3	pre-harvest, October 3

In the three-year research period, the average dates of corn, sunflower, and sugarbeet emergence were May 1, April 24, and April 21, respectively.

Plant materials sampled at the given five dates were analysed for the concentration of nitrogen, phosphorus, potassium and dry matter weight of individual plant organs and the whole plant.

Leaves of different age were analysed to find eventual differences in the efficiency of nitrogen, phosphorus, and potassium use in the synthesis of organic matter depending on the activity of leaves. Corn was analysed on July 15 (at the stage of flowering), and sugarbeet on July 16. The analyses were repeated four times to the end of the growth period. The first group represents the oldest leaves, the sixth group the youngest. The results discussed in this paper represent three-year averages values.

Results

Nitrogen, phosphorus, and potassium concentration in leaves and dry matter weight of leaves in the course of ontogenetic plant development

A comparison of the concentration of nitrogen, phosphorus, and potassium in leaves of the examined plant species showed that the average values for three elements were the lowest in corn (Fig. 1). Similar trends were observed also at individual stages of ontogenetic development. The superiority of corn was evident not only in view of the the efficieny of the use of nitrogen but also phosphorus and potassium. Furthermore, the weight of leaves per plant and total plant weight were always higher with corn than with sunflower and sugarbeet. In other words, corn used markedly lower amounts of the given three elements than the other two species for the formation of both a unit of organic matter and a unit of dry matter.

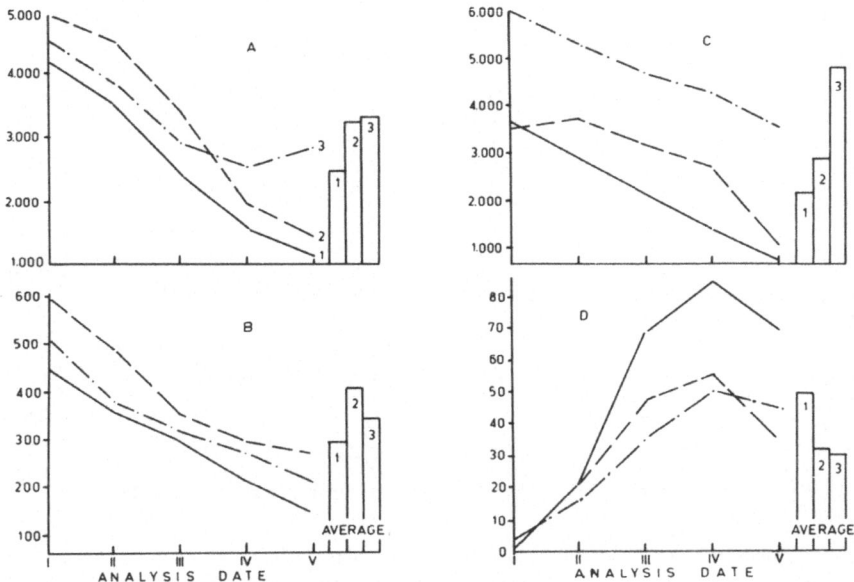

Fig. 1. Concentration of nitrogen (A), phosphorus (B) and potassium (C) in leaves (mg/100 g dry matter) and dry matter weight of leaves (D) (g/plant) in the course of ontogenetic development of corn (1), sunflower (2) and sugarbeet (3).

Nitrogen, phosphorus, and potassium concentration and dry matter weight in leaves of different age

Leaves of different age were analysed to determine the dependence of the efficiency of the use of the three elements on the physiological activity of leaves (Fig. 2). Corn was found to have lower concentrations of the elements than the other two plant species in all groups of the leaves analysed. However, N, P and K concentration varied with senescence. These data agree with the previous ones, pertaining to the concentration of the elements in leaves in the course of vegetation. Similar trend of average concentration in leaves of different age were obtained also at the other four dates.

N, P and K concentration depended on the specific patterns of development of the plant species studied. Each group of leaves had a specific concentration, especially the fourth group of corn leaves and the fifth group of sunflower leaves.

Differences in dry matter weight and concentration of the elements brought corresponding differences in the contents of these elements. The differences were also due to specific patterns of growth and development of the plant species, especially their generative parts. A detailed study of N, P and K concentration and dry matter weight of corn, sunflower, and sugarbeet plant organs was conducted by Jocić in 1974[2].

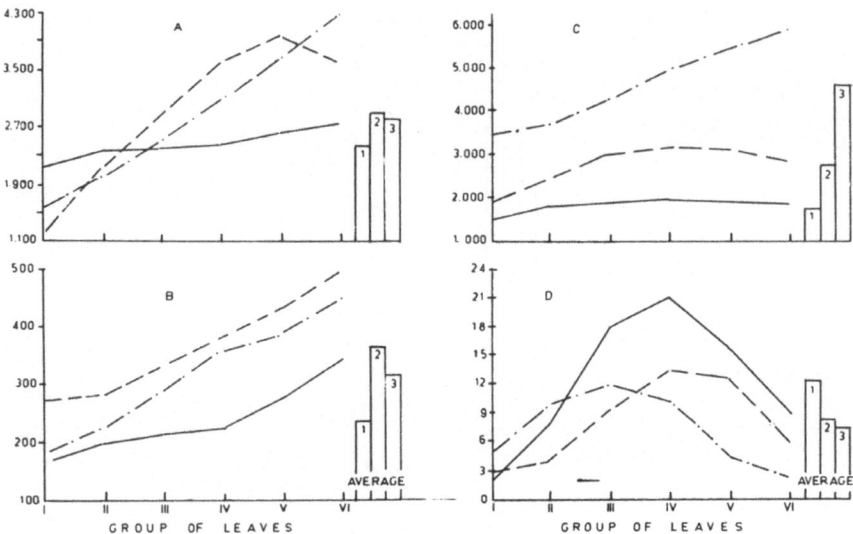

Fig. 2. Concentration of nitrogen (A), phosphorus (B) and potassium (C) in leaves of different age (mg/100 g dry matter) and dry matter weight of leaves (D) (g/leaf group) in corn (1), sunflower (2) and sugarbeet (3).

Discussion

It has been mentioned earlier that some recent studies differentiated C_3 and C_4 type plants with respect to the efficiency of nitrogen use[1,3] explaining the difference by a relatively lower participation of nitrogen in the enzyme of carboxylation in C_4 plants. Our results indicate that corn was a more efficient user of not only nitrogen but also phosphorus and potassium. Therefore, these differences are attributable to the characters of plant species.

An earlier study on the concentration of elements in various leaf segments showed that the concentration was lower, especially of nitrogen and potassium, in corn than in sugarbeet and sunflower[4].

In another study on the concentration of elements in seven plant species, including corn and sunflower, grown at 10 concentrations of nutritive solution, corn had considerably lower contents of N, K, Ca and Mg than sunflower[7].

It is well known that different varieties of the same species vary considerably in the concentration of elements. Our previous studies confirmed the above statement regarding the plant species examined. Differences between the minimum and maximum concentration (average for all organs) in 12 corn hybrids were 13, 20, 30, 29, and 18% for N, P, K, Ca and Mg, respectively in 10 sugarbeet varieties (in shoots), 13, 26, 22, 19, and 17% for N, P, K, Ca, and Mg respectively and in 22 sunflower inbreds (average for all organs) were 12, 36, 27, 47 and 33% for N, P, K, Ca and Mg, respectively[8]. These results show that plant species include varieties which efficiently use certain elements in the synthesis of organic matter. Moreover, differences between the varieties within one species are sometimes larger than differences between plant species themselves.

References

1 Brown R H 1978 A difference in N use efficiency in C_3 and C_4 plants and its implications in adaptation and evolution. Crop. Sci. 18, 93–98.
2 Jocić, B. 1974 Odnos izmedu lisne površine, sadržaja N, P, K i Ca u biljnom materijalu i prinosa u zavisnosti od mineralne ishrane kod kukuruza, suncokreta i šećerne repe. (Relations between leaf area, N, P, K and Ca contents in plant material and yield depending on mineral nutrition in corn, sunflower, and sugarbeet). Dodtorska disertacija, Poljoprivredni fakultet u Novom Sadu, pp 1–216.
3 Moore, R and Black C C 1979 Nitrogen assimilation pathways in leaf mesophyll and bundle sheath cells of C_4 photosynthesis plants formulated from comparative studies with *Digitaria sanguinalis* (L) Acop. Plant Physiol. 64, 309–313.
4 Sarić M and Jocić B 1978 Soderzhanie nekatoryh elementov v raznih segmentah lista kukuruzy, podsolnechika i saharnoi svekli. (Content of some elements in different leaf segments of maize, sunflower and sugarbeet). Fiziol. Biokhim. Kult Rast. 10, 125–131.
5 Sarić M and Krstić, B 1978 Ispitivanje zastupljenosti nekih elemenata mineralne ishrane u različitih hibrida kukuruza. (Studies of the amount of some elements of mineral nutrition in different maize hybrids). Arhiv za poljoprivredne nauke 113, 61–75.
6 Sarić M, Krstić, B, Petrović, M and Kastori, R 1979 Specifičnost reagovanja različitih sorta šećerne repe na ishranu azotom. (Specific Reaction of Different Sugarbeet Cultivars to Nitrogen Nutrition). Arhiv za poljoprivredne nauke 120, 25–40.
7 Sarić, M and Krstić B 1981 Uticaj različitih koncentracija hranljivog rastvora na masu suve materije i zastupljenost N, P, K, Ca i Mg u nekih biljnih vrsta. (Effect of different concentrations of nutritive solutions on the weight of dry matter and the content of N, P, K, Ca and Mg in some plant species). Arhiv za poljoprivredne nauke 147, 319–330.
8 Sarič M and Škorić D 1981 Relationship between the root and above ground parts of different sunflower inbreds regarding the content of some mineral elements. *In* Structure and Function of Plant Roots. Eds. R Brouwer *et al.* pp 399–401. Martinus Nijhoff/Dr. Junk Publishers, The Hague (Boston) London.

Differences in phosphate absorption in various barley genotypes

V. MEGO
Department of Biochemistry and Physiology, Research Institute of Plant Production, Piešt'any, 921 68, ČSSR

Introduction

The less-developed root system of short-stem barley varieties can limit yield formation through nutrition. These genotypes are not capable of utilizing higher rates of nutrients. Solution of the problem of intake increase and nutrient utilization by these varieties is considered by plant breeders to be one of the key tasks of the physiological-biochemical research of cereal crops. Kudrna [2] states (2) that the initiative work of (plant, stand) system manifesting itself in 3. thermodynamic phase by fast and vigorous growth of biomass is a consequence of utilization of internal energy that was accumulated due to energy absorption of existing factors, with mineral nutrients being among them. Dynamic observation of nutrient absorption processes enables the appraisal of genotype differences in absorption capacity of root tissues and thus to prepare conditions for selection of the varieties of higher performance. Varietal differences in capacities for ion absorption in cereals have been shown by Sarić and Kovačevič[4] and Glass et al.[1]

Materials and methods

Short-stem (SS) and long-stem (LS) types of barley, *Hordeum vulgare* L. var. Atem, GDR (LS) and Kŕ 999/7 (SS) were grown in Knopp's nutrient solution (0.25 concentration); granulated polyethylene (Bralen, Slovnaft, Bratislava) was used as compacting substrate allowing good aeration of solutions and excellent root extraction. Solution acidity was adjusted to pH 6. Plant sampling was carried out at the 3. leaf stage. The weight as well as the length of stem and roots and also their dry matter were determined. The minimal number of 70 individual plants from each variety were investigated. Absorption capacity of root tissues was measured by a dynamic radiometric method according to Mazel[3]. The amount of absorbed P was expressed in dpm, respectively, converted to micrograms of P per g of fresh biomass of roots after a 3 hours absorption period.

Results

Between the varieties with short stems and those with longer ones there are statistically significant differences not only in morphological characters, but also in the dynamics of P intake. As an example of these findings Table 1 and Figures 1

Table 1. Morphological characteristics of plants

Exp. No	Variety	Length cm		Fresh Wt g		Dry Wt %	
		stem	root	stem	root	stem	root
1.	LS	15.98	9.75	0.163	0.138	10.00	3.74
	SS	15.35	11.25	0.088	0.083	9.16	5.00
	t	0.94	3.35**	11.12**	9.76**	—	—
2.	LS	16.47	7.55	0.173	0.137	6.00	3.33
	SS	15.80	9.59	0.145	0.121	7.00	3.33
	t	1.34	5.85**	3.42**	2.17**	—	—

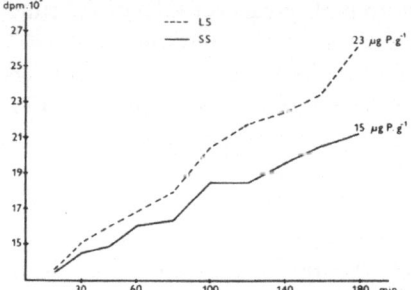

Fig. 1. Absorption of ^{32}P in barley root segments of two barley cultivars: LS = Atem GDR, SS = Kŕ 999/17 – Exp. No 1.

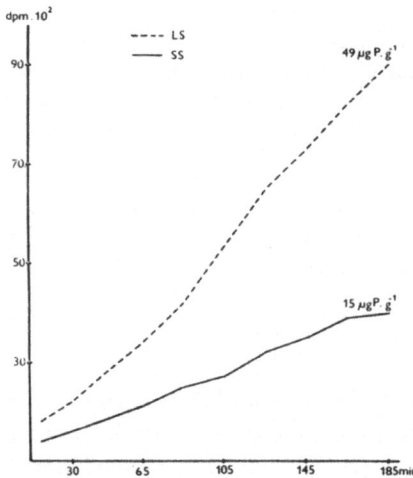

Fig. 2. Absorption of ^{32}P in barley root segments of two barley cultivars: LS = Atem GDR, SS = Kŕ 999/17 – Exp. No 2.

and 2 are given. From the results it is obvious that the root system of short-stem barley variety produces less biomass, but on the other hand it is significantly longer. Long-stem variety produces significantly more biomass, its aboveground part is of greater weight, but no differences appeared in length.

We observed marked intervarietal differences in the dynamics of P intake (Graphs 1 and 2). From Trial 2 it can be seen that absorption capacity of root tissues may be influenced in individual experiments by other factors.

References

1 Glass A D M *et al* 1981 Correlations between potassium uptake and hydrogen efflux in barley varieties. Plant. Physiol. 68, 457–459.
2 Kudrna K 1979 Zemědělské soustavy. SZN. Praha.
3 Mazel J J and Žitnewa N N 1973 Novyj dinamičeskij metod opredelenija poglotitel'noj sposobnosti rastitel'noj tkani. Fiziol. Rast. 20, 418–421.
4 Sarič M R and Kovačevič V 1981 Sortna specifičnost mineralne ischrane pšenice. *In* Fiziologija pšenice. Ed. J. Belič, SANU, Beograd.

Free-space binding and uptake of ions by excised roots of grapevines

A. MAGGIONI and Z. VARANINI
Institute of Agricultural Chemistry, University of Padua, I-35100 Padua, Italy

Key words Grapevine nutrition Potassium uptake Sulfate uptake Free space Ion binding Ion influx Ion efflux

Introduction

Mineral ions are taken up from the environment and transported to plant organs via several consecutive steps, namely 1) binding in the free space of roots, 2) crossing the plasma membrane of root cells, 3) radial transport in roots, 4) transfer to xylem sap, 5) translocation to shoots, 6) entry into leaf cells by crossing their plasma membrane, and 7) accumulation and or metabolic utilization. Except possibly the first, each step depends on metabolic activity and all of them are dependent on genetic characteristics for the metabolic structures (binding sites, carriers, enzymes etc.) accomplishing the transfer of ions from the outside to the interior of plants. In this way the whole process of plant nutrition is under genetic control as far as it is mediated by proteins or protein dependent structures, with multiple sites of genetically based variability of the overall efficiency. Therefore, analysis of each step is required to define the nutritional behaviour of genotypes different as to production aptitudes and adaptability to pedo-climatic environments.

On the basis of previous researches[6,7,10] the uptake of potassium and sulfate ions by excised roots of two species of *Vitis* (*V. vinifera* cv. Verduzzo trevigiano and the hybrid rootstock *V. berlandieri* × *V. rupestris* 1103 P) was studied and compared. This paper shows that important differences between genotypes can occur in the binding of ions in the free space.

Materials and methods

Roots were obtained as previously described[7]. K^+ and SO_4^{2-} uptake trials were carried out by using ^{86}Rb- and $^{35}SO_4^{2-}$ labelled solutions of KCl or K_2SO_4, according to the procedures already published[5,6] with minor changes. K^+ content of roots was measured by an EEL flame photometer. Root tissue was ashed with conc. H_2O_2 at 80°C and the residue dissolved with 1 M HCl.

Results

Potassium uptake

The time-course of K^+ (^{86}Rb) active uptake by excised roots of the two Vitis species is shown in Figure 1. In the first 5-min period the uptake rates were higher than in the subsequent (5–40 min) period. The amount of K^+ accumulated within the initial 5 min by 1103 P roots was about twice that of Verduzzo (88 and 42 n mol·g^{-1} fr·wt, respectively). In the subsequent period (5–40 min) the uptake occurred at the same rate of about 170 n mol·g^{-1}·h^{-1} for both species. Among the other factors, the concentration of K^+ in the root ($[K^+]_{root}$) regulates the uptake rate[2,4]. Therefore, the K^+ content was measured and $[K^+]_{root}$ calculated. Moreover, roots of both species were treated with distilled water at 30°C for 30

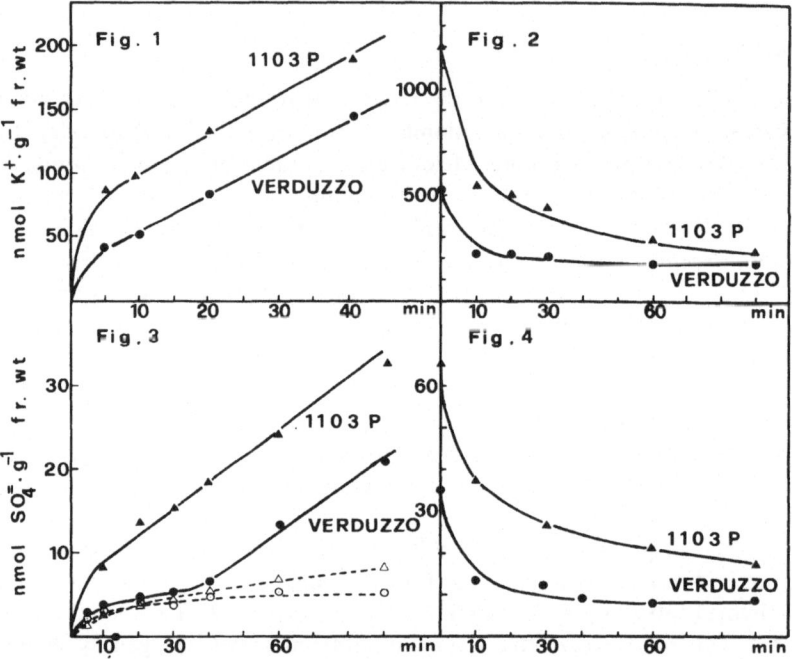

Fig. 1. The time course of K^+ (^{86}Rb) uptake by excised roots of Verduzzo and 1103 P grapevines. After contact with ^{86}Rb labelled $10^{-3}M$ $KCl + 10^{-3}M$ $CaSO_4$ solution roots were washed 60 min with cold non-radioactive solution of the same composition and submitted to liquid scintillation counting.

Fig. 2. Efflux curves of K^+ (^{86}Rb) from roots of Verduzzo and 1103 P grapevines. Roots were loaded for 30 min in ^{86}Rb labelled $10^{-3}M$ $KCl + 10^{-3}M$ $CaSO_4$ solution and the radioactivity remaining after increasing washing periods was measured.

Fig. 3. The time-course of SO_4^{2-} (^{35}S) uptake by excised roots of Verduzzo and 1103 P grapevines. Conditions as in Fig. 1, except that 2.10^{-3} M $K_2SO_4 + 10^{-3}M$ $CaCl_2$ was used. Filled symbols: experiments conducted at 30°C. Open symbols: experiments conducted at 0°C.

Fig. 4. Efflux curves of SO_4^{2-} (^{35}S) from roots of Verduzzo and 1103 P grapevines. Roots loaded for 30 min with ^{35}S labelled $2.10^{-3}M$ $K_2SO_4 + 10^{-3}M$ $CaCl_2$ were washed with cold, non radioactive solution of the same concentration and the radioactivity remaining after increasing washing periods was measured.

Mean values from two experiments, each with three parallels per experimental time.

Table 1. The effect of distilled water treatment (30 min at 30°C) on the uptake of K^+ by grapevine roots in the two (0–5 and 5–40 min) periods of uptake

Excised roots of	$[K^+]$ root $nmol \cdot g^{-1}$ fr. wt	Uptake rates ($n\,mol \cdot g^{-1} \cdot h^{-1}$)			
		control		pretr. dist. water	
		0–5	5–40	0–5	5–40
Verduzzo	25600	500	170	320	210
1103 P	28700	1060	180	540	300

min in order to lower $[K^+]_{root}$ by promoting K^+ efflux, and the capacity of these roots to take up K^+ was measured. Table 1 shows that $[K^+]_{root}$ of both *Vitis* species were rather similar, and the uptake rates of the second period were similar too. Distilled water treatment, intended to increase uptake rates by lowering $[K^+]_{root}$, changed the uptake rates in the expected way in the second period (5–40 min only. In the first 5 min the uptake rates were not related to $[K^+]_{root}$ and were decreased by the washing treatment.

Two different compartments appear to be consecutively involved in the two periods of uptake, the free space in the first period and the cell interior in the second one. This is confirmed by the efflux curves of total K^+ (^{86}Rb) absorbed (Figure 2), showing that the total amount taken up in 30 min by 1103 P roots was more than twice that of Verduzzo (1200 and 520 $n\,mol \cdot g^{-1} \cdot 30\,min^{-1}$, respectively). The difference principally concerned the amounts exchanged within 60 min from the free space (900 and 320 $n\,mol \cdot g^{-1}$). The residual difference between the non-exchanged amounts should concern the free space, since the uptake in the second period (active uptake to the cell interior) occurred at the same rate.

Sulfate uptake
The time-course is shown in Figure 3. The entry of SO_4^{2-} ions into 1103 P roots occurred at a higher rate in the first 20 min (about 40 $n\,mol \cdot g^{-1} \cdot h^{-1}$), then at about 16.5 $n\,mol \cdot g^{-1} \cdot h^{-1}$. Verduzzo roots behaved differently in that uptake rate declined progressively for 30–40 min, as if a closed compartment was saturated. Then, a second period of uptake started at a constant rate of about 17 $n\,mol \cdot g^{-1} \cdot h^{-1}$. To clarify this behaviour, uptake experiments were performed also at 0–2°C. They showed that the uptake rate of 1103 P roots decreased from 16.5 to 4 $n\,mol \cdot g^{-1} \cdot h^{-1}$ with an inhibition of about 75 per cent. Verduzzo roots behaved for 30 min in the same way as at 30°C, while in the subsequent time the uptake rate was close to zero. Showing that the first 30 min of SO_4^{2-} uptake by Verduzzo involves principally the free space of the root. Active uptake to the

interior of the cells started with a lag period of 15 to 20 min, as evaluated by extrapolating from the slope of the second period uptake. Efflux curves (Figure 4) of total SO_4^{2-} taken up in 30 min confirm the structure difference of the two Vitis roots. Total uptake (free space + cell interior) was higher for 1103 P (65 n mol·g^{-1}) than for Verduzzo (36 n mol·g^{-1}). Free space was also more ample in 1103 P roots than in Verduzzo (48 and 28 n mol·g^{-1}, respectively) and the amounts actively accumulated were different too (20 and 8 n mol·g^{-1} in 30 min). But these last figures were clearly affected by the different time-courses, since only after 30 min the active uptake rates were the same for both *Vitis* species.

Finally, the different shape of efflux curves shows that K^+ and SO_4^{2-} ions were quickly (10–20 min) exchanged and released from the free space of Verduzzo, while it took 60–90 min to complete the washing of the 1103 P free space.

Discussion

Excised roots of the two *Vitis* species transport ions through plasma membranes at the same rates, but important differences are noticeable in the early stages of uptake. In fact, in the free space of 1103 P roots, potassium and sulfate ions can be accumulated at levels which are respectively three and two times higher than in Verduzzo roots. Moreover, there seems to be a qualitative difference in the way these ions are retained. In fact, the exchange of labelled ions from the free space of Verduzzo roots requires no more than 20 min, while for 1103 P roots it takes at least 60 min. An interpretation can be attempted on the basis of Pettersson[8] and Ighe and Pettersson[2] as analyzed by Jensén[4]. Free space ions can be separated into three fractions: (i) ions freely permeating cell walls; (ii) ions electrostatically bound to charged groups; (iii) ions bound to specific metabolism-linked binding sites on the outer face of plasmalemma. The differences reported here can be attributed not only to the second fraction but probably also to the third one. By means of experiments with metabolic inhibitors, like 2,4-dinitrophenol[4], this hypothesis will be tested.

Another feature specific to sulfate uptake by Verduzzo roots is the 20 min lag period required for transmembrane transport to start. It seems that the saturation of sulfate binding sites in the free space ought to be accomplished before sulfate begins to be taken up, but this could depend on other metabolic facts, even complex ones as in the case of "aging" of storage tissues[9] or "augmentation" of maize roots[1].

Data reported in this work could be relevant for mineral nutrition in soil-plant relationships. In fact, ion binding in the free space can be considered as the uptake step intended to supply ions to membrane carriers. In soil most ions are bound to colloidal particles working as ion exchangers and the transfer of ions from the exchange sites of soil to plasmalemma carriers should occur down an "activity gradient" sustained by soil and free space binding sites, rather than through freely diffusing solutions[3]. In this view, the quantity and properties of sites determine the actual ion concentration (activity) at plasma membrane level. On the other hand, transmembrane transport, which is affected by the activity of site-bound ions, causes ions to move down the "activity gradient". Therefore, genetically determined differences of free space structures can modify the nutritional behaviour of soil grown plants by affecting ion availability for membrane carriers.

References

1   Frick H, Nicholson R and Bauman L F 1977 Characterization of augmented potassium uptake in the seedling primary root of inbred maize (*Zea mays*). Can. J. Bot. 55, 1128–1136.
2   Ighe U and Pettersson S 1974 Metabolism-linked binding of rubidium in the free space of wheat roots and its relation to active uptake. Physiol. Plant. 30, 24–29.
3   Jenny H 1966 Pathways of ions from soil into root according to diffusion models. Plant and Soil 25, 265–289.
4   Jensén P 1981 Separation of metabolic and non-metabolic steps of Rb uptake in spring wheat roots with different K^+ status. Physiol. Plant. 52, 437–441.
5   Maggioni A and Renosto F 1977 Cysteine and methionine regulation of sulfate uptake in potato tuber discs (*Solanum tuberosum*). Physiol. Plant. 39, 143–147.
6   Maggioni A 1979 Misure di assorbimento di ioni potassio in radici recise, per una migliore definizione delle caratteristiche intrinseche dell' apparato radicale. Riv. Vitic. Enol. (Conegliano) 32, 188–196.
7   Maggioni A 1980 Iron absorption by excised grapevine roots. Vitis 19, 105–112.
8   Pettersson S 1971 A labile-bound component of phosphate in the free space of sunflower plant roots. Physiol. Plant. 24, 485–490.
9   Van Steveninck R F M 1975 The "washing" or 'aging" phenomenon in plant tissues. Annu. Rev. Plant Physiol. 26, 237–258.
10 Varanini Z and Maggioni A 1982 Iron reduction and uptake by grapevine roots. J. Plant. Nutr. 5, 521–529.

The accumulation of phosphate in roots of different genotypes of maize

I. MICHALÍK
Department of Applied Chemistry and Biochemistry, Institute for Biological Rationalization of Agriculture, Agricultural University, Nitra, Czechoslovakia

Introduction

Plant productivity and crop quality depend on the properties of a variety, as well as on the agroecological conditions. The knowledge of differences among varieties regarding the influx of nutrients and molecular basis of the genetic dependence provides a goal-seeking rationalization of grain crops and food production. In general, it is supposed that the molecular mechanism of genetic heterogeneity of the nutrient influx is caused by differences occurring in the polygene structure, which is responsible for the synthesis of proteins—enzymes and mainly for the transport ATP-ases[4,5,10]. In this connection we have studied the influx, efflux and accumulation of phosphate, as well as the phosphatase activity of the roots of different lines and hybrids of maize (*Zea mays*).

Material and methods

We have used genotypes of maize which differ in their productivity and in the amount of mineral and organic substances contained in the grain. The following biological material has been investigated:
1. LSP—a simple paternal hybrid of LSP; LSP; 2. Kolektiv 440 Sc; 3. MVSc–429; 4. Pa—36; 5. WH; 6. SzTc—255; 7. Resi; 8. SzSc—369; 9. Blizard G—188/Tc/; 10. MVTc—269.

The influx of phosphate from 0.1 mM NaH$_2$PO$_4$ solution containing 0.001 mCi^{32}P.ml^{-1} has been studied in 4-cm segments of primary seminal roots of plants, 15 days old and grown in the water. Incubation of roots was made for 120 minutes in a radioactive solution. On the basis of the experimental values of the influx and desorption of phosphates the accumulation values have been calculated. In addition to this we have determined the activity of acid phosphatase by enzyme assay and electrophoretically in 10% polyacrylamide gel with subsequent detection Fast Blue B using α–naftyl phosphate substrate.

Results and discussion

From the works of many authors[2,6,7] it follows that the influx of nutrients and mainly of phosphates is mediated by metabolism. This is also demonstrated by the fact that kinetics of phosphate influx corresponds to that of enzyme reactions, which is of a multiphase character. Under the natural conditions of plant

M.R. Sarić and B.C. Loughman (eds.), Genetic Aspects of Plant Nutrition.
ISBN-13: 978-94-009-6838-7
©1983 Martinus Nijhoff/Dr W. Junk Publishers, The Hague/Boston/Lancaster.

Table 1. Values of the accumulation of phosphate in the roots of different genotypes of maize

Genetic material	Value of μ mol P.g^{-1} dry matter of root material			Accumulation (in %) vs influx (Ji = 100%)	Accumulation (in %) vs LSP	Total P content μmol. g^{-1}	Accumulation P (in %) vs total P (total P = 100%)
	influx (Ji)	efflux (Jo)	accumulation (J)				
1. LSP	5.15	4.37	0.78	15.4	100	96.16	0.81
2. Kolektiv	8.01	6.05	1.96	24.47	251.28	105.87	1.85
3. MVSc–429	7.72	6.24	1.48	19.17	189.74	91.16	1.62
4. Pa–36	6.99	3.70	3.29	47.07	421.79	95.84	3.43
5. WH	7.85	6.07	1.78	22.67	228.20	122.26	1.45
6. SzTc–255	5.85	4.79	1.06	18.12	135.89	92.10	1.15
7. Resi	6.67	6.15	0.52	7.79	66.67	87.06	0.60
8. SzSc–369	4.44	3.93	0.51	11.49	65.38	119.26	0.43
9. Blizard	5.51	4.98	0.53	9.62	67.95	102.58	0.52
10. MVTc–296	6.73	6.34	0.34	5.05	43.59	94.22	0.36

vegetation, the influx of phosphate is realized against a significant electrochemical gradient. From the thermodynamic aspect and under these conditions the influx of phosphates is realized only at the expense of metabolic energy consumption.

As the synthesis of proteins and enzymes is genetically determined, that of enzymes responsible for the ion absorption must also be undergenetic control. The results gained in yeast and bacteria cultures have confirmed the conclusions concerning the genetic dependence of the influx of nutrients[9].

Data given in Table 1 show that the roots of different maize lines and hybrids contain various amounts of total phosphorus (87–122 μmol.P.g^{-1}), the differences being up to 40%. However, the greatest differences, more than 9-fold, have been found in the accumulation of phosphates. The portion of phosphate accumulation in the total phosphorus influx in the PA–36 line is as high as 47% which demonstrates a high efficiency of phosphorus (P) influx. On the other hand, the lowest accumulation of only 5% has been determined in the roots of the MVTc–296 maize. In this sense our results are in accordance with those obtained by Clark and Brown[3]. The properties of a genotype are especially demonstrated by the realization of biochemical and physiological reactions, which are responsible for the transport of phosphates into the root cells and their metabolism. According to Schenk and Barber[13] the genotype differences in the influx of phosphate are connected with root morphology, while Barel and Peterson[1] assign it to different permeability of the plasmalemma. They have found a higher transport of foliarly applied phosphorus in the lines with male sterility compared to the maize izolines with standard cytoplasm[1]. We suppose that genotype differences in the influx of nutrients are produced by a complex of

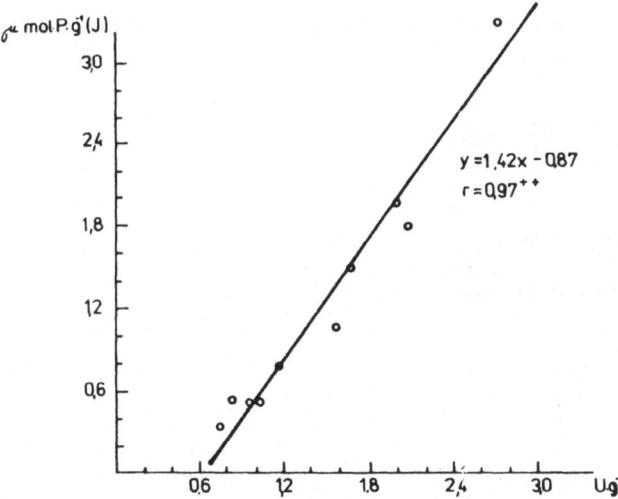

Fig. 1. Relationship between the activity of acid phosphatase and accumulation of phosphates.

Table 2. Values of the isoenzymes of phosphate isolated from maize roots by densitometric evaluation. Numbers in brackets refer to genetic material indicated in Table 1.

Isoenzymes	Genetic material; representation in %					
	LSP (1)	MVSc-429 (3)	Pa-36 (4)	WH (5)	S_zT_c-255 (6)	S_zS_c-36 (8)
Start						
1	6	3	—	—	—	—
2	7	13.5	10.5	8.5	3	—
3	30	15	50.5	4.0	26	57
4	21	44	17.5	48	25.5	
5	9	10	14.5	14.5	13	12
6	21	12	6.5	18	21	22
7	6	2.5	0.5	7	11.5	9

factors at the anatomical and morphological level, but mainly at the level of molecular mechanisms which are responsible for ion accumulation.

The calculated values of acid phosphatase activity (E.C: 3.1.3.3) show a direct dependence of the phosphate influx on the activity of phosphatase. The correlation coefficient is 0.97 (Fig. 1). From our results, obtained previously, it follows[7] that the activity of phosphatase is responsible for the phosphorus transport from roots into the above-ground organs, which can be adequately demonstrated by the stimulation of its influx too. Other authors have pointed out the correlation relationships between the influx of phosphates and the activity of phosphatase[12]. Despite the fact that phosphatase appears as an adaptive enzyme

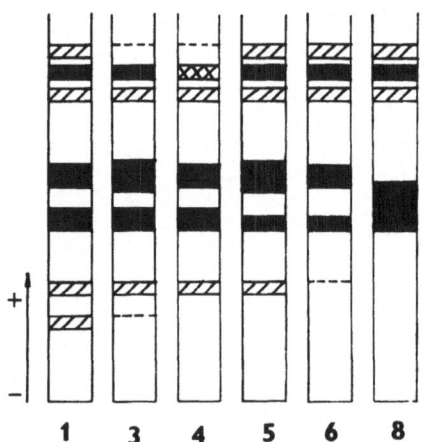

Fig. 2. Electrophoretic spectra of acid phosphatase.

the synthesis of which is repressed by phosphates, the properties of the genotype cannot be contested (Table 2).

The results of the electrophoretic spectra of phosphatase enzymes (Fig. 2) clearly show the existence of quantitative and qualitative differences. At the same time the spectra gives evidence about sensible phosphatase polymorphism. This is a consequence of individual plant adaptation to living conditions and the result of plant selection for yield under different soil and climatic conditions. As the productivity of plants also integrates processes responsible for the influx of nutrients and their utilization, it is very important to determine genetic anomalies of the transport system in cultivated plants. Positively demonstrated mutations of genes, controlling the mechanism of ion reception, prove the genetic dependence of mineral nutrition. It follows from the facts given above that the selection of plants can be taken as a method to produce new biological material with a high efficiency in the influx of nutrients.

References

1 Barel D and Peterson P A 1975 Variation in the differential leaf absorption of a high-molecular weight phosphate by plants of differing cytoplasms of three corn (*Zea mays*) inbreds. Can. J. Genet. Cytol. 17(2), 211–216.
2 Bieleski R L 1973 Phosphate pools, phosphate transport and phosphate availability. Ann. Rev. Plant Physiol. 24, 225–252.
3 Clark R B and Brown J C 1974 Differential phosphorus uptake by phosphorus-stressed corn inbreds. Crop Sci. 14(4), 505–508.
4 Epstein E 1972 Physiological genetics of plant nutrition. Mineral Nutrition of Plants. John Wiley and Sons, Inc., New York.
5 Feenstra W J and Oostinder Braaksma F J 1976 Genetic control of nitrate reduction in Arabidopsis. Arab. Inf. Serv. 13, 133–135.
6 Loughman B C 1980 Metabolic aspects of the transport of ions by cells and tissue of roots 2nd Internat. symposium "Structure and function of roots", Abstracts, Bratislava (Czechoslovakia) Sept. 1st–5th, 1980.
7 Michalík I 1978 The application of ^{32}P radionuclide in studying phosphorus transport mechanism in xylemic exudate. Radiochem. Radional. Letters, 34 (5–6), 371–382.
8 Michalík I 1982 The influence of phosphate concentration at the kinetics of uptake by maize roots. Biologia Plantarum (*in press*).
9 Nesmejanova M A Motloch O B Kolot M N and Kulajev I S 1979 Regul'acija ATP–aznoj aktivnosti kletok *E. coli* ekzogennym ortofosfatom. Biochimija, 44(7), 1212–1217.
10 Nielsen N E and Barber S A 1978 Differences among genotypes of corn in the kinetics of P uptake. Agron. J. 70, 695–698.
11 Nissen P 1973 Multiphasic uptake in plants. I. Phosphate and sulphate. Physiol. Plant. 28, 304–316.
12 Reid M S and Bieleski R L 1970 Response of *Spirodella oligorrhiza* to phosphorus deficiency. Plant Physiol. 46, 609–613.
13 Schenk M K and Barber S A 1978 Root characteristics of different corn genotypes related to P uptake. Agronomy abst. Ann. Meetings Amer. Society of Agronomy Crop Sci. Dec. 3–8, 1978 Chicago, Illinois.

Efficiency and kinetics of phosphorus uptake from soil by various barley genotypes

N. E. NIELSEN and J. K. SCHJØRRING
Department of Soil Fertility and Plant Nutrition, The Royal Veterinary and Agricultural University, Copenhagen, Denmark

Key words Barley Kinetic parameters Phosphorus Varietal effects

Summary Barley cultivars grown under field conditions of moderate deficiency of phosphorus (P) had great differences in P uptake and grain yields.

As the rate determining step in P uptake under these conditions is located in the root net influx of P ($\bar{I}_n L^*$) per g of dry matter of the plant can be expressed by

$$\bar{I}_n L^* = \bar{I}_{max} L^* \frac{c - c_{min}}{c - c_{min} + K_m}$$

where $\bar{I}_n$, L^*, $\bar{I}_{max}$, c, c_{min} and K_m denote mean net influx per unit length of the root, root length per unit weight of the plant, maximal mean net influx per unit length of the root, P concentration at the root surface, minimum concentration in solution of which net influx appears to be zero and Michaelis–Menten factor of P uptake, respectively. Studies of P uptake kinetics in water culture showed that the values of L^*, $\bar{I}_{max}$, K_m and c_{min} of P uptake varied considerably between barley cultivars. Furthermore, agreement was found between P uptake in the field and P uptake predicted from $\bar{I}_{max}$-, K_m-, c_{min}- and L^*-values observed in water culture experiments.

The data thus indicate that it should be possible to improve the efficiency by which plants utilize soil as a source of P by selecting and/or developing genotypes of barley with a smaller c_{min} and/or K_m and a greater $\bar{I}_{max}$ and/or L^* during the main period of growth.

The results suggest therefore that it should be feasible to adapt plants to a considerably lower soil P level.

Introduction

For many years the effort by man has been to fit the soil to the plant by application of fertilizers. We have only recently tried to fit the plant to the soil. Progress in the latter requires knowledge of plant parameters controlling the efficiency by which plants utilize soil as a source of nutrients.

The aim of the present work is:

1) to show how various genotypes of barley differ in their uptake of phosphorus from soil low in phosphorus,
2) to draw attention to four plant factors (kinetic parameters) important for the efficiency of nutrient uptake from soils under conditions in which the rate determining step in uptake of the nutrient is located in the root,
3) to give data for the range of the variation of these kinetic parameters between some genotypes of barley,

4) to establish the agreement between phosphorus uptake in the field and phosphorus uptake predicted from kinetic parameters observed in water culture experiments.

Results

The results in Table 1 show that barley varieties differ much in grain yields under conditions of moderate phosphorus deficiency, whereas grain yields were almost equal in experiments on Danish farms without phosphorus deficiency.

Fig. 1 illustrates how phosphorus uptake varied between two barley varieties representing extremes of 30 barley varieties grown at moderate deficiency of phosphorus in a field experiment.

The efficiency by which plants utilize soil and fertilizers as a source of nutrients is affected by several plant factors, *e.g.* radius, length, density and geometry of the roots in the soil; kinetic parameters of nutrient uptake, root exudates and the adaptability of the roots to symbiotic and non-symbiotic soil microorganisms.

The four plant factors (kinetic parameters) important for phosphorus influx per unit weight of the plant and the efficiency of phosphorus uptake from soils are given in Table 2. The latitude of variations between 6 barley varieties in the values of these kinetic parameters can be seen from the data in Table 3.

The net influx of phosphorus per g of dry matter of the plant can be expressed by

$$\bar{I}_n L^* = \bar{I}_{max} L^* \frac{c - c_{min}}{K_m + c - c_{min}} \quad (1)$$

where $\bar{I}_n L^*$ denotes mean net influx of phosphorus per g of dry matter of the plant and c the concentration of phosphorus at the root surface. Details of the

Table 1. Grain yields of 6 barley varieties, grown at moderate phosphorus deficiency or in experimental trials on Danish farms ($\times 100$ kg grain ha^{-1})

Barley variety	Low P level* in the soil	Average of Danish field trials**
Salka	49	48
Lofa	44	47
Rupal	41	48
Nürenberg	36	–
Mona	36	45
Zita	30	47

*Selected results from a field experiment with 30 barley varieties, Nielsen[2].
**Average yield in 1975 to 1979 of trials, conducted by the Danish Agricultural Advisory Centre.

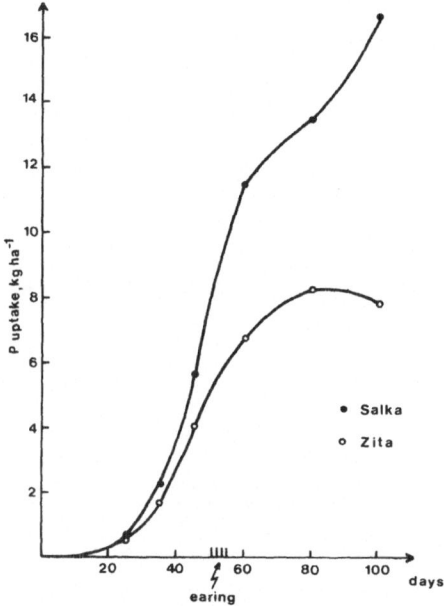

Fig. 1. Phosphorus (P) uptake by the barley varieties Salka and Zita during growth at moderate P deficiency under field conditions.

experimental procedure used to estimate L^*, $\bar{I}_{max}$, K_m and c_{min} has been given by Nielsen[1] and Schjørring and Nielsen[4]. Fig. 2 shows the predicted mean rate of P uptake per g DM, at varying P concentrations, c, calculated from Eq. 1 by use of values of L^*, $\bar{I}_{max}$, K_m and c_{min} in Table 3. From Fig. 2 it can be seen that Salka has the highest $\overline{In}L^*$-value (efficiency) of P-uptake per g DM of plant at c lower than 3 μM P, whereas Mona has the lowest efficiency.

Table 2. Plant factors (kinetic parameters) important for the efficiency of nutrient uptake from soils

Name	Symbol	Dimension
Root length per unit weight of the plant	L^*	mg^{-1}
Maximal mean net influx per unit length of the root	$\bar{I}_{max}$	pmole $cm^{-1}s^{-1}$
Michaelis–Menten factor of nutrient uptake	K_m	μM
Minimum concentration in solution at which net influx appears to be zero	c_{min}	μM

Table 3. Meters of root per g of dry matter (DM) of the plant (L*), mean maximal net influx ($\bar{I}_{max}$), Michaelis–Menten factor (K_m) and minimum concentration (c_{min}) for phosphorus (P) uptake by 35 days old barley plants, grown in water culture[4]

Barley variety	L* m root g DM	$\bar{I}_{max}$ pmole P cm^{-1}s^{-1}	K_m μM P	c_{min} μM P
Salka	65	0.08	2.9	0.02
Lofa	77	0.08	4.1	0.04
Rupal	46	0.10	3.6	0.04
Nürenberg	68	0.11	3.6	0.06
Mona	42	0.14	5.5	0.05
Zita	57	0.12	4.7	0.03
CV in %*	8.4	16	14	17

* Coefficient of Variance = 100 $s_{\bar{x}}/\bar{x}$

In order to study the agreement between P uptake in the field and P uptake expected according to the observed values of $\bar{I}_{max}$, K_m, c_{min} and L* in water culture experiments, transport kinetic models can be used as shown by Nielsen and Barber[3], Nielsen[1] and Schjørring and Nielsen[4]. By use of such a model and the kinetic parameters in Table 3 and estimated root lengths, Schjørring and

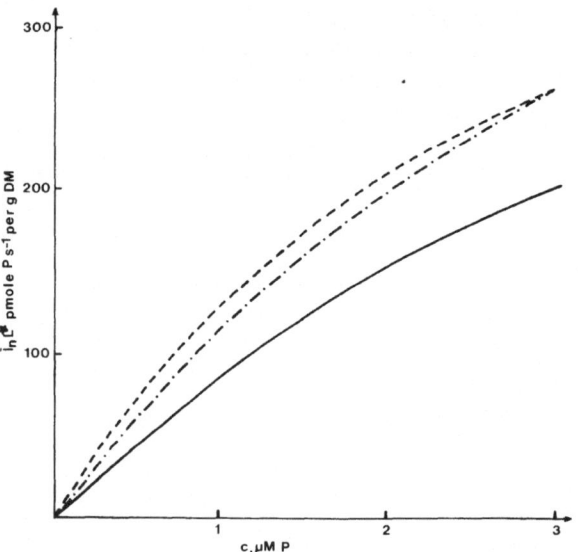

Fig. 2. Mean net influxes per g dry matter (DM) of plant, $I_n L^*$, at varying phosphorus (P) concentration, c, by the barley varieties Salka (– –), Zita (– · –) and Mona (——).

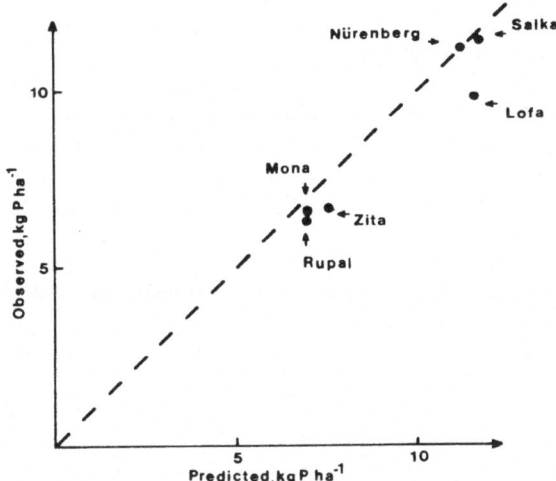

Fig. 3. The relation between observed phosphorus uptake in the field 60 days after emergence and phosphorus uptake predicted from estimated values of the kinetic parameters of 6 barley varieties in water culture experiments.

Nielsen[4] calculated the expected P uptake by some barley varieties. An example of the relation between observed and predicted P uptake is shown in Fig. 3. The agreement between observed P uptake in the field and the P uptake predicted by the model, Fig. 3, seems to justify the use of kinetic studies in water culture with the aim to characterize differences in P uptake efficiencies between plant varieties.

Conclusion

Root length, L, mean maximal net influx of nutrients into roots, $\bar{I}_{max}$, Michaelis–Menten factor for mean net influx, K_m, and the minimum concentration, c_{min}, at which mean net influx appears to be zero, are plant factors (parameters) that greatly affect the efficiency by which barley utilizes soil as a source of nutrients under conditions in which the rate determining step of uptake is located in the root.

Reliable procedures exist for estimating these plant parameters. The range of variation in the parameters between some barley genotypes are considerable. Improvement in the efficiency of phosphorus uptake may be possible by selecting from our barley varieties or by plant breeding to develop genotypes which have a smaller c_{min} and/or K_m, a higher $\bar{I}_{max}$ and/or L during the main growth period.

References

1   Nielsen N E 1979 Plant factors controlling the efficiency of nutrient uptake from soil and

genetics. Mineral nutrition of plants. Proceedings of the First International Symposium of Plant Nutrition. Varna, Bulgaria, Sept. 1979. No. 1, 203–220.

2   Nielsen N E 1981 Planteegenskaber (parametre), der påvirker effektiviteten af planters udnyttelse af jord som næringsstofkilde. I. Studier af forløbet af næringsoptagelsen hos 30 bygsorter ved moderat mangel på phosphor. Rapport No. 1122, 99 p. Department of Soil Fertility and Plant Nutrition, The Royal Vet. and Agric. Univ. Copenhagen.

3   Nielsen N E and Barber S A 1978 Differences between genotypes of corn in the kinetics of phosphorus uptake. Agron. J. 70: 695–698.

4   Schjørring J K and Nielsen N E 1982 Planteegenskaber (parametre) der påvirker effektiviteten af planters udnyttelse af jord som næringsstofkilde. II. Bestemmelse af de transportki netiske parametre for phosphoroptaglse hos 9 bygsorter, havre rug og hvede. Rapport No. 1123, 167 p. Department of Soil Fertility and Plant Nutrition, The Royal Vet. and Agric. Univ. Copenhagen.

Variation among species and varieties in uptake and utilization of potassium

S. PETTERSSON and P. JENSÉN
Department of Plant Physiology, Swedish University of Agricultural Sciences, S-750 07 Uppsala 7, Sweden and Department of Plant Physiology, University of Lund, Box 7007, S-220 07 Lund, Sweden

Key words Barley Potassium efflux Potassium influx Varietal differences

Summary In uptake experiments from water cultures K^+-influx in roots of sunflower (*Helianthus annuus* L.), cucumber (*Cucumis sativus* L.), birch (*Betula verrucosa* Ehrh.), lingonberry (*Vaccinium vitis-idaea* L.), and pine (*Pinus silvestris* L.) was related to the K^+-contents of the roots. However, due to genotypic variation, no universal "optimum" K^+-state of the roots for maximum K^+-influx could be defined.

Ranking of Rb^+ (K^+)-influxes into high K^+ and low K^+ roots of 11 cultivars of barley (*Hordeum vulgare* L.) brought the same sequence but the varietal differences were relatively greater in the high K^+ roots.

Net K^+ fluxes in barley roots were not related to K^+-influxes due apparently to varietal differences in K^+ effluxes from the roots.

Dry matter production per weight unit of K^+ present in the plants (K^+ use efficiency) was not related to the K^+-influxes of the roots in the barley cultivars.

It is concluded that several both morphological and physiological plant parameters must be evaluated and combined before selecting varieties for efficient mineral nutrient exploitation is possible.

Introduction

Differences among barley genotypes in K influx and efflux, transport to the shoot and K use efficiency (dry matter produced per K absorbed) were found by Pettersson[5], Jensén and Pettersson[4], and Glass and Perley[1]. Moreover the latter authors obtained a high correlation between growth rates and K influx in 10 tested cultivars. A screening technique for the determination of K uptake differences in genotypes using H^+ release by the roots was suggested by Glass *et al.*[2]

In the present paper data from our laboratory are put together which support the opinion that K^+ influx and efflux, translocation and K use efficiency are important factors which must be considered before genotypes with optimal K uptake and utilization can be described.

K^+ uptake (influx) controlled by K contents of roots

K^+-influx in barley roots is, among other factors, regulated by K^+-status of the plant and the ion concentration of the external solution[6]. Fig. 1 shows the

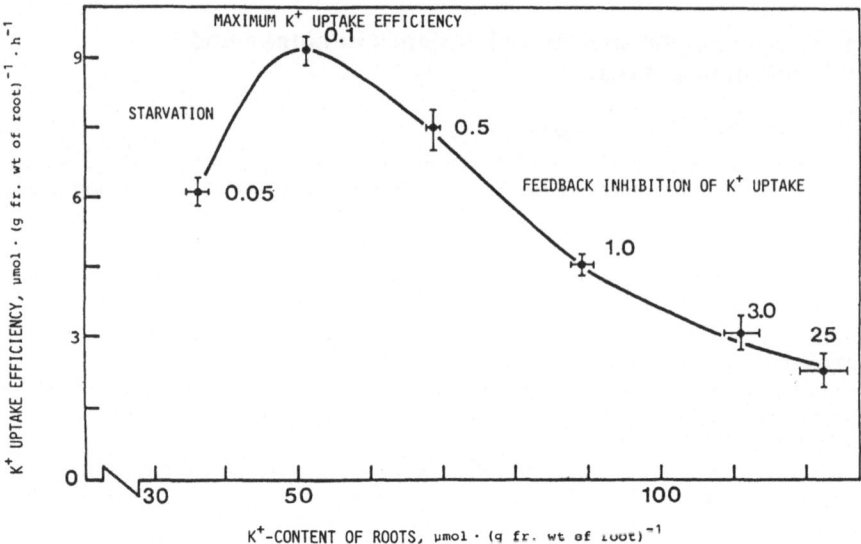

Fig. 1. K$^+$ (^{86}Rb) influx in roots of barley seedlings as a function of the K$^+$ concentration in the roots. Plants cultivated on complete nutrient solutions with concentrations of K$^+$ (mM) as indicated at symbols. Influx during 60 min from a solution with 2 mM K$^+$ (^{86}Rb) followed by 10 min rinse in inactive standard nutrient solution (3 mM K$^+$).

K$^+$(^{86}Rb) influx (uptake efficiency) into roots with different K$^+$-contents in a short term (1 h) experiment from an ^{86}Rb-labelled complete nutrient solution containing a standard concentration of K$^+$ (2 mM). Maximum K$^+$ influx is obtained within only a narrow interval of possible K contents of the roots.

It is clear that although they are apparently universal to their nature curves like that of Fig. 1 drawn for different species or varieties of species have each its specific characteristics (see ref. [3]). This observation points to the difficulty inherent in stating general conceptions like "optimum" conditions for nutrient uptake when treating more than one genotype in an experiment.

Maximum K$^+$ influx in relation to K contents of roots and relative growth rates

In most plant species investigated (sunflower, cucumber, birch, lingonberry, and pine) maximum K$^+$ influx was 3–6 μmol K$^+$ (^{86}Rb) g^{-1} fr. wt. of root · h^{-1} (Table 1). Two cereals (wheat and barley) has considerably higher values (10–15 μmol K$^+$ (^{86}Rb) g^{-1} fr. wt. of root · h^{-1}).

The data of Table 1 also visualize the statement made earlier that no simple relation exists between maximum K$^+$ influx and K contents of the roots. An "optimum" K$^+$ state of roots for maximal K$^+$ influx cannot be defined generally but must be related to the individual species or variety.

Nor was there a correlation between maximal K$^+$ influx and relative growth

Table 1. Maximum K^+ influx, K^+ contents of roots and relative growth rates (RGR) of seven plant species. K^+ uptake efficiency was determined as influx of K^+ (^{86}Rb) during 2 h from a complete ^{86}Rb-labelled nutrient solution with 2.0 mM K^+ followed by a 10 min rinse in inactive uptake solution to remove radio-labelled ions contained in surface film and the free space. (Mainly from Jensén and Pettersson[3])

Species	Maximum K^+ influx μmol · 9^{-1} FW · h^{-1}	K^+ contents of roots at max. K^+ influx μmol · g^{-1} FW	RGR % day^{-1}
Sunflower	4	50	25
Cucumber	6	30	22
Birch	4	30	22
Wheat	10	50	20
Barley	15	25	17
Lingonberry	6	35	4.5
Pine	3	40	4.0

rate of the plants (Table 1). The efficient K^+ absorbers wheat and barley did not grow particularly rapidly and contained no more internal K^+ than the other species. It appears that for a species like lingonberry a capacity for slow growth (low RGR) in combination with reasonably high ion uptake efficiency can have a survival and competitive value in nature on habitats poor in available mineral nutrients.

Influx, net flux and efflux of K^+ in cultivars of barley

Eleven cultivars of barley were ranked according to their Rb$^+$ influxes under conditions giving maximal K^+ influx (Table 2, low K^+ roots). Ranking of Rb$^+$ influxes into high K^+ roots (Table 2, high K^+ roots) brought the same sequence.

The decrease of Rb$^+$-influxes of high K^+ roots in comparison with Rb$^+$-influxes of low K^+ roots was considerable and in accordance with the feed-back control of ion uptake from the K^+ content of the roots as shown in Fig. 1. The varietal differences in Rb$^+$ influx were relatively greater in high K^+ roots than in low K^+ roots. For instance, for cv. Hellas Rb$^+$ influx was 71% in low K^+ roots but only 37% in high K^+ roots of the corresponding values for cv. Salve, the most efficient cultivar. Thus, relative differences in Rb$^+$-influxes among barley cultivars change with the K^+ status of the plants. This must be kept in mind when evaluating values from K^+ influx experiments performed under certain experimental conditions when selecting cultivars for efficient nutrient uptake. Plants with a high ion uptake efficiency at ample nutrient supply cannot *a priori* be expected to be effective also under restricted conditions.

Table 2. Rb$^+$ influxes and K$^+$ net flux for eleven cultivars of barley. Low K$^+$ plants cultivated in 0.1 mM K$^+$ and high K$^+$ plants cultivated in 3.0 mM K$^+$. Rb$^+$ influx from a complete nutrient solution with 2.0 mM Rb$^+$ followed by a 10 min rinse in inactive standard nutrient solution (3.0 mM K$^+$). For computation of net fluxes see text. Values for low K$^+$ roots from Pettersson[5]. Values for high K$^+$ roots from Jensén and Pettersson[4]

Cultivar	μmol · (g fr. wt. of root)$^{-1}$ · h^{-1}		
	Low K$^+$ roots	High K$^+$ roots	
	Rb$^+$ influx	Rb$^+$ influx	K$^+$ net flux (v)
1 Salve	20.1	4.1	2.6
2 Birgitta	17.8	3.3	3.5
3 Sv 67529	17.1	3.1	2.8
4 Bonus	16.9	2.9	2.7
5 Kristina	16.1	2.2	2.9
6 Gunilla	15.1	2.2	2.7
7 Eva	N.D.	2.2	2.9
8 Edda II	15.1	2.1	3.5
9 Hellas	14.3	1.5	3.3
10 Mona	N.D.	1.2	3.0
11 Ingrid	N.D.	0.8	3.1

Average net K$^+$ fluxes (v) over seven successive days on complete nutrient solutions with 3 mM K$^+$ was computed from the equation

$$v = \frac{(\ln W_2 - \ln W_1) \cdot (M_2 - M_1)}{(t_2 - t_1) \cdot (W_2 - W_1)}$$

as given by Williams[8]. W_1 and W_2 are the fresh weights of roots and M_1 and M_2 are the total quantities of K$^+$ in the plants at the start (t_1) and completion (t_2) of the uptake period. Values of v are shown in Table 2. It is evident that K$^+$ net fluxes are all close to 3 μmol K$^+$ · (g fr. wt. of root)$^{-1}$ · h^{-1} irrespective of variations in Rb$^+$-influxes among the cultivars. Its also clear that the maximal Rb$^+$ influx potential as shown for low K$^+$ roots is exploited to only a limited extent to supply the K$^+$ demand of the plants.

These results are different from those of Glass and Perley[1] who found a positive correlation between V_{max} and long-term growth performance in 10 cultivars of barley. However, the net K$^+$ uptake and growth figures of Glass and Perley[1] were from cultures with limited K$^+$ supply giving K$^+$ contents of 26–41 μmol K$^+$ · (g fr. wt. of root)$^{-1}$ only. In the present high K$^+$ plants K$^+$ content was 85–100 μmol K$^+$ · (g fr. wt of root)$^{-1}$ (see ref.[4], Fig. 2). In contrast to the results of Glass and Perley[1] for low salt barley plants our data seem to support the contention that the determination of kinetic parameters like V_{max} or maximal K^+ influx potential may be of limited value when trying to predict plant performance at ample nutrient supply, such as is usual in modern agriculture.

Although of the same order of magnitude Rb^+ influxes of high K^+ roots do not cope with net K^+ fluxes in cultivars 5–11 in our experiments (Table 2). This apparently anomalous result might be explained by the observation that K/Rb selectivity varies considerably between cultivars of barley[7] and that net K^+ fluxes were from the standard solution of cultivation with 3 mM K^+ while Rb^+-influxes were computed after uptake from a solution with 2 mM Rb^+. Under all circumstances the impression remains that Rb^+ influx studies of barley do not well reflect net K^+ uptake rates of high K^+ plants.

Non-correspondence between K^+-influx and net K^+-uptake will appear if K^+-effluxes are different. This becomes clear when comparing K^+-fluxes in cv. Salve and Hellas. The higher influx in cv. Salve is partly reduced to its net effect by K^+ (Rb^+) efflux from the roots (Table 3). The 30% lower K^+-influx of cv. Hellas is compensated for by a low efflux resulting in reduction of net K^+-content of 14% only.

K^+ use efficiency ratios in cultivars of barley

Mineral use efficiency can be defined as the ratio "weight units of plant dry matter produced/weight units of a certain mineral nutrient present in the plant". Great differences existed in K^+ use efficiency ratios among the 11 cultivars which were tested (Table 4). The highest production was found in cv. Mona (117 mg DM/mg K^+). Cv. Salve ranked as number 11 with 96 mg DM/mg K^+.

So, like net K^+-uptake, dry matter production was not related to K^+ influx of

Table 3. Influx and efflux of Rb^+ (^{86}Rb) and K^+ contents of two cultivars of barley. Rb^+ (^{86}Rb) influx determined during a 2 h uptake period from a complete ^{86}Rb-labelled nutrient solution with 2.0 mM Rb^+ followed by a 10 min rinse in inactive uptake solution. For efflux experiments plants were grown for six days on nutrient solutions with 3 mM K^+ and labelled with ^{86}Rb. Efflux to inactive nutrient solutions for different periods up to 6 h. Efflux was determined only as CPM of washing solutions (From Pettersson[5])

Cultivar	Influx		Efflux	K^+ content	
	$\mu mol \cdot g^{-1}$ FW $\cdot h^{-1}$	%	%	$\mu mol \cdot plant^{-1}$	%
Salve	19.9	100	100	860	100
Hellas	14.5	73	70	740	86

Table 4. K efficiency ratios for 11 cultivars of barley. Use efficiency ratios computed as mg dry weight/mg K^+ of plants or shoots only. Plants grown for 6 days on complete nutrient solutions with 3.0 mM K^+. (From Jensén and Pettersson[4]).

Cultivar	SK use efficiency ratio	
	Whole plant	Shoot
Mona	117	108
Kristina	110	102
Bonus	106	96
Hellas	105	94
Eva	104	92
Gunilla	101	89
Birgitta	101	95
Ingrid	100	90
Edda II	98	86
Sv 67529	97	86
Salve	96	89

the roots. Cv. Mona with rank 1 for K^+ use efficiency was ranked only as number 10 with respect to K^+ influx efficiency (Table 2).

It is noticeable that the ranking of K^+ use efficiency ratios for shoot production is with few exceptions similar to that for whole plants (Table 4).

Conclusions

Our experiments show that a high ion uptake efficiency (influx) is not necessarily correlated with a high dry matter production. Varying ion effluxes and mineral use efficiencies make the picture complicated. Perhaps a high ion uptake efficiency has a selective value more in the competition with neighbouring roots for sparsely occurring mineral nutrients in poor soils than for mineral supply for high production within the plant.

Breeders aiming consciously at a more efficient use of limited soil mineral resources must therefore consider and combine several both morphological and physiological plant characteristics, such as root growth and morphology, ion uptake efficiency, ion efflux, ion translocation and mineral use efficiency for the production of useful plant parts.

References

1 Glass A D M and Perley J E 1980 Varietal differences in potassium uptake by barley. Plant Physiol. 65, 160–164.

2   Glass A D M, Siddiqi H G and Giles K I 1981 Correlations between potassium uptake and hydrogen efflux in barley varieties. Plant Physiol. 68, 457–459.
3   Jensén P and Pettersson S 1978 Allosteric regulation of potassium uptake in plant roots. Physiol. Plant. 42, 207–213.
4   Jensén P and Pettersson S 1980 Varietal variation in uptake and utilization of potassium (rubidium) in high-salt seedlings of barley. Physiol. Plant. 48, 411–415.
5   Pettersson S 1978 Varietal differences in rubidium uptake efficiency of barley roots. Physiol. Plant. 44, 1–6.
6   Pettersson S and Jensén P 1978 Allosteric and non-allosteric regulation of rubidium influx in barley roots. Physiol. Plant. 44, 110–114.
7   Schimansky C and Marschner H 1971 Suitability of Rb-86 as a tracer for potassium in studies relating to potassium uptake by maize, sugar beet and four varieties of barley. Z. Pflanzenernaehr. Bodenkd. 129, 141–147.
8   Williams R F 1948 The effects of phosphorus supply on the rates of intake of phosphorus and nitrogen upon certain aspects of phosphorus metabolism in gramineous plants. Aust. J. Sci. Res. (B) 1, 333–361.

Application of root zone feeding for evaluation of ion uptake and efflux in soybean genotypes

J. WIENEKE
Institut für Radioagronomie, Kernforschungsanlage Jülich GmBH Postsach 1913, D-5170 Jülich, FRG

Key words Chloride   Root zone feeding   Sodium   Soybeans

Summary Preliminary results are presented of experiments on the application of a root zone feeding technique to pursue the pathway of $^{22}Na^+$ and $^{36}Cl^-$ in intact, differentially salt sensitive soybean varieties. In contrast to $^{36}Cl^-$, about 70% of the $^{22}Na^+$ taken up under saline conditions by the labelled apical 8 cm of the seminal root was released along the transport through the proximal root tissue into the ambient inactive nutrient solution. Additional investigations, including micro-autoradiography, are in progress to bring more light into the specific transport behaviour of the two ions.

Introduction

In long term salinity experiments with soybeans low Cl^- leaf includers were shown to take up not only smaller amounts of Cl^- but simultaneously to accumulate higher concentrations of this ion in the roots[6,11]. Greenway and Munns[4] questioned this because in other experiments the salt resistant plants always had lower Cl^- concentrations in the roots than the sensitive ones. The mechanism of this accumulation was suggested to be induced by the stress impact (Eggers, 1980). In general, problems of the regulation of Cl^- accumulation, transfer and translocation in the root are the subject of recent investigations (*e.g.* refs.[1,3,5]).

A number of investigations have indicated that not only the roots play an important role with respect to shoot metabolism. *Vice versa* the shoot interacts besides the supply of photosynthates and the maintenance of the transpiration stream by a feed back and signalling system which influences the function of the root including uptake processes[7]. Under stress conditions these interactions may be much affected. Thus it appears imperative to conduct, as well as compartment analysis with excised roots, experiments with intact plants when considering aspects of genotypic regulation of long distant ion transport under salt stress. However, even when using whole plants, information about processes related to specific root regions or occurring along the pathway of ions through the roots into the shoots may be masked. Root zone feeding with labelled nutrients is apparently a useful tool to elucidate the entrance of individual ions into and the translocation through the roots (*e.g.* refs. [8,9,10]).

M.R. Sarić and B.C. Loughman (eds.), Genetic Aspects of Plant Nutrition.
ISBN-13: 978-94-009-6838-7
©1983 Martinus Nijhoff/Dr W. Junk Publishers, The Hague/Boston/Lancaster.

Preliminary results are presented on uptake, translocation and loss of Na^+ and Cl^- in soybean varieties differing in salt resistance after labelling the apical 8 cm root zone.

Materials and methods

Soybeans (*Glycine max* L. Merrill) were cultivated in hydroculture as described previously[12]. The experiments were conducted in a growth room: daytime 12 h photoperiod plus 2 h each to adapt the light at the start and end, day/night temperature 23°/16°C and relative humidity 65%/85%. When 20 days old, 6 individual plants with good seminal roots were selected and transferred into plexiglas root boxes. Each box consisted of 3 containers (36 cm × 5 cm × 5cm) which were partitioned into three "cells". The separating plexiglas sheets (3 mm) contained two slits thus allowing to seal the seminal root of two plants into the 3 connected root cells: in cell A, filled with radioactive nutrient solution, the apical 8–9 cm root tip, in cell B, filled with identical but inactive solution, the proximal 8 cm of the seminal root (with small laterals) and in cell C, filled also with inactive solution, the basal part (20–25 cm) and the remaining bulk of roots. Between cell A and B a control cell (1 cm) was introduced. The cell solutions were circulated by a pump. Root sealing was performed with Xantopren® (Bayer Dental, Leverkusen), a sticky, flowing polysilicon material which polymerised by adding liquid elastromer activator to a silicon rubber within a few minutes. The plants were osmotically adjusted within 3 days to the final salinity level by gradually imposed NaCl.

The amount of radioactivity applied varied (0.2–0.6 μCi ml^{-1}) depending on the concentration of the corresponding ion in the solution and other factors. At the end of an experiment the solution in cell A was replaced by identical but inactive solution. After 3 minutes washing the plants were harvested and separated into 9 fractions: labelled root (1), root of cell B divided into parts of equal length B_1 (2) and B_2 (3) and that of cell C into part C_1 (4) and C_2 (5) (laterals cut off), remaining roots (6), primary leaves (7), trifoliate leaves (including plant top) (8) and stem (9). Fresh and dry weight determination allowed to follow the label on a concentration basis. The samples were dry ashed with caution for Cl^- loss and aliquots were counted as described previously[12]. After determination of the volume in cells B and C samples of the solutions were also counted, thus allowing to produce a balance of ion absorption, translocation and loss. For calculations the data of two corresponding plants of each root cell were pooled. Additional details of each experiment are given in the respective table.

Results and discussion

Sealing the roots into the slits of the plexiglas partitions at both sides with Xantopren was very efficient. Since the plants were transferred into the root boxes at noon the day before labelling and the level of the volume in cell B was kept higher than in the adjacent cells a possible leakage could easily be detected. Additional proof was possible due to the small control cell and at least by the use of double labelling.

The results of experiment (Table 1) show that at moderate low salinity the total absorption of Na^+ and its transport behaviour was very similar in both soybean varieties: Only 29% remained in the plants, mainly in the labelled root, 9–11% was retained along the pathway to the shoot with a strong gradient. In the shoot Na^+ was preferentially retained in the stem, and 70% of the total absorbed Na^+

Table 1. Uptake and distribution of Na and Cl in intact soybean plants after labelling the apical 8 cm of the seminal root with ^{22}Na and ^{36}Cl. Osmotically adjusted plants cultivated for 2 days in a root box with 3 separate cells: the first (A) containing labelled nutrient solution with 5 mM NaCl salinity, the others (B; C) identical but inactive solution

Soybean variety	Total uptake by lab. root (μ moles)	Distribution in % of total uptake						
		Total in plant	Lab. root (Root cell A)	Remaining roots	Stem	Leaves	Root cell B	Root cell C
				Na				
Lee	0.74±0.2	28.7±8.6	11.6	10.9	5.8	0.36	53.9	17.4
Bragg	0.74±0.3	28.6±9.4	17.9	8.6	1.8	0.34	64.8	6.7
				Cl				
Lee	0.82±0.1	97.0±1.8	91.3	4.0	0.4	0.3	3.0	N.D.
Bragg	0.87±0.2	95.6±1.9	80.4	2.6	4.0	8.6	4.0	0.3

was lost, most of it in cell B. Loss was detected in the control cell already 2 h after labelling. In cell B it increased after 6, 24 to 48 h with a ^{36}Cl/^{22}Na ratio different from that in the labelling solution.

In contrast to Na$^+$, more than 95% of the absorbed Cl$^-$ was retained in the plant, most of it in the root but more so in "Lee" than "Bragg". Higher amounts of Cl$^-$ were translocated into the shoot of "Bragg" and mainly deposited in the leaves. Only small losses were detected.

Table 2. Uptake and distribution of Na and Cl in intact soybean plants after labelling the apical 8 cm of the seminal root with ^{22}Na and ^{36}Cl. Osmotically adjusted plants cultivated for 2 days in a root box with 3 separate cells: the first (A) containing labelled nutrient solution with 20 mM NaCl salinity, the others (B; C) identical but inactive solution

Soybean variety	Total uptake by lab. root (μ moles)	Distribution in % of total uptake						
		Total in plant	Lab. root (Root cell A)	Remaining roots	Stem	Leaves	Root cell B	Root cell C
Lee	0.30±0.04	23.7±4.7	0.4	3.0	11.6	8.7	32.3	43.9
Jackson	0.79±0.2	26.6±4.1	0.9	5.4	13.2	7.1	24.7	48.7
Lee	0.88±0.01	80.0±2.6	0.9	21.2	13.7	45.2	16.1	4.2
Jackson	2.71±0.8	88.4±3.2	1.7	2.3	12.3	72.1	7.8	3.8

In another experiment (II) the salinity was increased to 20 mM NaCl and the variety "Bragg" was replaced by "Jackson". The uptake and transport behaviour differed remarkably from the previous data (Table 2): The total absorption of Na$^+$ and Cl$^-$ was about 3 times higher in "Jackson" as compared to "Lee". However, the percentage distribution was not varietally different. Again only 28% of the total Na$^+$ absorbed was present in the plant, about 72% was lost. The retention of Na$^+$ in the labelled root was severely reduced in favour of a greater translocation into the leaves. Again a gradual retention of Na$^+$ along its way through the root to the shoot was observed.

Contrary to Na$^+$, in both varieties more than 80% of the absorbed Cl$^-$ was retained in the plant, however, only to a very small amount in the labelled root. Loss of Cl$^-$ increased as compared to the findings in experiment (I). Considering the total uptake and the % – distribution, a pronounced varietally different Cl$^-$ in the leaves was revealed. This is due to the reduced Cl$^-$ loss by "Jackson", but rather more to the much smaller gradual Cl$^-$ retention in the roots of this variety (2.3%) as compared to "Lee" (21.2%), which at least resulted in a higher Cl$^-$ accumulation in the roots of "Lee" (0.19 μmoles against 0.06). This was also confirmed on a concentration basis.

Generally, the results of these preliminary experiments demonstrate that the total uptake of Na$^+$ and Cl$^-$ varied considerably between individual plants but the tendency appears to be well revealed (Tables 1 and 2). Thus the root zone feeding technique provided – particularly because multilabelling and plant fractionation was applied – valuable information on the distinguishing characteristics between the long distant transport of Na$^+$ and Cl$^-$. There is a capacity in the root tissue to retain both ions but to a much lower extent Na$^+$. After its entrance into the cortex cells Na$^+$ is supposed either to be less effectively absorbed into the vacuoles but easily transferred into the xylem and it is then reabsorbed and released into the free space. Alternatively it seems possible that a substantial amount of Na$^+$ moved longitudinally in the cortex tissue. In both instances most of the radioactive Na$^+$ is assumed to be exchanged in the apparent free space with inactive cations of the medium. The loss of Na$^+$ was not related to only a small area proximal to the labelled root as was observed for other ions by Shone et al.[10] This effect can be excluded due to the sealing technique used and the introduction of the control cell the counts of which were not included in the calculations. The exchange loss seems to be a prevailing feature of Na$^+$ transport in soybeans although the enhanced loss from the distal part of the seminal root under moderate high salinity (Table 2) may also indicate the extrusion of retranslocated Na$^+$.

Cl$^-$ appears to move across the root mainly in the symplasm, even under moderate high salinity. When the retention capacity in the uptake zone was saturated, Cl$^-$ was partly retained along the pathway through the root, but more effectively by the variety "Lee". Because of the insufficient control of the Cl$^-$ passage through the membranes by "Jackson"[12] the tissue retaining the ion along the way to the shoot may be earlier filled up in this variety. Further investigations are underway using microautoradiographic and root zone labelling techniques combined with the application of membrane protein blockers such as 4,4-diisothiocayno-2,2 – stilbenedisulfonic acid and hormones to bring more light into the pathway of Cl^-, particularly into the genotypic different accumulation in the root.

References

1 Alston A M and Miller M H 1978 Effect of water stress on subsequent uptake of chloride by wheat plants. Plant and Soil 49, 305–315.
2 Eggers H 1980 Untersuchungen zum Chlorid-Transport in Wurzeln von Sojabohnenvarietäten unterschiedlicher Salztoleranz, Tierärztl. Hochschule Hannover.
3 Cram W J 1980 Chloride accumulation as a homeostatic system: negative feedback signals for concentration and turgor maintenance differ in a glycophyte and a halophyte. Aust. J. Plant Physiol. 7, 237–249.
4 Greenway H and Munns R 1980 Mechanism of salt tolerance in nonhalophytes. Ann. Rev. Plant Physiol. 31, 149–190.
5 Hooymans I I M 1980 A lagphase in vacuolar accumulation of Cl^- ions in roots of intact barley plants. Z. Pflanzenphysiol. 100, 185–187.
6 Läuchli A and Wieneke I 1979 Studies on growth and distribution of Na^+, K^+ and Cl^- in soybean varieties differing in salt tolerance. Z. Pflanzenernaehr. Bodenkd. 142, 3–13.
7 Lüttge M and Higinbotham N 1979 Transport in plants. Springer-Verlag, New York, Heidelberg, Berlin.
8 Richter Ch 1972 Akkumulation und Translokation von Ionen bei Angebot zu verschiedenen Wurzelzonen von Mais und Bohnenkeimpflanzen. Diss. Nr. 20, Berlin.
9 Russell S R and Sanderson I 1967 Nutrient uptake by different parts of the intact roots of plants. J. Exp. Bot. 18, 491–508.
10 Shone T G M and Wood A V 1977 Longitudinal movement and loss of nutrients, pesticides and water in barley roots. J. Exp. Bot. 28, 872–885.
11 Wieneke J 1980 Responses of ion uptake and transport in plants in relation to salinity. Symp. on Plant-Soil-Water Interactions, NIAB, Faisalabad/Pakistan (*In press*).
12 Wieneke J and Läuchli A 1979 Short-term studies on the uptake and transport of Cl^- by soybean cultivars differing in salt tolerance. Z. Pflanzenernaehr Bodenkd. 142, 799–814.

Effect of genotype and iron, applied to soil, on the chemical composition and yield of corn plants

T. ZAHARIEVA
N. Poushkarov Institute of Soil Science and Yield Programming, Sofia, Bulgaria

Key words   Iron nutrition   Maize   Nutrient concentrations   Varietal differences

Introduction

It is common agronomical practice to overcome the insufficiency of a particular nutrient in soil by applying fertilizers. Intensive fertilization, an important tool in obtaining high yields, gives rise to new problems. These problems are both economic, in relation to the increasing cost of fertilizers, and ecological, in relation to the decline of natural soil fertility and pollution of soil with heavy metals.

Intensive fertilizer application in plant production in recent years has brought about physiological diseases related to the unbalanced mineral nutrition of agricultural crops. On calcareous soils development of Fe-chlorosis is becoming more and more widespread, a serious problem since these soils are widely distributed in Bulgaria. It is difficult to overcome Fe-chlorosis through applying Fe-fertilizers because of the low availability of inorganic Fe-compounds and the high cost of synthetic Fe-chelates. One means of solving these problems is to focus efforts on genetically determined differences in nutrient uptake by plants. It is known that some plant species and varieties adapt better to unfavourable soil and climatic conditions than others. This is explained by the fact that these plants either have a mechanism for a more efficient uptake of the forms of low availability of some elements, or tolerate toxic concentrations of elements in soil. Plant varieties with different sensitivity to Fe-insufficiency are comparatively well characterized[1,2].

In our previous studies a positive effect of Fe applied to soil on chlorotic corn plants grown on calcareous chernozem was shown[4] where it was found that the effect of FeEDDHA was different depending on the genotype of corn plants (unpublished).

The objective of the present study is to determine the effect of the genotype factor on the chemical composition and yield of *Zea mays* grown on calcareous chernozem. The study has also aimed at determining the effect of FeEDDHA applied to soil on nutrient concentrations and yield of a relatively large number of commercially popular maize hybrids.

M.R. Sarić and B.C. Loughman (eds.), Genetic Aspects of Plant Nutrition.
ISBN-13: 978-94-009-6838-7
©*1983 Martinus Nijhoff/Dr W. Junk Publishers, The Hague/Boston/Lancaster.*

Materials and methods

Abbreviations: EDDHA, ethylenediamine di (o-hydroxyphenyl-acetic acid).

A pot experiment with calcareous chernozem, representative of this soil type in North Bulgaria, was carried out. Fe was supplied as Fe EDDHA at a rate of 10 mg/kg soil. Twenty-three corn hybrids varying in pedigree and maturity classification were tested.

The plants were 33-day-old when harvested (6th–8th leaf stage). The concentrations of Ca, Mg, Mn, Zn, Cu and Fe in the dry plant material were determined by the atomic absorption method and K by a flame photometer. N was determined using the Kjeldahl method.

Results

A statistical analysis of the experimental data shows that varietal differences in the concentrations of P, K, Ca, Mg, Fe, Mn, Zn, Cu and in yield are statistically significant ($p = 0.1\%$). Correlation ratios are calculated using a dispersion analysis. They give a quantitative estimation of the size of the effect of the studied factors *i.e.* genotype, Fe applied to soil and their interaction, on chemical composition and yield of corn plants.

Data in Fig. 1 show that the genotype factor has a considerable effect on the concentrations of the elements studied. It is of interest that the genotype is found to exert a strong influence on Mn concentration in plants. It is noteworthy that with the majority of hybrids the concentrations of the elements do not vary within a wide range. With some hybrids, however, more significant deviations from the mean concentration are observed, which is illustrated by the distribution of Mn, K and Ca concentrations in different hybrids, given in Fig. 2. The concentrations of the other nutrients are analogously distributed.

The late hybrids of the BM group show more significant deviations from the mean concentration with respect to some nutrients. Fig. 3. shows plant nutrient concentration in BM 25 and BM 55 in percentage to those in BM 55–48. In

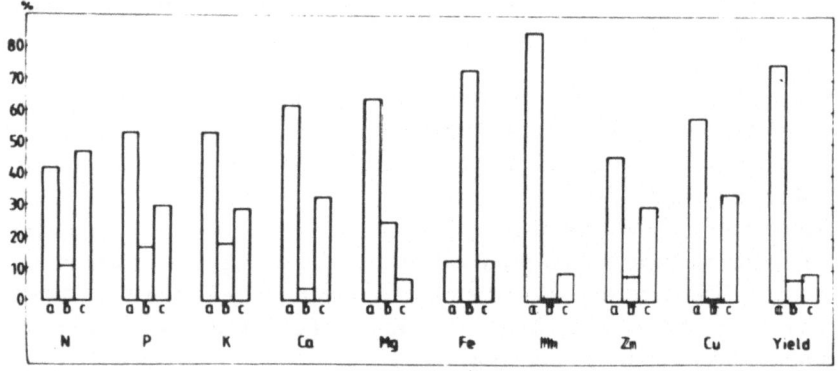

Fig. 1. Influence of genotype (a), Fe applied (b) and their interaction (c) on plant nutrient concentrations and yield.

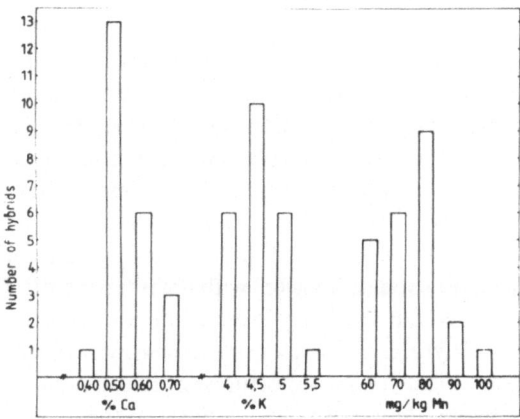

Fig. 2   Ca, K and Mn concentrations in 23 corn hybrids.

comparison with the BM 55–48 hybrid the BM 25 hybrid has higher concentrations of Ca, N, P and K and in this respect may be regarded as more efficient with respect to the uptake of these nutrients. The BM 55 hybrid tends to accumulate Fe and Mn. It is characterized by a very low concentration of Zn (6.5 ppm).

On averaging the data for all studied hybrids the numerical values of the independent effect of Fe applied to plant nutrient content were found. Fig. 1 indicates that Fe had a considerable influence only on Fe concentration in the plants.

The effect of Fe on N,P,K,Ca,Zn and Cu concentrations significantly increases in its interaction with the genotype factor. The numerical value for this effect is the highest with respect to N concentration in the plants. With 17 out of 23 hybrids studied the effect is proved to be positive.

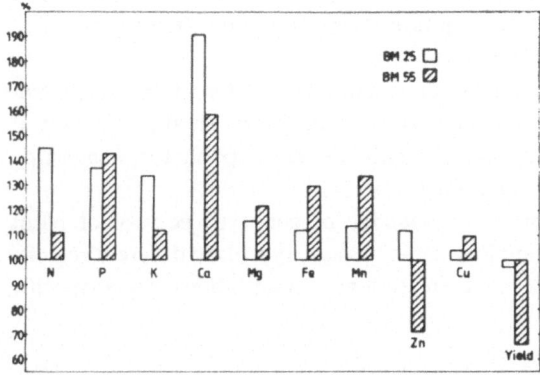

Fig. 3   Plant nutrient concentrations and yield of BM 25 and BM 55 in percent to those of BM 55–48.

Discussion

In connection with the variability of nutrient content between varieties of a given crop a reasonable question can be raised whether these differences do not challenge the validity of the use of plant analysis for diagnostic purposes. In spite of the statistical significance of varietal differences in plant nutrient concentrations, the latter do not vary within a wide range in this pot experiment. Data obtained agree with data reported by other authors and obtained in field experiments[3]. On the whole, genetically determined differences in nutrient concentrations (at the 6th–8th leaf stage) do not interfere with the interpretation of plant analysis data. As a result of a comparatively large number of hybrids analyzed some of them stand out as hybrids in which certain elements deviate from the mean value. This gives plant breeders the opportunity of selecting those varieties which "deviate from the norm" in the desired direction. In the case of plants grown on calcareous chernozem of interest are those hybrids which take up more Fe and N and do not accumulate P,Ca,Mn and others. However, the high total concentration of a given element in a variety is not always a proof of the efficiency with respect to this element. In this context examples are the Fe–efficient and Fe–inefficient plant varieties in which one and the same Fe-concentration is found. Considerable effect is exercised by the interaction of elements, which is supported by the experimental data obtained. With the BM 25 hybrid, which tends to accumulate Ca and K, no antagonistic interaction between Ca,K,Fe and Zn on their being taken up by the plant typical of calcareous soils is observed and hence no yield decline is reported. With BM 55 there is an antagonistic interaction between Fe and Mn on the one hand, and Zn on the other. Fe and Mn interfere with the uptake of Zn, which has a negative effect on the yield. With the ys_1 and WF9 hybrids different Fe–efficiency is a result of the changed ratios of K,Ca and Mn to other nutrients, (unpublished). Obviously, on assigning hybrids to different regions with calcareous soil not only the capacity of plants to accumulate some nutrients (Fe, Zn) should be taken into account, but also their capacity to maintain a favourable nutrient ratio, which ensures the balanced mineral nutrition of plants.

The present study shows that the independent effect of Fe on the concentrations of other nutrients in the plant is comparatively small, but that it substantially increases on Fe interaction with the genotype factor, suggesting that the Fe effect is genetically controlled.

Finally it can be stated that the knowledge of genotype control of plant mineral nutrition would help plant breeders to assign hybrids to different regions more rationally, on the basis of their different adaptability to a specific environment.

References

1 Brown J C 1979 Genetic improvement and nutrient uptake in plants. Bioscience, 29, 289–292.

2   Clark R B and Brown J C 1974 Differential mineral uptake by maize inbreds. Commun. Soil Sci. Plant. Anal 5, 213–227.
3   Rivard C E and Bandel V A 1974 Effect of variety on nutrient composition of field corn. Commun. Soil Sci. Plant Anal. 5, 229–242.
4   Zaharieva T 1976 Effect of increasing FeEDDHA levels on maize plants grown on calcareous chernozem. Soil Sci. and Agrochem. 11, 82–89.

SECTION III
THE INFLUENCE OF MINERAL NUTRITION ON PHYSIOLOGICAL AND BIOCHEMICAL PROCESSES OF GENOTYPES

Varietal differences in physiological and biochemical responses to changes in the ionic environment

B. C. LOUGHMAN, S. C. ROBERTS and C. I. GOODWIN-BAILEY
Department of Agricultural and Forest Sciences, University of Oxford, UK

Key words Armeria ecotypes *Armeria maritima* Maize Magnesium Manganese Mannose Phosphate Sodium *Zea mays*

Summary Experimental assessment of differences between cultivars of crop species or ecotypes of wild species from different localities in their capacities for ion absorption and transport is made difficult by the problem of obtaining seed material of comparable ionic content. When young seedlings are used this problem is particularly acute if the seeds of the different cultivars have not been raised under identical soil conditions. Propagation of material from ecotypes under controlled conditions is one approach to the solution of this problem. Six maize cultivars have been selected for similarity of phosphate content and the capacity for phosphate absorption from 5 μM KH_2PO_4 has been shown to vary by threefold whereas the proportion of the accumulated phosphate that reaches the shoot differs by much less. This level of phosphate supply approached that likely to induce deficiency and when the concentration is reduced to 1 μM differences in transport capacity of up to fourfold were observed when the rate of arrival at the tip of the first leaf was continuously monitored. The rapidity with which the transport is shut off by adding 1 mM D(+) mannose to the root environment also varies significantly indicating that sizeable differences in either the accumulation of mannose or the activity of phosphomannoisomerase exist in these cultivars.

Ecotypes of *Armeria maritima* collected from three sites, inland serpentine, inland mine dumps and coastal salt marsh were maintained as stock plants on the same peat mixture. Samples taken from these stocks were raised on a standard culture solution to provide genetically different material grown under constant conditions. The capacities for ion uptake were shown to differ very considerably and these differences were accentuated when the plants were grown in a range of concentrations of $MgSO_4$, NaCl and $MnSO_4$. The absorption of phosphate and its incorporation into nucleic acids were increased temporarily in the presence of 50 mM $MgSO_4$ but the pattern of these changes was different in the three ecotypes. The absorption of Na, Cl, and Rb was measured after treatment with a range of concentrations of NaCl and the effect of treatment with $MnSO_4$ on subsequent absorption of Mn and SO_4 was also measured. The coastal plants were significantly more efficient in their absorption of these ions when treated at the lower levels of NaCl (0.5 and 10.0 mM). The short term absorption rates were not reflected in the overall accumulation of sodium over periods of 10 weeks and the coastal plants appeared to reduce the root content of sodium by transfer to the shoot and by increased active pumping to the exterior.

Introduction

The way in which a plant responds to a deficiency of a single essential element is dependent on its stage of growth, the availability of water and nutrients and the particular environmental conditions under which it is growing. The element itself may be present in adequate amounts in some parts of the plant while other organs are deficient. Where the element is primarily concerned with the functioning of a particular organelle, *e.g.* manganese or chloride in the chloroplast, an inability to transport one of these elements from another part of the plant may be the crucial factor which determines whether or not the plant will survive. The stress of nutrient deficiency is caused by

(a) Low concentrations of the element in the soil and soil solution
(b) Inefficient absorptive mechanisms
(c) Binding of the element in an unavailable form in soil or within the plant
(d) Low rates of translocation to shoots from roots and vice versa.

Any method of overcoming the deficiency must involve the entry of the element via root or shoot and where the soil sequesters a particular ion the only means of relieving the deficiency is via the leaf. In general the mechanisms of ion absorption are similar whether the organ concerned is a root, stem or leaf and an understanding of the way in which minor nutrients exert their effect is helped by information concerning their processes of absorption. It is clear that genetic differences between species occur at least with regard to rate of absorption even if the mechanism is the same. We know relatively little about the characteristics of these absorption processes at the membrane and molecular level and it is perhaps presumptuous even to consider the factors controlling inherited differences with species.

A major problem in assessing the relative physiological capacities of a range of cultivars of a particular species is the difficulty of obtaining plants that are comparable in every respect even if grown from seed under absolutely controlled conditions. The only way in which the seeds can be guaranteed to contain similar mineral composition is for the parent plants to have all been grown to maturity under identical conditions. A single commercial seed variety, even though obtained from one supplier over a period of months or years, may have been grown at more than one site in more than one country. The ideal aim of comparing both the physiological and biochemical characteristic of a range of cultivars is therefore very difficult to achieve unless the experimentor has available the facilities of a large plant breeding station. However if a few varieties of seed can be obtained from a single harvest at one site it is possible to examine particular physiological and biochemical characteristics in order to attempt to correlate them.

Much of the relevant literature is concerned with differential susceptibilities of cultivars to deficiency or excess of a particular ionic species often associated with variations in the ability of the cultivar to absorb or transport it. Tolerance of copper deficiency varies widely between cultivars of wheat and at high nitrogen levels susceptible varieties produce no grain under conditions where others yield well[8]. Susceptibility is associated with increased tillering and the increase in the number of meristems possibly contributes to the immobilization of the limited amount of copper available. Zinc deficiency has been shown to depress growth within 7 days in two inbreds of maize. H.84 was more efficient at producing dry matter and showed fewer deficiency symptoms than A.63J. A greater proportion of the zinc moved to the tops in H.84 and there was less imbalance in other elements such as P, Ca, Mg, Mn and Fe that normally interact with zinc.

Sarić and Škorić[15] have shown that the ionic composition varies considerably between genotypes of sunflower. Apart from twofold differences in the content of specific ions such as calcium (569-1064 mg/100 g dry matter) the distribution of elements within the plant also varied indicating that cultivars differed in their capacity to absorb and transport them Zobel[17] has suggested that about one-third of the plant genome is concerned with growth, development and function of root systems. Capacities for transport and growth are genetically controlled even though environmental factors often overshadow the genetic differences between them. El Bassam[5] compared five barley varieties and found that the yield potential correlated best with the number of nodal roots rather than with root weight or length. At early stages of growth the yield potential of the cultivars could be correlated with the capacity of the root for water absorption at a fixed soil water potential.

Breeding of crop plants has been mainly for optimal yield or disease resistance although some high yielding strains resistant to fungal diseases may turn out to be very susceptible to Zn deficiency or to have an unusually high requirement for boron. Major nutrient stress conditions appear to be shortage of available iron and high salinity under alkaline conditions and P deficiency and Mn toxicity on acid soils. Plants adapted to grow well under these conditions must possess physiological and biochemical characteristics that enable them to extract or exclude more of one element than the less well adapted plants.

Deficiency of iron in soils is rare but alkaline conditions reduce its availability. Brown[2] has shown that tolerant species respond to iron stress by releasing protons, reducing compounds, chelating compounds such as citrate or by reducing Fe^{3+} to Fe^{2+} at the root surface. Selection of genotypes capable of some or all of these adaptations should enable the problem of iron stress to be countered under most agricultural conditions.

The mechanisms of ion absorption, transport through the root and into the xylem are all membrane controlled processes and as such depend on specific membrane carriers or result from the production of ion gradients or changes in electropotential. All these phenomena depend on the chemical constituents of the membrane components and the production of these components is likely to be genetically controlled. If different cultivars have differing capacities for absorption of nutrients or for exclusion of toxic ions then the presence in them of varying amounts of a particular ion will necessarily effect the activity of susceptible enzyme systems.

Caution is necessary concerning the relevance of the actual amount of the ion in the vicinity of a susceptible enzyme. It is possible that two cultivars containing identical amounts of an ion such as orthophosphate could exhibit significant differences in those biochemical processes in the cytoplasm or organelles known to be controlled by the level of inorganic phosphate. The relative amounts of inorganic and organic phosphates in the cytoplam may differ and the distribution of inorganic phosphate between vacuole and cytoplasm is particularly important. This distribution varies widely between different types of cells and it is important to know both the concentrations in cytoplasm and vacuole and their relative volumes before proper assessment can be made of the distribution of the inorganic phosphate within the cell. Estimation of pool sizes of phosphate is being aided by use of the ^{31}P – NMR spectrometer[9], but assessment of the distribution of other ions is more difficult. The effects of paramagnetic ions such as Fe^{2+} or Mn^{2+} on spectra of phosphorus compounds are helping to ascertain the location of such ions within cells.

The capacity for transfer of ions, both inorganic and organic across the tonoplast into the vacuole might be one of the most important factors in mediating the response of different cultivars to nutrient deficiencies. In addition, the resistance of particular cultivars to herbicides or insecticides may depend on the inherent ability to sequester the component in the vacuole by virtue of high capacity for transport across the tonoplast.

Materials and methods

(a) Maize

Maize seeds were obtained from Nickersons Seed Specialists, and after removal of excess fungicide by washing in running water were germinated on filter paper in the dark at 25°. After 48 hours they were either used directly for ^{31}P – NMR experiments[9] or transferred to containers of culture solution for growing at 25° at 12,700 lux, usually for 9–12 days. Transport experiments were carried out with individual plants or pairs of plants in 300 ml containers containing culture solution with 1 or 5 μM $KH_2\ ^{32}PO_4$ at pH 6.0 and 25°C at 12,700 lux. The top of the first leaf was lightly attached to the surface of a shielded Geiger Müller tube. The solution was magnetically stirred throughout the experiment and the output of the G. M. tube was connected via a ratemeter to a pen recorder. Additions were made to the solution by rapid injection into the stirred solution and 1.0 ml samples could also be rapidly withdrawn for assay of depletion rates. When changing the solution around the roots, the container was rapidly lowered, the roots rinsed and a new container of fresh solution immediately provided. It was possible to perform this operation without any fluctuation of the recorder trace.

Phosphate analyses were carried out by the method of Allen[1] after ashing the samples with a mixture of 60% $HClO_4$ and conc. HNO_3.

(b) Armeria maritima

The species *Armeria maritima* (Mill.) Willd is distributed thoughout the British Isles, on cliffs, salt marshes and inland sites that include serpentine soil and spoil tips of mineral ore mines where the levels of available ions, particularly Na, Mg and heavy metals vary enormously. Plants growing under constant environmental conditions on their native soils produce higher yields than plants from other sites. The rates of absorption of ions by plants from coastal sites are much higher than that of plants from inland serpentine sites. The former also show increased capacities for upward transport and outward efflux. The particular aspect of this type of research that is relevant to our studies here is that the genetic variation in these populations can be usefully studied by bringing material from the sites, taking cuttings and raising them on a standard nutrient solution.

Three major populations were used, Inland Serpentine, Meikle Kilrannoch (M. K.), Inland Non-Serpentine mine site, Garrigill (G) and Salt Marsh, Scolt Head (S). Individual rosettes were removed from the parent plant and propagated by root induction to allow root development in a sand-peat-loam mixture and then maintained as a stock. Roots of the same physiological age were obtained for experiments by stripping the lateral roots from single plants and allowing new roots to develop for 10 days in full culture solution. Prepared plants were treated with specific concentrations of ions for up to 96 hours. At appropriate times during the pretreatment, a number of 1 cm. root tips were removed for assay of their capacity to absorb and incorporate ions or amino acids[6].

The ability of the three populations to grow on a standard culture solution is shown in Fig. 1. It is clear that the capacity for growth is very different and reflects the influence of the ionic environment in which the plants occur naturally even though by the time of the assessment of growth capacity they have been exposed to the full culture medium for at least 10 days and have been growing for months prior to transfer on a standard sand-peat-loam mixture.

Experimental results

(a) Phosphate utilization by maize

It is generally agreed that roots take up phosphate faster than it can diffuse in

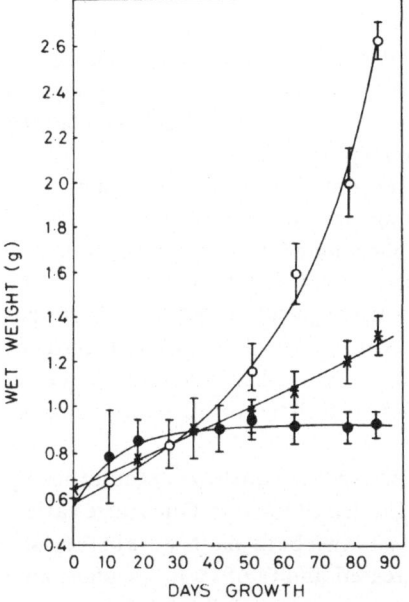

Fig. 1. Growth of three populations of *Armeria maritima* on a full culture solution. O—O Meikle Kilrannoch (inland serpentine) ×—× Garrigill (inland mine site) ●—● Scolt (coastal salt marsh). Concentration of elements in mmol/l. Na, 0.5; K, 0.2; Fe, 0.1; Ca, 0.1; Mg, 0.025; Mn, 0.01 B, 0.01; Cu. 0.001; Zn, 0.001; Mo, 0.00007; P, 0.001; S, 0.037; Cl, 0.10; N, 0.60. Each value is the mean of ten plants.

the soil solution to the root surface. Doubling or halving the capacity for absorption under standard water culture conditions might have absolutely no effect on the actual rate of absorption by the plant growing in the soil, where the diffusion rate is lower. The variation in ability, between and within species, to thrive on low phosphate soils is perhaps related to other factors such as mycorrhizal infection or by differences in proton extrusion. Whatever factors are involved, very significant differences have been observed when different varieties have been grown under standard conditions and the physiological and biochemical factors responsible for these differences are of great interest. The literature concerned with this aspect of plant nutrition is somewhat diffuse and caution is needed in interpreting the results. The major problem is, as has already been stressed, the source of seed, because of the wide differences in mineral composition of seeds of one variety grown at different sites with different manurial treatments. Such differences would affect the absorption and transport of ions from the soil solution and in addition modify the degree of mycorrhizal infection thus affecting those ions such as phosphate, copper and zinc with low diffusion rates. Water culture experiments would not necessarily show up these differences.

The concentration of inorganic orthophosphate in the solution in equilibrium with most soils is in the range 0.1 μM–10 μM and this means that all plants, whether growing in uncultivated or agricultural conditions, are absorbing from extremely low concentrations of phosphate that are sub optimal. Species differ in their ability to survive in these soils and differences between species in phosphorylative metabolism have been recorded and these may be related to the processes of absorption and transport of phosphate.

The well established effects of mannose on both the transport of phosphate into the xylem[10] and across the chloroplast membrane is due to sequestration of the cytoplasmic phosphate as mannose–6–phosphate (M–6–P) in species such as cereals where the phosphomannoisomerase activity is low. Consequently the M–6–P cannot be utilized via G–6–P and the usual glycolytic pathway. In species such as mung bean where the activity of this enzyme is higher, M–6–P does not build up and there is no effect on transport to the xylem[11]. Within the cereal species the effect of mannose is variable and may well be correlated with enzyme levels in the cultivars.

Evidence for differences in absorption of phosphate from low concentration by different varieties comes from the work of Clark and Brown[4]. One maize variety Pa36 absorbed more phosphate than WH even when both plants were in the same very dilute solution of phosphate. The increased ability of Pa36 for phosphate absorption appeared to be correlated with a higher level of phosphatase activity in the roots. At higher levels of phosphate WH outyielded Pa36 and the total phosphate contents of the two varieties were similar. It must be pointed out that the total phosphate content is unhelpful if no information concerning its distribution within the plant and between cellular organelles is available.

Following up the maize work we examined the capacities for absorption and transport of six varieties in which the total phosphate in the seeds was similar as was the distribution of this phosphate in 12-day-old plants grown in full culture solution without phosphate (Table 1). The rates of absorption by the roots of 12-day-old plants of the six varieties from 5 μM KH$_2$ ^{32}PO$_4$ differed very significantly but when the total content of the shoots was measured rates of transport were seen to be similar (Table 2). When the rate of transport from 1 μM KH$_2$^{32}PO$_4$ was assessed by measuring the rate of arrival of radioactive phosphate at the tip of the second leaf under constant absorption conditions[12] significant differences in the rate at which phosphate was transported were observed (Fig. 2). These differences at a concentration of phosphate in the root environment considered to be capable of leading to symptoms of phosphate deficiency possibly contribute significantly to the ability of these cultivars to survive on soils differing in available phosphate.

Differences in transport rates up to fourfold were measured and the rapidity of the response to mannose (inhibition of transport) also differed. The time taken to recover from mannose inhibition after transfer of the plants to water or 1 μM KH$_2$PO$_4$ also appeared to be characteristic of the variety.

These preliminary experiments indicate that varieties differ very considerably in their ability to transport phosphate to the shoot and also that the relative activities of phosphomannoisomerase and possibly of a mannose –6– phosphatase are also very different thus amplifying the observations of Clark and Brown[4] relating the activities of phosphatases with absorption rates.

^{31}P—NMR spectra of root tips (48 h) from these varieties are remarkably constant apart from differences in distribution of inorganic orthophosphate between cytoplasm and vacuole (Loughman and Rathcliffe unpublished).

Analyses of these varieties indicated a surprisingly similar level of phosphate in

Table 1. Distribution of phosphate in 12-day-old maize plants of six varieties. Values are the averages of 10 plants

Variety	Weight (mg)				Phosphate (mg)				mgP/g d·wt
	Shoot	Root	Seed	Total	Shoot	Root	Seed	Total	
Aurelia	47.2	10.1	184	241.3	0.83	0.07	0.98	1.88	7.79
W703	36.2	9.2	122	167.2	0.93	0.10	0.57	1.60	9.57
Fronica	51.9	12.2	214	278.2	1.01	0.11	1.13	2.25	8.09
Caldera	66.7	16.6	169	252.3	0.92	0.15	0.73	1.80	7.13
LG.11	79.2	19.2	180	278.4	1.03	0.14	0.62	1.79	6.43
Rubis	78.7	14.9	162	255.6	0.83	0.11	0.73	1.67	6.53

Table 2. Absorption and transport of phosphate from 12-day-old maize plants of six varieties. Absorption from 5 μM KH$_2$^{32}PO$_4$ (12.5 μ Ci/1) at pH 6.0, 25°C and 12.700 lux

Variety	Absorption ng g^{-1} h^{-1}		% transported to shoot in 4 h
	Roots	Shoots	
Aurelia	3.12	1.74	37
W 703	1.48	0.48	23
Fronica	2.64	0.89	26
Caldera	1.10	0.54	31
L. G. 11	2.20	1.09	32
Rubis	1.70	0.99	38

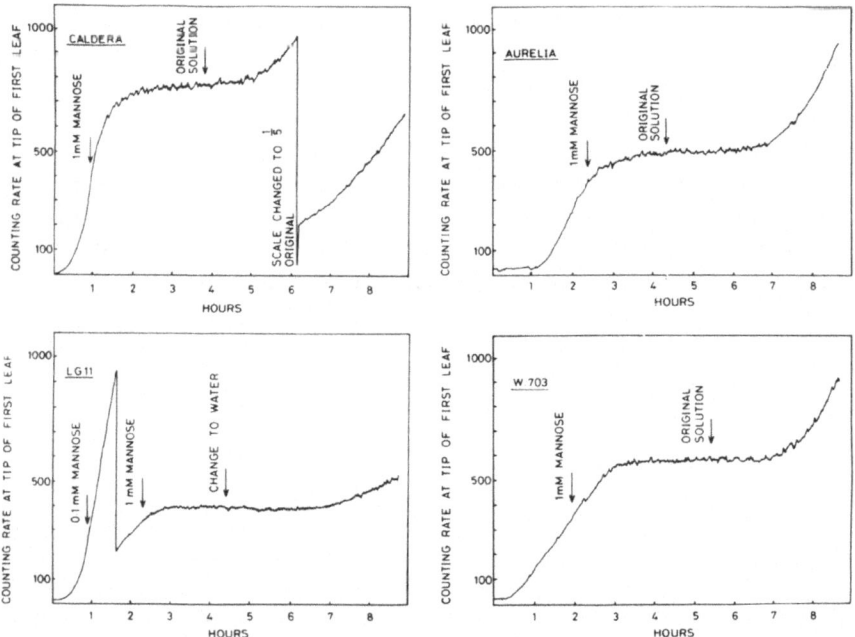

Fig. 2. The rates of transport of phosphate to the shoots of four 12-day-old maize cultivars exposed to 1 μM KH$_2$^{32}PO$_4$ at pH 6.0, 25°C and 12,700 lux. Plants were raised in full culture solution without phosphate.

the plants and in the distribution between seeds, roots and shoots (Table 3). Less than half the initial phosphate has left the seeds in 10 days and the greater part of that leaving the seed has moved to the shoot rather than the root. It is evident that varieties containing very similar amounts of phosphate transport the ion from the external solution to the leaf tip at very different rates even though the rate of absorption by the root differs relatively slightly (Table 3). It is of interest that levels of cis and trans of ABA in the first 1 mm apical segments of seven maize varieties have been shown to vary significantly and the effect of this hormone on proton transport could well be reflected in the absorption of phosphate[14].

Ion absorption by Armeria maritima

The rates of absorption of phosphate from low levels (10 μM NH$_4$H^{32}PO$_4$) by three populations was examined by removing plants at 12-hour intervals during treatment with 50 mM or 25 μM MgSO$_4$ for 80 hours and allowing them to absorb phosphate for 60 mins. Data for the high MgSO$_4$ treatment are presented in Fig. 3 in terms of phosphate soluble in 5% trichloracetic acid and the insoluble residue representing primarily nucleic acids.

The clear cut activation of phosphate incorporation into the soluble fraction by treatment with 50 mM MgSO$_4$ for about 50 hours occurs in all three populations but large differences are seen in the effect of MgSO$_4$ on incorporation of phosphate into the insoluble fraction. The capacity for absorption of phosphate maximizes in the three populations at 50 hours at 50 mM MgSO$_4$. Incorporation into the insoluble fraction also maximises at about 50 hours in the

Table 3. Rates of phosphate transport of six maize varieties. Individual plants of comparable size were allowed to absorb phosphate from 1 μM KH$_2$^{32}PO$_4$ at pH 6.0 at 25° and 12,700 lux. The counting rate of 0.5 cm^2 of the tip of the 1st leaf was continuously monitored. Relative transport rates are obtained from recorder charts (see Fig. 2)

Variety	Age (days)	Relative transport rate	Time taken for mannose to decrease rate (minutes)	Time to start of recovery in mannose free solution (minutes)
Aurelia	11	29	13	112
W 703	9	22	53	90
	12	26		
Fronica	11	50		
Caldera	9	102	15	60
L.G.11	9	107	8	–
	10	140		
Rubis	10	29		

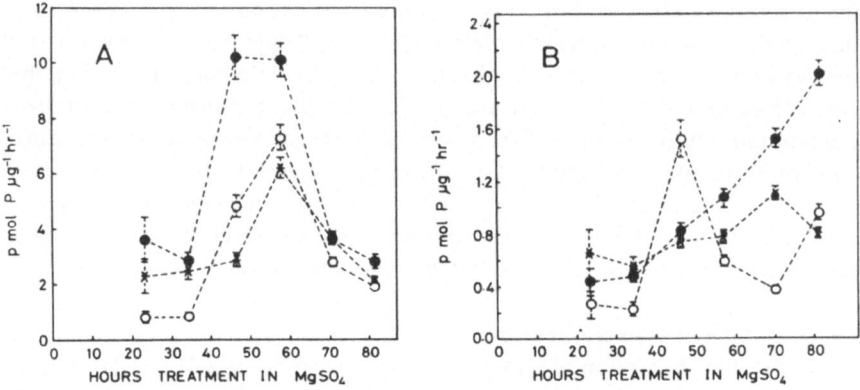

Fig. 3 The rates of absorption of phosphate from 10 μM NH$_4$H^{32}PO$_4$ and rates of incorporation into TCA soluble (A) and insoluble (B) fractions after treatment of three ecotypes of *Armeria maritima* with 50 mM MgSO$_4$. ○—○ (Mk); ×—× (9); ●—● (5).

Kilrannoch and Garigill plants while that of the Scolt plants continues to rise for 80 hours.

These responses are relatively minor in the low magnesium treatment but it is clear that the presence of high levels of magnesium has a major effect on the development of phosphate uptake capacity even through the increase in capacity is only transitory, returning to normal after about 80 hours. The mechanism of this readjustment of transport capacity is unknown but activations of between three- and ten-fold indicate a major reorganisation of the root membranes.

The relative rates of absorption of Cl$^-$, Na$^+$, Mn^{2+}, and SO$_4^{2-}$ ions after pretreatment of the three populations for 72 hours with NaCl or MnSO$_4$ are shown in Table 4. Large differences in the capacity for absorption of Na and Cl were observed, the coastal population (Scolt) being far more efficient in Na absorption at all levels of NaCl. The Scolt plants were more efficient than the inland levels in absorption of Na, Mn, Cl and SO$_4$ at the two lower levels of treatment.

Large differences in the rates of efflux of Na$^+$, Rb$^+$ and Cl$^-$ from plants of the three populations after treatment with full nutrient solution containing NaCl at the same three concentrations are shown in Fig. 4. The greatest differences between the three populations were seen at 50 mM NaCl for Na$^+$ efflux and 10 nM NaCl for Cl$^-$ efflux. An interesting aspect of these experiments is that in the presence of 2,4-dinitrophenol the differences between the populations were less marked. The long term accumulation of sodium by the three ecotypes does not necessarily reflect the short term absorption rates shown in Table 4. The relevant values for sodium absorption from 10 mM NaCl in Table 4 are compared with the total accumulation and distribution between shoot and root after treatment of the whole plant for 72 days in 10 mM NaCl (Table 5).

Table 4. The effect of NaCl and MnSO₄ on absorption of ions over 3 hours by roots of three populations of *Armeria maritima* after pretreatment for 72 hours with full nutrient solution containing NaCl or MnSO₄ at the specified concentrations. Values are the means of five replicates

Population	Treatment NaCl (mM)	Absorption rate (μmol g^{-1}h^{-1})	
		Cl	Na
M.K.		26.5	15.6
G	0.5	41.6	33.0
S		58.3	39.9
M.K.		26.7	47.8
G	10.0	53.5	62.8
S		52.5	101.1
M.K.		117.0	147.2
G	50.0	68.3	160.8
S		116.6	289.3
	MnSO₄ mM	SO₄	Mn
M.K.		1.92	1.71
G	0.05	1.72	2.04
S		2.42	2.09
M.K.		5.04	7.80
G	0.5	9.00	11.67
S		11.2	13.31
M.K.		27.4	35.40
G	2.50	38.0	40.61
S	41.0	40.10	

Discussion

Data relevant to the discussion concerning genetic differences have been presented for very different types of material. The very significant differences seen in some of the metabolic processes discussed indicate that more attention might be given to these capacities when considering the suitability of available cultivars for different soil types and nutrient availability.

The maize varieties used in these experiments vary in their size at maturity and differences are apparent even in the young seedlings in the growth rate over the

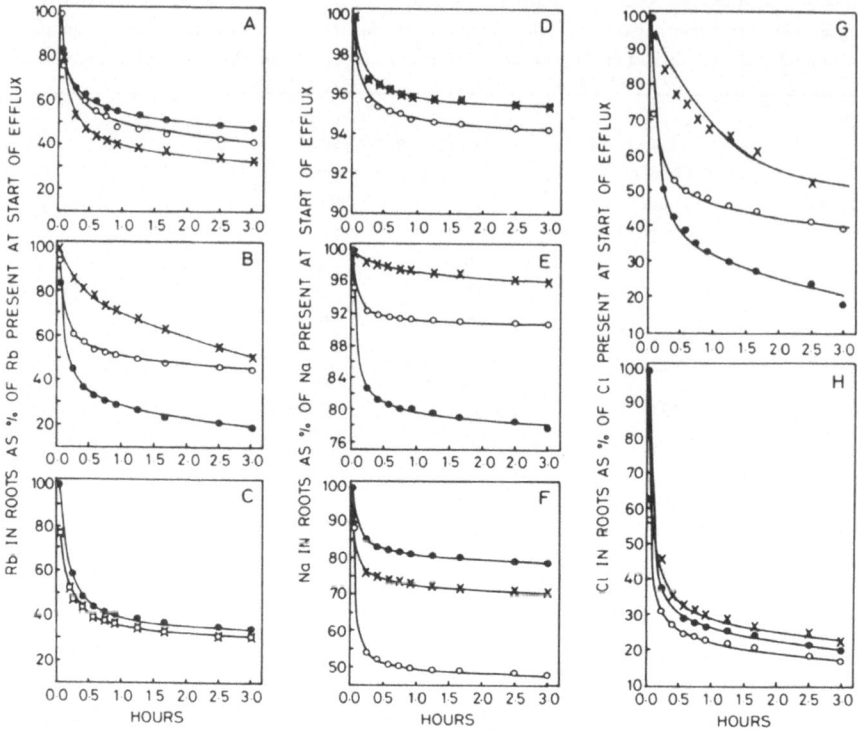

Fig. 4. Efflux of ions from root tips removed from three *Armeria maritima* ecotypes after pretreatment of the plants for 72 hours with a full culture solution containing NaCl at 0.5, 10.0 or 50 mM. Root tips were allowed to absorb ^{24}Na, ^{36}Cl or ^{86}Rb for 3 hours prior to transfer to unlabelled full culture solution for measurement of efflux expressed as % of the labelled ion in the root tips at the time of transfer. ^{36}RbCl supplied at 0.3 mM.
A and D Efflux from plants pretreated with 0.5 mM NaCl
B, E and G Efflux from plants pretreated with 10.0 mM NaCl.
C, F and H Efflux from plants pretreated with 50 mM NaCl
○—○ Meikle Kilrannoch. ×—× Garrigill. ●—● Scolt.

first two weeks and the final size of shoots and roots. When available phosphate is in limited supply, *e.g.* 1 uM, Aurelia and W703 transport slowly to the shoot whereas LG11 and Caldera possess the fastest transport. The latter varieties possess larger root systems than the others when the shoots are of equivalent length and the increased ability for transport appears to be correlated with root size. In this connection the smallest variety W703 also exhibits the slowest transport rate. The differences in the rapidity of the onset of inhibition of transport by mannose probably reflects the level of phosphomannoisomerase in the roots as well as the rate of mannose entry. In general, if transport is slow the response to mannose is slow and it also appears that recovery from mannose is faster in those varieties where the initial response is more rapid.

Table 5. Relative absorption rates (μmol g^{-1} h^1) and accumulation (μmol g^{-1}) of Na by Armeria plants grown on full culture solution containing 10 mM NaCl. Absorption measured over 3 hours with excised root tips after 72 hours treatment of whole plants. Accumulation measured after 71 days treatment of whole plants

Population	Absorption rate (μmol g^{-1} h^{-1})	Accumulation (μmol g^{-1})		% Transport
	Roots	Roots	Shoots	
M.K.	47.8	103.0	89.1	46.4
G	62.8	66.9	192.2	74.2
S	101.1	23.9	220.0	90.4

The results imply that some varieties are more efficient than others in absorbing and transporting phosphate from limiting concentrations of phosphate and that these differences are less marked when phosphate is in more abundant supply. There is little evidence so far that the varieties differ in their responses to shortage of other elements although there is some indication that the sensitivity to boron deficiency varies in these varieties suggesting that another membrane associated aspect is genetically determined. The differences in phosphatase activity in the roots of maize varieties and the correlation with absorption rates has already been referred to[4] and it is known that the decreased absorption associated with boron deficiency is correlated with reduced ATPase activity in the plasmalemma[13]. The fact that both the ion absorption rate and the ATPase activity can be restored almost immediately by the addition of 10 μMH$_3$BO$_3$ suggests that the two phenomena are linked and assay of ATPase activity of the maize varieties may also show differences between varieties that correlate with ion transport rates. Ecotypes of *Aegilops peregrina* (Hack.) Marie et Weil have been shown to differ in the pH optimum of the acid phosphatase developed in the roots as a result of growing in culture solution under conditions of phosphate deficiency (5 μM). The degree of stimulation of phosphatase activity brought about by growing on low phosphate soils also varied widely[16].

The Armeria populations are clearly adapted for growth on the soils of their own locality but even after they had been maintained under standard cultural conditions clear differences in absorption and efflux characteristics emerged. By raising plants in full culture solution and then stripping off the roots, regrowth of new roots in culture solution provided roots of similar physiological age and minimized differences in ion content carried over from the original sources of material. The genetic differences in these populations clearly show up in their ability to absorb ions, transport them to the shoots and in the capacity for incorporation of metabolisable ions such as phosphate and sulphate into organic forms. The plants from different sources have very different growth rates when raised on a standard culture solution. When grown at low levels of calcium, sodium and magnesium the plants from inland serpentine soils grow faster than coastal or mine site plants. The slower growing plants are clearly able to tolerate high concentrations of ions.

The coastal populations absorb ions at a much higher rate than plants from inland sites. However, the rates of efflux of ions both from the root to the external solution and by upward transport to the shoot can also be much greater. The genetically controlled ability of coastal plants to actively pump sodium, potassium and chloride ions to the soil solution enables them to cope with higher external levels of sodium chloride and thus confers on them a capacity for survival not possessed by the inland populations.

All these populations responded to the presence of high concentrations of cations for between 2 and 3 days by increasing their capacity to absorb phosphate and incorporate it into organic forms and at the same time the rates of respiration and protein synthesis increased. All these changes are transitory and the activities return to the original levels after a further period of a day or two. The maintenance of the osmotic potential of the tissues is of obvious inportance and the changes may reflect this. Alternatively the metabolic changes may be associated with the synthesis of new membrane components concerned with the development of mechanisms concerned with the active efflux of ions from the roots.

An interesting development in the search for the genetic basis of nutrient response is the screening of a series of Triticales containing various combinations of rye and wheat chromosomes for the capacity to grow on soils too low in copper to support other cereals. Genotypes carrying the rye chromosome 5R characterised by the presence of a hairy peduncle were most efficient. Wheat genotypes containing an extra pair of 5R chromosomes are the only additional lines of wheat that grow well on soils low in copper. It is clear that information on the segregation of this factor between genotypes of rye showing high and low efficiency will be of considerable value in selecting material for areas of copper deficiency[7].

In conclusion it must be said that much of the experimental data available on young seedlings is somewhat difficult to interpret without knowledge of the overall ionic composition of the plants and the seeds from which they were grown. More attention should be paid to the problem of raising absolutely comparable material before comparisons are made of possible physiological and biochemical characteristics of cultivars or ecotypes that contribute to their effectiveness under particular ionic environments. Reliable data of this type should help in the assessment of cultivars of crop plants for economic yields on soils where deficiencies of one element or surfeit of another have hitherto resulted in unacceptably low yields with the available cultivars.

Acknowledgements We are extremely grateful for the technical assistance of Mr. Brian Parker and for financial support from the Cecil Pilkington Charitable Trust.

References

1. Allen R J L 1940 The estimation of phosphorus. Biochem. J. 34, 858–865.
2. Brown J C 1977 Genetically controlled chemical factors involved in absorption and transport of iron by plants. Bioinorganic Chem. 11, 93–103.
3. Clark R B 1978 Differential response of maize inbreds to zinc. Agron. J. 20, 1057–1060.
4. Clark R B and Brown J C 1974 Differential phosphorus uptake by phosphorus stressed corn inbreds. Crop Science 14, 505–508.
5. El Bassam N 1981 Genetical variation in efficiency of plant root systems. *In* Structure and Function of Plant Roots. Eds. R Brouwer, O Gasparikova, J Kolek and B C Loughman. pp 237–245. Martinus Nijhoff, The Hague.
6. Goodwin-Bailey C I 1978 An Experimental Study of Inland and Maritime Populations of *Armeria maritima* (Mill.) Willd. D. Phil. Thesis. University of Oxford.
7. Graham R D 1981 Genetics of the copper efficiency factor in rye. *In* Copper in Soils and Plants. Eds. J F Loneragan, A D Robson and R D Graham. Academic Press, Sydney, New York.
8. Hill J, Robson A D and Loneragan J F 1978 The effects of copper and nitrogen supply on the retranslocation of copper in four cultivars of wheat. Aust. J. Agric. Res. 29, 925–939.
9. Kime M J, Longhman B C, Ratcliffe R G and Williams R J P 1982 The application of ^{31}P nuclear magnetic resonance to higher plant tissue. 1. Detection of spectra. J. Exp. Bot. 33, 665–669.
10. Loughman B C 1966 The mechanism of absorption and utilization of phosphate by barley plants in relation to subsequent transport to the shoot. New Phytol. 65, 388–397.
11. Loughman B C 1978 Metabolic factors and the utilization of phosphorus by plants. *In* Phosphorus in the Environment: its Chemistry and Biochemistry. Eds. R Porter and D W Fitzimons. pp 154–174. Elsevier, Amsterdam and New York.
12. Loughman B C 1981 Metabolic aspects of transport of ions by cells and tissues of roots. Plant and Soil 63, 47–55.
13. Pollard A S, Parr A J and Loughman B C 1977 Boron in relation to membrane function in higher plants. J. Exp. Bot 28, 831–839.
14. Rivier L and Pilet P E 1981 Absicic acid levels in the root tips of seven *Zea mays* varieties. Phytochemistry 20, 17–19.
15. Sarić M and Škorić D 1981 Relationship between the root and above ground parts of different sunflower inbreds regarding the content of some mineral elements. *In* Structure and Function of Plant Roots. Eds. R Brouwer, O. Gasparikova, J. Kolek and B C Loughman. pp 399–401. Martinus Nijhoff, The Hague.
16. Silberbush M, Shomer-Ilan A and Waisel Y 1981 Root surface phosphatase activities in ecotypes of *Aegilops peregrina* (Hack.) Marie et Weil. Physiol. Plant. 53, 501–504.
17. Zobel R W 1975 The genetics of root development. *In* The Development and Function of Roots. Eds. J G Torrey and D J Clarkson. pp 261–275. Academic Press, London, New York, San Francisco.

Genotypic differences in the mineral metabolism of plants adapted to extreme habitats

M. POPP
Institut für Pflanzenphysiologie der Universität Wien, Althanstrasse 14, Postfach 285, A-1091 Wien, Österreich

Key words Calcicole Calcifuge Calciphob Calcioltrophic Halophytes Metallophytes Salt accumulation Salt exclusion Serpentine plants

Summary Halophytes, metallophytes, serpentine plants, calcicoles, and calcifuges are adapted to soil conditions which are deleterious for other plants. Some mechanisms responsible for these adaptations are described. The respective changes in the genome for these alterations of the mineral metabolism are probably not too far-reaching, because already intraspecific differentiation may lead to resistant ecotypes. On the other hand it seems likely that some families or species are favoured in occupying extreme soil types in consequence of their basic physiological features (*i.e.* Chenopodiaceae, Poaceae in saline habitats, Brassicaceae, Caryophyllaceae on heavy metal soils).

Introduction

Extreme habitats will be confined in this paper to those environments which are special in their mineral composition and therefore only inhabited by particularly adapted plants: halophytes, metallophytes, serpentine plants, calcicoles, and calcifuges.

Emphasis will be put not only on intraspecific variations, which enable certain ecotypes of a species to cope with harsh edaphic conditions, but also on physiological features of broader taxonomic units (*i.e.* genus, family), which favour their adaptation to extreme soil types. For the entirety of the physiological characteristics common to the members of a taxonomic unit, Kinzel[29], Albert and Kinzel[2] proposed the term "physiotype". Concerning mineral metabolism physiotypical characteristics such as the dominant cation, K^+/Na^+ or K^+/Ca^{+2} ratios, overall low ion uptake, preference for NO_3^- or NH_4^+ are of some consequence for our understanding of how certain plants can survive under soil conditions which are deleterious for others.

To prevent any misunderstanding it should be pointed out here that the terms "tolerance" or "tolerant" will be not used according to Levitt's definition in stress physiology[37], but synonymous with "resistance" or "resistant", which do not indicate the mechanisms responsible for the adaptive effect.

M.R. Sarić and B.C. Loughman (eds.), Genetic Aspects of Plant Nutrition.
ISBN-13: 978-94-009-6838-7
©1983 Martinus Nijhoff/Dr W. Junk Publishers, The Hague/Boston/Lancaster.

Halophytes

Two ways in which halophytes cope with the high salt content of their habitats are shown in Fig. 1. 'Salt accumulators' such as *Salicornia rubra* (Chenopodiaceae) are resistant to high salt concentrations in their shoots. 'Salt excluders' such as *Puccinellia airoides* (Poaceae), which may grow beside *Salicornia rubra*, are able to keep the salt content of their shoots rather low. Both mechanisms demand considerable specialisation in ion uptake and transport.

In contrast to halophilic bacteria, in which the whole metabolic machinery is adapted to high salinity levels (*cf.* ref.[34]) a number of enzymes from halophytes have been shown to be as salt-sensitive as those from nonhalophytes[1,15,16,22,54]. Even when the kinetic properties of some enzymes of halophytes are positively influenced by high NaCl concentrations in the growth medium[45], it was assumed that approximately two thirds of the salt accumulated in these plants is sequestered within the vacuole[17,18]. Evidence for the validity of this assumption has now been obtained[7,25,55] and we may ask how can such concentration differences be achieved and maintained? In accumulator types this question concerns mainly the $K^+ - Na^+$ distribution between cytoplasm and vacuole in the shoots and translocation processes between roots and shoots, while within

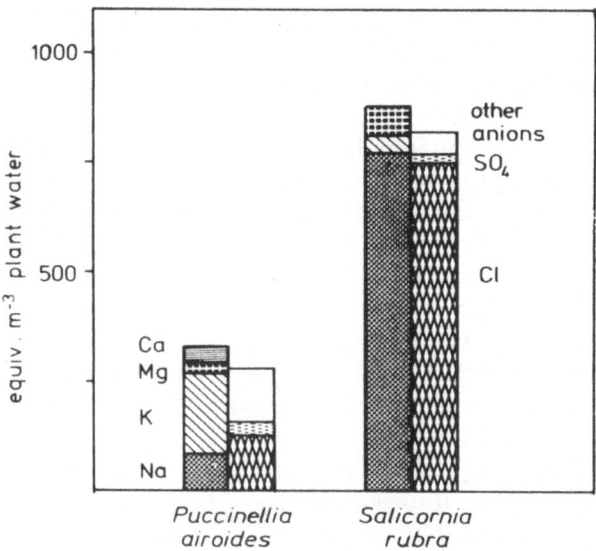

Fig. 1. Ion content (equiv. m^{-3} plant water) of two halophytes, *Puccinellia airoides* (Poaceae) and *Salicornia rubra* (Chenopodiaceae), growing side by side at the Great Salt Lake (USA) (Albert, R. unpublished results).

excluder types the ion uptake mechanisms in roots compared to those of nonhalophytes have to be altered, as well as the ion movement from the roots to the shoots.

Such alterations were first shown for leaf slices of the mangrove *Avicennia marina* by Rains and Epstein[48], who observed that K^+ uptake in the high concentration range was not decreased in the presence of concentrations of Na^+, which would have an inhibitory effect on K^+ uptake in nonhalophytes (*cf.* ref.[9]). Similar observations were reported by Jefferies[27] for K^+ uptake into roots of *Triglochin maritima*. In a solution containing 10 equiv $\cdot$ m^{-3} K^+ (high concentration range) 10 equiv $\cdot$ m^{-3} Na^+ had a slighly enhancing effect on K^+ uptake compared to the control at zero Na^+. Furthermore Jefferies[27] detected a positive influence of Na^+ on K^+ uptake even in the low concentration range. Roots of *Suaeda maritima* did not show such marked differences in their K^+ uptake properties Yeo[17]), but a special feature of this and other salt accumulating halophytes seems to be a preferential transport of Na^+ to the shoots. As pointed out by Osmond *et al.*[44] it is difficult to decide to what extent symplastic and apoplastic transport contribute to the high Na^+ concentrations in the shoots. In any case effective compartmentation of Na^+ between cytoplasm and vacuole requires unusual properties, especially of the tonoplast membranes. Investigations of strains of sugar beet[33] and *Plantago* species[11], differing in their salt tolerance, as well as *Halocnemum strobilaceum*[42] indicated marked changes in the $(Na^+ + K^+)$ ATPase activities in relation to the varying salt sensitivity of the organisms. And even the lipid composition of the membranes themselves seemed to be of importance for adaptation to salinity[11,32,33].

Although there are many factors involved in adjustment of plants to saline habitats, the respective changes for this purpose in the genome are probably not too far-reaching, because intraspecific differentiation can lead to salt tolerant ecotypes. Interestingly these ecotypes are mainly species belonging to the Poaceae or Juncaceae, families in which the halophytes are, with few exceptions, salt excluders characterized by K^+/Na^+ ratios greater than unity with low Cl^- and SO_4^{-2} uptake[4,21]. The adaptive significance of salt exclusion from the shoots becomes evident from experiments with *Agrostis stolonifera*, *Festuca rubra*[24,51] and *Juncus buffonius*[51], where the salt tolerant ecotypes always contained less salt in their shoots than the non tolerant ones.

Evidence for ecotypical differentiation in salt accumulator types is scarce. Fig. 2 shows the wide range of possible salt accumulation in the chenopod *Salicornia utahensis* in habitats of different salinity. Even under non saline conditions chenopods usually show a preponderance for Na^+ and are characterized by a high anion content due to high Cl^- uptake and storage of free oxalate. It remains to be determined whether or not the great variation in salt content seen in Fig. 2, and observed in other cases [44] is due to a wide physiological amplitude of one genotype or is provoked by intraspecific differentiation. If the latter is the case this would mean that in contrast to the ecological races of the Poaceae and

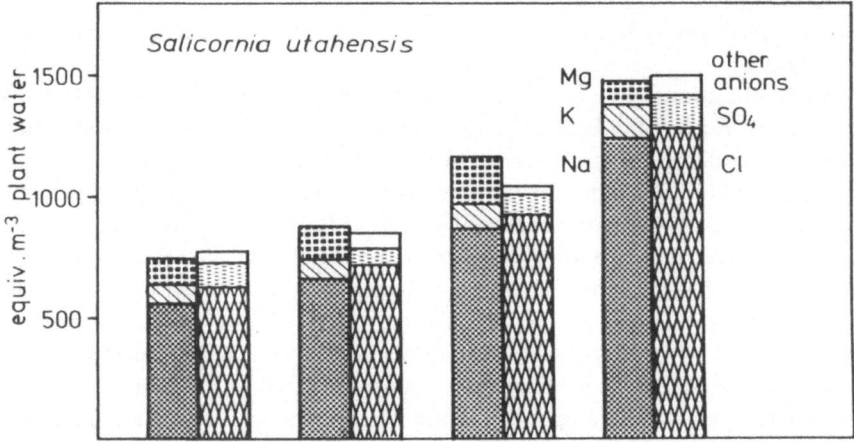

Fig. 2. Ion content (equiv. m^{-3} plant water) of the halophyte *Salicronia utahensis* (Chenopodiaceae) from several habitats throughout USA differing in soil salinity (Albert, R. unpublished results).

Juncaceae mentioned above, the more salt tolerant ecotypes of the Chenopodiaceae are the better salt accumulators.

Salt glands are of course another special feature in the mineral metabolism of halophytes and should be mentioned in the context of this brief review. They are not necessarily restricted to salt excluders, but seem to function also in maintaining the appropriate salt concentration in salt accumulators like many *Atriplex* species[44]. There is great variety in the structure and physiological properties throughout the halophytes (*cf.* ref.[38]). For example in some Cl$^-$ appears to be the actively transported ion while in others it is Na$^+$.

Metallophytes

Cu, Zn, Mn, Mo, and possibly Ni are necessary micronutrients for all plants, but when they occur in high concentrations such as in ore deposits or mining wastes, these and other heavy metals without known physiological function such as Cd, Pb, Cr become deleterious for most plants. Heavy metal tolerant plants are not adapted to all heavy metals, but only to those which occur in their natural habitats. For instance a plant from a Cu-rich soil is only Cu- and not Ni-, Zn-, or Mn-tolerant (*e.g.* different ecological races of *Agrostis tenuis*)[23]. From these and other observations it was assumed that for each heavy metal there exists a particular detoxifying mechanism in the resistant plants.

The shoots of heavy metal-tolerant plants may behave as accumulators or as excluders as was seen in halophytes. Exclusion may be achieved by reduced absorption by the roots and/or by lower translocation from the roots to the shoots[6]. In any case the basic mechanisms are not well understood. For Mn it is proposed that excluders are able to oxidise the available divalent form to the unavailable tetravalent ion[19]. Zn and Ni are found in high amounts bound to the cell walls (cf. ref[14]) and for different ecotypes of *Silene cucubalus* Ernst[13] observed a positive correlation between cation exchange capacity of the roots and the uptake of Zn and Cu.

Translocation to the shoots may be decreased either by reaction of heavy metals with phosphate to form sparingly soluble salts[12] or by increasing the Ca^{+2} support[53]. These two possibilities are not special features of heavy metal tolerant plants, but are generally of some consequence for the distribution of heavy metals between roots and shoots.

Again, as in the halophytes, even those metallophytes which accumulate heavy metals mainly in their shoots, possess no heavy metal resistant enzymes[40]. Possible mechanisms of preventing the damaging effects of heavy metals on

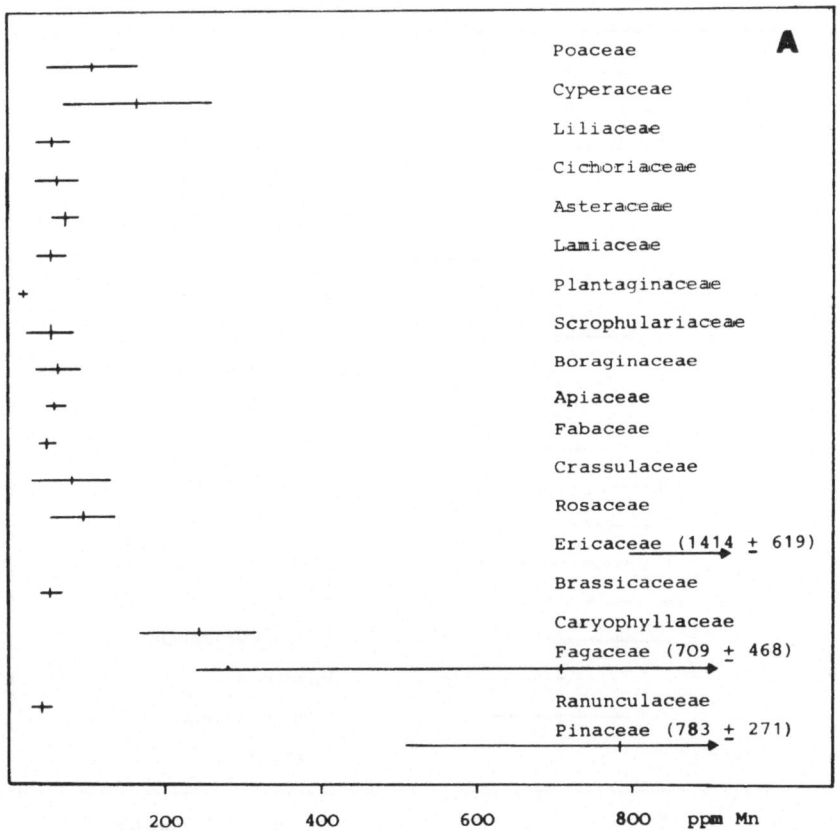

Fig. 3A The means ±95% confidence limits of Mn content (ppm dry substance) of different plant families collected from various habitats throughout Austria[43].

metabolism are i) complex formation and chelation with different organic compounds, ii) storage in the vacuole and iii) retention by the cell walls.

Organic compounds involved in detoxification are mainly organic acids (oxalate, malate, citrate), amino acids or mustard oil glucosides. Presumably members of the Caryophyllaceae, for which free oxalate content is typical, and members of the Brassicaceae, which are specialised in mustard oil glucoside production, may therefore be favoured in heavy metal habitats. In case of Zn and Ni the complexing mechanisms are highly effective, so that tolerant plants have a higher demand for these elements to reach optimal growth (*i.e.* to get the necessary amount of the micronutrient in an available form) than their sensitive counterparts.

There exists no special preference of certain plant families for particular heavy metals as was shown in an extensive study by Mutsch[43] on a great variety of plant species in many different habitats throughout Austria. Only in the case of

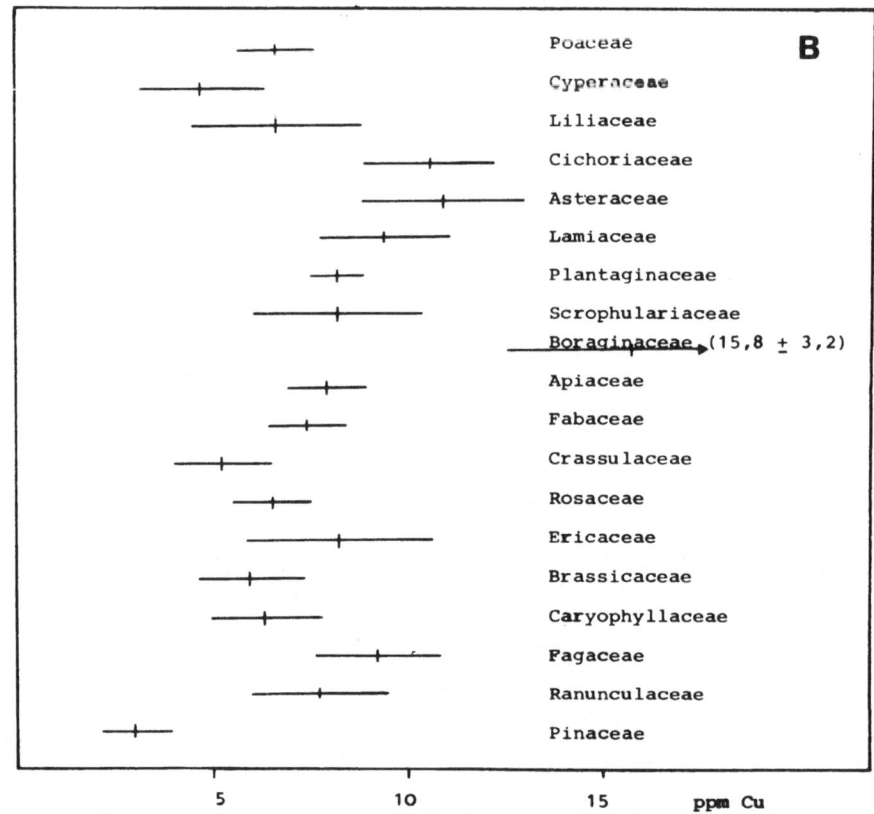

Fig. 3B. The means ±95% confidence limits of Cu content (ppm dry substance) of different plant families collected from various habitats throughout Austria[43].

Ericaceae (Fig. 3A) was there a trend for Mn accumulation. It should be pointed out that while these plants occur only in Mn-rich locations such as bogs some members such as *Vaccinium uliginosum* do not accumulate Mn beyond the range found in "normal" plants. Somewhat higher than normal levels of Mn are also found in members of Fagaceae and Pinaceae, but in this case accumulation might be a consequence of their tree-like structure.

With respect to Cu content (Fig. 3B) the Boraginaceae are slightly higher than all the other families but compared to Cu accumulators (see ref.[14]) their Cu storage is relatively low and they cannot be regarded as especially Cu tolerant.

Serpentine plants

A high Mg^{+2}/Ca^{+2} ratio, a high Ni content and rather small amounts of macro-nutrients are three main features of serpentine soils[46]. But there is a wide variation of these factors in serpentine habitats and Cr as well as Co may occur in relatively high concentrations (*cf.* ref.[31]). "Serpentine" plants therefore have not only to face heavy metal toxicity, but have also to cope with an extreme nutritional situation. For instance, in normal plants increasing amounts of Mg^{+2} reduce the uptake of K^+ (Fig. 4, *Helianthus annuus*) and Ca^{+2}. A serpentine tolerant plant such as *Helianthus bolanderi* var. *exilis* seems adapted in so far as its K^+ uptake is not influenced by Mg^{+2} levels up to $1\ M \cdot m^{-3}$ (ref.[39] and Fig. 4).

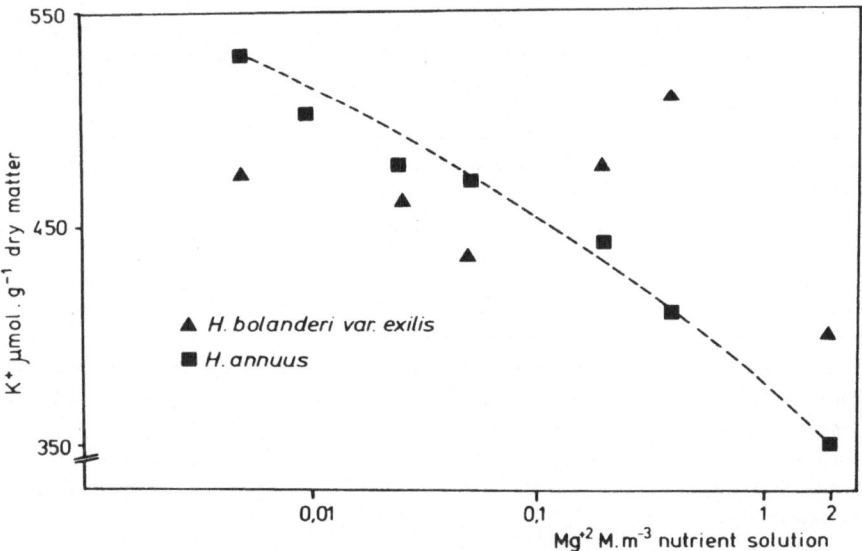

Fig. 4. K^+ content ($\mu mol \cdot g^{-1}$ dry matter) of the serpentine-tolerant *Helianthus bolanderi* var. *exilis* and the non-tolerant *Helianthus annuus* grown with various Mg^{+2} concentrations (logarithmic scale) in the nutrient solution (from Kinzel and Weber[31] after data from Madhok and Walker[39]).

This plant is also able to reduce Mg^{+2} accumulation in the shoots and this is of some consequence, because high levels of Mg^{+2} exhibit a negative effect on processes in the cytoplasm[31]. However, reduced Mg^{+2} storage in the shoots accompanied by a higher demand for Mg^{+2} to obtain optimal growth is not necessarily a common feature of all serpentine plants. As shown in Table 1 different species growing on the same serpentine soil vary widely in their K^+, Ca^{+2}, and Mg^{+2} content (heavy metals seem to be of minor importance in this habitat) and reflect their physiotypical characteristics. In *Festuca cinerea* (Poaceae) even on a serpentine soil K^+ is the dominant cation while in *Sedum album* (a member of the calciotrophic family of the Crassulaceae[28]) accumulates Ca^{+2} although soil content of this cation is very low. Because of their free oxalate content members of the Caryophyllaceae such as *Cerastium arvense* might detoxify excess levels of Mg as the sparingly soluble oxalate salt[31].

In contrast to heavy metal tolerant plants for which Antonovics[5] reported that ecological races occur in 30 different species very little is known about serpentine tolerant and non-tolerant ecotypes. Kruckeberg[47] reported such ecotypes of *Gilia capitata* and *Achillea borealis*, but the physiological consequences of the intraspecific differentiation were not revealed.

Calcicoles and calcifuges

An obvious difference between calcareous and silicate soils is of course their Ca content. But other soil factors such as HCO_3^- concentration, free Al ion content, Fe-availability, and pH also vary and are of considerably consequence for plant growth. As pointed out by Kinzel[30] for some plants it would be more correct to

Table 1. K^+, Na^+, Ca^{+2}, Mg^{+2} in $\mu mol \cdot g^{-1}$ dry matter, the molar K^+/Ca^{+2} ratio, and content of heavy metals in ppm dry matter from the shoots of plants from a serpentine soil (Gurhofgraben, Austria). Data for cations are from Horak[26], data for heavy metals are from Mutsch[43]

Plant species	K^+/Ca^{+2}	Na^+	K^+	Ca^{+2}	Mg^{+2}	Fe	Mn	Zn	Cu	Mo	Co	Ni	Cr
Cerastium arvense	307.0	4.3	799	2.6	325	226	283	74	4.9	0.04	2.10	9.4	0.94
Euphoriba cyparissias	6.3	1.7	276	44	344	60	35	41	5.3	0.05	2.37	33.9	0.21
Thlaspi montanum	9.1	3.1	453	50	404								
Biscutella laevigata	4.1	4.4	506	124	775	108	29	25	9.6	0.04	0.58	18.0	0.36
Sedum album	0.5	5.6	156	298	560	79	37	51	7.3	0.08	0.62	16.8	0.23
Potentilla arenaria	47.7	1.5	572	12	210								
Genista pilosa	2.8	1.7	59	21	126								
Dorycnium pentaphyllum agg.	6.0	2.4	199	33	300	88	21	17	5.2	0.05	0.14	21.3	0.23
Festuca cinerea	20.9	1.8	376	18	50	69	28	29	5.3	0.07	0.04	13.1	0.21

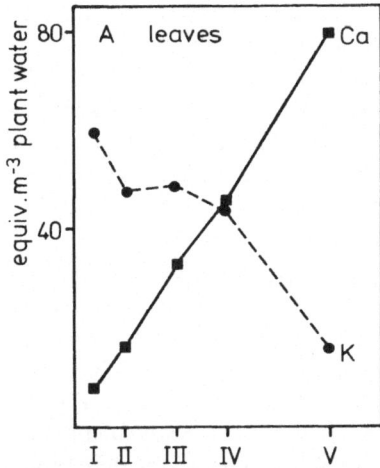

Fig 5 A K^+ and Ca^{+2} content (equiv. m^{-3} plant water) of leaves of *Kalanchoe daigremontiana* grown in nutrient solutions with different K^+/Ca^{+2} ratios. I 2.8/0.2; II 2.5/0.5; III 2.0/1.0; IV 1.5/1.5; V 0.5/2.5 K^+/Ca^{+2} equiv. m^{-3} nutrient solution. (Rößner, upublished results).

classify them as acidophilic or basiophilic, because their response to Ca^{+2} is not very pronounced.

Plants may vary appreciably in their Ca^{+2} content and Iljin[28] distinguished between calciotrophs which are rich in soluble Ca^{+2}, and calciphobes, the pressure sap of which contains almost no free Ca^{+2}. These terms for the

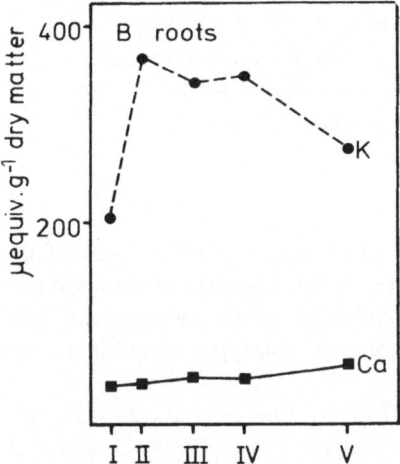

Fig 5 B K^+ and Ca^{+2} content (equiv. g^{-1} dry matter) of roots of *Kalanchoe daigremontiana* grown in nutrient solutions with different K^+/Ca^{+2} ratios. K^+/Ca^{+2} ratios are given in 5A. (Rößner, unpublished results).

physiological features concerning Ca content should not be confused with the specifications for the ecological distribution (calcicole, calcifuge). A calciotrophic plant need not necessarily be a calcicole and *vice versa*[3,28,30].

The low Ca^{+2} content found in physiology calciophobic plants like Chenopodiaceae, Caryophyllaceae, Polygonaceae and others depends on their storage of free oxalate. During extraction of the plant material oxalate from the vacuole mixes with Ca^{+2} from all other cell compartments and forms insoluble Ca-oxalate. Thus it is difficult to assess what the usual Ca^{+2} distribution *in vivo* is in these plants.

For a calciotriophic plant such as *Kalanchoe daigremontiana* (Crassulaceae) Rößner (in prep.) showed in a recent study that its preference for Ca^{+2} is restricted to the leaves, while the roots and the xylem sap contain more K^+ than Ca^{+2} even in nutrient solutions, where the K^+/Ca^{+2} ratio is below unity (Fig. 5). From this and other experiments it seems likely that *K. daigremontiana* is not only a better Ca^{+2} accumulator, but also a better K^+ retranslocator than other plants. A similar situation was observed in the calciotrophic members of the Brassicaceae, where the K/Ca^{+2} ratio of the leaves was below unity while the stems was markedly above unity[49]. As mentioned above these physiological features are not of very much consequence in determining the ecological distribution in relation to Ca^{+2}.

Possibly it is not the high absolute Ca^{+2} content of calcareous soils, but the very low K^+/Ca^{+2} ratio, which is unfavourable for certain plants. Silicate soils are not higher in K^+, but markedly lower in Ca^{+2} (ref.[20]) and the changed K^+/Ca^{+2} ratio seems to be of importance. Snaydon and Bradshaw[52] investigated ecotypes of *Festuca ovina* from habitats differing in their Ca^{+2} content. In the ecotype from the soil poor in Ca^{+2} K^+ uptake was reduced by amounts of Ca^{+2} (100 ppm), which had no effect in plants from a Ca^{+2}-rich habitat.

Although the calcicole–calcifuge problem should be not reduced to the Ca^{+2}-factor, it would lead too far to consider all the other soil conditions in the context of this paper (for further information see refs.[30,50]).

Concluding remarks

Since the review by Epstein and Jefferies[10] knowledge on genetic control of ion transport has increased significantly. While the varietal changes of ion uptake in cultivated plants are of practical interest, studies on edaphic ecotypes of wild plants are also helpful for our understanding of genotypic variation in ion transport[35].

Furthermore investigations of wild plants from extreme habitats illustrate also the great variability of mineral metabolism in higher plants. It may be assumed that some families are favoured to live on harsh soil conditions (*i.e.* Chenopodiaceae, Poaceae in saline habitats, Brassicaceae, Caryophyllaceae on heavy metal soils) because of their basic physiological constitution.

Acknowledgements The author is thankful to Doz. R. Albert and Prof. H. Kinzel for fruitful discussions. Special thanks are due to Dr. C. A. Atkins for reading the manuscript.

References

1 Albert R 1982 Halophyten. *In* Pflanzenökologie und Mineralstoffwechsel. Ed H Kinzel. Verlag Eugen Ulmer, Stuttgart.
2 Albert R and Kinzel H 1973 Unterscheidung von Physioltypen bei Halophyten des Neusiedlerseegebietes (Österreich). Z. Pflanzenphysiol. 70, 138–157.
3 Albert R, Königshofer H and Kinzel H 1980 Zur Osmoregulation einer physiologisch calciophoben und ökologish calcicolen Pflanze (*Dianthus lumnitzeri* WIESB.). Flora 169, 9–14.
4 Albert R and Popp M 1977 Chemical composition of halophytes from the Neusiedler lake region in Austria. Oecologia (Berl). 27, 157–170.
5 Antonovics J, Bradshaw A D and Turner R G 1971 Heavy metal tolerance in plants. Adv. Ecol. Res. 7, 1–85.
6 Baker A J M 1978 Ecophysiological aspects of zinc tolerance in *Silene mnaritime* With. New Phytol. 80, 635–642.
7 Beigl E 1981 Verteilung der Alkaliionen zwischen Cytoplasma und Vakuole in den Zellen höherer Pflanzen. Eine neuartige Untersuchsmethode. Phil. Diss. Universität Wien.
8 Cataldo D A, Gerland T R and Wildung R E 1978 Nickel in plants. Plant Physiol. 62, 563–570.
9 Epstein E 1972 Mineral nutrition of plants. Principles and perspectives. John Wiley and Sons, Inc. New York, London.
10 Epstein E and Jefferies R L 1964 The genetic basis of selective ion transport in plants. Annu. Rev. Plant Physiol. 15, 169–184.
11 Erdei L, Stuiver B and Kuiper P J C 1980 The effect of salinity on lipid composition and on activity of Ca^{+2} and Mg^{+2} stimulated ATPases in salt-sensitive and salt-tolerant *Plantago* species. Physiol. Plant. 49, 315–319.
12 Ernst W 1968 Der Einfluß der Phosphatversorgung sowie die Wirkung von ionogenem und chelatisiertum Zink auf die Zink und Phosphataufnahme einiger Schwermetallpflanzen. Physiol. Plant. 21, 323–333.
13 Ernst W 1972 Schwermetallresistenz und Mineralstoffhaushalt. Forschungsber. Land. Nordrhein. Westfalen. 2251, 1–38.
14 Ernst W 1982 Schwermetallpflanzen. *In* Pflanzenökologie und Mineralstoffwechsel. Ed. H Kinzel. Verlag Eugen Ulmer, Stuttgart.
15 Flowers T J 1972 Salt tolerance in *Suaede maritima* L. (Dum). The effect of sodium chloride on growth, respiration, and soluble enzymes in a comparative study with Pisum. J. Exp. Bot. 23, 310–321.
16 Flowers T J 1972 The effect of sodium chloride on enzyme activities from four halophyte species of Chenopodiaceae. Phytochem. 11, 1881–1886.
17 Flowers T J 1975 Halophytes. *In* Ion transport in plant cells and tissues. D A Baker and J L Hall. pp. 309–334. North Holland Amsterdam.
18 Flowers T J, Troke P F and Yeo A R 1977 The mechanism of salt tolerance in halophytes. Annu. Rev. Plant Physiol. 28, 89–121.
19 Foy C D, Chaney R L and White M C 1978 The physiology of metal toxicity in plants. Annu. Rev. Plant Physiol. 29, 511–566.
20 Gigon A 1971 Vergleich alpiner Rasen auf Silikat und Karbonatboden. Konkurrenz- und

Stickstoffformenversuche sowie standortskundliche Untersuchungen im Nardetum und im Seslerietum bei Davos. Phil. Diss. ETH Zürich. Veröff. Geobot. Institut. ETH, Stiftung Rübel.

21 Gorham J, Hughes L L and Wyn Jones R G 1980 Chemical composition of salt-marsh plants from Ynys-Mon (Anglesey) – the concept of physiotypes. Plant, Cell Environm. 3, 309–319.

22 Greenway H and Osmond C B 1972 Salt response of enzymes from species differing in salt tolerance. Plant Physiol. 49, 256–259.

23 Gregory R P G and Bradshaw A D 1965 Heavy metal tolerance in populations of *Agrostis tenuis* Sibth. and other grasses. New Phytol. 64, 131–143.

24 Hannon N J and Barber H N 1972 The mechanism of salt tolerance in naturally selected populations of grasses. Search. 3, 259–260.

25 Harvey D M R, Hall J L, Flowers T J and Kent B 1981 Quantitative ion localization within *Suaeda maritima* leaf mesophyll cells. Planta. 151, 555–560.

26 Horak O 1971 Vergleichende Untersuchungen zum Mineralstoffwechsel der Pflanzen. Diss. Univ. Wien, Band 60. Verlag Notring, Wien.

27 Jefferies R L 1973 The ionic relations of seedlings of the halophyte *Triglochin maritima* L. *In*: Ion Transport by Plants. Ed. W. P. Anderson. pp 297–321. Academic Press, London, New York

28 Kinzel, H 1963 Zellsaft-Analysen zum pflanzlichen Calcium- und Säurestoffwechsel und zum Problem der Kalk- und Silikatpflanzen. Protoplasma. 57, 522–555.

29 Kinzel H 1972 Biochemische Pflanzenökologie. Schriften d. Ver. z. Verbr. Naturwiss. Kenntn. in Wien. 112, 77–98.

30 Kinzel H 1982 Die calcicolen und calcifugen, basiphilen und acidophilen Pflanzen. *In*: Pflanzenökologie und Mineralstoffwechsel. Ed. H Kinzel. Verlag Eugen Ulmer, Stuttgart.

31 Kinzel H and Weber M 1982 Serpentine–Pflanzen. *In*: Pflanzenökologie und Mineralstoffwechsel. Ed. H Kinzel. Verlag Eugen Ulmer, Stuttgart.

32 Kuiper P J C 1968 Lipids in grape roots in relation to chloride transport. Plant Physiol. 43, 1367–1371.

33 Kylin A 1973 Adenosine triphosphatases stimulated by (sodium + potassium); biochemistry and possible significance for salt resistance. *In*: Ion transport in Plants. Ed. W P Anderson. Academic Press, London, New York.

34 Larson H 1967 Biochemical aspects of extreme halophilism. Adv. Microbiol. Physiol. 1, 97–132.

35 Läuchli, A 1976 Genotypic variation in transport. *In*: Encyclopedia of Plant Physiology. New Series, vol. 2, part B. Eds. U Lüttge and M G Pitman. Springer Verlag, Berlin, Heidelberg, New York.

36 Lee J, Reeves R D, Brooks R R and Jafffe T 1977 Isolation and identification of a citrato-complex of nickel from nickel accumulating plants. Phytochem. 16, 1503–1505.

37 Levitt J 1980 Responses of plants to environmental Stress. Academic Press, New York, London.

38 Lüttge U 1975 Salt glands. *In*: Ion Transport in Plant Cells and Tissues. Eds. D A Baker and J L Hall. North Holland Publishing Company.

39 Madhok O P and Walker R B 1969 Magnesium nutrition of two species of sunflower. Plant Physiol. 44, 1016–1022.

40 Mathys W 1975 Enzymes of heavy-metal-resistant and non-resistant populations of *Silene cucubalus* and their interaction with some heavy metals *in vitro* and *in vivo*. Physiol. Plant. 33, 161–165.

41 Mathys W 1977 The role of malate, oxalate, and mustard oil glucosides in the evolution of zinc-resistance in herbage plants. Physiol. Plant. 40, 130–136.

42 Mishustina N E, Tikhaya N I and Chaplygina N S 1979 ($Na^+ + K^+$) ATPase activity in membranes isolated from shoots of the halophyte *Halocnemum strobilaceum*. Fiziologiya Rastenii. 26, 541–547.
43 Mutsch F 1981 Schwermetallanalysen an Freilandpflanzen im Hinblick auf die natürliche Spurenelementversorgung und die Schwermetallintoxikation. Phil. Diss. Universität Wien.
44 Osmond C B, Björkman O and Anderson D J 1980 Physiological processes in plant ecology. Towards a synthesis with Atriplex. Springer Verlag, Berlin, Heidelberg, New York. Ecological Studies 36.
45 Priebe A and Jäger H J 1978 Responses of amino acid enzymes from plants differing in salt tolerance to NaCl. Oecologia (Berl.) 36, 307–315.
46 Proctor J, Johnston W R, Cottam D A and Wilson A B 1981 Field-capacity water extracts from serpentine soils. Nature, London 294, 245–246.
47 Proctor J and Woodell S R J 1975 The ecology of serpentine soils. Adv. Ecol. Res. 9, 255–366.
48 Rains D W and Epstein E 1967 Preferential absorption of potassium by leaf tissues of the mangrove *Avicennia marina*: an aspect of halophytic competence in coping with salt. Aust. J. Biol. Sci. 20, 847–857.
49 Rattenböck H 1978 Chemisch-physiologische Charakterisierung der Brassicaceae. Ein Beitrag zum Physiotypenkonzept. Phil. Diss. Universität Wien.
50 Rorison I H (ed) 1969 Ecological aspects of the mineral nutrition of plants. Blackwell Scientific Publications, Oxford, Edinburgh.
51 Rozema J 1978 On the ecology of some halophytes from a beach plain in the Netherlands. Ph. Th. Freie Universität, Amsterdam.
52 Snaydon R W and Bradshaw A D 1961 Differential response to calcium within the species *Festuca ovina* L. New Phytol. 60, 219–234.
53 Wainwright S J and Woolhouse H 1977 Some physiological aspects of copper and zinc tolerance in *Agrostis tenuis* Sibth: Cell elongation and membrane damage. J. Exp. Bot. 28, 1029–1036.
54 Weimberg R 1970 Enzyme levels in pea seedlings grown on highly saline media. Plant Physiol. 46, 466–470.
55 Yeo A R 1981 Salt tolerance in the halophyte *Suaeda maritima* L. Dum Intracellular compartmentation of irons. J. Exp. Bot. 32, 487–497.

NaCl—induced modifications of nitrogen absorption and assimilation in salt tolerant and salt resistant millet ecotypes (*Pennisetum typhoideum* L. Rich.)

A. BOTTACIN, M. SACCOMANI, P. SPETTOLI and G. CACCO
Institute of Agricultural Chemistry, Via Gradenigo 6, 35100 Padua, Italy

Key words Ecotype Enzyme activity Millet NaCl Nitrogen assimilation

Introduction

It has been estimated that one-third of the irrigated areas of the world is affected by salinity and increasing salts in the soil represent the major limitation to crop yield in many arid countries[5]. Although salinity is not incompatible with plant life, only a few crops are able to grow under high level of salts accumulated by fertilization practices or by use of saline water for irrigation. Since the salt-free water available for irrigation purpose is difficult and expensive to obtain, a promising approach to overcome the problem of salinity may be the genetic improvement of salt tolerance in crops by: 1) selection among agronomic cultivars; 2) hybridization and ploidy changes; 3) use of wild germplasm sources to develop new cultivated species. Promising studies have been undertaken with barley, tomato, wheat and wheatgrass[4,10,16,17]. Salt stress in plants can be accounted for an increase in sodium, chloride or both ion contents[9] and a decrease of the internal nutrient availability[2,6]. The consequent changes in plant metabolism are mainly on enzyme systems which can be repressed or inhibited, induced or activated[7,8,14,19]. Comparative studies on varieties of crop plants differing in salt tolerance appear useful. Millet (*Pennisetum typhoideum* L. Rich.) is one of the widely cultivated crops in African and Asian regions.[2]

The wild genotypes, growing in semi-arid and saline lands, are characterized by salt and drought resistance. In the present work we compared two millet ecotypes adapted to stress conditions in comparison with two varieties usually cultivated and non-adapted to stress. The effects of salt stress were evaluated on the nitrogen assimilatory pathway and particularly on the absorption of nitrate and ammonium and on the GDH/GS ratio.

Materials and methods

The four Tunisian ecotypes of millet (*Pennisetum typhoideum* L. Rich.) used in the present study, were obtained from the germ plasm bank of C.N.R., Bari, Italy; n.90.919 resistant to both drought and salinity (DSR); n.90.923 resistent to drought; n.90.915 (CV1) and n.90.916 (CV2) usually cultivated and non-adapted to stress. Millet seeds were germinated in quartz sand and moistened with Nitsch[13] nutrient solution. NaCl was supplied to plants at concentrations in accordance with the experimental schedule. Nitrogen uptake was evaluated by depletion; nitrate was tested according to McNamara et al.[11] and ammonium to Weatherburn[18]. Enzyme activities were assayed according to Neyra and Hageman[12] for nitrate reductase (E.C.1.6.6.1; NR), Bernt and Bergmeyer for glutamine dehydrogenase (E.C.1.4.1.3; GDH) and Rhodes et al.[15] for glutamine synthetase (E.C.6.3.1.2; GS).

Results and discussion

CV1 grew faster than DSR but after 5 weeks under 300 mM NaCl, the first failed, while the second exhibited growth one-half of the control (Fig. 1). At higher NaCl concentration the growth of DSR and DR ecotypes (g fresh weight/plant) was higher than that of CV1 and CV2, demonstrating their better salt resistance (Fig. 2). In all ecotypes nitrogen absorption was negatively affected by NaCl but the uptake inhibition was lower for ammonium than for nitrate (Table 1).

In all four ecotypes NaCl caused a decrease in NR level (Table 2) due, probably, either to salt toxicity[14], or to slower induction as a consequence of reduced NO_3^- absorption[12]. It has been demonstrated that salinity can modify the GDH/GS ratio both in halophyte and glycophyte shoots[3]: GS is induced in halophytes and repressed in glycophytes while GDH shows an opposite

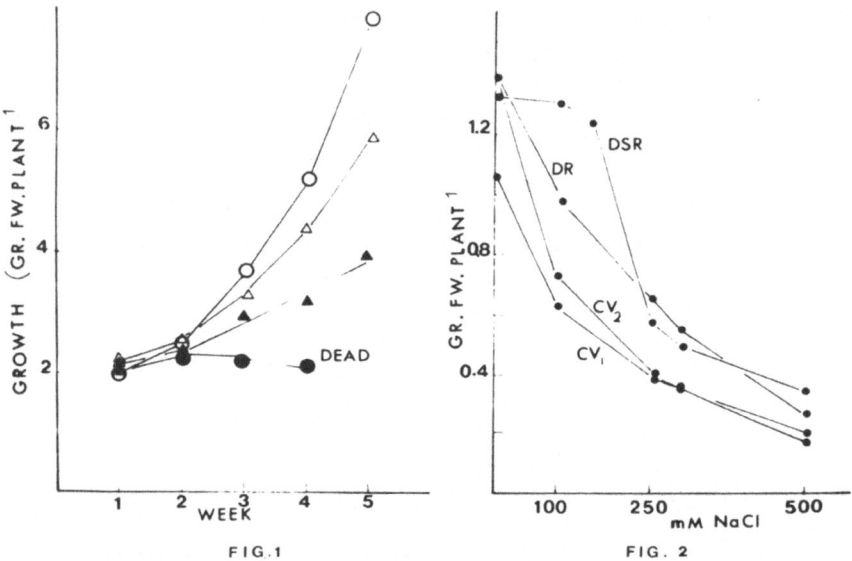

Fig. 1. Growth (g fresh weight/plant) of two millet ecotypes as function of time. DSR growth at O (control) (△) and 300 mM NaCl (▲). CV1 growth at O (control) (○) and 300 mM NaCl (●).

Fig. 2. Pattern of salt resistance (g fresh weight/plant) in four ecotypes of millet as function of increasing NaCl concentration.

Table 1. NO_3^- and NH_4^+ absorption by roots of four millet ecotypes grown at 0 (control) and 300 mM NaCl. Ion levels were assayed on 22-days old plants. Salt treatment was carried out for 7 days. Results are expressed in nmol. $\times$ min.$^{-1}$ $\times$ g d.m.$^{-1}$ Data reported are the average of three experiments $\pm$ standard error

	DSR	DR	CV1	CV2
NO_3^-				
control	91.0 ± 12.91	101.1 ± 11.63	74.7 ± 9.30	75.9 ± 10.29
300 mM	38.6 ± 6.06	65.6 ± 12.22	34.9 ± 3.35	43.7 ± 5.07
NH_4^+				
control	14.8 ± 0.99	16.9 ± 1.90	12.4 ± 2.13	12.3 ± 1.97
300 mM	12.1 ± 0.90	16.2 ± 1.70	14.3 ± 2.19	9.1 ± 1.25

behaviour[3]. Our experiments (Table 3) on millet plants demonstrate that GDH activity is stimulated by salinity as in glycophytes. On the contrary, as regards GS, DSR behaves like halophytes and DR, CV1 and CV2 ecotypes like glycophytes. This means that the GDH/GS ratio slightly decreased in DSR and increased in the other three ecotypes. The higher GDH/GS ratio observed in the three ecotypes, as a consequence of salinity, indicated that nitrogen was

Table 2. Specific activities of NR, GDH and GS in shoots of four millet ecotypes grown at 0 (control) and 300 mM NaCl. Enzymes were assayed on 22-days old plants. Salt treatment was carried out for 7 days. Results are expressed in nmol. $\times$ min.$^{-1}$ $\times$ mg prot.$^{-1}$. Data reported are the average of three experiments $\pm$ standard error

	DSR	DR	CV1	CV2
NR				
control	29.2 ± 4.38	19.0 ± 2.47	34.1 ± 4.6	23.5 ± 2.49
300 mM	3.7 ± 0.59	6.9 ± 1.24	3.9 ± 0.43	10.3 ± 1.95
GDH				
control	15.1 ± 1.34	13.6 ± 1.18	15.0 ± 1.48	12.3 ± 1.15
300 mM	17.7 ± 1.33	16.7 ± 1.23	13.4 ± 1.41	21.2 ± 2.50
GS				
control	43.9 ± 5.26	66.3 ± 6.46	56.8 ± 6.27	50.2 ± 8.15
300 mM	56.6 ± 3.87	43.1 ± 6.03	35.2 ± 3.08	28.4 ± 2.70
GDH/GS ratio				
control	0.34	0.20	0.26	0.24
300 mM	0.31	0.38	0.38	0.74

preferentially assimilated by GDH route, as in glycophytes. In the DSR ecotype the GS-GOGAT pathway represents the primary route for ammonium assimilation as in halophyte species. This could be an aspect of the adaptation to saline environmental conditions.

References

1. Bern E and Bergmeyer H U 1963 Methods of enzymatic analysis. Bergmeyer H U Ed, London, Academic Press.
2. Black R F 1956 Effect of sodium chloride in water culture on the ion uptake and growth of *Atriplex hastata*. Aust. J. Biol. Sci. 9, 20.
3. Boucaud J and Billard J P 1979 Etude comparée des activités glutamate deshydrogenasique et de glutamine synthetasique dans les racines et les parties aeriennes d'un alophyte obligatoire: le *Suaeda maritima* var macrocarpa et d'un glycophyte: le *Phaseolus vulgaris*, cultives en presence de differentes concentration en NaCl. C. R. Acad. S. C. Paris 289, 599–602.
4. Epstein E and Norlyn J D 1977 Toward sea water-based crop production. Science 197, 249–251.
5. Epstein E, Rush D W, Kingsbury R W, Kelley D B, Cunningham G A and Wrona A F 1980 Saline culture of crops: a genetic approach. Science 210, 399–404.
6. Greenway H 1962 Plant response to saline substrates. 1. Growth and iron uptake of several varieties of *Hordeum* during and after sodium chloride treatment. Aust. J. Biol. Sci. 15, 16–28.
7. Greenway H and Osmond C B 1972 Salt responses of enzymes from species differing in salt tolerance. Plant Physiol 49: 256–259.
8. Hason-Porath E and Poljakoff-Mayber A 1969 The effect of salinity on the malic dehydrogenase of pea roots. Plant Physiol. 44, 1031–1034.
9. Läuchli A and Wieneke J 1979 Studies on growth and distribution of Na^+, K^+ and Cl^- in soy-bean differing in salt tolerance. Pflanzenernachz. Bodenkd. 142, 3–13.
10. McGuire P E and Dvorak J 1981 High salt tolerance potential in wheatgrasses. Crop Sci. 21, 702–705.
11. McNamara A, Meeker L, Shaw G B and Hageman R H 1971 Use of a dissimilatory nitrate reductase from *Escherichia coli* and formate as reductive system for nitrate assays. J. Agr. Food Chem. 19, 229–231.
12. Neyra C A and Hageman R H 1975 Nitrate uptake and induction of nitrate reductase in excised corn roots. Plant Physiol. 56, 692–695.
13. Nitsch J P 1967 Enciclopedie de la Pleiade. Les phytotrons.
14. Rakova N M, Klyshev L K and Kasymbekov B K 1979 Effect of sodium salts on activity of nitrogen metabolism enzymes isolated from plants with different salt resistance. Fiziol. Rast. 25, 751–755.
15. Rhodes D, Rendon G A and Stewart G R 1975 The control of glutamine synthetase level in *Lemna minor* L. Planta, Berlin 125, 201–211.
16. Rush D W and Epstein E 1976 Genotypic responses to salinity: differences between salt sensitive and salt tolerant genotypes of the tomato. Plant Physiol 57, 162–166.
17. Tal M and Gardi I 1976 Physiology of polyploid plants: water balance in autotetraploid and diploid tomato under low and high salinity. Physiol. Plant. 38, 257–261.

18 Weatherburn M W 1967 Phenol-hypochlorite reaction for determination in ammonia. Anal. Chem. 39, 971–974.
19 Wyn Jones R G 1981 Salt tolerance. *In*: Physiological processes limiting plant productivity. (C. B. Johnson), Butterworths, Ed., p. 271–292. London, Boston, Sydney, Wellington, Durban, Toronto.

Soil preference of populations of genotypes of *Asplenium trichomanes* L. and *Polypodium vulgare* L. in Belgium as related to cation exchange capacity

P. BÜSCHER and N. KOEDAM
Dienst Plantenfysiologie, Vrije Universiteit Brussel, Paardenstraat 65–67, B-1640 St Genesius Rode, Belgium and Dienst Algemene Plantkunde en Natuurbeheer, Vrije Universiteit Brussel, Pleinlaan 2, B-1050 Brussel, Belgium

Key words Aluminium *Asplenium trichomanes* CaCO$_3$ Calcicole Calcifuge C.E.C. Fern Hydrogen pH *Polypodium vulgare* Soil preference Subspecies

Abstract A number of populations of *Polypodium vulgare* L. and *Asplenium trichomanes* L. were sampled with corresponding soils in Belgium in order to get an idea of their suitability for the investigation of related calcicole and calcifuge taxa. Morphology and cytology enabled us to distinguish the subspecies and the hybrids. Analyses of the soils for pH, CaCO$_3$, Al and H show that subspecies and hybrids have distinct soil preferences and can be characterised as calcicoles-neutrophile/basiphiles or calcifuge-acidiphiles. Physiological implications of the ecological status of the taxa are discussed in the light of their root cation exchange capacity.

Introduction

The major problem arising in the investigation of soil preference of plant species in their natural environment is the distinction between the responses to many interfering factors, at the ecological level (soil characteristics, humidity, light, temperature) as well as on the level of the plant itself (rooting depth, root mass, phenology). One way to cope with this problem is to choose the taxa so that their natural environment differs mainly in one soil characteristic or a group of related characteristics.

The purpose of this work was to determine whether the subspecies of *Asplenium trichomanes* L. and of *Polypodium vulgare* L., occurring in Belgium, represent a suitable tool for the investigation of the calcicole – neutrophile/basiphile and calcifuge – acidiphile habitat.

The subspecies and hybrids, listed in Table 1, may be found in Belgium. The nomenclature is according to De Langhe et al.[7]

This supposed soil preference is mainly based on vague accounts of the geology of the substrate rock. Acidification – decalcification processes may nevertheless alter the small scale root environment considerably and thus it seemed useful to determine both carbonate content and acidity of the substrate of a number of

The authors are granted a scholarship "aspirant-vorser" of the Belgian "Nationaal Fonds voor Wetenschappelijk Onderzoek".

M.R. Sarić and B.C. Loughman (eds.), *Genetic Aspects of Plant Nutrition*.
ISBN-13: 978-94-009-6838-7
©1983 *Martinus Nijhoff/Dr W. Junk Publishers, The Hague/Boston/Lancaster.*

Table 1. Subspecies and respective hybrids of *Asplenium trichomanes* and of *Polypodium vulgare*, occurring in Belgium

Supposed calcifuge – acidiphile	Supposed calcicole – basiphile/ neutrophile or indifferent
Asplenium trichomanes L. ssp. *trichomanes* (2n = 72)	*Asplenium trichomanes* L. ssp. *quadrivalens* D.E.Mey (2n = 144)
Polypodium vulgare L. ssp. *vulgare* (2n = 148)	*Polypodium vulgare* L. ssp. *prionodes* (Aschers)Rothm. (2n = 222)

hybrids: *Asplenium trichomanes* L. ssp. × *lusaticum* (D.E.Mey) Lawalrée; *Polypodium vulgare* L. ssp. × *mantoniae* (Rothm.) Schidlay

populations. In addition the exchangeable aluminium and hydrogen fractions have been determined, since the availability of these elements in soils is closely related to its acidity.

So far, numerous adaptive strategies of plants to acid or alkaline soil conditions have been proposed, cation exchange capacity (C.E.C.) being one of them, and we have examined the possible role of this factor.

Materials and methods

Plants from 21 populations of Asplenium and from 26 populations of Polypodium have been collected in February 1982, in Belgium, in the lower Ardennes and in the Meuse Valley. The substrate of each plant was taken as corresponding soil sample.

Plants were grown in the greenhouse in their native soils and irrigated with rainwater. Part of the soil was air-dried and sieved (ϕ 2 mm mesh) for analysis.

Subspecies were determined by the following features:
— cytology for Asplenium (chromosome numeration on acetocarmin squashes of sporangia);
— morphology for Polypodium (size of spores, number of indurated annulus cells, presence of connecting tissue between the leaf segments).

Hybrids were confirmed by an additional search for abortive sporangia and spores. The small number of *A. trichomanes* ssp. *trichomanes* sampled is probably the result of a sampling procedure, where more vigorous hybrids (due to heterosis effects) were favoured over the smaller diploid parent.

pH_{H2O} and pH_{CaCl2} (actual and potential acidity) were determined electrometrically in a H_2O (1:2.5 = soil:water ratio) or in a 0.02 N $CaCl_2$ (1:5 = soil:$CaCl_2$ ratio) suspension respectively, after 1 hour equilibration.

Carbonate was determined with a volumetric calcimeter. Since a slow increase of the gas volume has only been observed in one case, the carbonate in the studied soils may be assigned to calcium carbonate.

Exchangeable aluminium and hydrogen were determined titrimetrically in a 1 N KCl percolate after differentiation of the titration curves[12].

C.E.C. was determined by a modified Wacquant-method[15] for oven-dried (105°C) leaf or root material (newly unrolled leaf buds or newly developed roots).

Results

Soil factors

The subspecies of Polypodium have a distinct soil preference: *P. vulgare* ssp. *vulgare* behaves like a calcifuge – acidiphile, *P. vulgare* ssp. *prionodes* like a facultative calcicole – neutrophile/basiphile. The hybrid *P. vulgare* ssp. × *mantoniae* follows its calcifuge – acidiphile parent. The sampled soils of *A. trichomanes* ssp. *trichomanes* show a rather wide pH range, but they never contain detectable amounts of calcium carbonate and this subspecies appears to have, in consequence, a calcifuge tendency. *A. trichomanes* ssp. *quadrivalens* behaves like a facultative calcicole – neutrophile/basiphile. The hybrid *A. trichomanes* ssp. × *lusaticum* was sampled in acid, calcium carbonate-free soils, with one exception (Table 2).

These data confirm the soil preference of the subspecies and their hybrids as generally accepted, both for Polypodium and Asplenium (for Belgium: refs.[3,4]). According to Bouharmont[4] *A. trichomanes* ssp. *quadrivalens* is, however, known to colonize acid substrates as well.

The same distribution pattern of soil preference may be found when examining the results of exchangeable aluminium and, to a lesser extent, hydrogen in the soil. The facultative calcicoles – neutrophile/basiphiles were never found in soils containing detectable aluminium or hydrogen, while the calcifuges – acidiphiles tolerate sometimes large amounts of aluminium and hydrogen.

Cation exchange capacity

Root C.E.C. is generally lower in *P. vulgare* ssp. *vulgare* than in subsp. *prionodes*, *P. vulgare* ssp. × *mantoniae* being scattered also in the lower range. Root C.E.C. in *A. trichomanes* ssp. *trichomanes* is lower on average than in ssp. *quadrivalens*, yet the latter attains the lowest values. *A. trichomanes* ssp. × *lusaticum* covers the range of the diploid parent (Table 2).

Leaf C.E.C. on the contrary, allowed no differentiation between the subspecies (results not represented) and bears no relationship either to root C.E.C., or to soil preference. The relationship existing between soil preference and C.E.C. of whole plants, as established for certain moss species[9] is clearly relevant.

Discussion

Subspecies of Asplenium and Polypodium colonize different habitats with respect to their soil carbonate content and acidity, which is reflected in the analytical data listed above. The physiological properties, from which these tendencies originate, leave obviously enough room for a rather wide adaptability, since mixed subspecies stands were reported for both species[4,6]. Taking these general observations into consideration it seems highly probable that the ecological optimum of the subspecies results from a quantitatively differing

Table 2. Results of plant and soil analyses: minimum – mean – maximum(number of populations)

	Asplenium trichomanes	*quadrivalens*	× *lusaticum*
pH_{H_2O} soil	5.17	6.35	4.68
	6.133	7.298	5.752
	7.11 (3)	7.95 (10)	7.69 (8)
pH_{CaCl_2} soil	4.68	6.10	4.52
	5.830	7.027	5.454
	6.91 (3)	7.67 (10)	7.46 (8)
$CaCO_3$ (% dry soil)	0.00	0.00	0.00
	0.000	16.100	4.087
	0.00 (3)	61.90 (10)	32.70 (8)
Exchangeable aluminium	0.00	0.00	0.00
(meq/100 g dry soil)	0.301	0.000	0.265
	0.90 (3)	0.00 (10)	0.77 (8)
Exchangeable hydrogen	0.00	0.00	0.00
(meq/100 g dry soil)	0.143	0.000	0.020
	0.43 (3)	0.00 (10)	0.16 (8)
C.E.C. root	21.71	19.71	15.23
(meq/100 g dry matter)	24.583	27.554	20.466
	28.50 (3)	42.15 (9)	28.45 (8)
	Polypodium vulgare	*prionodes*	× *mantoniae*
pH_{H_2O} soil	3.49	6.02	3.57
	4.720	6.863	4.833
	5.98 (10)	7.32 (12)	5.96 (4)
pH_{CaCl_2}	3.02	5.68	3.23
	4.345	6.612	4.487
	5.60 (10)	7.06 (12)	5.73 (4)
$CaCO_3$ (% dry soil)	0.00	0.00	0.00
	0.000	6.742	0.000
	0.00 (10)	32.20 (12)	0.00 (4)
Exchangeable aluminium	0.03	0.00	0.57
(meq/100 g dry soil)	2.746	0.062	2.967
	6.25 (10)	0.43 (12)	9.65 (4)
Exchangeable hydrogen	0.00	0.00	0.00
(meq/100 g dry soil)	0.703	0.000	1.004
	5.43 (10)	0.00 (12)	3.34 (4)
C.E.C. root	15.30	19.65	17.40
(meq/100 g dry matter)	18.945	25.204	20.460
	24.32 (10)	31.22 (12)	24.52 (4)

property, enabling them to absorb or to reject certain amounts of ions, present in their environment.

A relation between cation exchange capacity and soil preference is observed by many authors (*e.g.* Wacquant[15]) and this property even proved to have competitive importance[10]. Since no satisfactory evidence could be provided for any hypothetical mechanism through which C.E.C. acts, the interest in this factor somewhat faded away. Nevertheless it is generally accepted that the density of cation adsorption sites, predominantly located in the cell wall, not only influences the concentration, but also the nature of the cations in the immediate plasmalemma environment.

The interaction between adsorption and availability of cations like Al, Fe or Mn might be responsible for the restriction of calcicoles – neutrophile/basiphiles or calcifuges – acidiphiles to their respective habitats. Deficiency of Fe or Mn in soils with low availability (in this case calcareous) may be enhanced by low C.E.C. of calcifuges, while higher C.E.C. of calcicoles grown in acid soils with a high Al-, Fe- or Mn-availability might favour intoxication[5,11,13,14].

However, some authors tend to revalue the role of Ca and H in the distribution of calcicoles and calcifuges. Bates and Brown[2], Ghorbal[8] and Bates[1] stressed the importance of Ca and H for the control of membrane permeability. Because of their higher adsorption capacity, calcicoles could maintain high Ca-levels, in accordance with their requirements, provided that enough Ca is available.

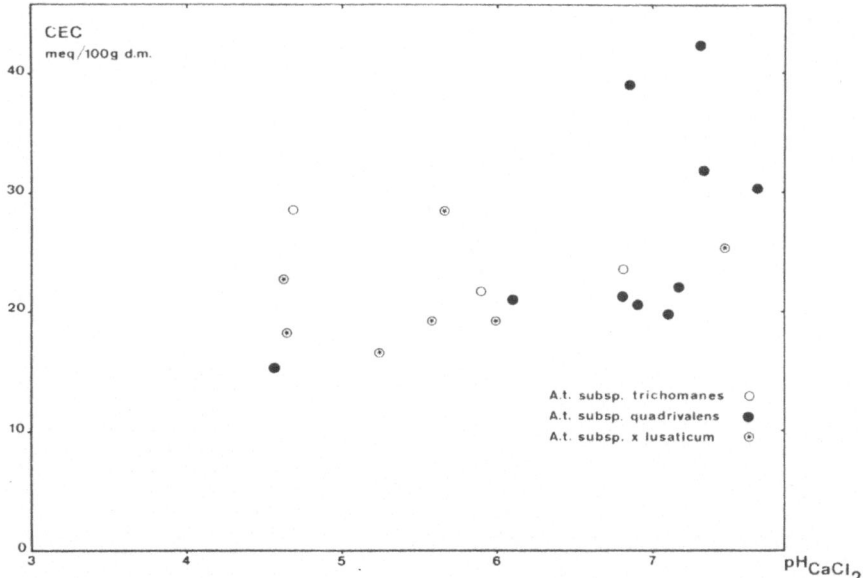

Fig. 1. *Asplenium trichomanes*. C.E.C. as a function of pH of the soil. Log (root C.E.C.)/pH: r = 0.52, P < 0.05

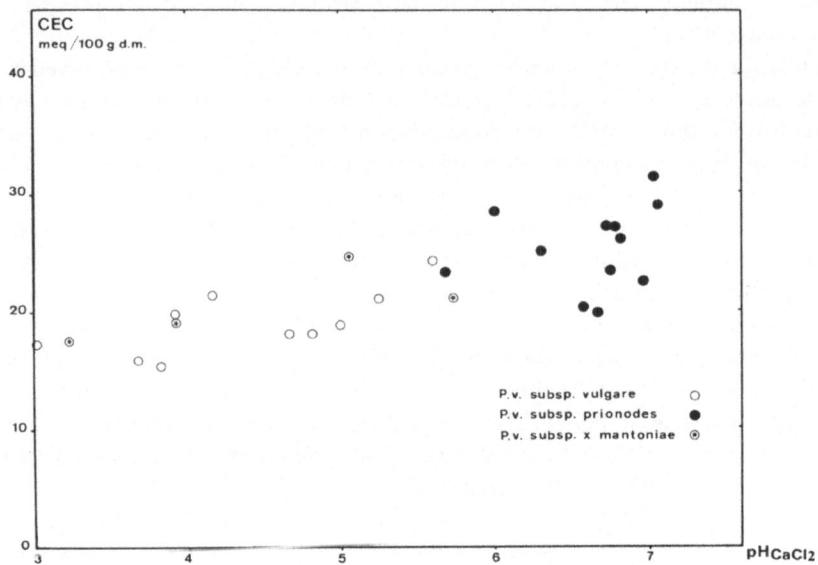

Fig. 2. *Polypodium vulgare*. C.E.C. as a function of pH of the soil. Log (root C.E.C.)/pH: r = 0.79, P < 0.001

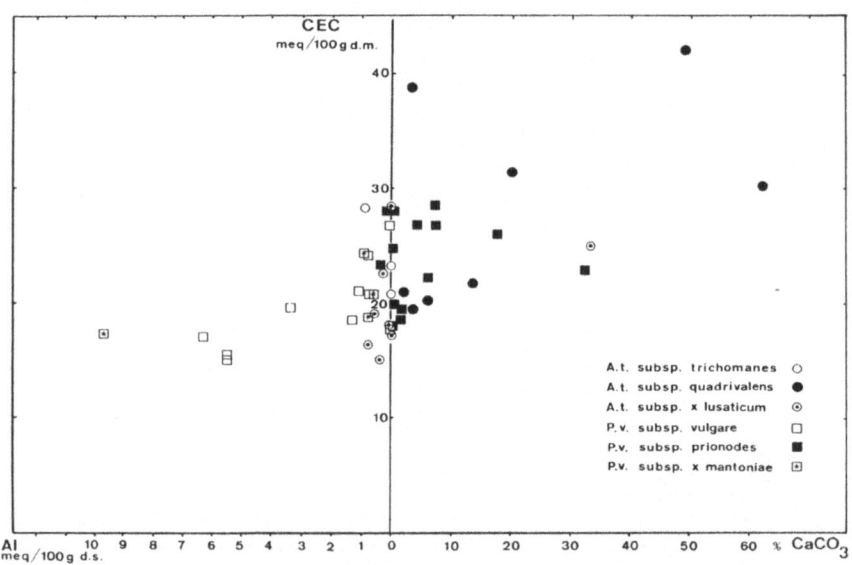

Fig. 3. *Asplenium trichomanes* and *Polypodium vulgare*. C.E.C. as a function of exchangeable Al, if present, or of CaCO$_3$, if present. Both species together: C.E.C./Al: r = −0.37, P < 0.01. C.E.C./CaCO$_3$: r = 0.51, P < 0.001

It remains impossible to endorse one hypothesis rather than another, because plants, collected in the environment they are so well adapted to, only yield indicative results. In the case of Polypodium and Asplenium it is important to draw the attention to the relations existing between root C.E.C. and pH_{CaCl_2} (Figs. 1 and 2), $CaCO_3$ (Fig. 3) and exchangeable Al (Fig. 3).

Well aware of the uncertainty concerning the Al- or Fe-forms in the soil complex or in the apoplast and the role of Ca and H in the membrane permeability of Asplenium and Polypodium, we can only see these results in the light of experimental data obtained by others and then, it is conceivable that C.E.C. may be partly responsible for the soil preference phenomenon.

Acknowledgements We thank our friends Dirk van Speybroeck for his help in sampling, Frank van Overwalle for data processing and Dave Evans for the correction of the translation.

References

1   Bates J W 1982 The role of exchangeable calcium in saxicolous calcicole and calcifuge mosses. New Phytol. 90, 239–252.
2   Bates, J W and Brown D H 1974 The control of cation levels in seashore and inland mosses. New Phytol. 73, 483–495.
3   Beckers B 1966 Bijdrage tot de biosystematiek van *Polypodium* L. in België en het Groothertogdom Luxemburg. Bull. Jard. Bot. Etat Brux. 36, 353–382.
4   Bouharmont J 1968 Les formes chromosomiques d'*Asplenium trichomanes* L. Bull. Jard. Bot. Natl. Belg. 38, 103–114.
5   Clarkson D T 1969 Metabolic aspects of aluminium toxicity and some possible mechanisms for resistance. *In* Ecological Aspects of the Mineral Nutrition of Plants. Ed. I H Rorison. Symposium no 9, pp 381–397. Blackwell Scientific Publications, Oxford & Edinburgh.
6   De Graaf D Th and Heukels P 1982 Over de eikvarens (*Polypodium vulgare* L. en *P. interjectum* Shivas) in het herbarium De Wever. Natuurhistorisch Maandblad 71, 81–84.
7   De Langhe J E, Delvosalle L, Duvigneaud J, Lambinon J and Vanden Berghen C 1978 Nouvelle flore de la Belgique, du Grand-Duché de Luxembourg, du Nord de la France et des Régions voisines (Ptéridophytes et spermatophytes). Deuxième édition. Jardin Botanique national de Belgique, Meise.
8   Ghorbal M H 1979 Absorption du calcium, localisation et rôle dans la perméabilité membranaire, relation avec le caractère calcicole ou calcifuge. Thèse de docteur-ès-sciences Université des sciences et techniques du Languedoc, Montpellier.
9   Koedam N and Büscher P 1983 Studies on the possible role of cation exchange capacity in the soil preference of mosses. Plant and Soil 70, 77–93.
10 Mouat M C H and Walker T W 1959 Competition for nutrients between grasses and white clover. II. Effect of root cation-exchange capacity and rate of emergence of associated species. Plant and Soil 11, 41–52.
11 Mugwira L M and Elgawhary S M 1979 Aluminium accumulation and tolerance of triticale and wheat in relation to root cation exchange capacity. Soil Sci. Soc. Am. Proc. 43, 736–740.
12 Penel M 1979 Caractérisation physico-chimique et classification des humus forestiers acides en relation avec la végétation et ses exigences écologiques. Thèse presentée à l'Université de Nancy. 1.

13 Straub R 1964 Untersuchungen über die Kationenaustauschkapazität der Wurzeln von Trockenrasen-und Wiesenpflanzen. Inaugural-Dissertation zur Erlangung des Doktorwürde der hohen naturwissenschaftlichen Fakultät der Julius-Maximilians-Universität zu Würzburg.
14 Vose P B and Randall P J 1962 Resistance to aluminium and manganese toxicities in plants related to variety and cation exchange capacity. Nature, London 196, 85–86.
15 Wacquant J P 1974 Recherches sur les propriétés d'adsorption cationique des racines (rôle physiologique et importance écologique). Thèse d'Etat-ès-Sciences. Université de Montpellier II.

Adaptation of selected trees and grasses to low availability of phosphorus

F. S. CHAPIN, III
Institute of Arctic Biology, University of Alaska, Fairbanks, AK 99701, USA

Key words Barley Chinochloa Growth rate Nutrient deficiency Nutrient stress Phosphorus fractions Root-shoot ratio Taiga

Abstract High-nutrient-adapted and low-nutrient-adapted species of New Zealand tussock grasses (*Chionochloa*), barley (*Hordeum*), and several taiga trees were grown at three rates of phosphorus supply. Low-nutrient-adapted species in each group of species had similar (grasses) or lower (trees) capacities for phosphate absorption, were less efficient in producing biomass (*i.e.* had higher nutrient concentrations), and grew more slowly than high-nutrient-adapted species. I conclude that the major adaptation to low nutrient availability in each of these comparisons is a slow growth rate that reduces the annual nutrient requirement.

Introduction

Low availability of nutrients, particularly nitrogen and phosphorus, is a major factor determining plant distribution and productivity of both natural and agricultural ecosystems. Some plants such as agricultural crops have a high nutrient requirement for effective growth and reproduction, whereas other species grow quite well and show no deficiency symptoms on infertile soils and normally occur only on such soils. Adaptations that could be important for effective growth in infertile soils include (a) a slow growth rate that minimizes annual nutrient requirement and prevents exhaustion of internal nutrient reserves, (b) a high capacity to extract nutrients from soil, and (c) differences in nutrient metabolism and efficiency in use of nutrients to produce new biomass.

This paper summarizes results of three studies comparing species that normally grow on either high-phosphorus (P) or low-P soils and were grown under controlled conditions at three rates of P supply. The species compared include (1) two New Zealand alpine tussock grasses, *Chionochloa pallens* from relatively high-P soils and *C. crassiuscula* from low-P soils[3], (2) commercial barley (*Hordeum vulgare*) which has a high fertilizer requirement and *H. leporinum* from Australian low-P soils[2], and (3) a series of several Alaskan taiga tree species that occur along a complex gradient of nitrogen and phosphorus availability and soil temperature[4]. Methods are described in the cited papers.

M.R. Sarić and B.C. Loughman (eds.), Genetic Aspects of Plant Nutrition.
ISBN-13: 978-94-009-6838-7
©1983 *Martinus Nijhoff/Dr W. Junk Publishers, The Hague/Boston/Lancaster.*

Results and discussion

Phosphate absorption capacity did not differ between *Chionochloa* species at low or moderate rates of phosphate supply (Table 1). In both these species P absorption capacity declined with increasing P availability, particularly in the high-P-adapted *C. pallens* so that at high P supply *C. pallens* had a lower P absorption capacity than *C. crassiuscula*. P absorption capacity did not differ significantly between low- and high-P-adapted *Hordeum* species. Among taiga trees P uptake capacity was highest among high-P-adapted trees and was lowest in those species characteristic of low-P soils. Thus, there was no evidence in these or other studies (see Chapin[5]) that low-P-adapted species have a high capacity to extract P from soils. A low P-absorption capacity might be expected in plants adapted to infertile soils because (1) diffusion to the root surface is the major

Table 1. Response of phosphate absorption (μmol g^{-1} h^{-1}) capacity and plant weight in grasses and trees to growth at three concentrations of phosphate. Within each group of comparisons high-P-adapted species are listed first and low-P-adapted species last. Data are mean$\pm$S.E., n=4.

	Growth phosphate concentration (μmol l^{-1})		
	1	10	100
Phosphate absorption (μmol g^{-1} h^{-1})			
Chionochloa pallens	1.47$\pm$0.13	0.78$\pm$0.08	0.45$\pm$0.05
C. crassiuscula	1.51$\pm$0.19	0.82$\pm$0.07	0.86$\pm$0.08
Hordeum vulgare	–	3.80$\pm$10.5	–
H. leporinum	–	17.60$\pm$12.6	–
Populus balsamifera	–	5.5$\pm$1.4	–
P. tremuloids	–	2.3$\pm$0.2	–
Betula papyrifera	–	2.6$\pm$0.8	–
Alnus crispa	–	1.3$\pm$0.2	–
Picea mariana	–	0.2$\pm$0.0	–
Plant mass (mg plant^{-1})			
Chionochloa pallens	53$\pm$9	1016$\pm$103	2200$\pm$135
C. crassiuscula	27$\pm$3	280$\pm$75	410$\pm$50
Hordeum vulgare	214$\pm$7	281$\pm$15	292$\pm$12
H. leporinum	38$\pm$4	47$\pm$7	44$\pm$3
Populus balsamifera	2057$\pm$211	4755$\pm$320	6444$\pm$551
P. tremuloides	3414$\pm$338	4486$\pm$550	5349$\pm$599
Betula papyrifera	2569$\pm$280	3019$\pm$475	5016$\pm$173
Alnus crispa	4605	4360	4810
Larix laricina	2166	2210	2664
Picea glauca	1835	2277	1582
P. mariana	850	673	909

limiting step in low-P soils and cannot be overcome by increased absorption capacity[8], and (2) P demand to support growth is the major determinant of P absorption rate[5] and would be highest in rapidly growing species.

Hordeum leporinum from Australian low-P soils had a higher root–shoot ratio than commercial barley[2], but the *Chionochloa* species and taiga trees showed no consistent species differences in root–shoot ratio[3] suggesting that some (but not all) species from infertile soils increase their capacity to extract nutrients from soil by increased allocation to roots.

Growth rate was the characteristic that differed most consistently among species. In each set of species comparisons, the high-P-adapted species grew most rapidly, but its growth was most strongly reduced by low P availability (Table 1). In contrast, the low-P-adapted species grew more slowly so that its P reserves were diluted over a smaller biomass, it experienced less P stress, and its growth was less strongly reduced by low P availability. A slow growth rate is advantageous in low-nutrient soils, because it minimizes annual nutrient

Table 2. Response of leaf phosphorus and nitrogen concentration in grasses and trees to growth at three concentrations of phosphate. Within each group of species comparisons, high-P-adapted species are listed first and low-P-adapted species last. Data are mean $\pm$ S.E., n = 4.

	Growth phosphate concentrations (μmol l^{-1})		
	1	10	100
Phosphorus concentration (% dry weight)			
Chionochloa pallens	0.058 ± 0.003	0.130 ± 0.006	0.383 ± 0.020
C. crassiuscula	0.055 ± 0.002	0.205 ± 0.011	0.830 ± 0.022
Hordeum vulgare	0.135 ± 0.002	0.315 ± 0.005	1.188 ± 0.013
H. leporinum	0.265 ± 0.010	0.682 ± 0.009	0.806 ± 0.007
Populus balsamifera	0.160 ± 0.009	0.135 ± 0.006	0.119 ± 0.010
P. tremuloides	0.128 ± 0.009	0.145 ± 0.009	0.137 ± 0.012
Betula papyrifera	0.152 ± 0.010	0.167 ± 0.011	0.139 ± 0.010
Alnus crispa	0.112 ± 0.012	0.109 ± 0.004	0.118 ± 0.008
Larix laricina	0.112 ± 0.016	0.087 ± 0.009	0.087 ± 0.006
Picea glauca	0.160 ± 0.013	0.147 ± 0.023	0.187 ± 0.017
P. mariana	0.124 ± 0.007	0.165 ± 0.011	0.143 ± 0.014
Nitrogen concentration (% dry weight)			
Populus balsamifera	0.64 ± 0.03	0.70 ± 0.01	0.97 ± 0.02
P. tremuloides	0.81 ± 0.03	0.99 ± 0.05	1.36 ± 0.10
Betula papyrifera	0.95 ± 0.03	1.12 ± 0.07	1.28 ± 0.07
Alnus crispa	0.76 ± 0.03	0.92 ± 0.02	1.38 ± 0.17
Larix laricina	0.68 ± 0.01	1.10 ± 0.11	1.18 ± 0.02
Picea glauca	1.09 ± 0.11	1.56 ± 0.23	2.04 ± 0.13
P. mariana	1.81 ± 0.15	2.35 ± 0.13	2.03 ± 0.07

requirement, prevents exhaustion of internal nutrient reserves, and thus minimizes nutient stress (*e.g.* refs.[1,7,9]). In contrast, a rapid growth rate that enables a plant to dominate available light, water, and mineral resources is clearly advantageous in a favorable environment.

In the *Chionochloa* and *Hordeum* comparisons, tissue phosphorus concentration was lower in the more rapidly growing, high-nutrient-adapted species (Table 2), indicating that the high-nutrient-adapted species produced more biomass per unit of tissue phosphorus (*i.e.* was more efficient in use of phosphorus to produce biomass). Among taiga trees there was no consistent difference among species in tissue phosphorus concentration, but the high-nutrient-adapted species had lower tissue nitrogen concentrations (*i.e.* were more efficient in producing biomass per unit tissue nitrogen.) Thus in all species' comparisons, the low-P-adapted species were less efficient at producing biomass per unit plant nutrient. The high tissue phosphorus concentration (low efficiency of P use) in low-P-adapted species is a logical and necessary consequence of the fact that these species have a much slower growth rate and similar (or slightly lower) rate of P absorption compared with high-P-adapted species.

In summary, a slow growth rate (and consequently a low annual P requirement and prevention of depletion of tissue preserves) is the major adaptation of wild plants to infertile soils, and among the species examined low-P-adapted plants are not more effective at extracting P from soils nor are they more efficient in producing biomass with the nutrient which they acquire. This contrasts strikingly with crop species in which it is possible to select for high P use efficiency or high P uptake capacity (*e.g.* Gerloff[6]). I suggest that these characteristics have not been selected for in wild plants on infertile soils because (1) a high P uptake capacity is of little use in an infertile soil, as described above, (2) criteria for fitness under natural selection (*e.g.* low mortality) may be quite different from the high productivity selected by agronomists, and (3) multiple nutritional stresses that occur naturally may preclude more efficient use of a particular mineral.

Acknowledgments Studies were supported by the Guggenheim Foundation and by NSF grant DEB 78-11594 to the University of Alaska.

References

1 Chapin F S III 1980 The mineral nutrition of wild plants. Annu. Rev. Ecol. Syst. 11, 233–260.
2 Chapin F S III and Bieleski R L 1982 Mild phosphorus stress in barley and a related low-phosphorus-adapted barleygrass. Phosphorus fractions and phosphate absorption in relation to growth. Physiol. Plant. 54, 309–317.
3 Chapin F S III, Follett J M and O'Connor K F 1982 Growth, phosphate absorption and phosphorus chemical fractions of two *Chionochloa* species. J. Ecol. 70, 305–321.
4 Chapin F S III, Tryon P R and Van Cleve K 1983 Influence of phosphorus supply on the growth and biomass allocation of Alaskan taiga tree seedlings. Can. J. For. Res. *In press*.

5 Clarkson D T and Hanson J B 1980 The mineral nutrition of higher plants. Annu. Rev. Plant Physiol. 31, 239–298.
6 Gerloff G C 1976 Plant efficiences in the use of nitrogen, phosphorus and potassium. *In* Plant Adaptation to Mineral Stress in Problem Soils. Ed. M J Wright. pp 161–169. Cornell University Agricultural Experiment Station, Ithaca, New York.
7 Grime J P and Hunt R 1975 Relative growth rate: its range and adaptive significance in a local flora. J. Ecol. 63, 393–422.
8 Nye P H 1977 The rate-limiting step in plant nutrient absorption from soil. Soil Science 123, 292–297.
9 Rorison I H 1968 The response to phosphorus of some ecologically distinct plant species. I. Growth rates and phosphorus absorption. New Phytol. 67, 913–923.

Comparison of $K^+ - Na^+$ selectivity mechanisms in roots of Fagopyrum and Triticum

H. EGGERS and W. D. JESCHKE
Lehrstuhl für Botanik I, Universität Würzbug, D-8700 Würzburg, FRG

Key words Buckwheat Flux Analysis Influx Potassium Selectivity Sodium Wheat

Introduction

Higher plants show great variation in their selectivity of K^+ and Na^+ uptake and transport[7,12]. According to these authors, several mechanisms can be responsible for the selective uptake of alkali cations. These include influx selectivity and $K^+ - Na^+$ exchange at the plasmalemma, selective accumulation across the tonoplast and reabsorption of sodium from the xylem sap in exchange for K^+. These mechanisms are often combined and variably realized in different species and proved to be genetically determined[9] as could be shown *e.g.* for soybean varieties (ref.[10] and Eggers, unpublished results).

Amongst all plants investigated by Collander[1], buckwheat (*Fagopyrum esculentum*) had the highest K/Na ratio in the shoot, suggesting a high $K^+ - Na^+$ selectivity of uptake and transport to the shoot. It seemed to be of interest to study in detail by which strategies this species achieves the high degree of discrimination. For the sake of comparison, *Triticum* was chosen as another species since this cereal grass was shown to be highly efficient in $K^+ - Na^+$ exchange[8], but little more data about wheat were available.

Material and methods

a) Plants

Seedlings of buckwheat (*Fagopyrum esculentum* Moench) and winter wheat (*Triticum aestivum* L. cv. Carstacht) were germinated and then grown in aerated 0.5 mM CaSO$_4$ solution at day-night temperatures of 23 and 20°C under continuous light conditions. After 3 (Fagop.) or 4 (Tritic.) days the resulting low-salt roots were used.

b) Influx measurements

The apical 4 cm of the roots were excised and were preincubated in an aerated 0.5 mM CaSO$_4$ solution for three hours to reduce wounding effects. 6 to 10 roots at a time were then suspended in an aerated experimental solution, containing in meq. l.$^{-1}$: Ca(NO$_3$)$_2$: 6, MgSO$_4$: 1 (basal solution) and in addition Na-phosphate buffer (pH 5.8): 0.02 to 1 (labelled with ^{22}Na) for measuring Na$^+$ influx, or K-phosphate buffer (pH 5.8): 0.01 to 0.5 (labelled with ^{86}Rb) for measuring K^+ influx. K^+ or Na$^+$ were added as chlorides to some samples in order to study their mutual competition. After 20 min, the roots were washed 3 × short and 3 × 30s with cold (4°C) inactive experimental solution. The radioactivity found in the roots was related to the fresh weight and the rate of K^+ or Na^+ influx was calculated. $K_{m\,app.}$ and V_{max} were determined as described by Cornish-Bowden[2].

c) Transport and flux measurements

For transport measurements, excised roots (3.5 cm) were mounted in two-chambered vessels (see Jeschke[5]). The apical part (2.5–3 cm) was in chamber S (containing the experimental solution), the excised basal part protruded into chamber X (containing basal solution). The solutions in the chambers were sampled frequently and K^+ and Na^+ were determined by flame photometry and related to the fresh weight of the apical root parts in chamber S. Where indicated, $10^{-6}\,M$ CCCP was added; a stock solution either in NaOH (final concentration 0.1 mM, Fagopyrum) or in methanol (final concentration 0.03%, Triticum) was used.

Flux analysis was carried out using the same apparatus and was described in detail earlier[5]. During a loading period of 21 to 22 h the roots were equilibrated with Na^+ (^{22}Na), they were then washed out with unlabelled solution; in a subsequent period of "re-elution" the washing solutions in chamber S additionally contained potassium. Unidirectional sodium fluxes were determined as described earlier[5,6,11].

The experimental solutions had the same composition as described for influx measurements. Na^+ was given as 0.2 mM Na-phosphate buffer (pH 5.8) with additional NaCl.

Results and discussion

a) Influx measurements

Table 1 shows the kinetic parameters of K^+ and Na^+ short-time uptake (= influx) in a concentration range of ion uptake by system 1 (Epstein *et al.*[4]). For buckwheat and wheat half-maximal rates of influx for K^+ uptake were reached at very low and almost similar K^+ concentrations ($K_{m\,app}$), indicating a high affinity of system 1 for potassium in both species. Addition of sodium had only a slight influence on K^+ influx, possibly a weak stimulation was detectable. The maximal influx rates (V_{max}) of wheat and buckwheat differed remarkably but were not affected by sodium.

Table 1. Kinetic constants* of Na^+ (^{22}Na) and K^+ (^{86}Rb) influx in excised roots of Fagopyrum and Triticum. Duration of uptake: 20 min. Concentration range for K^+: 0.01–0.5 mM, for Na^+: 0.02–1 mM

	Fagopyrum[a]		Triticum[b]	
	$K_{m\,app.}$	V_{max}	$K_{m\,app.}$	V_{max}
K^+	0.020	3.2	0.023	6.7
(+1 mM Na^+)	0.015	3.0	0.015	6.7
Na^+	1.56	0.63	0.40	1.28
(+0.2 mM K^+)	1.53	0.52	2.00	1.21

* Apparent Michaelis constant $K_{m\,app}$ in mM, maximal velocity V_{max} in μmol g^{-1} fr. wt. h^{-1}.
[a] Mean values of 2 (Na^+) or 3 (K^+) replications.
[b] Data by S. Kolibius (unpublished), mean values of 6 replications.

A preference for potassium became evident when the uptake kinetics of Na^+ was studied in comparison. Both plants – particularly Fagopyrum – reached half maximal Na^+ influx rates at much higher concentrations than with potassium and V_{max} was relatively low in particular for Fagopyrum. These observations agree with earlier measurements on many other species[3].

When potassium was added, the Na^+ influx in buckwheat and wheat roots responded quite differently. The $K_{m\,app}$ for Triticum was drastically increased but V_{max} remained nearly unchanged. This indicates a strong competition of K^+ with Na^+, apparently for the same binding sites. By contrast, Na^+ influx in Fagopyrum roots was almost unaffected by the presence of K^+ (Table 2). This behaviour is quite different also from that reported for barley[13] and appears to indicate that Na^+ influx in Fagopyrum roots does not proceed by a system-1 type of ion uptake. This is suggested also by the high value for $K_{m\,app.}$ (1.56 mM) that lies beyond the concentrations of system 1 (up to 0.5 mM, see Epstein[3]).

Table 2. Unidirectional steady-state fluxes and compartmental contents of Na^+* in Fagopyrum** and Triticum*** roots and the effect of K^+ on Na^+ efflux. In addition, some fluxes and the vacuolar content are related to the cytoplasmic sodium content

	Fagopyrum		Triticum
	1 mM Na^+	10 mM Na^+	1 mM Na^+
ϕ_{oc}	0.085	0.47	3.0
ϕ_{co}	0.074	0.38	2.0
ϕ_{co}/Q_c	0.6	0.7	0.9
ϕ_{cv} ($=\phi_{vc}$)	0.024	0.069	2.1
ϕ_{cx}	0.011	0.083	0.95
ϕ_{cx}/Q_c	0.09	0.15	0.43
Q_c	0.13	0.54	2.22
Q_v	0.49	4.13	51.1
Q_v/Q_c	3.9	7.7	23
ϕ_{co} (K^+-dep.)†	n.d.	n.d.	7.0

* All fluxes are given in μmol Na^+ g^{-1} fr. wt. h^{-1}, the Na^+ contents in μmol g^{-1} fr. wt. Abbreviations: ϕ_{oc}, ϕ_{co}: plasmalemma influx and efflux; ϕ_{cv},ϕ_{vc}: tonoplast influx and efflux; ϕ_{cx}: efflux from the xylem parenchyma cells to the xylem vessels (corresponding to transport through the xylem vessels); Q_c, Q_v: cytoplasmic and vacuolar ion content; ϕ_{co} (K^+-dep.): K^+-dependent sodium efflux across the plasmalemma.

** Mean values of 4 replications, SEM for all data was about 10%.

*** Data by Jeschke and Nassery[8].

† K^+-dependent Na^+ efflux across the plasmalemma induced by addition of 0.2 mM K^+ or 1 mM K^+ when 10 mM Na^+ was present. n.d. = not detectable.

b) Flux analysis

In order to estimate individual sodium fluxes, roots had to be equilibrated with Na$^+$ (^{22}Na). As shown by independent measurements – not presented here – Fagopyrum roots reached saturation after about 8 hours. The rate of transport was relatively high after 4 to 6 hours but decreased later (see Fig. 2) and reached relatively stable values after 20 hours. The flux measurements done after 21–22 h loading, therefore, yielded these low transport rates.

In Table 2 flux data of Fagopyrum are shown in comparison to data of Triticum[8]. With Fagopyrum all fluxes – even at 10 mM Na$^+$ – were conspicuously smaller than the fluxes of Triticum at 1 mM Na$^+$, for both species the influx ϕ_{oc} obtained by compartmental analysis and influx measurements were in good agreement (see Tables 1 and 2). Nevertheless, for Triticum and for Fagopyrum ϕ_{co} seems similarly related to the cytoplasmic sodium content as can be seen by the comparable ratios ϕ_{co}/Q_c. Furthermore, the ability of buckwheat to accumulate sodium in the root cell vacuoles was noticeably low, as is demonstrated by the low ratio Q_v/Q_c between vacuolar and cytoplasmic Na$^+$ contents (see Table 2).

Since the discrimination between K$^+$ and Na$^+$ is high in Fagopyrum[1] it was anticipated that this pertains also to K$^+$ – Na$^+$ exchange at the plasmalemma.

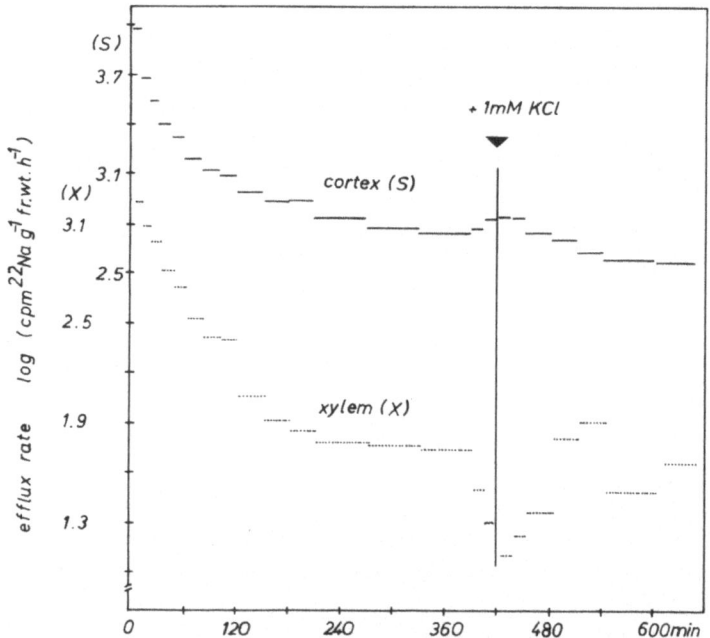

Fig. 1. Rate of ^{22}Na efflux from excised buckwheat roots across the plasmalemma of root cortex cells (S) and from the xylem vessels (X) during efflux analysis and during re-elution in the presence of 1 mM K$^+$. The roots were equilibrated with 10 mM Na$^+$ (^{22}Na) for 21 h (see Material and methods).

Unexpectedly this was not true and Fagopyrum did not show any significant short-time effect following a treatment with potassium, only a small decrease of tracer efflux after about 1 hour was detectable (see Fig. 1 and Table 2). This suggests that no or only a very limited $K^+ - Na^+$ exchange at the plasmalemma of the root cortical cells does occur.

In this way, buckwheat contrasts strongly to wheat which showed a highly efficient $K^+ - Na^+$ exchange[8] and to a various degree to other species that have been studied until now (see Jeschke, this volume).

c) Transport

Fig. 2 shows rates of Na^+ transport in roots of Fagopyrum and Triticum under 10 mM Na^+ conditions. Sodium transport by wheat roots was selectively inhibited by an addition of potassium (Fig. 2). The resulting low rate of transport then was unaffected by CCCP.

In contrast to wheat, sodium transport in buckwheat roots showed a strong time-dependence. After high, transient rates the transport decreased after about 6 h. When K^+ was added, there was some additional decrease but Na^+ transport in the presence of potassium could further be decreased by an addition of CCCP. Whatever this CCCP-inhibition indicates, it appears to show that in buckwheat Na^+ transport cannot be inhibited by an addition of K^+ as efficiently as in wheat. This observation appears to be consistent with the absence of $K^+ - Na^+$ exchange in buckwheat roots (see above). Furthermore, Na^+ apparently is less efficiently removed from the cytoplasm to vacuoles—see the low ratio Q_v/Q_c of

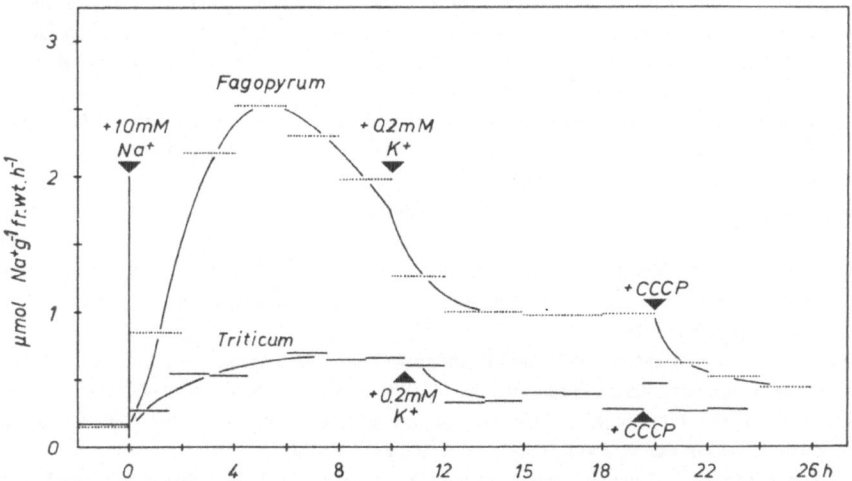

Fig. 2. Rates of Na^+ transport through excised roots of buckweat and wheat. At the time $t=0$, 10 mM Na^+ was added to basal solution (methods) in chamber S. Where indicated, 0.2 mM K^+ or 10^{-6} M CCCP respectively were given in addition. The transport rates before $t=0$ represent transport of endogenous Na^+.

3.9 in Fagopyrum roots which is much smaller than the corresponding ratio in wheat roots ($Q_v/Q_c = 23$).

Conclusions

Triticum resembles barley in showing an efficient influx *and* exchange selectivity at the plasmalemma and in being capable of accumulating Na^+ in root vacuoles.

In contrast, the main source of selectivity in Fagopyrum appears to be the discrimination between K^+ and Na^+ during influx. Remarkably, however, Na^+ uptake in these roots was almost unaffected by external K^+ and does not appear to proceed by a system-1 type of ion uptake. Neither $K^+ - Na^+$ exchange at the plasmalemma nor selective accumulation of Na^+ in vacuoles contribute noticeably to the overall selectivity of the root. According to these results Fagopyrum may be termed a highly efficient Na^+ excluder.

References

1 Collander R 1941 Selective absorption of cations by higher plants. Plant Physiol. 16, 691–721.
2 Cornish-Bowden A 1976 Principles of Enzyme Kinetics. Butterworths, London, Boston.
3 Epstein E 1972 Mineral Nutrition of Plants: Priniciples and Perspectives. Wiley and Sons, New York. ISBN 0–471–24340–X.
4 Epstein E, Rains D W and Elzam O E 1963 Resolution of dual mechanisms of potassium absorption by barley roots. Proc. Natl. Acad. Sci. U.S.A. 49, 684–892.
5 Jeschke W D 1972 Wirkung von K^+ auf die Fluxe und den Transport von Na^+ in Gerstenwurzeln, K^+-stimulierter Na^+-Efflux in der Wurzelrinde. Planta (Berl) 106, 73–90.
6 Jeschke W D 1977 $K^+ - Na^+$ selectivity in roots, localization of selective fluxes and their regulation. *In* Regulation of Cell Membrane Activities in Plants (E. Marrè and O Cifferi eds), pp 63–78. Elsevier/North Holland, Amsterdam.
7 Jeschke W D 1979 Univalent cation slectivity and compartmentation in cereals. *In* Recent Advances in the Biochemistry of Cereals (D L Laidman and R G Wyn Jones eds) pp 37–61. Academic Press, London, New York.
8 Jeschke W D and Nassery H 1981 $K^+ - Na^+$ selectivity in roots of Triticum, Helianthus and Allium. Physiol. Plant. 52, 217–224.
9 Läuchli A 1976 Genotypic variation in transport. *In* Encyclopedia of Plant Physiology, New Series (U. Lüttge and M G Pitman, eds), Vol. 2B, pp 372–393. Springer Verlag, Berlin, Heidelberg, New York.
10 Läuchli A and Wieneke J 1979 Studies on growth and distribution of Na^+, K^+ and Cl^- in soybean varieties differing in salt tolerance. Z. Pflanzenernaehr. Bodenkd. 142, 3–13.
11 Pitman M G 1971 Uptake and transport of ions in barley seedlings. I. Estimation of chloride fluxes in cells of excised roots. Aust. J. Biol. Sci. 24, 407–421.
12 Pitman M G 1976 Ion uptake by plant roots. *In* Encyclopedia of Plant Physiology, New Series (U Lüttge and M G Pitman eds), Vol. 2B, pp 95–128. Springer Verlag, Berlin, Heidelberg, New York.
13 Rains H D W and Epstein E 1967 Sodium absorption by barley roots: Role of the dual mechanisms of alkali cation transport. Plant Physiol. 42, 319–323.

Genetic differentiation in Taraxacum and its relation to mineral nutrition

C. H. HOMMELS, A. A. STERK and O. Gy. TANCZOS
University of Groningen, Dept. of Plant Physiology, P.O. Box 14 9750 AA Haren (GN), The Netherlands and University of Amsterdam, Hugo de Vries Laboratory, Plantage Middenlaan 2a, 1018 DD Amsterdam, The Netherlands (A. A. Sterk)

Key words Microspecies Nitrogen Salinity Species *Taraxacum*

Introduction

In western Europe the genus *Taraxacum* is rich in species. Twenty-one sections containing about 1200 microspecies have been described[14]. Most of the microspecies are obligate agamospermous and polyploid. Triploids are most common. Recently diploid sexual species have been found in Central Europe[12]. Further north, these species are gradually replaced by the agamospermous triploid microspecies. It is very likely that the diploid sexual species have participated in the evolution of the agamospermous triploid microspecies by hybridization. It is supposed that the driving force behind the colonization of the northern parts of Europe is the high "immediate fitness' of the microspecies.

About 200 microspecies have been found in the Netherlands, within the following sections: Obliqua (1), Spectibilia (22), Palustria (10), Erythrosperma (9) and Vulgaria (157). The richness of these forms is probably due partly to hybridization, and partly to adaptation to different levels of mineral nutrition, moisture content and changes in both factors.

In this paper, two important physiological parameters will be discussed, growth and the activity of membrane-bound ATPases in roots as carriers of the ion transport mechanism, with a view to determining in what respects the microspecies differ and their bearing on such processes as ecological differentiation and evolution.

Nutrient-rich soils enable plants to grow quickly and fast growth may be advantageous[6]. Nutrient-poor soils force plants to grow less quickly. Plants may adapt to both conditions, resulting in a fixed way of growth. The longer a microspecies occurs in a specific habitat, a higher degree of adaptation will be expected, resulting in more rigid genotypes. Possible differences in growth rate and in flexibility of the genotypes, regarding this parameter, will be examined in this publication. Growth is a resultant of many physiological processes in the plant[2]; one being the uptake and transport of minerals, in which the membrane-bound ATPases of the microsomal fraction are supposed to play a key-role. It is suggested that fast-growing plants have an ATPase enzyme characterized by a high capacity, *i.e.* a high V_{max} value. The opposite will be true for slow-growing plants, adapted to poor soils, that will have a high affinity to the substrate cation-ATP, *i.e.* a low K_m value, in order to generate sufficient energy for the uptake of minerals, at low levels of mineral nutrition. This necessity does not exist for fast-growing plants on nutrient-rich soils.

Results

I Distribution of microspecies in the Netherlands

Results of field research on the habitat of a number of *Taraxacum* microspecies are summarized in Table 1.

A second method of determining the ecological preference of the microspecies is given by Ellenbergs average nitrogen number (ANN). This ANN is a measure of preference of a microspecies for a certain nitrogen level of the soil. Table 2 summarizes the range of ANN of a few relevant microspecies.

T. corynodes, *G. ekmanii* and *T. lancidens* were not mentioned in the results, because the field research was still in progress at that time. Preliminary results indicated that *T. ekmanii* and *T. corynodes* prefer high nutrient levels while *T. lancidens* prefer moderate nutrient levels.

The results in Table 1 and Table 2 can be interpreted in terms of ecological

Table 1. Habitats of *Taraxacum* species in the Netherlands

Taraxacum species	Frequency of presence in a number of:				
	Dune grasslands (23)		Hayfields (16)	Hayfields/pastures (12)	Pastures (11)
	Ungrazed Unfertilized Dry	Grazed Lightly fertilized Dry	Ungrazed Non/lightly fertilized Moist	Variedly grazed Moderately fertilized Moderate moist	Heavily grazed Heavily fertilized Moderate moist
Obliquum (O)	91.7%	9.1%	–	–	–
Taeniatun (E)	91.7%	–	–	–	–
Nordstedtii (S)	–	–	81.3%	35.7%	–
Hollandium (P)	–	–	65.0%	–	–
Adamii (V)	–	–	75.0%	64.3%	–
Sellandii (V)	–	–	6.3%	45.7%	72.7%
Eudontum (V)	–	–	–	–	90.9%

(O) s. Oblique; (E) s. Erythrosperma; (S) s. Spectibilia; (P) s. Palustria; (V) s. Vulgaria; the Figures 23, 16, 12 and 11 are numbers of investigated plots.

Table 2. The average nitrogen numbers of a number of *Taraxacum* microspecies in the Netherlands

Taraxacum species	Range of ANN
Taeniatum	2.7–4.2
Obliquum	3.2–4.2
Adamii	4.2–5.8
Nordstedtii	4.2–6.5
Sellandii	4.4–6.5
Eudontum	5.2–6.4

preferences. Going from unfertilized, nitrogen-poor soils to heavily fertilized, nitrogen-rich soils, the following sequence of sections can be made: Obliqua—Erythrosperma—Spectibilia—Palustria—Vulgaria. It is also known that the first four sections have been settled in the Netherlands for centuries, while section Vulgaria has extended greatly during the last fifty years.

On the basis of the above results microspecies were choosen, using at least one microspecies from each section. *T. nordstedtii* and *T. hollandicum* were needed as examples of plants from poor nutrient soils because seeds of *T. obliqua* and *T. taeniatum* were not available until very recently. *T. sellandii*, *T. eudontum*, *T. ekmanii* and *T. corynodes* were used as examples from nutrient-rich soils and *T. adamii* and *T. lancidens* from intermediate soils.

II Growth analysis

The experiments were carried out during the winter, in a glasshouse with supplementary light to a maximum of 60 W/m^2, for 16 hours. The temperature was kept about 25°C and the relative humidity was 60–70%. Seeds of a single motherplant were sown in vermiculite. After 14 days the seedlings were transferred to waterculture with a nutrient level of 10% Hoagland[9]. After 28 days, the plants of each microspecies were divided into two groups and placed on waterculture with 1% and 25% Hoagland respectively (high and low salt conditions). The solution was changed weekly. Every week 6 plants of each microspecies were harvested randomly from each treatment. The average dry weight of root and shoot were determined (Table 3). Microspecies of nutrient-rich soils grew relatively fast under both high and low salt conditions, with some minor shifts while microspecies of nutrient-poor soils grew relatively slowly in both high and low salt conditions, although the shifts were greater. These shifts in sequences were caused by differences in plasticity of the genotypes, *i.e.* the degree of change in response to alterations in the conditions. A minor change in response, expressed on the HS/LS ratio, points to a rigid genotype: *T. eudontum*, *T. adamii* and *T. sellandii*. Large changes in response point to a flexible genotype: *T. corynodes*, *T. hollandicum* and *T. nordstedtii*. These results do not agree with

Table 3. The average dry weights of the total plants, after 70 days, in water culture under high and low salt conditions

Species	Dry weights of total plants		HS/LS ratio
	Low salt	High salt	
Eudontum	1.19 (1)	1.40 (4)	1.2
Corynodes	0.98 (2)	3.10 (1)	3.2
Ekmanii	0.98 (2)	1.74 (2)	1.8
Sellandii	0.96 (3)	1.25 (5)	1.3
Adamii	0.71 (4)	0.97 (8)	1.4
Hollandicum	0.64 (5)	1.65 (3)	2.6
Lancidens	0.56 (6)	1.05 (7)	1.9
Nordstedtii	0.48 (7)	1.22 (6)	2.6

The dry weights are given in grams; the figures between the parenthesis are the ordinal numbers.

the hypothesis, that the most flexible microspecies were to be found in the section Vulgaria. Also, the second part of this hypothesis, that the section Palustria and Spectibilia have more rigid microspecies, is not confirmed by our results.

One of the concepts developed for use in growth analysis is the efficiency index, later called the relative growth rate (RGR). This is the increase in dry weight per unit of original dry weight over a time interval t, or in formula: $RGR = \ln N(t_2) - \ln N(t_1)/t_2 - t_1$. Using the seedweight as $N(t_1)$, and the dry weight of the total plant on $t = 70$ days as $N(t_2)$, the average RGR over a period of 70 days was calculated (Table 4). One must bear in mind that germination takes time, about 5 days elapsing before photosynthetic parts form and dry weight increases. After a correction for the 5 days, the average RGR will be higher than those given in Table 4.

Comparing the results given in Table 3 and Table 4, it is obvious that the average RGR values show less variation than the dry weights of the total plant. This deviation was caused by the $N(t_1)$ values, the seed weights. There are considerable differences in seed weights between the investigated microspecies, up to 225%. The fact that the fast-growing plants have the heaviest seeds causes a smaller variation in the RGR values. It is remarkable that a typical representative of the rich-soil microspecies, *T. sellandii*, showed the lowest efficiency index or average RGR under high salt conditions. *T. hollandicum* on the other hand, a microspecies of relatively poor soils, showed a very high average RGR under high salt conditions. The average RGR calculated over a number of 3-week periods, showed a very steep decline. Under low salt conditions this decline was about 0.12 g/g/day during the first period to a level of about 0.04 g/g/day during

Table 4. The average RGR, expressed in g/g/day, of a number of *Taraxacum* microspecies over a period of 70 days under low and high salt conditions

Taraxacum species	The relative growth rates	
	Low salt	High salt
Eudontum	10.5×10^{-2}	10.7×10^{-2}
Ekmanii	10.3	11.1
Corynodes	9.9	11.6
Sellandii	10.1	10.5
Adamii	10.5	10.9
Hollandicum	10.2	11.6
Lancidens	9.9	10.8
Nordstedtii	9.5	10.8

the last period. Under high salt conditions this decline was less steep to a level of about 0.06 g/g/day during the last period. Possible causes for this decline are (i) lack of nutrition, especially under low salt conditions, (ii) mutual shading, especially under high salt conditions and (iii) the fallen leaves were not systematically taken into account.

The shoot-root ratio (S/R) gave information over the distribution of dry matter in the plant (Fig. 1). Under high salt conditions the S/R of the 4 microspecies showed a clear optimum. The fast-growing plants reached an earlier optimum and at a higher level. A possible explanation is that under high salt conditions all plants invest far more energy in the shoot than in the root causing accelarated growth of the shoot and subsequently a high production of photosynthetic products. Not all of the products will be used for growth, so some must be stored in the root. As a consequence of this storage the S/R will decline. Fast-growing plants will reach this moment of storage sooner. The difference in S/R level is probably due to a genetic fixed allocation of energy and in fast-growing plants this allocation will be more in favour of the shoot. The S/R under low salt conditions gave a different picture. During the first weeks the roots of all plants were more developed, compared with the plants at high salt treatment. The slow-growing plants showed a constant decline in S/R while the fast-growing plants showed a minor rise during the first weeks. This phenomenon can also be explained by the genetic fixation of the allocation. The rise was followed by an accelerated decline and both the fast- and slow-growing plants ended up at the same S/R level.

III Ion-stimulated ATPases in the microsomal fraction

The preliminary results of experiments with ion-stimulated ATPases in the

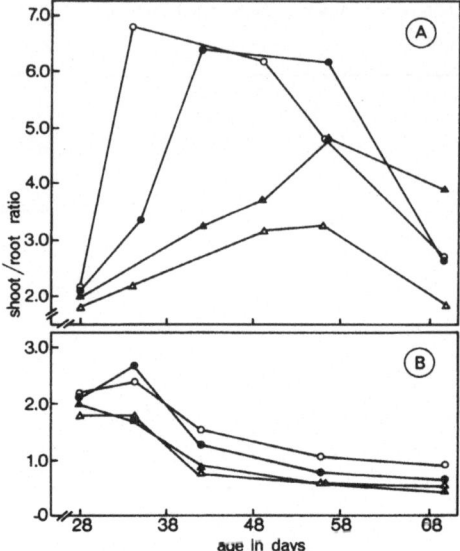

Fig. 1. The shoot-root ratio of a number of *Taraxacum* microspecies over a period of 70 days under low and high salt conditions. ○ *T. corynodes*, ● *T. sellandii*, △ *T. lancidens*, ▲ *T. nordstedtii*.

microsomal fraction of the roots will be summarized in the following section. The ion-stimulated activity of the ATPase declined in the presence of the following cations: Mg-Mn-Ca-Fe-Zn-Al-K-Na. None of these cations inhibited activity while that of the ATPase with MnATP as substrate was about 60% compared with MgATP as substrate. Saturation experiments showed only one type of ATPase as an acceptor of substrate. Earlier statements about the existence of an optimal Mg/ATP ratio of 1 (Christiansen and Lindberg[3]), and the inhibiting effect of free ATP, free Mg^{2+}-ions, and MgATP in high concentrations on the enzyme activity (Balke and Hodges[1]) were confirmed. K^+-ions in the presence of Mg^{2+}-ions produced an enzyme activity higher than the sum of the independent activities of both ions. This was most obvious in cases where the Mg/ATP was less than 1. The presence of Na^+ in the incubation should be noted. An optimum pH of 7.0–8.5 was determined. The V_{max} and K_m have been calculated (Table 5).

The data of Table 5 are preliminary and are discussed only in terms of tendencies. The data of Tables 3 and 5 show that there is no simple correlation between growth and the V_{max}-values of the ATP-ases. Furthermore those of Table 5 exhibit the lack of response of the activity of the ATP-ases on the level of mineral nutrition. As far as this is concerned the genetical make up of the investigated microspecies is well fixed.

For K_m-values conclusions cannot yet be made although there is a declining tendency going from high salt to low salt conditions, indicating the necessity for a high affinity of ATP-ases under nutrient-poor conditions.

Table 5. The V_{max}-values of the ATP-ases of a number of microspecies after 42 days, under high salt and low salt conditions

	V_{max}	
	High salt	Low salt
T. sellandii (V)	17.2	17.0
T. eudontum (V)	19.9	19.9
T. eckmanii (V)	19.9	18.3
T. adamii (V)	18.8	18.8
T. lancidens (V)	18.8	17.3
T. hollandicum (P)	18.8	19.4
T. taeniatum (E)	14.2	13.2
T. nordstedtii (S)	11.2	13.0
T. limburgense (—)	14.2	10.7

* V_{max} is expressed in micromol Pi/h/mg protein. The protein content of the enzyme extracts was determined with a slightly altered Lowry method.

References

1. Balke N E and Hodges T K 1975 Plasma membrane adenosine triphosphatase of oat roots. Plant Physiol. 55, 83–86.
2. Chapin F Stuart 1980 The mineral nutrition of wild plants. Ann. Rev. Ecol. Syst. 11, 223–260.
3. Christiansen J L and Lindberg S (1976) Kinetic studies of (Mg^{2+}) ATP-ase in oat roots. Physiol. Plantarum 36, 110–112.
4. Ellenberg H 1974 Zeigerwerte der Gefässpflanzen Mitteleuropas. Göttingen.
5. Evans G C 1972 The Quantitative Analysis of Plant Growth. Oxford (E.).
6. Grime J P 1977 Evidence for the existence of three primary strateigies in plants and its relevance to ecological and evolutionary theory. Am. Nat. 111, 1189–1194.
7. Hagendijk A et al 1975 Taraxacum (behalve sectie Vulgaria). Flora Neerl. IV (9), 1–52.
8. Hagendijk A (in publ) Taraxacum section Vulgaria. Flora Neerl.
9. Hoagland D R and Snijder W C 1933 Nutrition of strawberry plants under controlled conditions. Proc. Am. Soc. Sci. 30, 288–296.
10. Hodges T K 1976 ATPases associated with membrane of plant cells. In Encyclop. Plant Physiol. New Series, Vol 2, part A (U Lüttge and M G Pitman, Eds.), pp 260–283. Springer Verlag, ISBN 3-540-07452-8.
11. Lehninger A L 1965 Bioenergetics. W A Benjamin, Inc.
12. Nijs den J C M and Sterk A A 1980 Cytogeographical studies of Taraxacum section Taraxacum (= section Vulgaria) in Central Europe. Bot. Jahrb. Syst. 101, 527–554.
13. Sterk A A and den Nijs J C M 1978 Classificatie en sexualiteit. Gorteria 9(5), 178–188.
14. Richards A J and Sell P D 1976 Taraxacum Wigg. in Flora Europaea 4, 332–343.
15. Veltrup W 1980 Die Wirkung von Schwermetallen auf ATPasen. Ber. Deutsch. Bot. Ges. 93, 659–666.

The effect of nitrogen on the activity of some enzymes of nitrogen metabolism during ontogenesis of maize kernel hybrids

D. JELENIĆ and V. HADŽI-TAŠKOVIĆ ŠUKALOVIĆ
Maize Research Institute Zemun Polje, Beograd, Yugoslavia

Key words: Alanine aminotransferase Aspartate aminotransferase Glutamate dehydrogenase Maize Nitrate reductase Nitrogen Protein Top-dressing Yield

Introduction

The absorption by maize of NO_3^- ions and distribution of absorbed and transformed nitrogen between individual plant organs[1,3,7,9,10] have been well studied. However, it is very important to study why different maize genotypes, under the same conditions, have different amounts of nitrogen compounds and, at the same time, different build-up of organic matter. It is necessary to observe the intensity of physiological and biochemical processes during kernel ontogenesis in order to study genotypical specificity concerning nitrogen absorption, as well as its further transformation into organic compounds.

Materials and methods

Maize hybrids ZPL-773 × Mo 17 Ht and ZPL-233 × K-708 were grown in the trial field of the Maize Research Institute in Zemun Polje, 1981. These two hybrids belong to FAO maturity group 700, but have considerably different yield productions. The trial was conducted in three replications. The elementary plot was 49 m². Phosphorus and potassium were used in the form of combined fertilizer 8:24:16 before planting. Nitrogen was introduced in the KAN form (27% N).

Table 1. Scheme for distribution of nutritive elements of kg/ha

Variants	Basic			N top-dressing		
	N	P_2O_5	K_2O	in the 3–4 leaf stage	in the 8–9 leaf stage	before tasseling
A	133	116	77	53.2	46.7	33.2
B	266	236	157	106.4	93.1	66.5

Control variant without use of nutritive elements (Ø)

M.R. Sarić and B.C. Loughman (eds.), *Genetic Aspects of Plant Nutrition.*
ISBN-13: 978-94-009-6838-7
©1983 Martinus Nijhoff/Dr W. Junk Publishers, The Hague/Boston/Lancaster.

NR activity was determined according to Hageman et al.[5] and Schrader et al.[11], GS activity according to Elliot[4], GOT and GPT activity according to Tonhazy[13], GDH activity according to Bergmayer[2], and total nitrogen according to the standard Kjeldahl method. For the total protein content the factor N × 6.25 was used. Enzyme activity was determined only in the leaves above the ear, and protein content in the leaves above and below the ear.

Results and discussion

Enzyme activity in the leaf

NR activity (Fig. 1) decreased with leaf ageing, but was the highest in the treatment without top-dressing, followed by A variant in both hybrids. Total NR activity in control variant (Ø) was 16.3% higher in the leaf of hybrid ZPL-773 × Mo 17 Ht than in the leaf ZPL-233 × K-708, which also had lower grain yield. The same was reflected in top-dressing variants. Nitrogen top-dressing was reflected in higher physiological activity of the leaf. Therefore total NR activity was 36.1% higher in A variant and 74.8% higher in B variant in hybrid ZPL-773 × Mo 17 Ht, and in the leaf of hybrid ZPL-233 × K-708 the increase of total NR activity amounted to 81.4% in A variant and 89% in B variant compared to (Ø) variant.

GS activity (Fig. 1) ranged from 39.3 to 121.2 μmol/g fresh matter·hour. Nitrogen top-dressing brought about increased GS activity, which coincided with increased NR activity in the same variants. These results contribute to the explanation of a role for GS in primary nitrogen assimilation[8]. The increase of total GS activity was 14.5% in A variant and 22.2% in B variant in the leaf of hybrid ZPL-773 × Mo 17 Ht, whereas in the leaf of ZPL-233 × K-708 it amounted to 32.9% in A variant and 55.0% in B variant compared to control variant.

GDH activity in the maize leaf was very high during kernel development. However, the role of GDH has been unexplained, because the recent studies have confirmed the results of Miflin et al.[8], who stated that the principal way of NH_4^+ ion assimilation in the tissue performing photosynthesis is conducted through glutamine synthetase and glutamate synthase. GDH activity ranged from 19.2 to 42.5 μmol/g fresh matter·hour. There were no differences in total GDH activity (Fig. 1) between variants, i.e. no specific influence of nitrogen top-dressing was evident, because GDH activity varied regardless of nitrogen quantity in nutrition. The differences between studied hybrids were also small—total GDH activity was only 15% higher in Ø variant in the hybrid of ZPL-773 × Mo 17 Ht. However, high GDH activity in leaf, when protein degradation processes in the leaf are intensified, indicate that GDH enzyme is involved into the mechanism of protein degradation during kernel development, when assimilative transport from leaf to ear is intensified.

Total protein quantity in the leaf decreased from the beginning of kernel formation to the end of the study (10th to 65th day after pollination) and that decrease was greater in the leaf under the ear. The better the plant was supplied

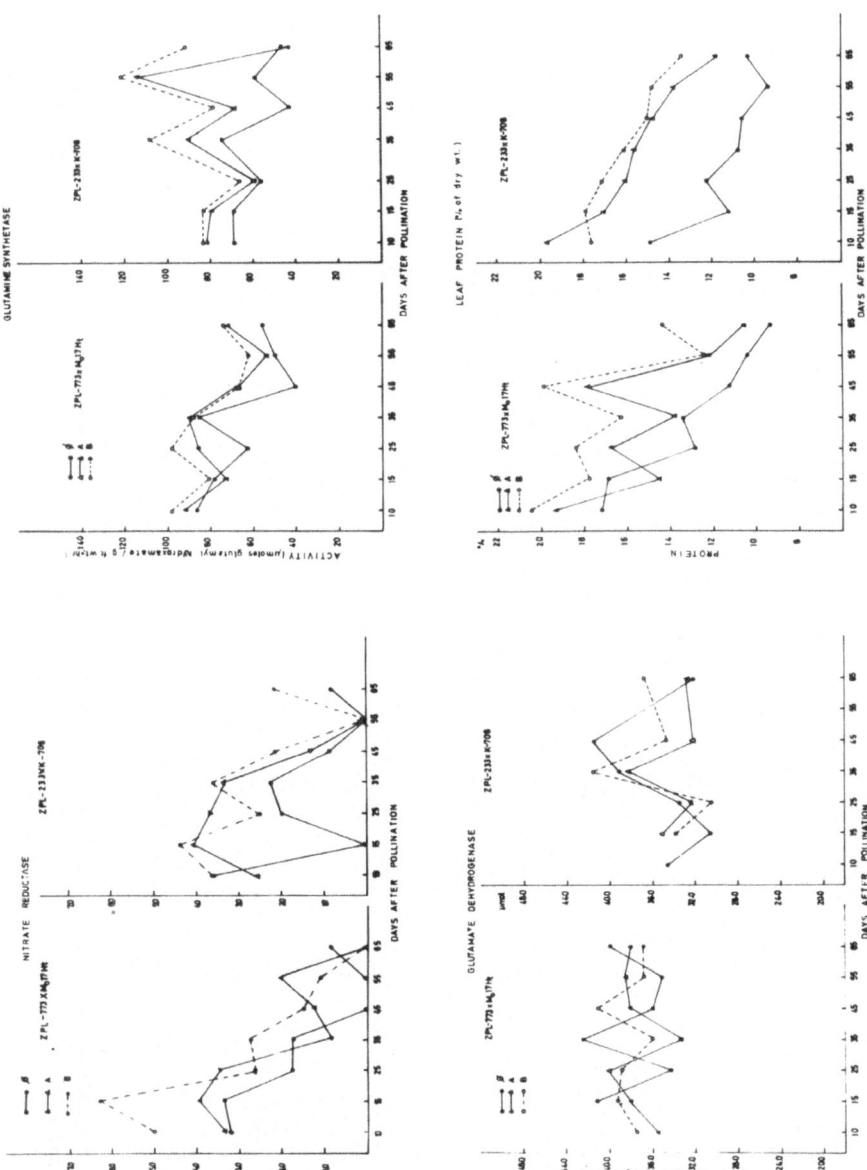

Fig. 1. Effect of the nitrogen top-dressing on NR, GS, GDH activities and on total protein content in the maize leaf (Activity in μmol/g fresh weight·hour).

240

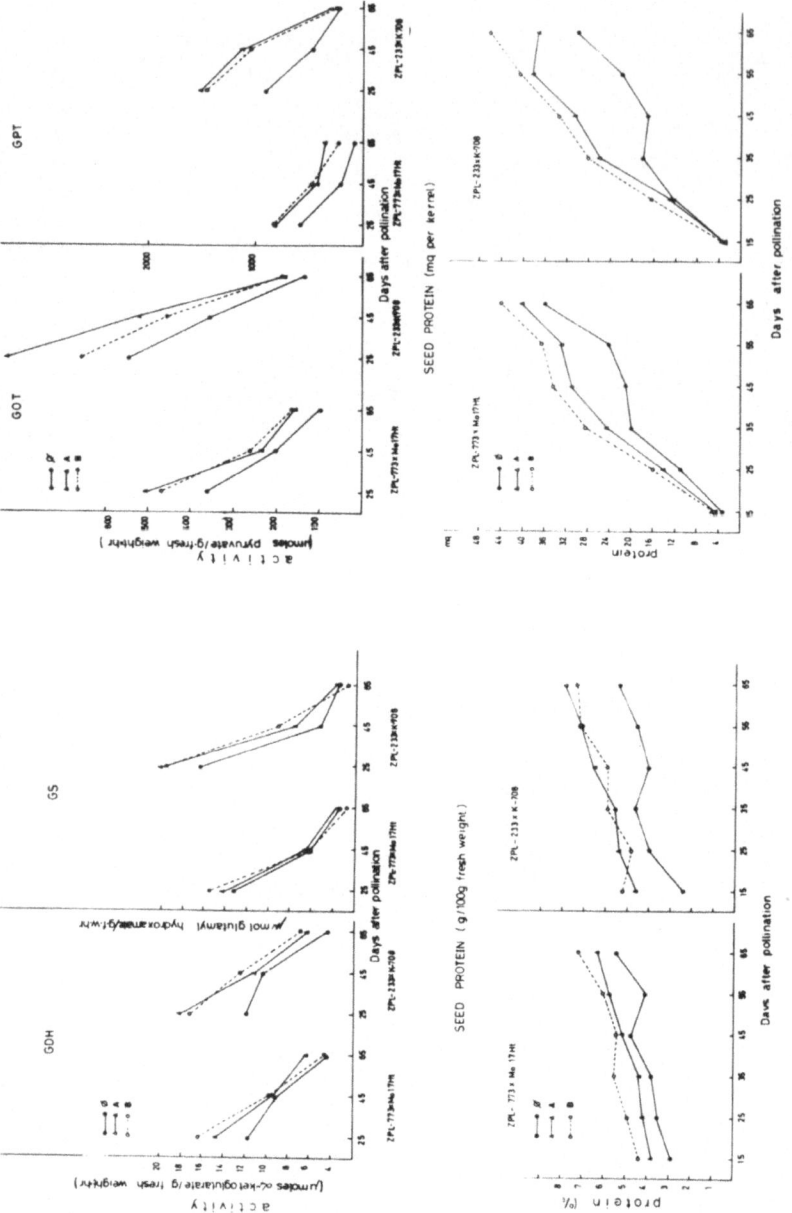

Fig. 2. Effect of the nitrogen top-dressing on GS, GDH, GOT, GPT activities and on total protein in the maize kernel (Activity in µmol/g fresh weight·hour).

with nitrogen, the lower was the total protein decrease in the leaf. Nitrogen top-dressing was reflected in the increase of protein quality of the leaf. On the average, it amounted to 15.3% in A variant and 31.1% in B variant in the leaf above the ear of the hybrid ZPL-773 × Mo 17 Ht, and to 34.1% in A and 44.2% in B in the leaf below the ear. In the leaf above the ear of hybrid ZPL-233 × K-708, protein quantity increase amounted to 31.1% in A 40.3% in B; in the leaf below the ear it amounted to 52.2% in A and 57.2% in B. These results indicate the well-known fact that the transport process is more efficient from the leaves above the ear, or rather, that they are physiologically more active during kernel ontogenesis.

Enzyme activity in maize kernel

In order to determine whether there are any differences on the level of NH_4^+ assimilation and transamination, which would be conditioned by genotypes and nitrogen concentration applied, the activity of GS, GDH, GOT, and GPT enzymes during development was studied in the endosperm (Fig. 2).

GS enzyme, which catalyzes ammonia incorporation into the amide position of glutamine, was active in the stages of endosperm development studied, *i.e.* during intensive synthesis of reserve proteins. GS activities were the highest at the initial stages of endosperm development and decreased at the end of kernel formation. Maintaining relatively high GS activity during reserve protein synthesis could mean that GS participates in ammonia assimilation from sources other than nitrate assimilation. Similar results were reported by Scott *et al.*[12] and they show that GS activity does not decrease according to the decrease in nitrate absorption. The most pronounced differences were in GS activity between the control variant and the variant with top-dressing on the 25th day after pollination.

GDH enzyme was present in the developing endosperm. This enzyme catalyzes ammonia incorporation into ketoglutaric acid in order to obtain glutamic acid. The highest GS activities were measured on the 25th day after pollination in both genotypes. There was an increased GDH activity in variants with top-dressing, but there were no specific differences between genotypes.

The activity of GS and GDH enzymes in the kernel was considerably lower than their activity in the leaf, which would conform to the reported results that the activity of these enzymes is not essential for endosperm development.

The activity of transaminases, which include glutamate as the product of the reaction of glutamate synthase, was proved in endosperm of the developing maize kernel. Aspartate aminotransferase enzyme (GOT) activity was higher during intensive synthesis of endosperm reserve proteins and was highest on the 25th day after pollination in both genotypes and in variants with nitrogen top-dressing compared to control variant. Alanine aminotransferase enzyme (GPT) had higher activities than GOT during intensive synthesis of reserve proteins. The sum of GPT activities was greater in the endosperm of the hybrid

ZPL-233 × K-708. The increase in GPT activity was approximate in variants with nitrogen side-dressing and was the highest on the 45th day after pollination.

Top-dressing also influenced the increase in the quantity of total kernel protein. This increase was the highest after the 35th day after pollination in both hybrids. Differences in the quantity of total protein in the kernel are insignificant between hybrids. However, great differences were found per unit of area. Grain yield for hybrid ZPL-733 × Mo 17 Ht amounted to 10136 kg/ha and protein yield to 1256 kg/ha. Grain yield for hybrid ZPL-233 × K-708 amounted to 7771 kg/ha and protein yield to 955 kg/ha in B variant, where the yield increase, compared to Ø variant, amounted to 52% and 23% respectively.

References

1   Arnold J M, Bauman L F and Aycock H S 1977 Interrelation among protein, lysin, oil, certain mineral element concentrations and physical kernel characteristics in two maize populations. Crop Sci. 17, 421–426.
2   Bergmeyer H I 1970 Glutamate Dehygrogenase. *In* Bergmeyer H I, Methoden der enzymatischen Analyse. Academic Verlag, Berlin. pp 613–616.
3   Chevalier D S and Schrader L E 1977 Genotypic differences in nitrate absorption and partitioning of N among plant parts in maize. Crop Sci. 17, 897–902.
4   Elliot W H 1953 Isolation of glutamine synthetase and glutamate transferase from green peas. J. Biol. Chem. 201:661.
5   Hageman R H and Flesh D 1960 Nitrate reductase activity in corn seedlings as affected by light and nitrate content of nutrition media. Plant Physiol. 35, 700.
6   Hadži-Tašković Sukalović V 1981 Study of the activity of more important enzymes from the oxydoreductase and transaminase system and protein changes in normal and O_2 maize genotypes. Doctoral thesis, Agricultural faculty, Beograd.
7   Kovačević V 1977 Particularity of contents of some elements in the leaf, pollen and style of inbred lines of corn. J. Sci. Agr. Res. 120, 129–136.
8   Miflin B and Lea P 1976 The pathway of nitrogen assimilation in plants. Phytochemistry 15, 873–885.
9   Sarić M and Krstić B 1978 Studies of the amount of some elements of mineral nutrition in different maize hybrids. J. Sci. Agr. Res. 113, 61–75.
10 Sarić M and Krstić B 1981 Effect of different concentrations of nutritive solution on the weight of dry matter and the content of N, P, K, Ca and Mg in some plant species. J. Sci. Agr. Res. 147, 319–329.
11 Schrader L E, Cataldo D A and Peterson D M 1974 The use of protein in extraction and stabilization of nitrate reductase. Plant Physiol. 53, 688.
12 Scott B D and Neyra A C 1979 Glutamate synthetase and nitrate assimilation in Sorghum (*Sorghum vulgare*) leaves. J. Botany 57, 754.
13 Tonhazy N E 1960 Glutamat-oxalacetat-transaminase und glutamat pyruvat-transaminase. *In* H I. Bergmayer, Methoden der enzymatischen Analyse. Academice Verlag, Berlin, pp. 695, pp 723.

Changes in ion composition and hexitol content of different *Plantago* species under the influence of salt stress

H. KÖNIGSHOFER
Botanisches Institut, Universität für Bodenkultur, Gregor Mendel-Straße 33, A-1180 Wien, Austria

Key words Anions Cations Culture experiment Hexitols Ionic balance Inorganic anions Organic anions Osmoregulation *Plantago* spp. Salinity Salt tolerance

Summary The effect of salinity on the ionic balance and the hexitol content of the halophyte *P. maritima* L. and the nonhalophytes *P. major* L. ssp. *major*, *P. lanceolata* L., and *P. media* L. was studied in a culture experiment. In response to salt application the nonhalophilous species increased their internal electrolyte content to a considerably greater extent at high transpiration than *P. maritima*. Excessive uptake of Cl and SO_4 following salt-treatment was in most cases compensated by a decline in the total organic anion content in all species under investigation. Na was strongly accumulated in the shoots of *P. maritima* when subjected to salt-stress, while the nonhalophytic species tended to exclude this ion from leaf tissue enhancing Mg-uptake for charge balance. Acyclic polyhydric alcohols (sorbitol and mannitol), the dominant soluble carbohydrates in all *Plantago* species studied, increased, with one exception, in all plants under saline conditions. The results indicate marked physiological differences between the halophyte *P. maritima* and the nonhalophytic members of the genus *Plantago* in their reaction to salinity being perhaps partially responsible for differences in the degree of salt tolerance.

Introduction

Adaptive response of plants to salinity involves various aspects including maintenance of turgor (osmotic adjustment) regulation of the ion content (ionic balance), adequate distribution of the accumulated ions within the whole plant as well as at sub-cellular level (compartmentation), and synthesis of compatible solutes[3,10,13]. Though physiological differences in the reaction of plants to saline substrate depend largely on factors such as type, level and duration of salinity, general environmental conditions, and developmental stage of plants, comparative investigations have also revealed the importance of genetic control of salt transport and salt tolerance in halophytes and nonhalophytes (cf.[6,18]).

In this paper four species of the same genus differing in their ecological and nutritional demands, the halophyte *Plantago maritima* L. and the nonhalophytic members *Plantago major* L. ssp. *major*, *Plantago lanceolata* L. and *Plantago media* L., were studied in their response to saline conditions with regard to changes in the ion composition and hexitol content. *P. major* ssp. *major* is known to prefer relatively nutrient-rich habitats; in addition this species is considered to be somewhat more salt tolerant than *P. lanceolata* and *P. media* which grow on soils generally low in salt and nutrients[5,26].

M.R. Sarić and B.C. Loughman (eds.), Genetic Aspects of Plant Nutrition.
ISBN-13: 978-94-009-6838-7
©1983 *Martinus Nijhoff/Dr W. Junk Publishers, The Hague/Boston/Lancaster.*

Materials and methods

Growth conditions. Culture experiments on four *Plantago* species (*Plantago maritima* L., *Plantago major* L. ssp. *major*, *Plantago lanceolata* L., and *Plantago media* L.) were carried out in 1976, between mid-March and mid-October. Vegetative rosettes collected from their natural habitats (*P. maritima* from saline soils near the Neusiedler Lake in Austria, the nonhalophytic species from circumneutral soils near Vienna) were planted in plastic vessels filled with quartz sand. The pots were watered daily for 2–4 h with a nutrient solution composed of (in equiv. m^{-3}): K, 8.0; Ca, 4.5; Mg, 2.2; NH_4, 1.1; NO_3, 6.7; H_2PO_4, 1.7; HPO_4, 1.1; SO_4, 5.2; Cl, 1.1 either with or without addition of appropriate salt concentrations (50 equiv. m^{-3} NaCl + 25 equiv. m^{-3} Na_2SO_4 + 25 equiv. m^{-3} $MgSO_4$ in the case of *P. maritima* and 25 equiv. m^{-3} NaCl + 12.5 equiv. m^{-3} Na_2SO_4 + 12.5 equiv. m^{-3} $MgSO_4$ in the case of the nonhalophytic species). Salt-treated plants of *P. maritima* were grown under saline conditions continuously from the beginning until the end of the experiment, whereas the nonhalophilous species were imposed to salt-stress only twice (for two weeks in July and for three weeks from the end of September until mid-October). After each period of salt-treatment leaf material was harvested. The experiments were conducted in the open air; the culture vessels were protected from rain by a mobile roof. Culture solutions were renewed fortnightly. Chemical analyses have been carried out on hot-water extracts from freeze dried and ground plant samples. Inorganic ions were measured by standard methods, organic anions and hexitols were determined by gas liquid chromatography (for details see refs.[16,22]).

Results

Plants of *P. maritima* subjected to saline conditions were reduced in growth and appeared more succulent than the control plants but did not exhibit any visible salt-injury symptoms, whereas the nonhalophytic species (especially young leaves in July) became slightly chlorotic during exposure to salinity. Shoots of *P. major* ssp. *major* were infected by a fungus during the culture period and could therefore be harvested only once.

Fig. 1 and Fig. 2 show the effect of salinity on the ionic balance of the *Plantago* species investigated. The total anion level indicates that *P. maritima* increased its internal electrolyte content under saline conditions (100 equiv. m^{-3}) only from 480 to 594 equiv. m^{-3} at the first harvest. Even though the nonhalophilous species were subjected to salt concentrations half that applied to *P. maritima*, increments in the total anion content of 240 (*P. major* ssp. *major*), 254 (*P. media*) and 359 equiv. m^{-3} (*P. lanceolata*) are probably the result from exposure to salt-stress during the hot culture period in July. However, the second salt-treatment at the end of the season enhanced the internal electrolyte levels significantly less in the nonhalophytic species.

The quantitative changes in the inorganic anion content indicate that all species adapted their osmotic potential to the salinity of the medium to a great extent by massive uptake of Cl and SO_4 in more or less equal amounts according to the composition of the culture solution. Parallel to the total anion content the sum of Cl and SO_4 (in equiv. m^{-3}) increased in July to considerably higher values in the nonhalophilous species treated with only 50 equiv. m^{-3} salt concentration

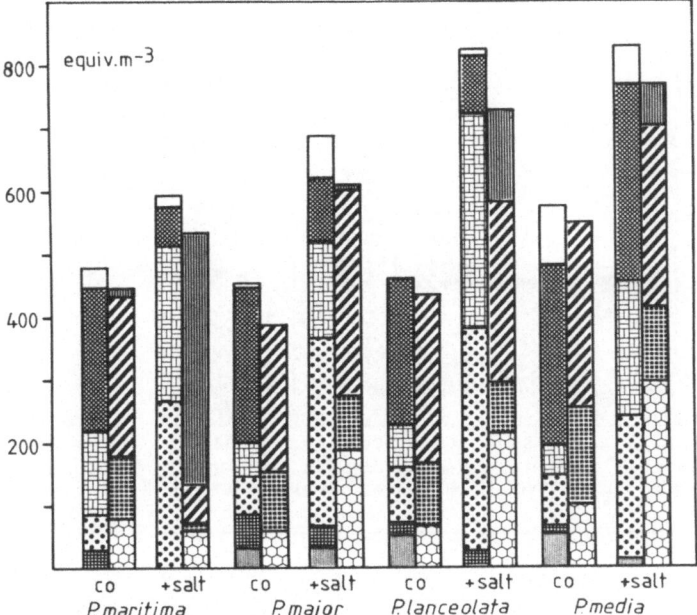

Fig. 1. Effect of salinity on the ionic concentrations, based on tissue water content, of the plants under investigation. First harvest: mid-July. co = control. Key to the symbols used see Fig. 2.

(about 3.4-fold in *P. media*, 4.2-fold in *P. major* ssp. *major* and 4.5-fold in *P. lanceolata*) than in *P. maritima* treated with 100 equiv. m^{-3} (about 2.5-fold). At the end of the vegetation period the nonhalophytic *Plantago* species accumulated both anions to a markedly lesser degree under saline conditions.

Alterations in the NO_3 and PO_4 content were not uniform; in some cases the tissue concentration of these anions was reduced under salt-stress.

Striking differences between the species investigated could be detected with respect to the effect of salinity on the cation content. In *P. maritima* Na was transported preferentially to the shoot, whereas the nonhalophilous species tended to exclude Na from the aerial parts of the plants. The most effective exclusion of Na from the leaf tissue was achieved by *P. major* ssp. *major*. In *P. maritima* accumulation of Na was offset by a drastic decline in K, while in the salt-treated nonhalophytic species K remained either nearly at the same level compared with the controls or decreased only slightly (in *P. lanceolata*, second harvest, old leaves) respectively increased even in *P. major* ssp. *major*. The Ca content, too, was generally lowered under saline conditions in the nonhalophytic species to a lesser degree than in *P. maritima*. Lack of high levels of Na in the leaf tissue of the salt-treated nonhalophilous *Plantago* species was compensated partially by a remarkable high intake of Mg, whereas in *P. maritima* the Mg content was hardly affected by salinity.

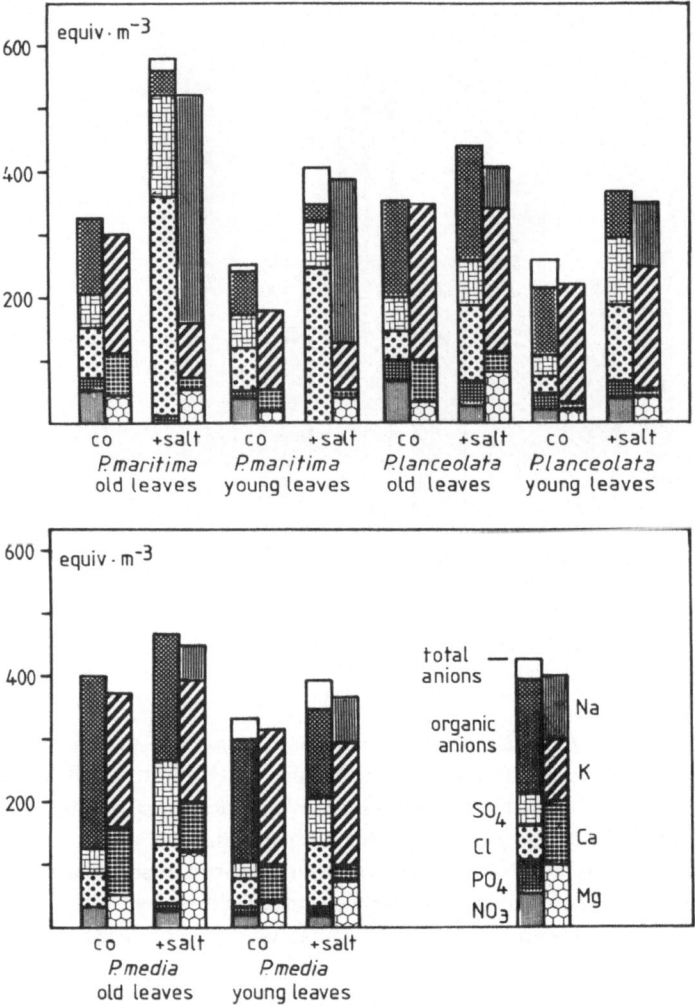

Fig. 2. Effect of salinity on the ionic concentrations, based on tissue water content, of the plants under investigation. Second harvest: mid-October. co = control.

With two exceptions (*P. media*, first harvest and *P. lanceolata*, second harvest, old leaves), the total amount of organic anions was more or less diminished in all species subjected to saline environment regardless of the qualitatively different organic acid patterns. The decline in the organic anion content under salt-stress was mostly pronounced in *P. maritima* and *P. major* spp. *major*.

All four species studied contained large quantities of acyclic C_6-sugar alcohols. Separation of the acyclic polyols revealed that sorbitol contributes 80–90% and mannitol correspondingly 10–20% to the total amount of acyclic hexitols. With

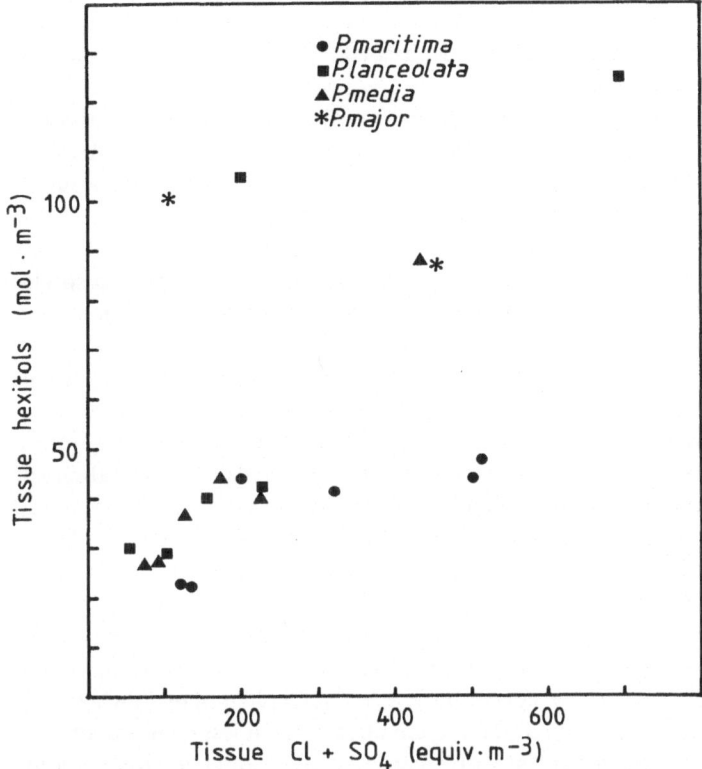

Fig. 3. Relationships between hexitol and electrolyte accumulation. Data were taken from both series of the experiment.

one exception (*P. major* ssp. *major*), the hexitol level increased in all species upon exposure to salinity (up to 3.5-fold: *P. lanceolata*, second harvest, old leaves). As can be seen in Fig. 3 a quite good relationship between hexitol and tissue salt (Cl + SO$_4$ in equiv. m^{-3}) concentrations could be found in *P. media* and *P. lanceolata*, while in *P. maritima* and *P. major* ssp. *major* such distinct correlations were not evident. But the results for *P. major* ssp. *major* are certainly of minor validity, as this species could be investigated only once.

Discussion

Comparative studies with related genotypes have revealed that it is generally not possible to establish a clear correlation between the internal electrolyte concentration and the degree of salt tolerance of either halophytes or nonhalophytes. For instance wild species of *Lycopersicon* increase their tissue level of Na and Cl to a considerably greater extent under salt-stress and prove to be much more salt tolerant than the cultivated tomato[25,27]. On the other hand, in soybean mutants[20], in varieties of grapes[4], in different species of *Juncus*[24] or in ecotypes of *Agrostis stolonifera*[2], salt tolerance is associated with relatively low salt concentrations in the shoot.

In the *Plantago* species investigated the most striking differences in response to salinity between the halophyte *P. maritima* and the nonhalophilous members were obtained after the first salt-treatment of the nonhalophytes which occurred during a hot period in July. The results concerning ion content and ion composition indicate that *P. maritima* was able to restrict salt intake to some extent when exposed to salinity while the salt-stressed nonhalophytic species were obviously not capable to prevent excessive electrolyte accumulation (mainly Cl and SO_4) in the shoots of the plants at high transpiration when passive movement of mineral ions with the water flow may be enhanced. This led perhaps to internal osmotic imbalances resulting from a buildup of high salt concentrations in the cell walls of the nonhalophilous species at the end of the transpiration stream which would affect the water relations of the leaf cells[13,21,23]. The ability of *P. maritima* to limit ion uptake under saline conditions may be partially due to reduced transpiration rates of its small succulent leaves while the broader leaves of the nonhalophytic species certainly have a higher water turn-over. In addition, differences between salt-sensitive (*P. media*) and salt-tolerant (among them *P. maritima*) *Plantago* species were observed concerning changes in the lipid levels in roots and shoots upon exposure to NaCl[8]. In roots of *P. media* the lipid level decreased strongly even at low concentrations of NaCl (50 $mol.m^{-3}$) whereas in *P. maritima* the level of most lipid classes remained more or less constant up to 75 $mol.m^{-3}$. This could be important with regard to control and regulation of ionic permeability of the root cell membranes.

The results obtained concerning the effect of salinity on the cation content are consistent with several other findings. *P. maritima* accumulated high quantities of Na in the shoot associated with a decrease in K and Ca as do many "salt accumulators" among halophytes (cf.[3,9,10]). In contrast, the nonhalophilous *Plantago* species kept the Na content rather low in the leaf tissue under saline conditions. Perhaps Na is retained in basal plant parts (roots or rhizomes) by the nonhalophytic species as described for some crop plants[14,19]. In this connection a comparative study of Na-uptake and transport has shown that in fact Na-translocation from the root to the shoot is more pronounced in *P. maritima* than in *P. media*, while no differences could be detected concerning the primary uptake process through the plasmalemma of the cortical root cells between these two species[7].

Considerable accumulation of Cl and SO_4 in the leaves of the salt-treated species was balanced in most cases by a decline in the organic anion content. Similar shifts in the ionic balance due to salinity were found in halophytic and nonhalophytic *Juncus* species[24] and in some varieties of *Centaurium littorale*[11].

The role of sorbitol as a possible compatible cytoplasmic solute in higher plants subjected to salt-stress or low external osmotic potentials has been discussed for *P. maritima* and *P. coronopus* in former reports[1,12,15,17]. The present results show that alterations in the hexitol content in response to salinity also occur in nonhalophytic members of the genus *Plantago*. In *P. media* and *P. lanceolata* enhanced polyol content was even better correlated with increasing internal salt concentrations than in *P. maritima*. However, it is certainly difficult to estimate adequately the significance of the acyclic polyhydric alcohols for osmoregulation in *Plantago* in the absence of information about the sub-cellular localization of these polyols.

Acknowledgement The author is grateful to Doz. R. Albert and Dr. M. Popp for critical reading of the manuscript.

References

1 Ahmad I, Larher F and Stewart G R 1979 Sorbitol, a compatible osmotic solute in *Plantago maritima*. New Phytol. 82, 671–678.
2 Ahmad I, Wainwright S J and Stewart G R 1981 The solute and water relations of *Agrostis stolonifera* ecotypes differing in their salt tolerance. New Phytol. 87, 615–629.
3 Albert R 1982 Halophyten. *In* Pflanzenökologie und Mineralstoffwechsel. Ed. H Kinzel. Verlag Eugen Ulmer, Stuttgart.
4 Bernstein L, Ehlig C F and Clark R A 1969 Effect of grape rootstocks on chloride accumulation in leaves. J. Am. Soc. Hortic. Sci. 94, 584–590.
5 Blom C W P M 1976 Effects of trampling and soil compaction on the occurence of some *Plantago* species in coastal sand dunes. I. Soil compaction, soil moisture and seedling emergence. Oecol. Plant. 11, 225–241.
6 Epstein E and Jefferies R L 1964 The genetic basis of selective ion transport in plants. Annu. Rev. Plant Physiol. 15, 169–184.
7 Erdei L and Kuiper P J C 1979 The effect of salinity on growth, cation content, Na^+-uptake and translocation in salt-sensitive and salt-tolerant *Plantago* species. Physiol. Plant. 47, 95–99.
8 Erdei L, Stuiver C E E and Kuiper P J C 1980 The effect of salinity on lipid composition and on activity of Ca^{2+}- and Mg^{2+}-stimulated ATPases in salt-sensitive and salt-tolerant *Plantago* species. Physiol. Plant. 49, 315–319.
9 Flowers T J 1975 Halophytes. *In* Ion Transport in Plant Cells and Tissues. Eds. D A Baker and J L Hall. pp. 309–334. North Holland, Amsterdam.
10 Flowers T J, Troke P F and Yeo A R 1977 The mechanism of salt tolerance in halophytes. Annu. Rev. Plant Physiol. 28, 89–121.
11 Freijsen A H J and Van Dijk A 1975 Differences in growth rate and salt tolerance between varieties of the halophyte *Centaurium littorale* (Turner) Gilmour, and their ecological significance. Acta Bot. Neerl. 24, 7–22.
12 Gorham J, Hughes L I and Wyn Jones R G 1981 Low-molecular-weight carbohydrates in some salt-stressed plants. Physiol. Plant. 53, 27–33.
13 Greenway H and Munns R 1980 Mechanisms of salt tolerance in nonhalophytes. Annu. Rev. Plant Physiol. 31, 149–190.
14 Jacoby B 1964 Function of bean roots and stems in sodium retention. Plant Physiol. 39, 445–449.

15 Jefferies R L, Rudmik T and Dillon E M 1979 Responses of halophytes to high salinities and low water potentials. Plant Physiol. 64, 989–994.

16 Königshofer H 1981 Stoffwechselphysiologische Untersuchungen an Plantago-Arten unter besonderer Berücksichtigung ökologischer Aspekte. Phil. Diss. Universität Wien.

17 Lambers H, Blacquière T and Stuiver C E E 1981 Interactions between osmoregulation and the alternative respiratory pathway in *Plantago coronopus* as affected by salinity. Physiol. Plant. 51, 63–68.

18 Läuchli A 1976 Genotypic variation in transport. *In* Encyclopedia of Plant Physiology. New Series, vol. 2, part B. Eds. U Lüttge and M Pitman. Springer Verlag, Berlin-Heidelberg-New York.

19 Läuchli A 1976 Symplasmic transport and ion release to the xylem. *In* Transport and Transfer Processes in Plants. Eds. I F Wardlaw and J B Passioura. Academic Press, London, New York.

20 Läuchli A 1979 Regulation des Salztransportes und Salzausschließung in Glykophyten und Halophyten. Ber Deutsch. Bot. Ges. 92, 87–94.

21 Oertli J J 1968 Extracellular salt accumulation; a possible mechanism of salt injury in plants. Agrochimica 12, 461–469.

22 Popp M and Kinzel H 1981 Changes in the organic acid content of some cultivated plants induced by mineral ion deficiency. J. Exp. Bot. 32, 1–8.

23 Ramati A, Liphschitz N and Waisel Y 1979 Osmotic adaptation in *Panicum repens*. Differences between organ, cellular and subcellular levels. Physiol. Plant. 45, 325–331.

24 Rozema J 1976 An ecophysiological study on the response to salt of four halophytic and glycophytic *Juncus* species. Flora 165, 197–209.

25 Rush D W and Epstein E 1976 Genotypic responses to salinity. Differences between salt-sensitive and salt-tolerant genotypes of the tomato. Plant Physiol. 57, 162–166.

26 Sagar J R and Harper J L 1964 *Plantago major*, *Plantago lanceolata* and *Plantago media*. J. Ecol. 52, 189–221.

27 1971 Salt tolerance in the wild relatives of the cultivated tomato: responses of *Lycopersicon esculentum*, *L. peruvianum* and *L. esculentum* minor to sodium chloride solution. Aust. J. Agric. Res. 22, 631–638.

Influence of chlorate ions on some characteristics of maize seedlings with different protein content

K. KONSTANTINOV, V. LAZIĆ, M. DENIĆ, Č. RADENOVIĆ and V. FURTULA
*Maize Research Institute, Zemun, Yugoslavia
and Faculty of Science, University of Belgrade, Belgrade, Yugoslavia (V.F.)*

Key words Ions Maize Roots Proteins Genetic control

Introduction

Nitrate reductase plays a major role in regulating nitrogen metabolism[2] and its activity can be related to the yield capacity of the crop[3,7]. According to data reported before[12,14] there is a competition between chlorate and nitrate for the same enzyme. The behaviour of chlorate suggests that chlorate and nitrate are transported by the same carrier system[6]. It is assumed that a direct effect of chlorate on the membranes cannot be excluded[6]. This paper deals with a study on the effect of chlorate ions on the plant height, weight of root system, and water content in roots of different genotypes of maize.

Materials and methods

For the investigation of the effect of chlorate ions on plant height two genotypes with different protein content were grown in nutrient solution with different chlorate ion concentrations and other inbred lines have been treated in the same way. For the determination of the effect of chlorate on the water content two inbred lines and their hybrid combination have been used. Plants were measured seven days after planting.

Results and discussion

To find out if any correlation exists between the protein content in the kernel and the effect of chlorate ions on plant height two genotypes with extremely high and low protein content (ILL HP, Illinois High Protein Corn and ILL LP, Illinois Low Protein Corn) were grown in different concentrations of chlorate ions.

The effect of chlorate on plant height in different genotypes is presented in Fig. 1. Chlorate ions have different effect on plants with different content of protein. Increased chlorate concentration caused a decrease of plant height. Similar results were obtained when three inbred lines with different protein content in the kernel were treated with chlorate are presented in Fig. 2. Since protein content is

M.R. Sarić and B.C. Loughman (eds.), Genetic Aspects of Plant Nutrition.
ISBN-13: 978-94-009-6838-7
©1983 *Martinus Nijhoff/Dr W. Junk Publishers, The Hague/Boston/Lancaster.*

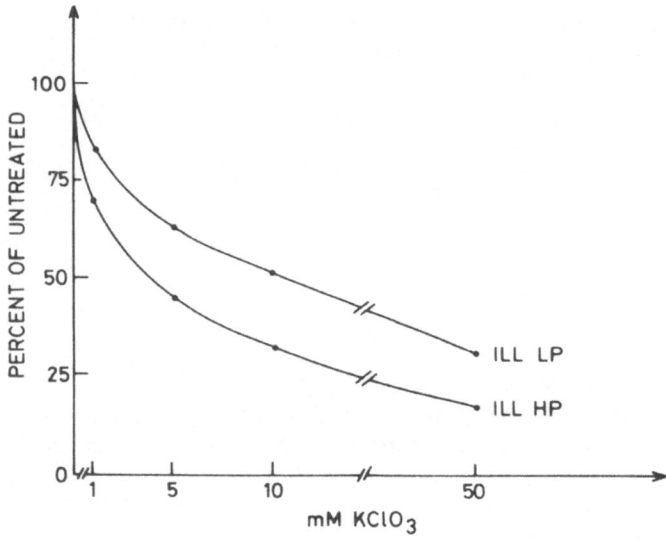

Fig. 1. A – Effect of different KClO₃ concentrations on plant height.

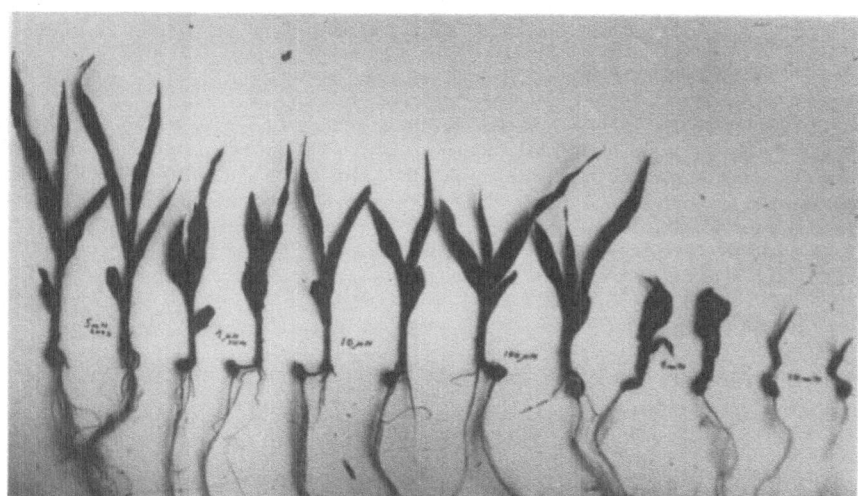

Fig. 1. B – Appearance of the seedlings obtained fron treated seeds with different chlorate concentrations.

genetically determined it can be concluded that the chlorate effect on plant height was dependent on genotype.

As the root system is reponsible for the absorption of ions and water the effect of chlorate ion on the water content in different genotypes (two inbred lines and their hybrid combination was examined). Results are presented in Table 1.

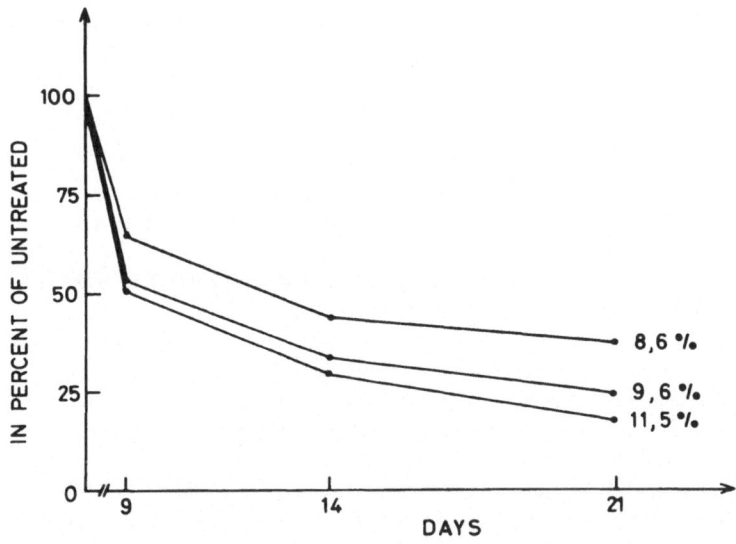

Fig. 2. Influence of chlorate ions on plant height of the genotypes with different protein content in kernel.

Table 1. Effect of chlorate on water content in primary roots of different maize genotypes

Concentration of chlorate mM	Water content		
	L_1	L_2	F_1
0	93.2	94.2	92.3
1.0	93.5	94.2	93.0
5.0	93.0	94.3	92.7
10.0	95.4	92.3	91.2

Reduction of water content was obtained in inbred line L_2 and the hybrid combination. The inbred line L_1 reacts differently.

It could be concluded that the effect of chlorate on the water content was different in different genotypes and that hybrid plants were less sensitive than inbred lines.

In order to determine the mechanism by which ions pass through biological membranes and have a particular effect on the whole plant, one would first have to identify the molecules involved and to determine whether the carrier system is either a lipid–soluble ionophore, or a membrane-spanning polypeptide[8]. Experiments on the protein and lipo-protein complex of the plasma membranes of maize genotypes, on which the effect of chlorate ions investigated, are in progress (manuscript in preparation).

References and literature

1. Bartlett G R 1959 Phosphorus assays in column chromatography. J. Biol. Chem. 234, 466–468.
2. Beevers L D and Hageman R H 1969 Nitrate reduction in higher plants. Annu. Rev. Plant. Physiol. 20, 45.
3. Deckard E L 1970 Nitrate reductase activity and its relationships to yields of grain and grain protein in normal opaque-2 corn (*Zea mays* L.). Ph. D. Thesis, University of Illinois.
4. Denić M 1968 An improved method for determination of amino acid composition in bulk protein of individual maize kernals. Acta Chem. Scand. 22, 1809.
5. Denić M, Konstantinov K, Radenović Č and Furtula V 1981 A study of the amino acid composition of protein fractions in plasma membranes from primary maize roots. Studia Biophysica 83, 3, 173–180.
6. Doddema H, Van de Dijk S I, Hofstra J J, Lambers H, Lauting L and Stulen G 1978 Nitrate uptake and metabolism in relation to ecological factors. Abstract poster. Federation of European Societies of Plant Physiology, Edinburgh.
7. Eilrich G H 1968 Nitrate reductase in wheat (*T. aestivum* L) and is relationship to grain protein production as affected by genotype and spring application of calcium nitrate. Ph. D. Thesis, University of Illinois.
8. Fisher K A 1980 Split membrane analysis. Ann. Rev. Physiology 42, 261–273.
9. Hall J L and Roberts R M 1975 Biochemical characteristics of membrane fractions isolated from maize (*Zea mays*, L) roots. Ann. Bot. 39, 983–993.
10. Hodges T K and Leonard R T 1974 Purification of a plasma membrane bound adenosine triphosphatase from plant roots. Methods Enzymol. 32, 392–406.
11. Hodges T K, Leonard R T, Bracker C E and Keenan T W 1972 Purification of an ion-stimulated adenosine triphosphatase from plant roots associated with plant roots. Proc. Natl. Acad. Sci. USA 69, 3307–3311.
12. Hofstra J J 1977 Chlorate toxicity and nitrate reductase activity in tomato plants. Physiol. Plant. 41, 65–69.
13. Leonard R T and Van Der Wande W J 1976 Isolation of plasma membranes from corn roots by sucrose density centrifugation. An anomalous effect of Ficoll. Plant Physiol. 57, 105–114.
14. Liljestrom S and Åberg B 1966 Studies on the mechanism of chlorate toxicity. K Lautbrukskögsk. Ann 32, 93–107.
15. Popović V and Denić M 1978 Activity of nitrate reductase in some genotypes of maize. Genetika 10 (1), 123–129.
16. Saxton M J, Breidenbach R W and Lyons J M 1980 Membrane dynamics: Effect of environmental stress. *In* Genetic Engineering of osmoregulation. Impact on Plant Productivity Food, Chemicals and Energy. Plenum Press, New York and London, 1980.
17. Solomonson L P and Vennesland B 1972 Nitrate reductase and chlorate toxicity in *Chlorella vulgaris* Beyerinck. Plant Physiol. 50, 421–424.
18. Spackman D H, Stein W H and Moore S 1958 Automatic recording apparatus for use in the chromatography of amino acids. Anal. Chem. 30, 1190.

Efficiency of nitrogen utilization and photosynthetic rate in C_3 and C_4 plants

B. KRSTIĆ and M. R. SARIĆ
Institute of Biology, Faculty of Natural Science, University of Novi Sad, Yugoslavia

Key words C_3 and C_4 plants   N utilization   Photosynthesis

Introduction

It was postulated by the theoretical studies of Brown[1,3] that, due to specific metabolism in C_3 and C_4 plants, the latter utilize nitrogen more efficiently than C_3. According to these authors, such efficiency is primarily the result of a decrease in RuDP-carboxylase content in C_4 plants. Since this enzyme may contain as much as 50% of total nitrogen[9], nitrogen concentration in C_4 plants must be lower than that of C_3.

It is worth noting that there are almost no comparative investigations of both the efficiency of nitrogen utilization and the photosynthetic rate in these two plant types. Furthermore, the evidence concerning this question has been obtained from experiments undertaken in various agroecological conditions. Since not only plant species but cultivars as well differ in the concentration of mineral elements, adequate numbers of both plant types must be analyzed. Therefore, distinctions in the efficiency of utilization of various mineral elements, nitrogen particularly, as well as the photosynthetic rate in a number of plant species belonging to C_3 and C_4 plants were chosen for the subject of the present paper.

Material and methods

A number of the C_4 plants (maize, sweet corn, broom corn, and millet) and some C_3 species (sunflower, barley, and soybean) were included. The plants were grown in water culture for 25 days on complete nutrient solution[4], and on distilled water. Dry weight, leaf area, CO_2 absorption rate, and nitrogen concentration were determined.

Results

Dry weight (Table 1)

Dry weight was higher in plants grown on nutrient solution than in those on distilled water. Due to the contents of individual substances of the nutritive tissue of seeds different dry weights were obtained in the plant species grown on

Table 1. Dry weight of above-ground parts and roots of different plant species grown on distilled water and nutrient solution (mg/plant)

Plant species	Distilled water			Nutrient solution			Ratio Root on nutrient solution/ Root on distilled water	Ratio Above-ground part on nutrient solution/ Above-ground part on distilled water
	Above-ground part	Root	Ratio Above-ground part/ Root	Above-ground part	Root	Ratio Above-ground part/ Root		
Maize	11.67	86.00	1.29	369.00	95.33	3.87	1.10	3.30
Broom corn	16.33	21.00	0.81	62.00	33.00	1.87	1.57	3.79
Sweet corn	20.00	37.67	0.53	83.00	41.33	2.00	1.11	4.15
Millet	5.00	1.33	3.75	57.53	23.53	2.86	4.51	11.40
Sunflower	58.00	33.00	1.56	255.33	37.00	7.72	1.12	4.39
Barley	19.00	19.33	0.98	110.67	21.00	5.26	0.92	5.78
Soybean	127.33	33.33	3.82	130.67	55.67	2.34	1.67	1.02
LSD 5%	9.37	4.11		44.03	4.22			
1%	13.14	5.76		61.74	5.92			

distilled water. Thus, for instance, the highest dry weight of both the aboveground part and root was recorded in maize, the lowest in common millet. Highly significant differences in dry weight of the same two plant species were obtained when they were grown on nutrient solution and the sequence of recorded dry weights was almost the same as in the plants grown on distilled water. It should be emphasized that after 25 days no differences in the synthesis of organic matter between the investigated groups of C_3 and C_4 plants were observed, whereas great differences were recorded among the plant species disregarding the plant type. In addition, according to the ratios obtained for the above-ground parts and roots of plants grown on distilled water and nutrient solution, no relationship was found between the two plant types in synthesizing the organic matter. It should be also emphasized that the ratios obtained for the aboveground parts and roots of plants grown on distilled water were evidently smaller than in those grown on nutrient solution. Such a result may be explained by nutrition conditions. Finally, when the ratios obtained for plant roots, grown on nutrient solution and distilled water were compared, as well as those for the aboveground parts grown on distilled water and nutrient solution, no relationship between the two types under consideration was recorded. The relationship was found only among the plant species.

Leaf area (Table 2)

The leaf area was considerably smaller in the plants grown on distilled water than in those on nutrient solution, ranging from 1.27 in soybean to 25.41 in common millet. Significant differences in leaf area were found among the plant species grown on nutrient solution, and highly significant differences were

Table 2. Leaf areas of different plant species grown on distilled water and nutrient solution (cm^2/plant)

Plant species	Distilled water	Nutrient solution	*Ratio* Nutrient solution / Distilled water
Maize	28.36	166.29	5.85
Broom corn	5.11	35.98	7.62
Sweet corn	5.87	35.35	5.67
Millet	1.34	34.06	25.41
Sunflower	7.88	84.45	10.73
Barley	4.77	47.45	10.00
Soybean	34.79	44.41	1.27
LSD 5%	2.02	14.75	
1%	2.83	20.68	

recorded between the maize and sunflower plants, whereas no significant differences were recorded among either sorghum species, common millet, soybean and barley. In the plant species grown on distilled water, no differences were established between either sorghum species or barley.

CO_2 absorption rate (Table 3)

The amount of absorbed $^{14}CO_2$ depended upon both the growing conditions and plant species under investigation. Considerably lower $^{14}CO_2$ absorption rates were obtained in the plants grown on distilled water than in those grown on nutrient solution. Generally, the C_3 plants absorbed smaller amounts of $^{14}CO_2$ than the C_4 species. However, highly significant differences in $^{14}CO_2$ absorption rate were found among the plant species grown on both nutrient solution and distilled water with the exception of barley and sorghum.

Nitrogen concentration (Table 4)

As has been already quoted, C_4 plants are characterized by lower nitrogen concentration than the C_3. In our experiments plants grown on distilled water, as well as those on nutrient solution, showed different nitrogen content. Mostly, significant or highly significant differences were obtained among the plant species under investigation. The maize plants, grown on nutrient solution, were characterized by the smallest nitrogen concentration in both the aboveground part and root. However, the other C_4 plants showed higher nitrogen content in the aboveground parts than the sunflower. In the roots of broom corn and soybean the nitrogen concentration was almost the same. The same was for sweet corn and barley. Different nitrogen concentrations were obtained in true tissues of the plants grown on distilled water and they depended primarily upon the content of nitrogen stored in kernels.

Table 3. $^{14}CO_2$ absorption rate in different plant species grown on distilled water and nutrient solution (Bq/10 mg dry weight)

Plant species	Distilled water	Nutrient solution
Maize	29.1	95.2
Broom corn	40.1	145.7
Sweet corn	79.1	115.4
Millet	56.2	151.2
Sunflower	6.9	19.7
Barley	38.6	50.2
Soybean	13.9	27.1
LSD 5%	3.15	2.81
1%	4.41	3.96

Table 4. The nitrogen concentration in aboveground aparts and roots of different plant species grown on distilled water and nutrient solution (mg/100 g dry weight)

Plant species	Distilled water		Nutrient solution	
	Above-ground part	Root	Above-ground part	Root
Maize	3066	1547	4039	3073
Broom corn	1379	739	5056	4245
Sweet corn	2042	1021	5126	3539
Millet	2427	736	5300	4047
Sunflower	2605	731	4994	4098
Barley	2214	1004	5334	3514
Soybean	4149	2494	5999	4274
LSD 5%	72	23	88	103
1%	101	33	124	145

Discussion

It is difficult to state that the C_4 plants are characterized by lower nitrogen concentrations than the C_3 plants. According to our results, the C_4 plants grown on nutrient solution absorbed greater amounts of $^{14}CO_2$ than C_3. Such results are in agreement with those obtained by Das et al.[2] However, we found no relationship between CO_2 absorption and dry weight as related to plant type (Tables 1 and 3). No relationship between the nitrogen content and the CO_2 absorption rate was found. Also, Smith and Boutton[8], working on communities of both C_3 and C_4 plants, found that the nitrogen concentration was higher in the C_4 than in the C_3 species. On the other hand, Moore and Black[3] state that the C_4 plants utilize nitrogen more efficiently than the C_3 ones due to the "division of labor" between mesophyll and bundle sheath cells.

Taking into consideration the fact that even greater differences in nitrogen concentration and concentrations of other elements occur among the cultivars of a single plant species[6] than among the species[5,7], not only plant species belonging to the C_3 and C_4 plant types but also a great number of cultivars belonging to the species of the two considered types should be investigated. Only in such a case, we can consider a higher or lower concentration of nitrogen or some other mineral element, i.e. a higher or lower efficiency of a certain element in synthesis of organic matter in the plant species analyzed.

References

1 Brown R H 1978 Difference in N use efficiency in C_3 and C_4 plants and its implications in adaption and evolution. Crop. Sci. 18, 93–98.

2 Das V S R, Veeranjaneyulu K and Ramachandra Reddy A 1981 Photosynthesis and bioproductivity in some crop and weed species. *In* Photosynthesis VI, Photosynthesis and Productivity, Photosynthesis and Environment (G Akoyunoglou, eds.) pp. 63–72. Balaban International Science Services, Phyladelphia.

3 Moore R and Black C C 1979 Nitrogen assimilation pathways in leaf mesophyll and bundle sheath cells of C_4 photosynthesis plants formulated from comparative studies with Digitaria sanguinalis (L) Acop. Plant Physiol. 64, 309–313.

4 Reid P H and York E T 1958 Effect of nutrient deficiencies on growth and fruiting characteristics of pean tus in sand cultures. Agronomy J. 50, 37–63.

5 Sarić M and Petrović M 1976 Uticaj totalne deficijencije odedenih jona u hranljivom rastvoru na sadržaj N, P, K i Ca kod nekih biljnih vrsta (Effect of total deficiency of certain ions in the medium on N, P, K and Ca content in some plant species). Zbornik za prirodne nauke. Matica srpska 51, 47–61.

6 Sarić M and Škorić D 1981 Relationship between the root and above ground parts of different sunflower inbreds regarding the content of some mineral elements. *In* Structure and Function of Plant Roots (R Brouwer *et al.*, Eds). Martinus Nijhoff/Dr W Junk, The Hague, Boston, London p 399–401.

7 Sarić M and Krstić B 1981 Uticaj različitih knocentracija hranljivog rastvora na masu suve materije i zastupljen ost N,P,K,Ca i Mg u nekih biljnih vrsta. (Effect of different concentrations of nutritive solutions on the weight of dry matter and the content of N,P,K, Ca and Mg in some plant species). Arhiv za poljoprivredne nauke 147, 319–330.

8 Smith B N and Boutton T W 1981 Environmental influences on $^{13}C/^{12}C$ ratios and C_4 photosynthesis. *In* Photosynthesis VI, Photosynthesis and Productivity, Photosynthesis and Environment (G Akoyunoglou, eds) pp 255–262. Balaban International Science Services, Philadelphia.

9 Sytnik K M, Musatenko L I and Bogdanova T L 1978 Leaf Physiology, pp 229–234. Naukova dumka, Kiev.

Genetic differentiation of various physiological parameters of *Plantago major* and their role in strategies of adaptation to different levels of mineral nutrition*

D. KUIPER*
University of Groningen, Dept. of Plant Physiology, Haren (Gr), The Netherlands

Key words ATPase Genetic differences Inbred lines Plantago Plasticity Respiration Shoot/root ratio

Introduction

Strategies of a particular plant species for survival are essential to the total sum of processes leading to maintenance of the species in its ecological niche. The characteristics of the various strategies are especially determined by regular and irregular changes in environmental factors. Two mechanisms are available for the plant species to cope with these environmental changes, *viz* genetic variability and phenotypic flexibility of the individual plants. The genetic variability of a species offers the possibility of individual plants which are better adapted to a new environment. Genetic differentiation is a long-term response to environmental changes.

Phenotypic flexibility is the ability of the species to alter the characteristics of the individual plants in the population to a changed environment. Phenotypic flexibility may differ widely between genotypes within a population, depending upon the number of available plastic physiological characteristics and upon the extent of the plasticity of the physiological factor in question. Phenotypic flexibility forms a short-term response to a changed environment.

The genetic differentiation of plant characteristic is well known, especially their morphological features. However, genetic differentiations of physiological properties have also often been demonstrated: *e.g.* in sodium accumulation[14], in sodium chloride extrusion[13], and in potassium uptake[2]. In most cases crop plants were used but only a few reports give information on genetic differentiation of physiological characteristics in wild plants, as *e.g.* heavy metal tolerance[1] and sulphur dioxide tolerance[15,18].

The purpose of this investigation is to test whether populations of *P. major* survive by the combination of a high genetic variability, which is due to highly homozygoteous genotypes and a low physiological plasticity of the individual plants. Inbred lines (G_1, K_1, A_3 and Z_2, Van Dijk and Van Delden[19]) characterized by a genetically determined number of seeds per capsule were used (Table 1).

* Grassland species research group, Publication no 58.

M.R. Sarić and B.C. Loughman (eds.), Genetic Aspects of Plant Nutrition.
ISBN-13: 978-94-009-6838-7
©1983 Martinus Nijhoff/Dr W. Junk Publishers, The Hague/Boston/Lancaster.

Table 1. Mean number of seeds per capsule of the four selected inbred lines of P. major

	Inbred line	Mean number of seeds per capsule	
1	G_1	10	(ssp. major)
2	K_1	18	(average ssp. pleiosperma)
3	A_3	19	(average ssp. pleiosperma)
4	Z_2	33	(extreme ssp. pleiosperma)

Previous experiments revealed that in contrast to P. lanceolata, P. major is unable to change the activity of ATPases of the microsomal fraction from roots after alteration in the nutrient level[8]. Also, several parameters of the energy and nitrogen metabolism did not show any response, P. lanceolata, on the contrary, is characterized by a certain plasticity of the above physiological characteristics as far as mineral nutration is concerned[9,10,16].

Material and methods

Plants from selected inbred lines were subjected to a full (100%) and a diluted (2%) nutrient solution, and to transfers from one condition to the other (100%–2% and 2%–100%)[5]. The activity of the ATPase of the microsomal fraction from the root was determined[6] and root respiration was separated into the activity of the cytochrome system and of the alternative pathway. Also, several growth parameters were determined[7]. Description of the used methods may be found for ATPase determination in Kuiper and Kuiper[8], and for root respiration in Lambers et al.[12]

Results

A large difference in sensitivity of the two extreme lines 1 and 4 to alterations in nutrient solutions was observed. Figs. 1A and 1B show the shoot to root ratios of lines 1 and 4. Differences in nutrient solution on transfers from one to the other condition had no influence on the shoot/root ratio in line 1. Obviously it was more or less fixed in plants of line 1, caused by a rather rigid growth pattern. The only influence at a low nutrient level was a dramatic decline in growth rate of shoot and root after a while (not shown). Loss of dry matter due to mineral deficiency occurred but anyhow no regulatory response to low nutrition was noted.

On the contrary, the shoot to root ratio of plants of line 4 was very plastic with regard to mineral nutrition and a lower ratio was observed at the low mineral level. The ratios of the transferred plants were not significantly different from those of the plants grown continuously on the new nutrient solution.

Figs. 2A and 2B show the response of the Mg^{2+}-stimulated ATPase activity of the microsomal fraction from roots of transferred and control plants of line 1 and

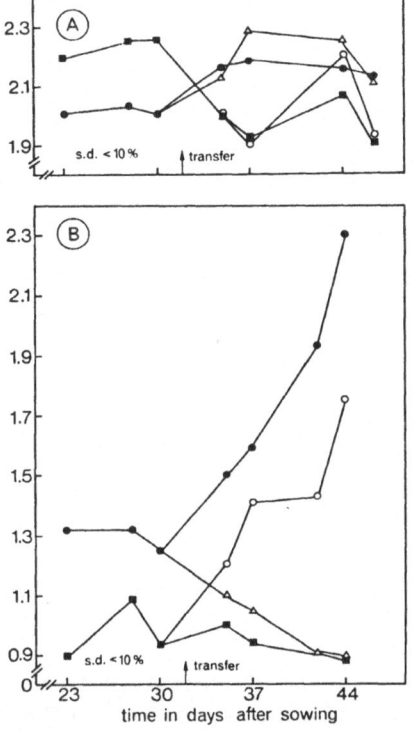

Fig. 1. Shoot to root ratios from two inbred lines. A. line 1, B. line 4. ●--● 100%, △--△ 100%-2% ■--■ 2%, ○--○ 2%-100%.

line 4, respectively. The strength of the mineral nutrient solution had no influence on the maximum velocity (V_{max}) of the root ATPases from plants of line 1, while a low nutrient level decreased the V_{max} values in plants of line 4. The V_{max} value of the activity of root ATPase preparations of plants of line 1 probably was genetically fixed. In plants of line 4 the ATPase activity was sensitive to mineral nutrition, not only at the beginning of the experiment but also later on when transfer of the plants to the other mineral condition took place.

The behaviour of the ATPase activity from roots of plants of line 2 and line 3, the lines with an intermediate number of seeds, was also intermediate in all respects. The absolute value of the activity was intermediate and the response of the ATPase on transfer of the plant to the other mineral condition was intermediate. Plants of line 2 were sensitive to mineral nutrition at the beginning of the experiment but these plants were fixed as far as ATPases are concerned at the time of transfer of the plants. As a consequence, the V_{max} value of the ATPase activity of 100%–2% plants resembled that of the 100% plants and the same holds true for 2%–100% and 2% plants, respectively.

Figs. 3A and 3B show the root respiration of line 1 and line 4, respectively.

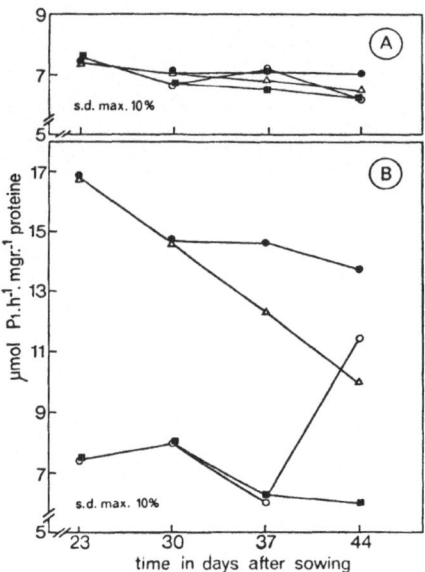

Fig. 2. The V_{max} values of the Mg^{2+}-stimulated ATPase activity of roots from two inbred lines during the experiment. A. line 1, B. line 4. See Fig. 1. Arrow (↑) indicates transfer.

Neither the strength of the initial nutrient solution nor the strength of the new nutrient solution after the transfer influenced the root respiration of plants of line 1 (Fig. 3A) up to day 35.

After day 35 the root respiration of 2% plants declined and after day 42 the 100%–2% plants reacted similarly.

The nutrient level determined the height of the root respiration in plants of line 4. Transferred plants resembled very much, with respect to root respiration, the plants grown continuously on the new nutrient solution. The difference in values of root respiration activity of plants of line 1 and line 4, comparing 100% and 2% plants of each line, is a clear indication of genetic differentiation. The results of line 2 and line 3 are not shown but they behaved in an intermediate manner. The root respiration of plants of line 2 behaved similarly to the ATPase activity; the nutrient level affected root respiration at the beginning of the experiment but transferred plants did not possess any plasticity in this respect.

Discussion

A clear genetic differentiation of physiological parameters in *Plantago major* is seen in Table 2. A different genetic background with respect to the extent of phenotypic plasticity seems to be present in this species also. The arrangement of the inbred lines in number of seeds per capsule leads to a correlation between seed number and dry weight of the plants, respiratory activity and ATPase activity of

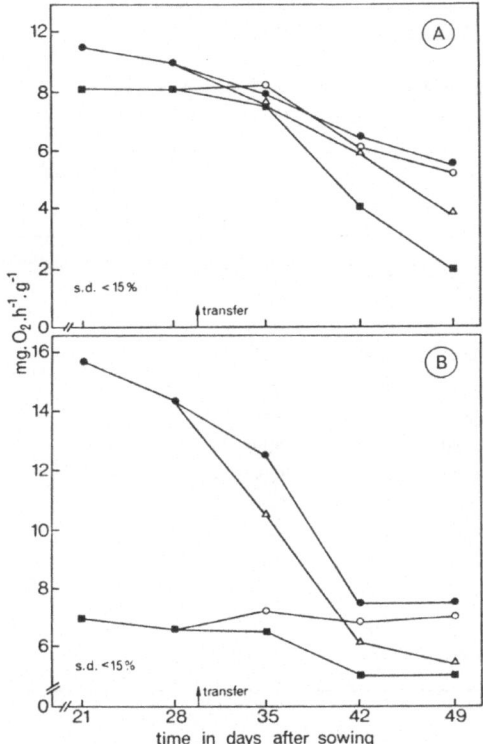

Fig. 3. Total root respiration from two inbred lines during the experiment. A. line 1, B. line 4. See Fig. 1.

the roots. The presence of plasticity and the quickness of response time to a changed situation also showed a direct correlation with the number of seeds per capsule.

It is doubtful, whether the plasticity in the different genotypes has survival value as far as physiological flexibility is concerned. The time needed to respond to a change in the mineral level, in line 4 about 14 days, seems rather long for an adequate reaction to a changed environment. The other three lines had no plasticity or needed still more time for a plastic response. So if the plasticity has any survival value in line 4, and if it is truly correlated with the number of seeds, then only a small proportion of the genotypes of this species possess useful plasticity. In conclusion, the physiological flexibility at the species as a whole is rather poor.

According to Grime[3] and Grime and Hunt[4] *P. major* has a ruderal strategy. Characteristics of a ruderal plant species are a high relative growth rate, life-time of one year or less and a high seed production. Physiological flexibility is low; plants of line 4 had the highest growth rate and seed production, a high degree of

Table 2. The properties of the four inbred lines of *Plantago major* at 100% nutrient level

Line	ssp. *major*		ssp. *pleiosperma*	
	1	2	2	4
Mean number of seeds per capsule	11	18	19	33
100% plants				
Dry weight-shoot*	low			high
Dry weight-root*	low			high
Total root respiration	low	high		high
Activity of the cytochrome mechanism*	low	high		high
Activity of the alternative mechanism*	high			low
Plasticity of shoot to root ratio	absent	low		high
Response time of this plasticity	absent	slow		less slow
Plasticity of root respiration	absent	low		high
Response time of this plasticity	absent	slow		less slow
$V_{max}Mg^{2+}$—ATPase	low			high
$V_{max}Ca^{2+}$—ATPase*	low			high
Plasticity of ATPase activity	absent	low		high
Response time of this plasticity	absent	slow		less slow

* not shown

physiological flexibility may contribute little to the survival value in short-living annual plants. The plants of line 1 show a more competitive strategy for survival.

References

1 Antonovics J, Bradshaw A D and Turner R G 1975 Heavy metal tolerance in plants. Adv. Ecol. Res. 7, 2–85.
2 Glass A D M and Perley J E 1980 Varietal differences in potassium uptake by barley. Plant Physiol. 65, 160–164.
3 Grime J P 1977 Evidence for the existence of three primary strategies in plants and its relevance to ecological and evolutionary theories. Am. Nat. 111, 1169–1194.
4 Grime J P and Hunt R 1975 Relative growth rate, its range and adaptive significance in local flora. J. Ecol. 63, 393–422.
5 Hoagland D P and Snijder W C 1933 Nutrition of strawberry plants under controlled conditions. Proc. Am. Soc. Hort. Sci. 30, 288–296.
6 Kuiper D 1982 Genetic differentiation in *Plantago major*: Ca^{2+}—and Mg^{2+} stimulated ATPases from roots and their role in phenotypic adaptation. Physiol. Plant. 56, 436–443.
7 Kuiper D 1983 Genetic differentiation in *Plantago major*: Growth and root respiration and their role in phenotypic adaptation. Physiol. Plant. 57, 222–230.

8 Kuiper D and Kuiper P J C 1979 Ca^{2+}—and Mg^{2+} stimulated ATPases from roots of *Plantago major* and *Plantago maritima*: Response to alterations of the level of mineral nutrition and ecological significance. Physiol. Plant. 45, 1–6.
9 Kuiper D and Kuiper P J C 1979 Ca^{2+}– and Mg^{2+}–stimulated ATPases from roots of *Plantago lanceolata, Plantago media* and *Plantago coronopus*. Response to alterations of the level of mineral nutrition and ecological significance. Physiol. Plant. 45, 240–244.
10 Lambers H, Posthumus F, Stulen J, Lanting L, Van de Dijk S and Hofstra R 1981 Energy metabolism of *Plantago major* ssp. *major* as dependent on the supply of mineral nutrients. Physiol. Plant. 51, 234–238.
11 Lambers H, Postumus F, Stulen I, Lanting L, Van de Dijk S and Hofstra R 1981 Energy metabolism of *Plantago lanceolata* as dependent on the supply of mineral nutrients. Physiol. Plant. 51, 85–92.
12 Lambers H, Blacquière Tj and Stuiver C E E 1981 Interactions between osmoregulation and the alternative respiratory pathway in *Plantago coronopus* as affected by salinity. Physiol. Plant. 51, 63–68.
13 Läuchli A 1976 Genotypic variation in transport, *from* Transport in Plants II, part B, pp 372–393. Encyclop. Plant Physiol.
14 Marschner, H, Kylin A and Kuiper P J C 1981 Differences in salt tolerance of three sugarbeet genotypes. Physiol. Plant. 51, 234–238.
15 Pollard A J 1980 Diversity of metal tolerance in *Plantago lanceolata* L. from the south eastern United States. New Phytol. 86, 109–117.
16 Stulen I, Lanting L, Lambers H, Posthumus F, Van de Dijk S and Hofstra R 1981 Nitrogen metabolism of *Plantago lanceolata* as dependent on the supply of mineral nutrients. Physiol. Plant. 51, 93–98.
17 Stulen I, Lanting L, Lambers H, Posthumus F, Van de Dijk S and Hofstra R 1981 Nitrogen metabolism of *Plantago major* ssp. major as dependent on the supply of mineral nutrients. Physiol. Plant. 52, 108–114.
18 Taylor G E and Tingey D T 1981 Physiology of ecotypic plant response to sulfurdioxide on *Geranium carolinianum* L. Oecologia 49, 76–82.
19 Van Dijk H and Van Delden W 1981 Genetic variability in Plantago species in relation to their ecology. Theor. Appl. Genet. 60, 205–290.

Genotypic differences in growth and photosynthesis of young barley plants

L. NÁTR
Department of Plant Physiology, Charles University, Praha, Czechoslovakia

Key words Barley Genotype Nitrogen Nutrient stress Phosphorus Photosynthesis

Introduction

Both mineral nutrition and dry matter production in plants are very complex. The analysis of genotypic differences of several parameters are very important for understanding the basic processes of plant growth as well as for improving crop productivity[17].

For the following reasons there is an increasing interest in the effects of mineral nutrients on photosynthesis in different species and genotypes (*e.g.* refs. [2,6,8]).

(1) The understanding of photosynthetic processes and structures should enable an optimization of crop yield.

(2) The use of mineral fertilizers represents a very effective means for the regulation of photosynthesis[7,15].

In this contribution, we first summarize some results obtained in our department in experiments carried out to examine the very different direct and indirect effects of nutrient stress on photosynthetic characteristics (Part A). In Part B, the effects of nitrogen stress on the growth of 8 barley genotypes are described. Some general conclusions as to the varietal specificity of mineral nutrient effects are discussed.

Materials and methods

A. Using spring barley the changes brought about by nitrogen or phosphorus deficiencies were studied and attention was focussed on the effects on:

1. Leaf structure: The 1st to 3rd leaves of young barley plants grown in full Richter's nutrient solution (K) or in solutions without phosphorus (−P) or nitrogen (−N) were analysed by stereological procedures—for details see Pazourek and Nátr[14].

2. Leaf energy budget: On leaves of plants as in 1 we measured the dependence of transpiration on water saturation deficit as well as optical properties. The individual components of leaf energy budget[13] were calculated[3].

3. Nitrogen and phosphorus utilization efficiency: Barley plants were grown in nutrient solutions. The unique source of nitrogen or phosphorus was that of the caryopsis. There were

constant growing conditions with high light (HL: 122 $W \cdot m^{-2}$) or low light (LL: 28 $W \cdot m^{-2}$). For details see Nátr et al.[10].

4. Rate of photosynthesis: In a closed system, the changes of the rate of photosynthesis (P_N) caused by nitrogen supply for 24 hours to previously-N-deficient young barley plants were investigated[5].

5. In an open system for gas exchange measurements the light curves of the rate of photosynthesis and transpiration were determined. Leaves of young barley plants grown in nutrient solutions K, $-N$, $-P$ (see ad 1) of the following varieties were measured[8].

Selekční hanácký (a local variety from Haná region),
Valtický (a standard variety of the 60's),
Diamant (a highly productive variety of the 70's).

B. Under partially constant temperature (day/night 22°C/18°C $\pm$ 2°C) and constant illuminance (8 k lux) the barley plants were cultivated in the following Hoagland nutrient solutions:

H1 (0.014 mg $N \cdot l^{-1}$ corresponding to $1 \cdot 10^{-3} NO_3^-$ meq.)
H2 (0.070 mg $N \cdot l^{-1} \sim 5 \cdot 10^{-3} NO_3^-$ meq.) and
H3 (0.240 mg $N \cdot l^{-1} \sim 15 \cdot 10^{-3} NO_3$ meq.).

The following genotypes were investigated:

1. Selekční hanácký II.   5. Korál
2. Proskovcův hanácký   6. HE 2553/1.8
3. Valtický   7. HE 2107
4. Favorit   8. HE 1526

Seeds were kindly supplied by the Breeding Station Hrubčice.

Results and discussion

The purpose of the experiments in A was to analyse the manifold effects of N or P on several photosynthetic characteristics. The results should also have helped to distinguish between the direct and indirect effects of the particular nutrient on photosynthetic structures and processes.

1. Both N and P deficiency induced profound changes in the leaf anatomical structures (Table 1). There are reasons for believing that some changes of P_N

Table 1. The effect of N or P deficiency on barley leaf structure (from Pazourek and Nátr[14])

	1st leaf			2nd leaf		
	Control	$-P$	$-N$	Control	$-P$	$-N$
Leaf volume [mm^3]	60.9	73.1	46.5	106.9	117.7	49.1
Number of stomata per leaf volume [mm^{-3}]	587	452	784	578	506	990
Specific volume of photosynth. tissue [$\times 10^{-3} mm^3 mm^{-2}$]	59.4	89.7	57.2	73.5	80.5	50.2

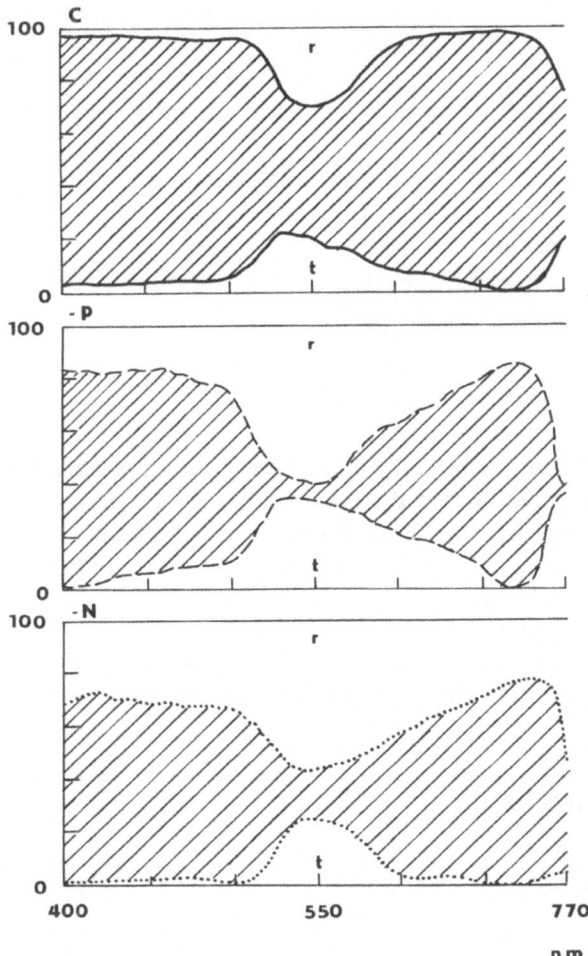

Fig. 1. Reflection (r), transmission (t) and absorption (/////) characteristics in leaves of control (———), N (– – –) or P (· · · ·) deficient barley plants in the spectral region of photosynthetic active radiation. (Unpublished data of Hák.)

brought about by nutrient deficiency are due to changes of leaf anatomy[11]. However, more quantitative data on tissue structures are needed for explaining the effects of alterations in anatomy of nutrient stressed plants on individual processes of photosynthesis.

2. Absorbed solar radiation, absorbed infrared radiation, convection heat transfer and latent heat transfer-transpiration are the main components of the leaf energy budget. Nitrogen or phosphorus deficiency altered the rates of the various components of energy balance. Due to changes in the optimal properties

Table 2. Nitrogen and phosphorus untilization efficiency [mg dry weight·(mg N)$^{-1}$ or (mgP)$^{-1}$] in barley plants grown without P (−P) or N (−N) supply under high (HL) and low (LL) illuminance (from Nátr et al.[10])

	Control	−P	−N
Phosphorus efficiency			
HL	79	849	106
LL	82	551	66
Nitrogen efficiency			
HL	29.4	23.0	66.7
LL	20.7	21.8	48.9

of the leaf (Fig. 1) the absorption of radiation energy differs among the treatments. Considerable differences were also found in the rate of transpiration under various water deficiencies[3]. It may be concluded that changes in the energy budget of the leaves taken from N or P deficient plants will cause additional changes in the rate of photosynthesis.

3. Nitrogen or phosphorus deficiency increased the efficiency of the particular nutrient, *i.e.* the amount of total dry matter per unit of N or P of the plant (Table 2). This efficiency is considerably modified by light and most probably by other external factors as well.

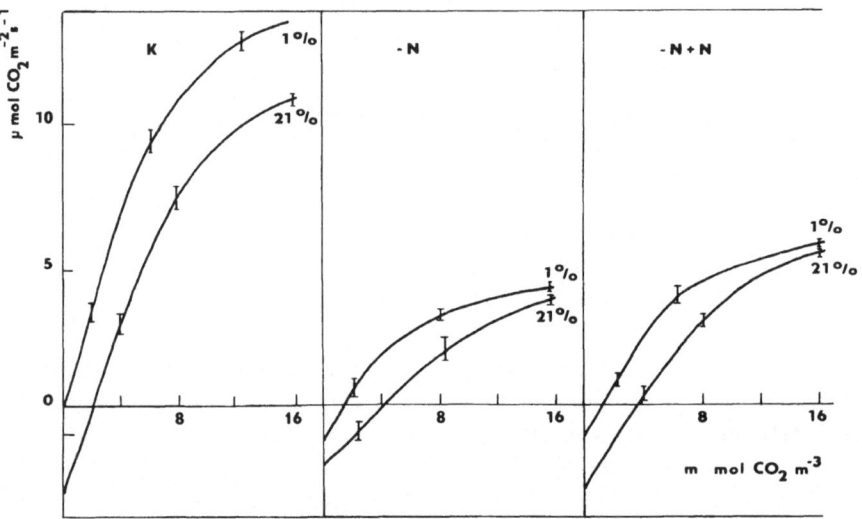

Fig. 2. The dependence of rate of photosynthesis (P_N) on CO_2 concentration in 1% or 21% O_2 of young barley plants grown in nutrient solution (K), without nitrogen (−N) and with N for only 24 hours before measurements (−N +N). (Unpublished data of Marek and Frank.)

4. Several photosynthetic characteristics have been proposed as criteria for the evaluation of the status of mineral nutrition[1,16]. Based on previous experiments[7] further investigations have been carried out in order to support or abandon the idea of such a biological test. Supply of N for 24 hours to previously N deficient plants induced changes in all measured and calculated parameters that are commonly used to describe photosynthetic activity in higher plants (Fig 2).

5. The light curves of the rate of photosynthesis (Fig 3) show significant varietal differences in the effect of a nutrient deficiency. The more productive the variety, the more sensitive to N stress when evaluated by the decrease in the rate of photosynthesis. However, such a conclusion is not likely to be generally valid for all genotypes (see Part B).

B. The eight barley genotypes represent typical varieties grown in the Haná region (Czechoslovakia) during the last 100 years. In this period, breeding

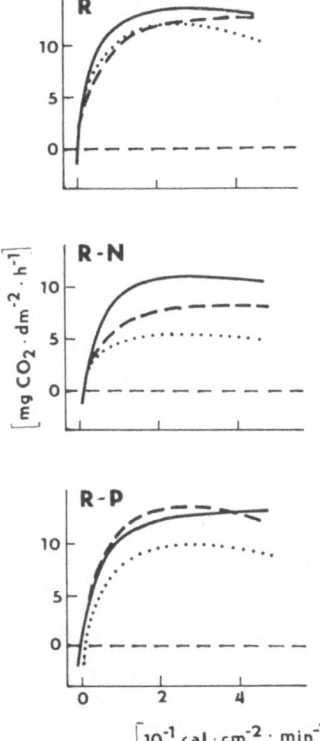

Fig. 3. The dependence of the rate of photosynthesis (P_N) on photosynthetic active radiation (PhAR) of three spring barley varieties (Selekční (——), Valtický (– – –), Diamant (· · · ·) grown in full nutrient solution (R) or without P (R−P) or N (R−N)). (From Nátr[8].)

succeeded in increasing the yield from about 1 t·ha^{-1} to about 8 t·ha^{-1}. The higher productivity is due to

increasing grain/straw ratio,
lowering the plant height,
increasing the ear number per hectare[4].

Information on grain yield formation indicates[12,18] that there will be several ideotypes of high-yielding varieties, *i.e.* the same degree of productivity could be achieved by different combinations of structures and functions at the whole plant, organ, tissue, cell and molecular levels. Hence, it cannot be expected that the genotypes used represent the unique and generally valid way of progress in breeding barley varieties in the last century. The same is true of the dependence of these varieties on mineral fertilizers. However, these genotypes are likely to react in very different ways to changes in nutrient supply.

The growth of the eight barley genotypes showed a characteristic exponential course (Fig. 4). The time dependence of total dry weight W_T could be expressed by the formula $W_T = a \cdot e^{b \cdot t}$, where a, b, are parameters and t indicates days after planting. Because of the exponential growth the parameter b equals the relative growth rate. Calculating the parameters, the fitted curves are indicated (Fig. 4) and the significant interactions between genotypes and N level in the nutrient solutions are obvious. The decrease in N supply decreased the relative growth rates to very different degrees in the genotypes.

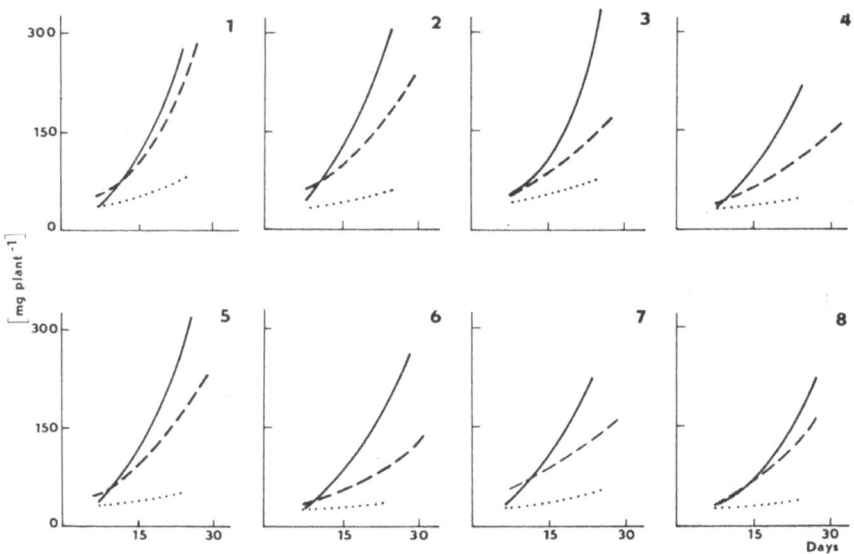

Fig. 4. Plant growth (total dry weight) of 8 spring barley genotypes as influenced by nitrogen concentration in nutrient solution. H 1 (····): 10^{-3} NO_3^- meq, H 2 (---): $5 \cdot 10^{-3}$ NO_3^- meq, H 3 (——): $15 \cdot 10^{-3}$ NO_3^- meq. The numbers refer to the genotypes (see Materials and methods).

There were also significant interactions between varieties and N nutrition in the distribution of dry matter as shown by the time course of the root/plant dry weight ratio (Fig. 5). Derived and calculated data commonly used for describing growth and photosynthesis of higher plants (to be published elsewhere) confirm considerable genotypic differences in the effects of N stress.

It is very important to emphasise the fact that changes of plant growth, relative growth rate, root/plant ratio and other parameters brought about by N deficiency do not indicate any unique tendency in either old or recent varieties[9]. Hence, any generalisations about the effects of nutrient deficiencies on varieties differing in their productivity deduced from only two genotypes should be made with caution.

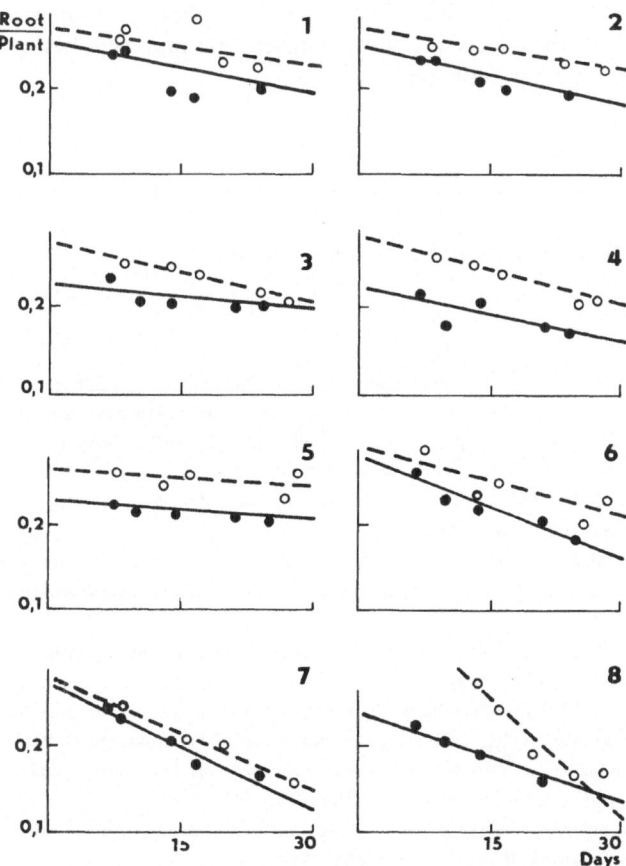

Fig. 5. The root plant dry weight ratio of 8 barley genotypes cultivated in H2 (– – –) and H3 (——) nutrient solutions (see caption to Fig. 4).

Conclusion

Both mineral nutrition and photosynthetic dry matter production are very complex processes. There is an increasing amount of data showing genotypic differences in various processes or structures determining or influencing assimilation of nutrients and their functioning in dry matter production and distribution.

The descriptions of genotypic differences of parameters as complex as nutrient efficiency (dry matter produced per unit of a particular nutrient) or nutrient content in the plants are not likely to bring new information about the mechanisms that reflect the appropriate genetic information. Similarly, results of the effects of nutrient stress on two genotypes with different productivity should be generalised only with caution because of the varying combinations of both processes and structures producing the same degree of plant productivity.

The use of mathematical models will be of great use in handling the complexity of genotypic specificity of the effects of mineral nutrients on photosynthesis. It seems useful to concentrate especially on

(1) the mechanisms regulating the distribution of dry matter between shoot and root;

(2) leaf anatomy and its relationship with both light energy interception and gas exchange;

(3) cell structures including chloroplast number and activity.

References

1. Bouma D, Spencer K and Dowling E J 1969 Assessment of the phosphorus and sulphur status of subterranean clover pastures. 3. Plant tests. Aust. J. Exp. Agr. Animal. Husb. 9, 329–340.
2. Bravdo B and Pallas J E jr 1982 Photosynthesis, photorespiraton and RuBP carboxylase/oxygenase activity in selected peanut genotypes. Photosynthetica 16, 36–42.
3. Hák R and Nátr L. 1982 The dependence of transpiration on water deficiency in nitrogen or phosphorus stressed plants. Biol. Plant. (In press).
4. Kopecký M, Nátr L, Pešík J, Zemánek M and Zeniščeva L 1975 Physiological characteristics and varietal growing technology of high yielding spring barley and winter wheat varieties. Rost. Výroba (Praha) 21, 861–873.
5. Marek M and Frank R 1983 The influence of nitrogen nutrition on the rate of photosynthesis in barley leaves. Photosynthetica 17, (In press).
6. Murthy K K and Singh M 1979 Photosynthesis, chlorophyll content and ribulose diphosphate carboxylase activity in relation to yield in wheat genotypes. J. Agr. Sci. Cambridge 93, 7–11.
7. Nátr L 1970 The influence of removal of mineral deficiency on dry weight, rate of photosynthesis and N, P, K concentration in barley. Flora 159, 589–599.
8. Nátr L 1971 Varietal differences in the effect of mineral nutrient deficiency on the rate of photosynthesis and transpiration. Rost. Výroba (Praha) 17, 411–418.
9. Nátr L 1972 Influence of mineral nutrients on photosynthesis of higher plants. Photosynthetica 6, 80–99.

10. Nátr L, Apel P and Fialová S 1982 The influence of P or N deficiency on growth of young barley plants as dependent on illuminance. Biol. Plantarum (*In press*).
11. Nobel P S 1977 Internal leaf area and cellular CO_2 resistance: photosynthetic implicatons of variations with growth conditions and plant species. Physiol. Plant. 40, 137–144.
12. Nichiporovich A A 1964 Photosynthesis and mineral fertilizers. (*In Russ.*) Agrokhimiya 1, 40–52.
13. Parkinson K J, Day W and Leach J E 1980 A portable system for measuring the photosynthesis and transpiration of graminaceous leaves. J. Exptl. Bot. 31, 1441–1453.
14. Pazourek J and Nátr L 1981 Changes in the anatomical structure of the first two leaves of barley caused by the absence of nitrogen or phosphorus in the nutrient medium. Biol. Plantarum 23, 296–301.
15. Pham Thi Nhu Nghia, Nátr L and Fialová S 1981 Changes in photosynthetic rate of spring barley induced by removal of nitrogen or phosphorus deficiency. Photosynthetica 15, 216–220.
16. Reuther W and Burrows F W 1942 Effect of manganese sulfate on the photosynthetic activity of frenched tung foliage. Proc. Am. Soc Hort. Sci. 40, 73–76.
17. Sarić R M 1981 Genetic specificity in relation to plant mineral nutrition. J. Plant Nutr. 3:743–766.
18. Tooming Kh.G 1977 Sun radiation and yield formation. (*In Russ.*) Gidrometeoizdat, Leningrad.

The effects of mineral nutrition on the photosynthetic and respiratory activity of leaves of winter wheat and maize varieties

J. REPKA
University of Agriculture, Nitra, Czechoslovakia

Key words Maize Mineral nutrition Photosynthesis Respiration Wheat Yield

Introduction

Specific reactions of varieties to the factors of environment and especially to mineral nutrition provide a reason for a study of genetic conditioning of the uptake and utilization of mineral elements. Genetic specificity is conditioned by morphological, physiological and biochemical factors[5,8,9,13].

Similarly, genetic control of differences in the rate of photosynthesis has been confirmed not only between species but also within species[2,14]. Nátr[6,7] has shown significant differences in the P_N of the flag leaf of winter wheat, as well as a great dependence of the leaf on the P_N under varying conditions of mineral nutrition as well as on the light regime during cultivation.

Young plants are normally used in the study of the genetic specificity of the uptake and utilization of nutrients. There are fewer data on genetically determined accumulation and utilization of nutrients during ontogenesis in relation to the final yield. In this paper some data are given on the content of mineral elements in the leaf of two maize hybrids and four cultivars of winter wheat in the period of generative organ formation and plant ripening.

Material and methods

Maize hybrids CE-250 and CE-330 were grown under field conditions with 134 kg N, 60 kg P and 124 kg K $\cdot$ ha^{-1}. Winter wheat cultivars Slávia, Grana, MW$_2$ and BU-11 were grown at nutrient doses of 150 kg N, 64 kg P and 150 kg K $\cdot$ ha^{-1}. In maize the 8th and 9th leaves were taken for analyses in the period from June 22nd to August 7th; in winter wheat varieties the flag leaf was analysed in the period from May 23rd to June 27th. The following values were estimated:

The content of mineral elements after digesting dried leaf blades with sulphuric acid: N (with Nessler reagent), P (with ammonium molybdate) colorimetrically; K, Ca, Mg by means of flame photometry.

The net photosynthetic rate (P_{Ns}) in mg $\cdot$ dm^{-2} $\cdot$ h^{-1} according to the dry matter increase in leaf segments with a total area of 20 cm^2 after a 5 h exposure in rotation device (Šestak and Čatský 1966) at 22°C and irradiance 26 W $\cdot$ m^{-2}.

The content of chlorophyll a+b spectrophotometrically in acetone extracts (Šestak and Čatsky 1966).

M.R. Sarić and B.C. Loughman (eds.), Genetic Aspects of Plant Nutrition.
ISBN-13: 978-94-009-6838-7
©1983 Martinus Nijhoff/Dr W. Junk Publishers, The Hague/Boston/Lancaster.

The respiration rate (R_D) in $\mu l\ O_2 \cdot g^{-1}$ dry matter $\cdot 30\ min^{-1}$ was determined by the Warburg manometric method.

Results

The content of elements and the rate of photosynthesis and respiration confirm that significant changes occur during the life of the flag leaf (Fig. 1). Differences in cultivars are shown by changes in quantitative values of the accumulation of elements in different periods of growth of the flag leaf. This variability is more significant than the mean content of nutrients over the whole period of life of the flag leaf. The cultivars compared from this point of view can be characterized by the data shown in Table 1.

Differences between minimal and maximal values of mean concentrations of elements between cultivars are 9% in nitrogen, 5.5% in P, 22% in K and Ca, 16.1% in Mg and 9.2% in the sum of nutrients. At some periods of life the flag leaf differs in concentration of elements by 25% higher in N and P, 50–70% in K, Ca, and Mg. It is not possible to judge from our data whether these differences are more significant for the evaluation of the functional activity of the leaf than the mean concentration of elements over the whole period of life of the flag leaf.

P_N and R_D changes do not follow the curves of mineral elements and are subject normally to the influence of other factors. Differences in P_N values between the cultivars vary within a range of 20.3%, R_D 9.5% and chlorophyll content 9.5%.

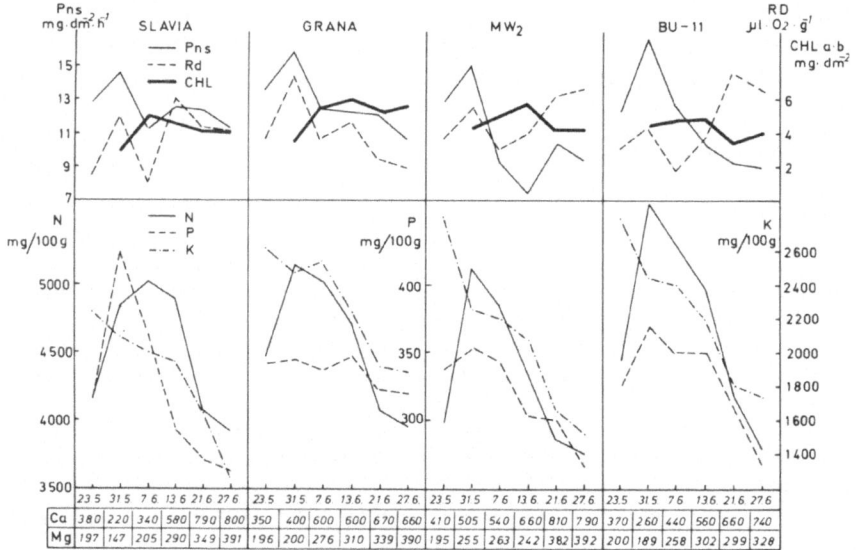

Fig. 1. Changes of mineral elements content (mg $\cdot$ 100 g^{-1}), rate of photosynthesis (P_{Ns}) and respiration (R_D), and the content of chlorophyll in the flag leaf of winter wheat varieties.

Table 1. Mean contents of mineral nutrients of the different cultivars and other characteristics

Cultivar	Mean content of mineral nutrients in leaf dry matter in %						P_{Ns}	R_D	Chl. a+b	Grain yield t. ha^{-1}
	N	P	K	Ca	Mg	total				
Grana	4.55	0.33	2.27	0.54	0.28	7.99	12.78	1092	5.02	8.8
Slávia	4.48	0.32	1.86	0.51	0.26	7.45	12.50	1062	4.26	7.5
MW$_2$	4.29	0.31	2.01	0.61	0.61	7.30	7.54	1162	4.69	6.1
BU-11	4.70	0.32	2.22	0.50	0.26	7.59	11.99	1146	4.20	5.6

The differences in grain yields exceed the differences in the concentration of elements as well as those of the physiological activity of the flag leaf, when the difference between the minimum and maximum yields is 57.5%. The flag leaf does not fully reflect the processes responsible for the full extent of the yield. The cultivar with the highest production (Grana) showed a significantly higher P content during the whole period and maintained also a higher K concentration for longer periods. Under these concentration conditions the flag leaf kept a higher physiological activity, had a higher P_N value and a lower R_D, especially by the end of its functional activity.

Since the first demonstration of the specific role of the flag leaf on wheat yield[4,11], many investigators studied the importance of the flag leaf from various aspects. This paper examines certain parameters concerning the flag leaf and the results obtained show that the parameters under investigation are cultivar-specific. Therefore, when we discuss the specificity of cultivars in mineral nutrition, a particular plant organ used as a criterion for evaluating this specificity should be taken into account. It seems that in wheat the flag leaf represents the organ of particular importance to which attention should be paid.

It is possible to say from the dynamics of P_N and R_D, as well as from that of the chlorophyll content that the cultivars which maintained higher mean P_N values and lower R_D of the flag leaf (Grana, Slávia) also gave a higher grain yield. The most productive cultivar (Grana) had a higher total content of elements and particularly a higher concentration of P and K.

Comparison of the changes of the content of mineral elements P_N and R_D in the 8th and 9th maize leaves (Fig. 2) confirms quantitative differences in various phases of ontogenesis between the hybrids. The mean concentration of elements in the leaves characterizes the hybrids by the values shown in Table 2.

In comparison with the hybrid CE-250, the hybrid CE-330 had a higher mean concentration of P and K and the total content of nutrients. The higher accumulation of all elements at initial growth stages and a higher N, P and K content at the terminal stage of the life of leaves of the hybrid CE-330 should be

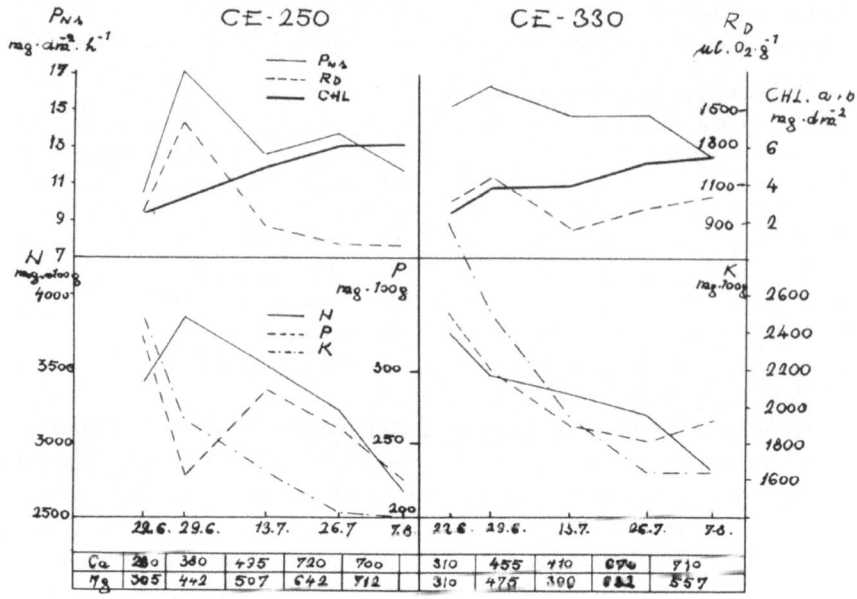

Fig. 2. Changes of mineral element content (mg·100 g^{-1}), rate of photosynthesis (P_{Ns}) and respiration (R_D) and the content of chlorophyll in the 8th and 9th leaves of maize hybrids.

regarded as specific. The concentration conditions were more favourable to a higher photosynthetic activity of leaves with higher values of P_N by 13.7% at a lower chlorophyll content by 10%.

The hybrid CE-330 produced 163.3 t·ha^{-1} biomass and 8.16 t·ha^{-1} grain, *i.e.* 21.0% and 15.8% more than the hybrid CE-250.

The content of nutrients as well as the P_N and R_D values of the flag leaf of winter wheat varieties and of the leaves of maize hybrids confirm differences in the accumulation of the elements and in the physiological activity of the analyzed leaves. A higher production in both species was obtained in the variety and in the

Table 2. Element concentrations in different hybrids

Hybrid	Mean concentration of elements in % of leaf dry matter					
	N	P	K	Mg	Ca	Total
CE-250	3.34	0.26	1.83	0.55	0.51	6.37
CE-330	−3.32	0.28	2.14	0.47	0.51	6.73
±%	−0.7	+7.5	+16.9	−14.5	0	+5.6

hybrid with a higher total concentration of mineral elements in leaves at a higher P and K concentration. In view of the fact that differences in yield are higher than the differences in the concentration of nutrients and in the photosynthetic and respiratory activity of the analyzed leaves, it is not possible to regard the data on these leaves as a sufficient proof of the varietal specificity of accumulation and utilization of ions in relation to the yield obtained. Klimashevsky[5], Sarić[8] and Vose[13] indicate difficulties in the physiological and genetical interpretation of the effect of nutrients in one organ in relation to the yield. Similarly, the problem of the relationship between the concentration of nutrients in leaves, their photosynthetic activity and the yield of plants has not been solved[7].

References

1   Aslam N and Hunt L A 1978 Photosynthesis and transpiration of the flag leaf in four spring–wheat cultivars. Planta, Berlin 141, 23–28.
2   Avratovščukova N 1977 Genetics of Photosynthesis. Stud. informace UVTI, Praha, 1, 96 p (*In Czech*).
3   Dantuma G 1973 Rates of photosynthesis in leaves of wheat and barley varieties. Neth. J. Agr. Sci. 21, 188–198.
4   Lupton F G H 1966 Translocation of photosynthetic assimilates in wheat. Ann. Appl. Biol. 57, 355–364.
5   Klimashevsky E L 1974 The problem of genotypic specificity of plant root nutrition. *In* Variety and Nutrition, Irkutsk, pp 11–53 (*In Russian*).
6   Nátr L 1966 Odrudove rozdily v intenzite fotosyntezy. Rostl. vyroba 12, 163–178 (*In Czech*).
7   Nátr L 1972 Influence of mineral nutrients on photosynthesis of higher plants. Photosynthetica 6, 80–99.
8   Sarić M R 1974 The importance of the problem of variety specificity of mineral nutrition. *In* Variety and Nutrition, Irkutsk, pp 54–60 (*In Russian*)
9   Sarić M R and Kovačević V 1981 Varietal specifity of wheat mineral nutrition. *In* Physiology of wheat, Beograd, pp 61–77 (*In Serbian*).
10 Sarić M, Jocić B and Krstić B 1982 The effect of mineral nutrition upon leaf area and yield in some wheat varieties. Zbornik za prirodne nauke, Matica srpska Novi Sad 62, 1–14 (*In Serbian*)
11 Stoy V 1963 The translokation of ^{14}C-labelled photosynthetic products from the leaf to the ear in wheat. Physiol. Plantarum 16, 851–866.
12 Tarčevski I A, Čikov vV I, Ivanova A P, Suleimanova A Yu and Andrianova Yu E 1977 Osobenosti fotosinteza i ottoka asimilyatov u različnyh sortov yarovoi pšenicy *In* Genetika fotosinteza, Dušanbe, 234–237 (*In Russian*).
13 Vose P B 1974 Utilization and recognition of nutritional varieties in crop plants. *In* Variety and Nutrition, Irkutsk, pp 61–70 (*In Russian*).
14 Zelitch J 1971 Photosynthesis, photorespiration and plant productivity. Academic Press, New York.

Nutritional differentiation within the species *Dittrichia viscosa* W. Greuter, between a population from a calcareous habitat and another from an acidic habitat

J.-P. WACQUANT and N. BOUAB
Département de Physiologie écologique, Centre Emberger, C.N.R.S., F-34033 Montpellier Cédex, France

Key words   Calcicole   Calcifuge   Calcium   *Dittrichia viscosa*   Magnesium   Potassium   Sodium

Summary   *Dittrichia viscosa* is a Mediterranean bush widespread on various soil types. It is shown that the plants from a calcareous habitat (G plants) and those from an acidic habitat (M plants) differ in their ability to accumulate various cations when growing in the same experimental conditions. On this acidic soil the G plants accumulate more Ca and less K than the M plants. On a calcareous soil the response is reversed; it is the M plants which contain more Ca and often less K. This behaviour on each of the soils is typically that of a calcifuge for M and a calcicole for G. The two types of plants also differ in their affinity for Mg and Na. The ubiquity of the species could well be explained at least partially by genetic differentiation in the ability of plants to select ions.

Introduction

Dittrichia viscosa W. Greuter (ex *Inula viscosa* Ait.) is a bush widely spread in the Mediterranean region on recently abandoned lands. This post-cultural species is a remarkable pioneer able to colonize various chemical types of soils such as calcareous as well as acidic or even slightly salty. How can such an ubiquitous species adapt to a wide range of edaphic conditions? Is it by wide physiological plasticity of individual plants, or rather through a physiological polymorphism of the species composed of different populations each closely adapted to a narrow range of soil conditions[8,10,11,12].

In order to answer the question for the species *Dittrichia viscosa* we shall see if plants from a calcareous habitat and those from an acidic habitat differ in their ability to accumulate various cations and if the eventual differences are of the same order as those described for calcicole and calcifuge species.

Materials and methods

Two contrasted habitats of *Dittrichia viscosa* were chosen in the Mediterranean area of France near Montpellier; one with a typical calcareous soil will be called Saint-Gély or G; the other with a very acidic soil will be named Mauguio or M.

The plants were compared in their habitat and in cultivation as well. Rooting of transplants was stimulated by using hormones. The germination of the seeds collected in the habitats was facilitated by stratification in low temperature and moist conditions. The soils used for experimental cultures were collected in the habitats (Table 1). All plants grown in pots were irrigated with deionised water

M.R. Sarić and B.C. Loughman (eds.), Genetic Aspects of Plant Nutrition.
ISBN-13: 978-94-009-6838-7
©1983 Martinus Nijhoff/Dr W. Junk Publishers, The Hague/Boston/Lancaster.

Table 1. Analysis of the calcareous soil of the Saint-Gély population (G) and the acidic soil of the Mauguio population (M)

Soil	pH	CaCO$_3$ g. 100 g^{-1}	meq. 100 g^{-1}**				
			C.E.C.	Ca	Mg	K	Na
Calcareous (G)	7.7	55.2	10.8	49.8*	0.8	0.2	0.1
Acidic (M)	5.4	0	13.8	9.8	2.1	0.2	0.1

** Exchangeable cations extracted with 1N CH$_3$COONH$_4$ * Free CaCO$_3$

and received a diluted nutrient solution once a week during the first four weeks in order to stimulate the growth which was otherwise poor on such soils.

The content in Ca, Mg, K and Na of plant shoots was determined in a 0.1N HNO$_3$ extract at 28°C for 36 hours (50 ml per g of fresh weight). Cations were assayed by flame emission (Ca, K and Na) and atomic absorption (Mg) spectrophotometry. For each ion the mean value of three replicate assays is provided (Tables 2 to 4).

The mineral content of the plant is given in equivalent fractions in order to express the proportion (as a percentage) of each ion Ca, Mg and K relative to their total amount S (Tables 2 to 4). The proportion for Na is given relative to S' which is the total content of cations including Na. This equivalent fraction is a way to express ion interactions that occurred during accumulation; it is used to compare the ability of plants to select ions[12].

Results

The G and M populations in their habitat

In the calcareous habitat G, *Dittrichia viscosa* is the dominant species (70% of

Table 2. Comparison of proportions of Ca, Mg and K (% S) and Na (% S') in shoot tips between G and M transplants collected as adults in their habitat and cultivated on the same acidic or calcareous soil

Transplants	Culture soil	% S			% S'	µeq g^{-1}	
		Ca	Mg	K	Na	S	S'
G	Acidic	51.2	17.0	31.9	29.7	284	404
M		50.8	15.6	33.6	29.1	297	419
G	Calcareous	60.8	15.7	23.5	26.3	320	434
M		63.6	14.2	22.2	24.6	372	493

S, amount of Ca, Mg and K per g of fresh weight; S' including Na. Each analysed sample refers to 4 genotypes (12 shoot tips)

Table 3. Proportions of Ca, Mg and K (% S) and Na (%S') in shoots of G and M seedlings when germinated on the same acidic or calcareous soil. For S and S' see Table 2

Seedlings	Germination soil	% S			% S'	$\mu eq \cdot g^{-1}$	
		Ca	Mg	K	Na	S	S'
G	Acidic	46.5	14.3	39.3	49.0	102	201
M		40.9	12.9	46.2	48.3	103	199
G	Calcareous	53.2	14.0	32.8	45.9	119	221
M		54.9	12.1	33.1	43.0	127	234

Each analysed shoot sample refers to about 100 seedlings

the cover) in a very dense stand of a ten year old fallow. In the acidic habitat M located in a gravel pit abandoned about 10 years ago, the species while not dominant (30%) has taller plants than those of the G habitat.

Five adult plants (estimated age: 4 to 6 years) representative of the stand were chosen in each habitat to compare size and shoot tips mineral content. Fig. 1 describes how the plant types differ in their size. Fig. 2 shows that they also differ mainly in their Ca and K content since there is no difference for Mg and a tendency for the G plants to contain more Na than the M plants. The G plants growing in their calcareous habitat accumulate more Ca and less K than the M

Table 4. Proportions of Ca, Mg and K (% S) and Na (% S') in shoots of G and M plants grown from seeds, germinated and cultivated on acidic (A) and calcareous (C) soils. For S and S' see Table 2

Plants	G		M		G		M	
Germination soil	A	C	A	C	A	C	A	C
Culture soil	Acidic				Calcareous			
Ca % S	45.3	47.1	43.7	43.8	62.0	61.6	65.7	65.3
Mg	17.2	16.8	15.2	14.8	13.8	14.7	12.1	12.8
K	37.4	36.0	41.1	41.1	24.2	23.7	22.2	21.9
Na % S'	42.4	41.9	37.7	39.0	34.1	34.2	30.9	31.3
S $\mu eq\ g^{-1}$	180	178	196	202	251	233	250	248
S'	312	307	315	331	380	354	362	361

Each analysed sample refers to 17 genotypes (17 shoots)

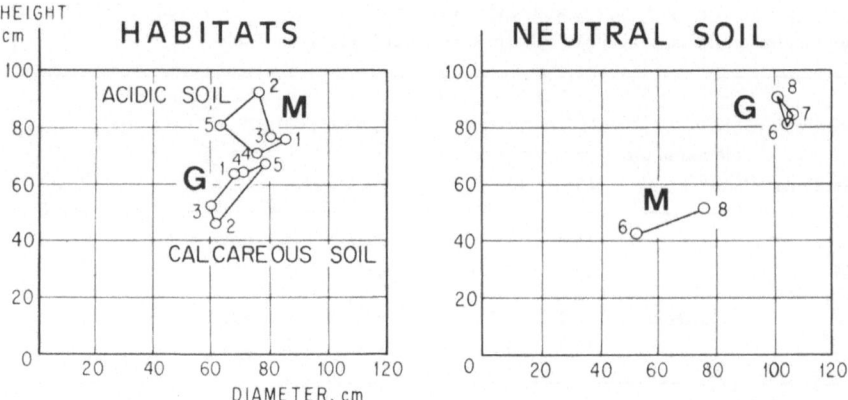

Fig. 1. Size of plant individuals of the Saint-Gély (G) and Mauguio (M) populations in their own habitat (*left*) and when transplanted and grown for 4 months on a neutral calcimorphic soil (*right*). Numbers are to identify individuals in each population.

plants in their acidic habitat. Is this because the calcareous habitat is richer in calcium than the acidic? To check this assumption we studied the response of the two plant types after transplanting into the same soil.

Adult plants transplanted on a neutral substratum

Transplantation of three adult plants of each habitat into a neutral (pH 7.1) calcimorphic soil showed that the M plants grow poorly compared to the G plants (Fig. 1 *right*) and one of the M plants did not survive the transplantation. All this suggests that the M plants, like *calcifuge* species, do not thrive as well on a neutral calcimorphic soil as the G plants, which in turn behave like calcicoles. Since there were no differences in the mineral content of the shoot tips between

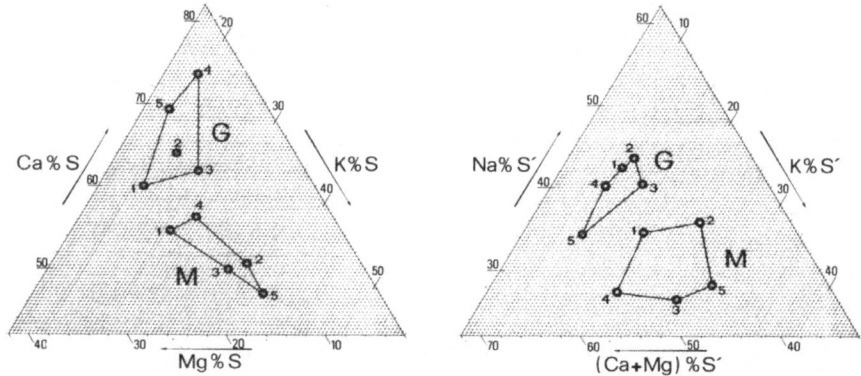

Fig. 2. Proportions of Ca, K and Mg % S (*left*) and Na % S' (*right*) in the shoot tips of five G plants and five M plants in their own habitat, *i.e.* the G on the calcareous soil and M on the acidic. For S and S' see Table 2.

the plants (data not presented) we ran another experiment with more stressing soil conditions.

Divided adult plants transplanted on acidic and calcareous soils

Five adult plants of *Dittrichia viscosa* (4 to 6 years old) were collected in March in each of the habitats and cultivated in large clay pots exposed in the open. The stem and root of each plant were divided longitudinally into two parts. One was planted in a pot with the acidic soil the other in a pot with the calcareous soil. At the end of July while no morphological difference appeared between plants, the shoot tips of each plant type were collected and analysed. The transplants of the two populations have a distinctive response in their mineral content in each soil (Table 2). *In the same acidic soil*, the G plants accumulate more Ca and less K than the M plants as we noticed before in the habitat study. Is this because the G plants have a high affinity for Ca, like the *calcicolous* species, and the M for K, like the *calcifuges*[4,6], or because of a carry-over effect of the oldest tissues of the transplant? *In the same calcareous soil*, the two types of plants have a reversed behaviour, excluding a carry-over effect. It is the M plants which accumulate more Ca and less K. Is this because, like *calcifuge* species, the M plants, contrary to the G, do not control in calcareous soil the entrance of calcium at the expense of potassium[4,9]?

The two populations also differ in their Mg and Na amounts. The G transplants contain proportionately more Mg and Na than the M transplants in both soils. It will be of interest to see if plants grown from seeds confirm the physiological tendency observed with transplants.

Seedlings on the same acidic or calcareous soil

The seeds collected from one plant of each habitat were sown on both types of soil contained in trays (about 200 seeds in each). The seedlings did not present any visible morphological difference, whatever the population or the soil; however the differences appearing in the mineral content of the shoots (Table 3) are quite similar to those observed in the transplant experiment except in the case of M on the calcareous soil which contains more K than G. Let us see now how the seedlings respond as plants a few months later.

Plants grown from seeds, germinated and cultivated on acidic and calcareous soils

The seeds of each population were sown (Table 4) in April on both acidic and calcareous soils. The seedlings of each type were then transplanted, one per pot, in either acidic or calcareous soil and grown in a greenhouse. Three and a half months later there were no morphological differences in size or colour between the G and M plants growing on the same soil whatever the soil of germination or culture. The shoots of seventeen plants of each type were collected as a sample and analysed (Table 4).

No difference appears in the mineral content within each population between the plants germinated on the acidic and on the calcareous soils, whatever the culture soil. Does that mean that each of the populations is not physiologically polymorphic? Nevertheless the differences between the populations still remain as described in the other experiments. *On the acidic soil* the G plants (germinated on either A or C soils) contain more Ca and less K than the M plants (on A or C). *On the calcareous soil* the M plants contain more Ca and less K than the G plants on both germination conditions. It is also clear with this study that in any of the soil conditions the G plants always accumulate more Mg and Na than do the M plants.

Discussion and conclusion

Several workers have demonstrated that *calcicole* and *calcifuge* species differ mainly in their ability to accumulate calcium and potassium[2,4,5,6,9]. The present work has shown that the plants of *Dittrichia viscosa* from calcareous and acidic habitats also differ similarly in their response with respect to Ca and K. *In their own habitat i.e.* the G plants on the calcareous soil and the M on the acidic soil, the G contain more Ca and less K than the M. At first it seems to be a soil effect since the G plants grow on a soil very rich in calcium in contrast to the M plants. However if we look at the behaviour of the two types of plants when grown *on the same acidic soil*, very poor in calcium (Table 1), in all cases (Tables 2 and 4) the G plants still accumulate more Ca and less K than the M plants. Such results cannot be explained by a carry-over effect as mentioned previously. They suggest that the G plants, like *calcicole* species, have a high affinity for calcium and the M plants, like *calcifuge* species, for potassium[2,3,4,5,6,9]. On the same calcareous soil the G and M plants had their behaviour reversed. This is again another typical feature described for *calcicole* and *calcifuge* species on calcareous soils[3,4,7]. It is the M plants, like a *calcifuge* species which accumulate more Ca and generally less K (Tables 2, 3 and 4). This seems to be due less to a change in their affinity for Ca than to their lack of ability to limit Ca entrance into the plant[9]. All of our experiments strongly suggest that the plants of the acidic habitat behave physiologicaly as *calcifuges* and that of the calcareous habitat as calcicoles.

The differences between the two populations for Ca and K accumulation appear to have a physiological adaptational significance. At present this is not so for Mg and Na even though these ions allow clear differentiation of the two populations. Thus in all cases (Tables 2, 3 and 4) on acidic as well as on calcareous soil the G plants always accumulate proportionately more Mg and Na than the M plants.

As a conclusion this work did not show any visible *morphological differentiation* between the plants of the two populations when compared under the same growing conditions. However it provided evidence for *physiological differentiation* since the plants of the two habitats differed in their ability to accumulate Ca,

K, Mg and Na cations. The ubiquity of *Dittrichia viscosa* could well be explained at least partially by *genetic differentiation* in the ability of plants to select ions.

Acknowledgement Thanks are due to Mr. Raymond Marican for his technical assistance.

References

1 Bradshaw A D 1965 Adv. Genet. 13, 115–155.
2 Duvigneaud P and Denaeyer-de Smet S 1964 Lejeunia 28, 1–148.
3 Hamzé M 1973 Thèse spéc. Université Montpellier II, 110 p.
4 Hamzé M, Salsac L. and Wacquant J P 1980 Agrochimica, XXIV, 5–6, 432–442.
5 Kinzel H 1969 *In* A Läuchli 1976. Encyclop. Plant Physiol. Lüttge *et al.* Ed. 2(B), 372–393.
6 Passama L 1970 Oecol. Plant. 5, 225–246.
7 Passama L, Ghorbal H, Hamże M, Salsac L and Wacquant J P 1975 Rev. Ecol. Biol. Sol. 12, 295–313.
8 Ramakrishnan P S 1969. Can. J. Bot. 47, 175–181.
9 Salsac L 1980 Bull. A.F.E.S. Sci. du Sol 1, 45–77.
10 Snaydon R W 1970 Evolution 24, 2, 257–269.
11 Snaydon R W and Bradshaw A D 1961 New Phytol. 60, 219–235.
12 Wacquant J P, Hochepot M and Valdeyron G 1981 C. R. Acad. Sci. Paris 293, III, 813–816.

SECTION IV
THE INFLUENCE OF MINERAL NUTRITION ON YIELD AND QUALITY OF DIFFERENT GENOTYPES

Responses of various crop species and cultivars to fertilizer application

K. MENGEL
Institut f. Pflanzenernährung a. Justus-Liebig-Universität, D-6300 Giessen, FRG

Key words Cation excess Fertilizer response Hydrogen release Root length Root morphology Soil pH Wheat cultivars

Summary Crop response to fertilizer application depends not only on the level of available plant nutrients in the soil but is also related to crop physiology and morphology. For a well balanced nutrition the rate of nutrient supply to the roots must correspond with the rate of nutrient required for growth. Species or cultivars with a high growth rate generally respond more favourably to fertilizer application than those with low growth rates. An analogous relationship holds for the biomass produced per unit soil surface. Thus modern rice and wheat cultivars tolerate a more dense spacing than older ones. Due to the dense stand the yield and particularly the grain yield of the modern varieties may be several times higher than those of older cultivars, and therefore also the nutrient requirement, especially the demand for N and P, is higher for the modern cultivars.

Modern cereal cultivars are characterized by a high crop index which means that after flowering a high proportion of grain filling material must be produced by photosynthesis. Assimilation and translocation of photosynthates are favoured by K^+. Thus in particular modern cultivars require a high K^+ content for optimum grain filling.

Nutrient exploitation of soils by plant roots depends on root morphology and root physiology. Grasses generally have much longer roots than dicots. Thus the rate of K^+ and phosphate uptake per unit root length is lower for grasses than for dicots. It is for this reason that dicots respond earlier to a K^+ and phosphate dressing than grasses.

Species living symbiotically with Rhizobium may depress the rhizosphere pH considerably and thus promote the dissolution of phosphate rock.

Introduction

It has been well known since the 18th century that plants require inorganic nutrients for normal growth. Liebig[28] stated that the primary plant nutrients are exclusively of "inorganic nature". In those days in Central Europe the soils were much depleted of plant nutrients and thus often gave very poor yields. In maintaining and improving soil fertility Leibig[28] wrote: "It must be borne in mind that as a principle of husbandry, what is taken from the soil must be returned to it in full measure". Liebig's advice has been followed and as a consequence soil fertility has been much improved and crop yields have approximately tripled. However, although much research work has been carried out since the time of Liebig, there is still uncertainty whether fertilizer application will result in crop response under given conditions.

Crop response as a consequence of fertilizer application should be higher the more the soil is depleted of plant nutrients. This relationship has been found by various researchers, *e.g.* for nitrogen by Wehrmann and Scharpf[47] or for potassium by von Braunschweig[7] in Germany and Loué[29] in France. In many cases, however, the use of soil nutrient status as assessed by soil tests for fertilizer recommendations is unsatisfactory, because weather conditions[26], soil structure[11] and soil type[19] also have an impact on fertilizer response.

The main emphasis of this paper is focussed on the question of why, when grown under the same conditions, some species or cultivars respond to fertilizer application whereas others do not. The reasons for this differential behaviour are mainly related to physiological and morphological plant properties. These will be considered in relation to the utilization of N, P and K.

Growth rate

Numerous cultivars have been bred with the main object of developing crops which produce more plant material per unit area and time. This is particularly true for forage crops where the total plant matter above ground is harvested and regarded as economic yield. For optimum nutrition soils have to provide plant nutrients at rates which are adequate to the growth rates of the crop. Thus a cultivar with a low growth rate may draw sufficient nutrients from a soil with a low or medium nutrient status so that fertilizer application will have little effect on yield production. In contrast, species or cultivars with high growth rates are unable to attain plant nutrients in adequate amounts under low soil fertility conditions, as they require a higher rate of nutrients per unit soil surface (or volume). It is mainly for this reason that the higher yielding variety or crop species responds more readily to fertilizers. This is a general relationship which always applies provided that in the process of breeding for higher yields of vegetative material root growth and root metabolism have not been significantly altered.

Table 1. Effect of K^+ fertilizer application on the grain yield (ton/ha) on two different K^+ fixing sites

K rate (kg K_2O/ha)	Dornwag		Weng	
	Spring wheat 1972	Maize 1973	Maize 1972	Spring wheat 1973
0	3.27	2.48	5.34	4.83
300	3.96	3.88	5.63	4.62
600	6.16	5.04	8.66	5.07
900	4.48	5.48	9.37	5.21

A spectacular example of plants with a high growth rate are tropical grasses which produce large amounts of plant material per unit time and thus produce high fertilizer responses[17]. In Table 1 the effect of K^+ application on maize and spring wheat is shown. The crops were grown on K^+ fixing soils and for this reason high K^+ fertilizer rates were required. It can be seen that maize gave a greater response than spring wheat[8], because maize required higher rates of K^+ per unit time.

In inbred maize lines known for a high yield potential the high grain yield is associated with a high production of stalks and leaves and it is for this reason that these lines require high nutrient rates for obtaining maximum yields. Agboola[1] in testing 8 different maize cultivars found that the N fertilizer rates for obtaining maximum grain yield ranged between 45 and 220 kg N/ha. The grain yields varied from 2.30 to 5.52 t/ha with a clear tendency for the higher yielding cultivars to require the higher fertilizer rate.

Plant morphology

In contrast to inbred maize cultivars, modern high yielding wheat and rice varieties give high grain yields associated with low straw production. These cultivars are characterized by short, stiff culms, which are very resistant to lodging; and by small erect leaves which allow ample utilization of solar energy in a dense plant stand[43]. The single plant of these new cultivars requires lower amounts of P and particularly lower amounts of K than a single plant of the old long culm varieties. In these older varieties a substantial amount of P and K is required for the production of vegetative plant material. Modern high yielding varieties, however, allow a much denser plant stand and for this reason the production of vegetative plant matter per unit soil surface, e.g. ha, is higher than in stands of older varieties. New short culm rice cultivars tolerate a spacing as dense as 20×20 cm per plant while the older varieties need 50×50 cm[45]. Because of their droopy leaves, mutual shading is high in stands of older rice or wheat cultivars and their tall culms are very susceptible to lodging. The high yields obtained from modern rice and wheat cultivars are primarily a consequence of the dense stand with 350 to 400 panicles (rice) or 600 to 700 ears (wheat) per m². Such dense stands produce grain yields in the range of 8 to 10 t/ha which contain a protein-N quantity equivalent to about 160 to 200 kg N. Most soils are not capable of providing such high quantities of N to the crop during the growth period. It is for this reason that the modern cultivars generally need fertilizer N in order to obtain maximum yield. Undeveloped local varieties with a yield potential of about 2 to 3 t/ha require about 40 to 60 kg N/ha for grain protein production. They therefore respond only marginally to fertilizer application. This relationship is shown in Fig. 1[9]. The typical tall local Indian wheat cultivar C-306 only responded to a low rate of nitrogen application and maximum grain yields were obtained at about 80 kg·N ha^{-1}. In contrast the short stiff straw

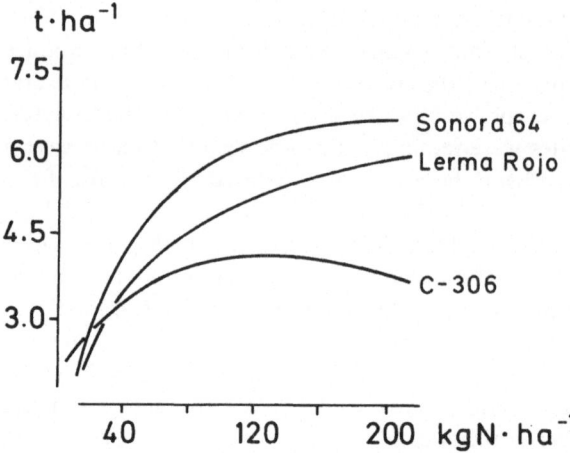

Fig. 1. Response to nitrogen of one old (C-306) and two modern wheat cultivars (after Chandler[9]).

Mexican cultivars Sonora 64 and Lerma Rojo gave a much greater response to N and about twice the grain yields were obtained with a nitrogen application rate of 160 kg N · ha^{-1}. NAS et al.[36] in testing 6 different spring wheat cultivars under field conditions found that in most cases a rate of 90 kg N · ha^{-1} gave the highest yield. The cultivar with the lowest grain yield reached its maximum grain yield (1 t · ha^{-1}) with a fertilizer rate as low as 45 kg N ha^{-1}. Similar results have been obtained for maize and sorghum[10].

As modern cereal cultivars tolerate high rates of N application their N nutritional status and especially their content of soluble amino acids is high. It is possible that this has a bearing on the synthesis of cytokinins[46] since the purine ring, a component of cytokinins, is synthesized from amino acids. Cytokinins may influence grain yield in different ways[35]. They may stimulate tillering and thus increase the number of ears per m^2, they may stimulate ear growth and thus

Table 2. Yield and yield characteristics of an older and a modern wheat cultivar; **significant difference at 1%, ***significant difference at 0.1% level (after Haeder et al.[48])

	Hohenheimer Franken, old cv	Kolibri modern cv
Duration from anthesis to maturation, d	56	47
Grain yield/ear, g	1.12	1.24
Single kernel weight, mg	48.2	44.2**
No of kernels/ear	23.0	28.0**
Crop index	29.3	37.7***

increase the number of grains per ear. They are also known to retard the senescence of the flag leaf and thus promote grain filling[23]. These additional positive effects are only implemented, if the plants tolerate a high N nutritional level which is the case for modern cereal cultivars. Local older cultivars are more sensitive to a high N nutritional status.

Modern cereal cultivars are frequently characterized by a long period of ear formation which means that the beginning of anthesis is relatively late[13]. This longer period of ear formation results in long ears with a high number of spikelets. Because of the later anthesis, the period of grain filling is generally shorter in modern cultivars as compared with older ones. For this reason modern cultivars often possess more kernels per ear, but the single kernel weight is lower than that of older varieties. These characteristics of a local spring wheat cultivar "Hohenheimer Franken" and a modern cultivar "Kolibri" are shown in Table 2. The "crop index", which is the % of grain weight of the total dry weight of the upper plant material at harvest, was significantly higher in the modern varieity as compared with the older one. This means that the production of grains per leaf and culm unit was much higher in the modern cultivar. Reserves of assimilates produced before anthesis are relatively higher in older varieties than in modern ones[16]. These reserves contribute to grain filling. Modern varieities must thus produce relatively high amounts of assimilates in the period between anthesis and maturation. Photosynthesis and particularly assimilate translocation from the leaves towards the ears depend on the K^+ nutritional status of the plants[14,31,34]. It is for this reason, that modern cultivars frequently need a higher K^+ status than older ones in order to obtain satisfactory grain filling. This relationship is shown in the data of Table 3, in which the relative grain filling of an old and a modern spring wheat cultivar is presented[16]. It is obvious that the modern cultivar (Kolibri) responded much better to increasing K^+ supply than the older one.

Grain filling is not only a question of assimilate translocation towards the grains, but also depends much on grain metabolism[32]. Often the metabolic activity in the grain may be the limiting factor. This has been shown for the barley cultivar "Bomi" and its isogenic lysine rich mutant "Risø" by Beringer and

Table 3. Effect of K^+ nutrition on the single kernel weight (relative data) of an old and modern cultivar of spring wheat (after Forster and Mengel[16])

Fertilizer rate	Old Hohenheimer Franken	Modern cv Kolibri
K_1	100	100
K_2	121	139
K_3	122	145

Koch[5]. The cultivar "Bomi" responded to a higher rate of K^+ application, whilst "Risø" did not. Even at maturation "Risø" still had a high content of soluble amino acids in the grains, and the authors suppose that in this cultivar protein synthesis rather than photosynthates supply to the grains was the process limiting grain filling. Thus the typical K^+ effect, the promotion of assimilate transport, was of no major importance for this cultivar.

Root growth, root morphology and root metabolism

Whether or not a crop species or a cultivar responds to fertilizer application the extent to which plant roots are capable of exploiting the soil for plant nutrients is important. In this respect considerable differences between species and even cultivars may occur. Steffens and Mengel[42] found that *Lolium perenne* could well feed from the fraction of non exchangeable soil K^+, while *Trifolium pratense* could not. This difference in the capability to exploit soil K^+ was particularly evident, when both species were grown together. Under conditions of low K^+ availability the clover was almost completely replaced by the grass. Potassium application resulted in a marked increase of clover yield while the yield of grass was only marginally influenced. This effect of K^+ on grass and clover yield is shown in Fig. 2. Steffens[41] in studying the question why rye-grass as compared with clover was such a potent K^+ extractor, found that both species were capable of depressing the K^+ concentration in the soil solution to a rather

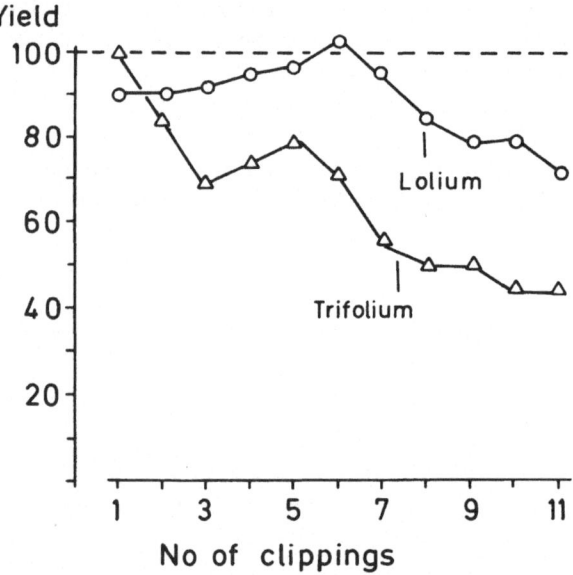

Fig. 2. Relative yield of *Lolium* and *Trifolium* in the treatment not fertilized with K. Yield of the K fertilized treatment = 100 (after Steffens and Mengel[42]).

low level (2 μM) and that in this respect the species did not differ fundamentally. Under K^+ deficiency conditions, however, rye-grass, was able to take up larger amounts of Na^+ as compared with clover and produced more plant matter per unit K^+ than did clover. One may suppose that Na^+ is better able to substitute for K^+ in rye-grass than in clover.

The most marked difference between both species, however, was root growth. This was investigated by Steffens[41] throughout a growth period under field conditions on a gray brown podsolic soil derived from loess. It was found that root parameters measured and listed in Table 4 were better for rye-grass than for red clover. A most spectacular difference was found in root length being about six times greater in grass than in clover. In Fig. 3 the relationship between root length and K^+ uptake is shown for both species. It is evident that the slope of the curve for clover is steeper than that for grass. In calculating the K^+ requirement of clover and grass per unit surface of crop stand, Steffens[41] found that the K^+ uptake per unit root length must be about three times as high in clover as in grass in order to maintain the K^+ demand of the crop.

Potassium uptake rate vs K^+ concentration in the nutrient or soil solution is described by a saturation type of curve[3]. Taking this into account, the K^+ concentration in the soil solution must be about 4 to 5 times higher for an adequate K^+ supply to clover compared with rye-grass. This relationship is shown schematically in Fig. 4 in which the K^+ uptake rates for grass and clover are shown on the y-axis, the latter being 3 times as great as that of rye-grass. From the K^+ uptake rates indicated on the y-axis the corresponding K^+ concentrations shown on the x-axis can be obtained. It is obvious that the K^+ concentration in the soil solution must be several times higher for clover as compared with rye-grass, in order to assure an adequate K^+ supply to the crop.

Table 4. Above ground yield and root characteristics of red clover and rye-grass grown on the same site under field conditions. The yield relates to a soil surface of 25 × 25 cm, the root distribution to a profile surface of 25 × 100 cm. All other root characteristics relate to a soil volume of 25 × 25 × 100. 25 × 25 cm = soil surface. 100 cm = soil depth.* ** *** significant differences at 5%, 1% and 0.1% level respectively (after Steftens[41])

	Clover	Grass
Yield, g DM	24.0	29.6
Root fresh weight	63.7	179.5***
Root distribution, no. of cut roots unit surface	489	845*
Cation exchange cap., meq. unit soil volume	1.47	2.77**
Root length, m	510	3477***
Rooth depth, cm	81.7	70.0
Root surface, m^2	0.88	3.13***

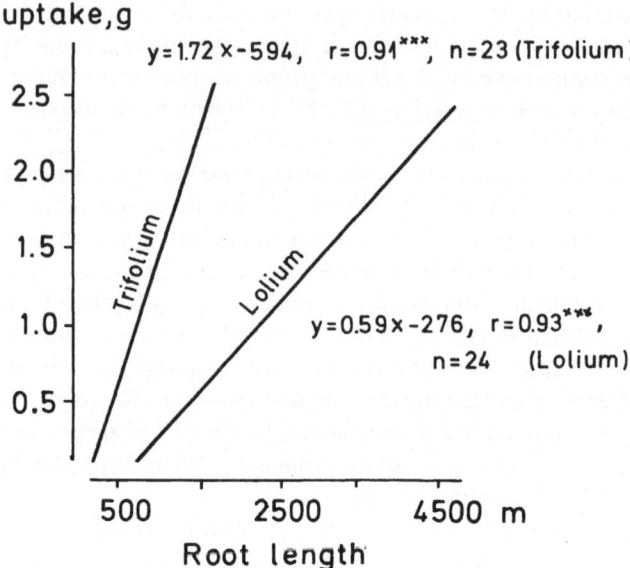

Fig. 3. Relationship between root length and K uptake of *Trifolium pratense* and *Lolium perenne* grown in open field. The data relate to a soil surface of 25 × 25 cm (after Steffens[41]).

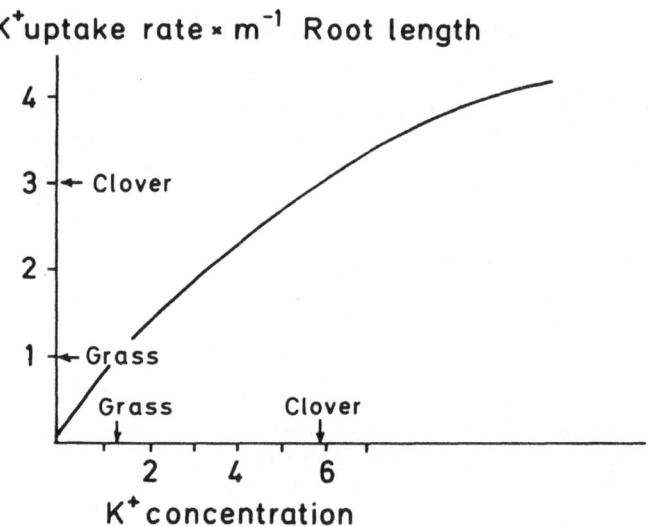

Fig. 4. Relationship between K^+ uptake rate per unit root length and K^+ concentration in the soil solution. Scales on the axes are relative.

What has been said for K^+, is also true for phosphate. Plant species with a high root length per unit soil surface may feed well from relatively low levels of available P in the soil in comparison with species with shorter root lengths or lower root weights. It is also for this reason that under conditions of low soil P availability grasses frequently hardly respond to a P application, whereas legumes will respond by marked yield increases[39]. Besides the root length, root mass, root suface, and root hairs also may play an important role in mining the soil for plant nutrients. Generally grasses have longer root hairs than legumes[21,41] and for this reason are more capable of exploiting the soil adjacent to the root for plant nutrients. This is particularly true for nutrients with a low diffusion rate in the soil medium[4] such as phosphate. Thus Hendriks et al.[22] reported that rape (*Brassica napus*) had longer root hairs than maize and therefore exploited a larger soil volume around the root than wheat. Differences in root hair density also exist between the cultivars of one species as has been shown by Bole[6] for wheat. Whether these differences are of major importance for fertilizer response is not yet clear.

Monocots and dicots differ considerably in their root morphology, the former developing a fibrous root system with numerous slender long roots, while dicots possess a tap root with laterals. The soil depth penetrated by the roots may differ considerably for the various crop species[44] and this may have an impact on fertilizer response. Generally, however, the soil depth penetrated by a root system is of greater importance for water uptake and of minor importance for the uptake of nutrients.

Screening the K^+ fertilizer response of various crop species, one frequently finds a fair response with dicots but not with grasses. This is shown in the data of Table 5 obtained from a long term field experiment carried out by Schön et al.[40]. It can be seen that the response of the cereals to a K application was (compare column NP with column NPK) not dramatic and in a range of about 10% while the response of the dicots (potatoes, broad beans, clover/grass mixture) was several times higher. Van der Paauw[37] also reported that potatoes responded much better to K^+ fertilizer dressings than did wheat. Hulpoi et al.[24], carrying out field experiments on chernozem soils in Roumania found no K^+ response for maize and wheat, while sugar beets responded favourably to K^+ application. The reason for these differences in behaviour is not yet understood. It is tempting to suppose, however, that grasses, because of their extended root system (root length, root density, root mass, root hairs), are able to obtain adequate amounts of K^+ from the soil under conditions under which the dicots can not.

Differences in root length do not only exist between widely differing species such as monocots and dicots but also between more closely related species. Forster[15] reported that under the same growing conditions oats developed a root mass which was about twice as high as the root mass of spring wheat. Consequently spring wheat responded in its production of vegetative plant material to K^+ application while oats did not. The high nutrient exploitation

Table 5. Response of various crops to K fertilizer application. Average yield data from an experiment lasting 20 years[40]

Crop	Yield, t/ha of tubers, grains, seeds or fresh matter		
	No	NP	NPK
Winter wheat	3	4.21	4.55 (108)
Spring barley	5	2.84	3.21 (113)
Oats	3	3.54	3.80 (107)
Winter rye	2	3.08	3.15 (102)
Spring rye	1	2.26	2.49 (115)
Broad beans	2	1.27	2.46 (194)
Clover-grass	1	38.3	45.8 (120)
Potato	4	23.8	32.6 (137)

The figures in parentheses are relative values as compared with the control (NP = 100)

power of oats was also evident in experiments of van Praag et al.[38], who reported that oats were capable of taking up large amounts of NH_4^+ tightly bound in the interlayer clay fraction.

Finally it should be stressed that the degree of root development during the growth period may also have an impact on fertilizer response. Halevy[20] reported that the cotton cultivar "Acala 1517-C" was more susceptible to K^+ deficiency than the cultivar "Acala 4-42". Investigations into the root development revealed that the root growth of the K susceptible cultivar (Acala 1517-C) terminated earlier than root growth of "Acala 4-42". It is assumed that the prolonged root growth enabled the cultivar to feed from soil K^+ for a longer time and thus was less sensitive to low availability of soil K^+ than the cultivar with the earlier termination of root growth.

Uptake of the major plant nutrients is in most cases an active process and thus directly related to root metabolism and the supply of the root system with carbohydrates. Drews[12] reported that *Lolium perenne* plants growing under high light intensity were more capable of utilizing the non exchangeable soil K^+ fraction than plants which were shaded. It is probable that also between species and cultivars differences in root metabolism do exist which may affect their response to fertilizer application. As yet, however, this question has not been investigated.

Differential root metabolism may also result from the form of nitrogen which is supplied to a plant species. Fig. 5 shows the pH development in a gray brown podzolic soil under *Lolium perenne* and under *Trifolium pratense*, both species

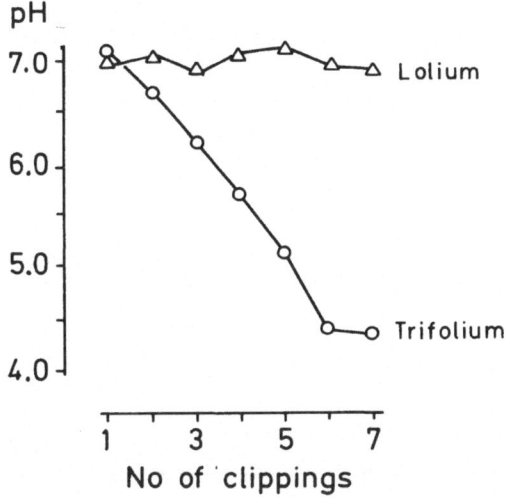

Fig. 5. Development of pH under *Lolium* and *Trifolium* throughout a period of clippings (after Mengel and Steffens[33]).

grown in a pot culture simultaneously under the same environmental conditions. The clover received only a small nitrogen dressing at the beginning of the experiment and had thus been supplied largely with N fixed by Rhizobium bacteria. The rye-grass was supplied with an adequate quantity of NH_4NO_3 after each clipping. The dramatic pH decrease under clover was the result of the high excess of cation over anion uptake. This excess of cation uptake was quantitatively almost as high as the amount of protons released by the roots into the soil[33]. The conclusion is justified that leguminous species living from symbiotically fixed nitrogen have to excrete large amounts of H^+ in order to maintain their cation anion balance. This may lead to marked pH decreases in the rhizosphere and thus have an impact on plant nutrient availability, especially on the availability of phosphate[2]. The general observation that on slightly acid to neutral soils (pH 6 to 7) leguminous species still respond to rock phosphate while grasses do not[25,30] may be explained by the solubilization effect of H^+ excreted by the leguminous root. This assumption has been substantiated by a recent, as yet unpublished experiment of Hauter (Inst. of Plant Nutrition, Giessen), in which clover was grown in one treatment with symbiotically fixed N and in the other treatment with NH_4NO_3 and later with $Ca(NO_3)_2$. Phosphate was applied exclusively in the form of rock phosphate. The most important result of this investigation is shown in Table 6. During the first period of the experiment the soil pH declined in both treatments in the same manner and the P-uptake of the crop was lower in the treatment with symbiotically fixed N. During the last two clippings, however, the pH under the symbiotically living clover dropped dramatically and this soil pH decrease was accompanied by a significantly higher

Table 6. pH alteration under *Trifolium pratense* and P uptake of the crop as related to symbiotically and mineral N nutrition; ***significant difference in uptake at 0.1% level (after Hauter, unpublished data)

No of clippings	pH symb.	pH min.	P uptake, mg P/pot symb.	P uptake, mg P/pot min.
1.	7.0	6.9	69	73***
2.	6.4	6.3	103	112***
3.	5.9	6.0	83	78***
4.	6.0	6.2	89	67
5.	5.3	6.0	55	48***
6.	5.2	6.2	33	26***

P uptake of the crop. The higher P uptake of the plants harvested at the first two clippings in the treatment with mineral N was due to a better growth during the early stage of development during which the plants of the symbiotically fed treatment still had to establish their root nodules.

Concluding remarks

In view of the rapidly declining resources of raw materials and energy the importance of crop production for food, fuel and fibre in the near future cannot be overestimated. In order to meet this challenge it is important to increase the yield per unit area even more than it has been increased in recent years. This yield increase obtained in the past was mainly a consequence of an improved plant nutrition and partially also a result of plant breeding. In the near future plants must be developed which utilize solar energy and atmospheric CO_2 with a higher efficiency in the production of the economic yield, *i.e.* grains, roots, tubers *etc*. These new plant types will also require optimum nutrition, for it is only wishful thinking to imagine that high yielding crop cultivars or species can be developed which need no fertilizer. The high amounts of plant nutrients which are removed from the field by high crop yields, must be replenished, either by the soil, by fertilizer or by both in order to maintain soil fertility, a prerequisite for an efficient utilization of solar radiation and atmospheric CO_2. Generally the regeneration of available plant nutrients by weathering of soil minerals and microbial activity in the soil is a process with narrow limits. Its importance declines the higher the crop yield level. It is for this reason that fertilizer application will also increase in importance in the near future. From a consideration of the energy input/output ratio as well appropriate fertilizer dressing is a powerful means in net energy production[27].

It is, however, also of importance to develop new crops and cultivars which are highly efficient in the utilization of plant nutrients. This is already the case, if new wheat or rice cultivars are compared with older ones, the latter requiring more P and K and producing less grain per plant. The partial substitution of Na^+ for K^+ may also be a means of improving the efficiency of K^+ (ref.[18]). Finally it should be stressed that plants with a powerful root system have a higher efficiency in fertilizer utilization. Such plants are able to utilize more strongly bound nutrients *e.g.* K^+ and NH_4^+ and thus the level of available nutrients in the soil can be kept relatively low which reduces the risk of nutrient losses as a result of leaching volatilization or denitrification.

Acknowledgment The author is much obliged to Mr. E. A. Kirkby for his valuable assistance in preparing the English text.

References

1 Agboola A A 1972 The relationship between the yields of eight varieties of Nigerian maize and content of nitrogen, phosphorus and potassium in the leaf at flowering stage. J. Agric. Sci. 79, 391–396.
2 Aguilar A and van Diest A 1981 Rock phosphate mobilization induced by the alkaline uptake pattern of legumes utilizing symbiotically fixed nitrogen. Plant and Soil 61, 27–42.
3 Barber S A 1979 Growth requirements of nutrients in relation to demand at the root surface. *In:* The Soil Root Interface. Eds. J L Harley and R Scott Russel. pp 5–20. Academic Press, London, New York, San Francisco.
4 Barley K P 1970 The configuration of the root system in relation to nutrient uptake. Adv. Agron. 22, 159–201.
5 Beringer H and Koch K 1977 Stickstoffmetabolismus in einer normalen und in einer lysinreichen Gerste bei unterschiedlicher Kaliumernährung. Landw. Forsch. Sonderh. 34/II: 36–44. Kongreßband.
6 Bole J B 1973 Influence of root hairs in supplying soil phosphorus to wheat. Can. J. Soil. Sci. 53, 169–175.
7 Braunschweig v L C 1978 Ergebnisse aus mehrjährigen Feldversuchen zur Überprüfung der optimalen Kaliversorgung des Bodens. Landw. Forsch. Sonderh. 35, 219–231.
8 Burkart N 1975 Kaliumdynamik und Ertragsbildung Kalium fixierender Böden Südbayerns. Dissertation Fachbereich für Landwirtschaft und Gartenbau der Technischen Universität München.
9 Chandler R F 1970 Overcoming physiological barriers to higher yields through plant breeding. *In* Role of Fertilization in the Intensification of Agricultural Production". pp 421–434. Intern. Potash Institute Berne.
10 De R 1974 Cultural practices for maize, sorghum and millets. 1. FAO/SIDA Seminar for plant scientists from Africa and the Near East. pp 440–451. FAO Rome.
11 Diez Th and Hege J 1980 Stickstoffdüngung des Weizens nach Bodenuntersuchung (N_{min}) in Abhängigkeit von den Standortverhältnissen. Bayr. Landw. Jahrbuch 57, 944–951.
12 Drews J U 1978 Die Aufnahme von Kalium aus der nichtaustauschbaren Kaliumfraktion des

Bodens in Abhängigkeit von der Belichtung der Pflanzen. Dissertation Fachbereich 19, Justus Leibig-Universität Gießen.
13 Evans L T, Wardlaw I F and Fischer R A 1975 Wheat. *In* Crop Physiology. Ed. L T Evans. pp 101–149. Cambridge University Press, Cambridge.
14 Forster H 1976 Einfluß der Kaliumernährung auf Ausbildung und Chlorophyllgehalt des Fahnenblattes und auf die Ertragskomponenten von Sommerweizen – Untersuchungen an fünf verschiedenen Sommerweizensorten. Z. Acker Pflanzenbau 143, 169–178.
15 Forster H 1980 K-Aneignungsvermögen verschiedener Pflanzenarten im Gefäßversuch. Landw. Forsch. Sonderh. 37, 645–652. Kongreßband.
16 Forster H and Mengel K 1974 Einflüsse der Kaliumernährung auf die Ertragsbildung verschiedener Sommerweizensorten (*Triticum aestivum* L.). Z. Acker Pflanzenbau 139, 146–156.
17 Gartner J A 1969 Effect of fertilizer nitrogen on a dense sward of Kikuyu. Paspalum and carpet grass. 2. Interactions with phosphorus and potassium. Q. J. Agric. Anim. Sci. 26, 365–372.
18 Gerloff G. C. 1976 Plant efficiencies in the use of nitrogen, phosphorus, and potassium. *In* Plant Adaption to Mineral Stress in Problem Soils. Eds. M J Wright and S A Ferrari, pp 161–173. Beltsville Maryland.
19 Grass K 1972 Bodenuntersuchung und Düngungswirkung auf Grund von Feldversuchen. Landw. Forsch. 23, Sonderh. 27/11, 146–158.
20 Halevy J 1977 Wachstumsrate und Nährstoffaufnahme von zwei Baumwollsorten bei Bewässerung. Kali-Briefe (Bern) Fachgeb. 27, 79. Folge, No. 5, 1–13.
21 Haynes R J 1980 Competitive aspects of the grass-legume association. Adv. Agron. 33, 227–261.
22 Hendriks L, Claassen N and Jungk A 1981 Phosphatverarmung des wurzelnahen Bodens und Phosphataufnahme von Mais und Raps. Z. Pflanenernaehr. Bodenkd. 144, 486–499.
23 Herzog H and Geisler G 1977 Der Einfluß von Cytokininapplikation auf die Assimilateinlagerung und die endogene Cytokininaktivität der Karyopsen bei 2 Sommerweizensorten. Z. Acker Pflanzenbau 144, 230–242.
24 Hulpoi N Picu I and Tianu A 1971 Researches concerning the application of fertilizers to irrigated field crops. Probleme Agricole No. 8, Ministerul Agriculturii, Industriei Alimentare, Silviculturii Si Apeor, Roumania.
25 Khasawneh F E and Doll E C 1978 The use of phosphate rock for direct applications to soils. Adv. Agron. 30, 159–206.
26 Kuntze H and Bartels R 1975 Nährstoffversorgung und Leistung von Hochmoorgrünland. Landw. Forsch. 31, Sonderheft I, 208–219.
27 Lewis D A and Tatchell J A 1979 Energy in UK agriculture. J. Sc. Food Agric. 30, 449–457.
28 Liebig v J 1841 Die organische Chemie in ihrer Anwendung auf Agrikultur und Physiologie. Verlag Vieweg, Braunschweig.
29 Loué A 1979 Die durchschnittliche Wirkung der Kalidüngung auf Großkulturen in Dauerversuchen. Kali-Briefe (Bern) Fachgeb. 16, 79. Folge, No. 4.
30 Maloth S and Prasad R 1976 Relative efficiency of rock phosphate and superphosphate for cowpea (*Vigna sinensis* Savi) fodder. Plant and Soil 45, 295–300.
31 Mengel K and Haeder H E 1974 Photosynthese und Assimilattransport bei Weizen während der Kornausbildung bei unterschiedlicher Kaliumernährung. Z. Acker Pflanzenbau 140, 206–213.
32 Mengel K and Judel G K 1981 Effect of light intensity on the activity of starch synthesizing enzymes and starch synthesis in developing wheat grains. Physiol. Plant 51, 13–18.

33 Mengel K and Steffens D 1982 Beziehung zwischen der Kationen/Anionen-Aufnahme und der Protonenabscheidung der Wurzeln bei Rotklee. Z. Pflanzenernaehr. Bodenkd. 145, 229–236.
34 Mengel K, Seçer M. and Koch K 1981 Potassium effect on protein formation and amino acid turnover in developing wheat grain. Agron. J. 73, 74–78.
35 Michael G and Beringer H 1980 The role of hormones in yield formation. In: Physiological Aspects of Crop Productivity. pp 85–116. 15th Colloq. Intern. Potash Institute Berne.
36 Nas H G, MacLeod J A and Suzuki M 1976 Effects of nitrogen application on yield, plant characters, and N levels in grain of six spring wheat cultivars. Crop Sci. 16, 877–879.
37 Paauw van der F 1958 Relations between the potash requirements of crops and meterological conditions. Plant and Soil 9, 254–268.
38 Praag van H J, Fischer V and Riga A 1980 Fate of fertilizer nitrogen applied to winter wheat as Na^{15}NO$_3$ and (^{15}NH$_4$)$_2$SO$_4$ studied in microplots through a fourcourse rotation: 2. Fixed ammonium turnover and nitrogen reversion. Soil Sci. 130, 100–105.
39 Schmitt L and Brauer A 1979 75 Jahre Darmstädter Wiesendüngungsversuche. J D Sauerländer's Verlag, Frankfurt/M.
40 Schön M, Niederbudde, E A and Mahkorn A 1976 Ergebnisse eines 20-jährigen Versuches mit Mineral-und Stallmistdüngung im Lößgebiet bei Landsberg (Lech). Z. Acker Pflanzenbau 143, 27–37.
41 Steffens D 1982 Vergleichende Untersuchungen über das Kaliumaufnahmevermögen und die Entwicklung des Wurzelsystems von *Lolium perenne* und *Trifolium pratense*. Diss. Fachbereich 19, Justus Liebig-Universität Gießen.
42 Steffens D and Mengel K 1979 Das Aneignungsvermögen von *Lolium perenne* im Vergleich zu *Trifolium pratense* für Zwischenschicht Kalium der Tonminerale. Landw. Forsch. Sonderh. 36, 120–127.
43 Tanaka A 1973 Influence of special ecological conditions on growth, metabolism and potassium nutrition of tropical crops (as exemplified by the case of rice). In: Potassium in Tropical Crops and Soils. pp 147–167. Proc. 10th Colloq. Intern. Potash Institute, Berne.
44 Taylor H M and Klepper B 1978 The role of rooting characteristics in the supply of water to plants. Adv. Agron. 30, 99–128.
45 Toriyama K 1974 Development of agronomic practices for production of field crops under irrigated conditions – rice. In: 1. FAO/SIDA Seminar for Plant Scientists from Africa and the Near East. pp 452–456. FAO, Rome.
46 Wagner H and Michael G 1971 Der Einfluß unterschiedlicher Stickstoffversorgung auf die Cytokininbildung in Wurzeln von Sonnenblumenpflanzen. Biochem. Physiol. Pflanzen 162, 147–158.
47 Wehrmann J and Scharpf H C 1979 Der Mineralstoffgehalt des Bodens als Maßstab für den Stickstoffdüngerbedarf (N$_{min}$-Methode). Plant and Soil 52, 109–126.
48 Haeder H E, Beringer H and Mengel K 1977 Assimilateinlagerung in das Korn bei zwei Sommerweizensorten. Z. Pflanzenernaehr. Bodenkd. 140, 409–419.

Genetic specificity of nitrogen nutrition in leguminous plants

G. SCHILLING
Martin-Luther-Universität Halle-Wittenberg, Sektion Pflanzenproduktion, Wissenschaftsbereich Agrochemie, DDR-4020 Halle/S., Adam-Kuckhoff-Str. 17b, Deutsche Demokratische Republik

Key words Leguminous plants   Nitrogen fixation   Protein   Seed   Yield

Summary Mineral nitrogen did not increase grain yield and seed protein levels of *Vicia faba* L. and *Lupinus luteus* L. in field trials and pot experiments. Fixed N_2 was substituted by mineral nitrogen in these cases because of inhibition of N_2 fixation by mineral nitrogen. Contrary to these results mineral nitrogen increased grain yields and seed protein amounts of *Lupinus albus* L., *Pisum sativum* L., and *Glycine max.* (L.) Merr. The nitrogen effect was caused at an early stage by saving energy due to inhibition of N_2 fixation (measurement of gas exchange by means of IRGA). In case of the N application after flowering grain, yields and seed protein levels increased because the mineral N was an additional nitrogen source for plants. At this stage the plants had ceased fixing atmospheric nitrogen. The high sink activity of growing fruits induced a lack of assimilates in nodules (determined by means of $^{14}CO_2$ application). The N effect was therefore the consequence of the lower assimilate pool for supplying root nodules in these plants in comparison with *Vicia faba* L. and *Lupinus luteus* L. Hence it follows that response to mineral nitrogen can be a criterion for discovering more effective Rhizobium- host combinations.

Introduction

In recent years the nitrogen metabolism of leguminous plants has become of increasing interest[1,15,16] because the seeds of different species have a high protein content with a high lysine level. Therefore they are suitable as food for pigs and poultry in addition to cereal proteins. On the other hand leguminous plants use atmospheric nitrogen by symbiosis with Rhizobia. Therefore knowledge of their nitrogen metabolism can reveal possibilities for creation of new organism associations able to fix more atmospheric nitrogen.

Nitrogen metabolism was investigated in an attempt to recognize generally acceptable physiological relationships in N_2 fixing systems and to draw conclusions for new systems which increase the utilisation of atmospheric nitrogen.

Abbreviations

ATP = adenosine triphosphate
ADP = adenosine diphosphate P_i = inorganic phosphate
N_i = inorganic nitrogen   at-%-^{15}N-exc. = ^{15}N frequency as atom-% ^{15}N-excess
^{15}N amount = N amount $\dfrac{\text{at. \% }^{15}\text{N exc.}}{100}$

M.R. Sarić and B.C. Loughman (eds.), Genetic Aspects of Plant Nutrition.
ISBN-13: 978-94-009-6838-7
©1983 Martinus Nijhoff/Dr W. Junk Publishers, The Hague/Boston/Lancaster.

Materials and methods

The investigations were carried out using the following species: *Lupinus albus* L. (cv. Kiewskij mutant and Drushba), *Lupinus luteus* L. (cv. Gülzower Süße Gelbe), *Pisum sativum* L. (cv. Dornburger Gelbe Speiseerbse and Auralia), *Vicia faba* L. (especially cv. Fribo), *Glycine max.* (L.) Merr. (cv. Fiskeby V). All plants were cultivated in "Mitscherlich pots" containing 5–6 kg quartz sand (Hohenbocka) or soil[11]. Some replicates (3–5 per treatment) were harvested at early stages and other ones later in order to study the nitrogen metabolism at different stages of plant development. In addition field experiments (mostly 13 m^2 per plot, 4 replicates) were carried out in order to determine the effect of mineral N on grain yields and seed protein composition.

In pot experiments plants were fertilized as follows (amounts per pot): 0.52 g P as $CaHPO_4 \cdot 2H_2O$, 0.83 g K as K_2SO_4, 0.3 g Mg as $MgSO_4 \cdot 7H_2O$ and micronutrients as 2 ml Hoagland AZ solution a + b. Nitrogen was given as NH_4NO_3 or $^{15}NH_4\ ^{15}NO_3$ depending on the aim of the trial (for details see figures and tables). The harvested plants were cut into organs, lyophilized, weighed and analysed for nitrogen concentration[3] and at-% ^{15}N-exc.[4]. In some cases seed N fractions were separated using fractional extraction and dialysis[10,13]. The seed globulins legumin and vicilin were characterized by means of disc electrophoresis[18]. Amino acid estimations in the different proteins were peformed by means of an amino acid analyzer (Hd 1200E, CSSR) after acid hydrolysis[12]. For the determination of the origin of nitrogen in the samples mineral N and/or soil N were used in ^{15}N labelled form in some experiments (3–5 mg N/100 g soil as soluble inorganic N ($= N_i$) according to Bremner[2], ^{15}N frequency was uniform in all soil N fractions). The following equations were used in order to estimate the origin of N found in plants:

$$\frac{\text{fixed N}}{\text{in mg/pot}} = \frac{\text{total N}}{\text{in mg/pot}} - \left[\frac{\text{seed-grain N}}{\text{in mg/pot}} + \frac{\text{fertilizer N}}{\text{in mg/pot}} + \frac{\text{soil borne N}}{\text{in mg/pot}} \right]$$

and

a) Sand culture (^{15}N labelled mineral fertilizer)

$$\frac{\text{Fertilizer N}}{\text{in mg/pot}} = \frac{^{15}\text{N amount in plants}}{\text{given in mg/pot}} \cdot \frac{100}{\text{at-\% }^{15}\text{N-exc. in fertilizer}}$$

b) ^{15}N labelled soil without mineral N or with ^{14}N application

$$\frac{\text{Soil borne N}}{\text{in mg/pot}} = \frac{^{15}\text{N amount in plants}}{\text{given in mg/pot}} \cdot \frac{100}{\text{at-\% }^{15}\text{N-exc. in soil}}$$

c) ^{15}N labelled soil, ^{15}N labelled fertilizer

$$\frac{\text{Fertilizer N}}{\text{in mg/pot}} = \left[\frac{^{15}\text{N amount in plants}}{\text{given in mg/pot}} - \frac{\text{soil borne }^{15}\text{N}}{\text{in plants (mg/pot)}} \right] \cdot \frac{100}{\text{at-\% }^{15}\text{N-exc. in fertlizer}}$$

The amount of soil-borne ^{15}N was determined in a special trial according to b).

In some cases the net–CO_2-assimilation of whole plants or of particular leaves *in situ* was proved. Therefore we used an open system of gas exchange analysis including IRGA and partial $^{14}CO_2$ application (estimation of radioactivity by means of flow counter, for details see Freye[5]).

All results were tested for statistical significance using variance analysis and Tukey test.

Results and discussion

Table 1 shows results of field trials. Nitrogen application prior to sowing or in the period of flowering increased the yields of seeds and the amounts of seed

Table 1. Effect of nitrogen fertilization on yields of seeds and amounts of seed proteins. Results of field experiments: Seeds were inoculated with suitable Rhizobium preparations before sowing (unpublished)

Nitrogen fertilization	Lupinus albus L. (cv. Kiewskij mutant)		Glycine max. (L.) Merr. (cv. Fiskeby V)		Vicia faba L. (different cv.)	
	n = 4 Seeds dt/ha	Crude protein in seeds dt/ha	n = 3 Seeds dt/ha	Crude protein in seeds dt/ha	n = 9 Seeds dt/ha	Crude protein in seeds dt/ha
a) Soil content < 7 mg N_i/100 g						
Without N	18.4 (100)	5.9 (100)	12.3 (100)	4.2 (100)	25.4 (100)	6.8 (100)
+ 60 kg N/ha prior to sowing	21.2 (115)	7.3 (124)	14.0 (114)	5.4 (129)	25.4 (100)	6.9 (101)
+ 60 kg N/ha in the period of flowering	20.9 (114)	7.1 (120)	13.8 (112)	5.4 (129)	25.2 (99)	6.8 (100)
b) Soil content > 7 mg N_i/100g	No differences between "without N" and + 60 kg/ha N prior to sowing or in the period of flowering					

The soil content was not determined in each case in the field experiments with *Vicia faba* L.

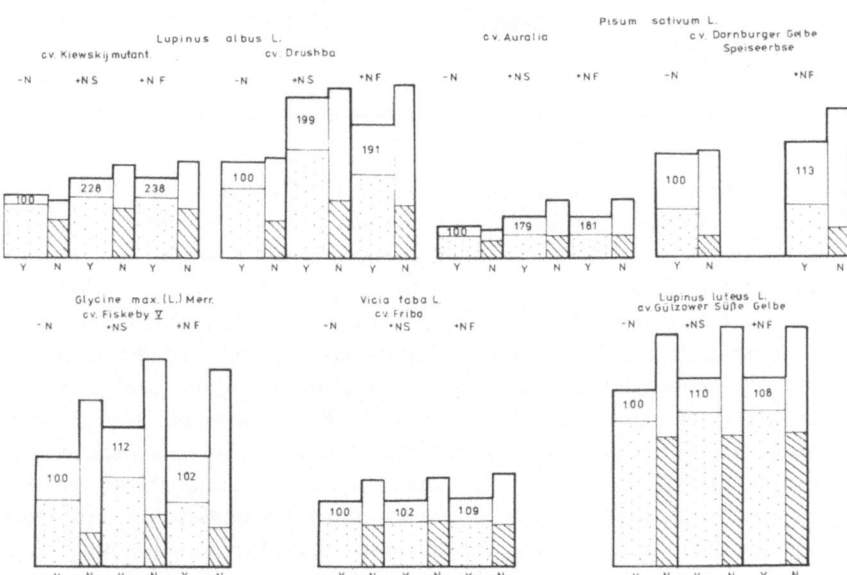

Fig. 1. The influence of mineral N on yield (g dry matter per pot) and nitrogen amount in plants (mg N per pot). Experiments in Mitscherlich pots. Broad columns = yield (pointed part = roots and shoots without grains, empty part = grains); narrow columns = nitrogen amount in plants (hatched part = roots and shoots without grains empty part = grains); −N = without N, +NS = +N prior to sowing, +NF = +N after flowering, substrate sand or N deficient soil, unpublished.

Table 2. Uptake of ^{15}N labelled mineral nitrogen by different leguminous species and its distribution in the plants. Pot experiments with 0.6 g N as $^{15}NH_4^{15}NO_3$ (10 at % ^{15}N-exc. = 60 mg ^{15}N) applied to *Lupinus albus* L. and *Vicia faba* L. in the period of flowering. 0.5 g N in the same form were applied to *Pisum sativum* L. Inoculaton by suitable preparations of Rhizobium. All plants were harvested when ripe. In the case of Pisum mean values of 4 experiments[13]. The other experiments were performed by Dr W. Merbach

Analytical characteristics	*Lupinus albus* L. (cv. Kiewskij mutant) Soil culture	*Pisum sativum* L. (Dornburger Gelbe Speiseerbse) Sand culture	*Vicia faba* L. (cv. Fribo) Sand culture
Additional N found in plants in comparison with control in mg N/pot	375 (=100)	192 (=100)	50 (=100) not significant
N uptake from fertilizer (^{15}N frequency) in mg/pot and (% of added amount)	363 (61)	361 (72)	309 (52)
Deficit in fixed N_2 in plants caused by N fertilizing in mg/pot and (%)	none	169 (=18%)	259 (=42%)
Proportion of additional nitrogen localized in seeds in mg/pot and (%)	277 (74)	157 (82)	48 (96) not significant
Fertilizer N (^{15}N frequency) in seeds in mg/pot and (% of fertilizer N taken up)	219 (69)	312 (86)	203 (66)

proteins of *Lupinus albus* L. and *Glycine max.* (L.) Merr. all cases with < 7 mg N_i/100 g soil. Contrary to this result *Vicia faba* L. never produced additional yields. In order to analyse these findings more precisely we carried out pot experiments with (Fig. 1) and the results confirmed the findings of the field trials. In addition they demonstrated that Pisum reacts similarly to *Lupinus albus* L. and *Glycine max.* (L.) Merr. However, *Lupinus luteus* L. behaved like *Vicia faba* L. and the cultivar did not play a role. Therefore, we found two groups of leguminous species: one group (*Lupinus albus* L., *Pisum sativum* L., *Glycine max.* (L) Merr.) transformed mineral nitrogen into additional proteins (especially seed proteins) but the other (*Vicia faba* L., *Lupinus luteus* L.) was not able to perform this process. In experiments with ^{15}N labelled mineral N we tested the uptake of nitrogen by the different species. Table 2 shows that the plants of both groups absorbed mineral nitrogen to a similar extent. But *Vicia faba* L. plants contained no additional nitrogen because the fixed N_2 decreased to nearly the same extent

Table 3. The fate of mineral nitrogen ($^{15}NH_4^{15}NO_3$) applied in the period of flowering to plants of *Pisum sativum* L. (cv. Dornburger Gelbe Speiseerbse) and the transport of different amino acids in leguminous plants. Pot experiments in quarts sand (Pisum) and soil (Lupinus). All seeds were inoculated with suitable Rhizobium preparations. The experimental data were obtained in co-operation with Dr. Merbach and Dr. Schalldach

Tested characteristics	Total mg fertilizer N per pot	^{15}N labelled compounds in mg per pot or (% of total)	
^{15}N labelled soluble N compounds in stems of Pisum 10 h after nitrogen application	8.4	Glutamic acid 0.8, alanine 2.0, serine + threonine 0.7, amide N 2.4, aspartic acid 0.5, 12 other nitrogen compounds in total 2.0	
Fertilizer N in seeds (ripeness)	326	Soluble N compounds (28) Proteins (72)	
Translocation of ^{15}N labelled amino acids from older leaves to pods (*Lupinus albus* L.)		Protein N in pods	Soluble N in pods
L-Alanine	7.57 mg ^{15}N	3.9 mg ^{15}N (51)	3.67 mg ^{15}N (49)
α-Aminoisobutyric acid	7.36 mg ^{15}N	0.29 mg ^{15}N (4)	7.07 mg ^{15}N (96)
other organs			
L-Alanine	2.26 mg ^{15}N	(23)	
α-Aminoisobutyric acid	2.59 mg ^{15}N	(26)	

as the absorption of mineral nitrogen increased. The absorbed mineral N was preferentially transported in all cases tested into the seeds. Only in *Lupinus albus* L. (and to a lesser extent in *Pisum sativum* L.) was this nitrogen deposited as additional N. Its route to seeds seemed to be characterized by incorporation into some transport amino acids (already in roots, Table 3), by translocation of these compounds to fruits which were strong sinks in this period, and by converting the nitrogen into seed proteins. Obviously, uptake into the pods was not dependent on the consumption of amino acids by protein biosynthesis because the non-protein amino acid α-aminoisobutyric acid was "pumped" into the pods to the same extent as L-alanine ^{15}N (Table 3). Such results were obtained also in *Sinapis alba* L.[14]. Table 4 demonstrates that the additional nitrogen was incorporated preferentially into legumin, the reserve protein which is formed at the end of vegetative growth. Its composition was not changed in consequence of the increased formation (Table 5). But legumins of the different species had a varying structure (Fig. 2, Table 5). Additional nitrogen in seeds originating almost solely from mineral N (see ^{15}N distribution) was incorporated into the proteins vicilin and albumin also. But the amounts were smaller. The results

Table 4. Concentration of different seed proteins and distribution of additional seed nitrogen found in consequence of nitrogen fertilization carried out in the period of flowering. Pot experiments in quartz sand (Pisum) or soil (Lupinus, Glycine). N as $^{15}NH_4^{15}NO_3$ (10 at % ^{15}N-exc., in case of Glycine 13.1 at % ^{15}N-exc.). All seeds were inoculated with suitable Rhizobium preparations prior to sowing. The experiments were performed by Dr. Merbach

Plant species, treatment and analytical characteristics	Total N in seeds mg/pot	Nitrogen in seeds*				
		Soluble	Albumin	Legumin	Vicilin	Rest
Lupinus albus L. (cv. Kiewskij mutant)						
Without Ni +0.6 g *N	179 (100)	(4)	(37)	(27)	(4)	(28)
·additional total N	276 (100)	(3)	(22)	(46)	(8)	(21)
·additional ^{15}N	22 (100)	(3)	(28)	(41)	(7)	(21)
Pisum sativum L. (cv. Auralia)						
Without Ni + 0.6 g *Ni	134 (100)	(9)	(28)	(23)	(11)	(29)
·additional total N	217 (100)	(10)	(6)	(43)	(16)	(25)
·additional ^{15}N	28 (100)	(8)	(12)	(44)	(13)	(23)
Glycine max. (L.) Merr. (cv. Fiskeby V)						
Without Ni + 0.8 g *Ni	1293 (100)	(4)	(21)	(35)	(19)	(21)
·additional total N	298 (100)	(−4)	(−2)	(65)	(18)	(23)
·additional ^{15}N	48 (100)	(3)	(9)	(48)	(18)	(22)

* Values in per cent of total N in seeds

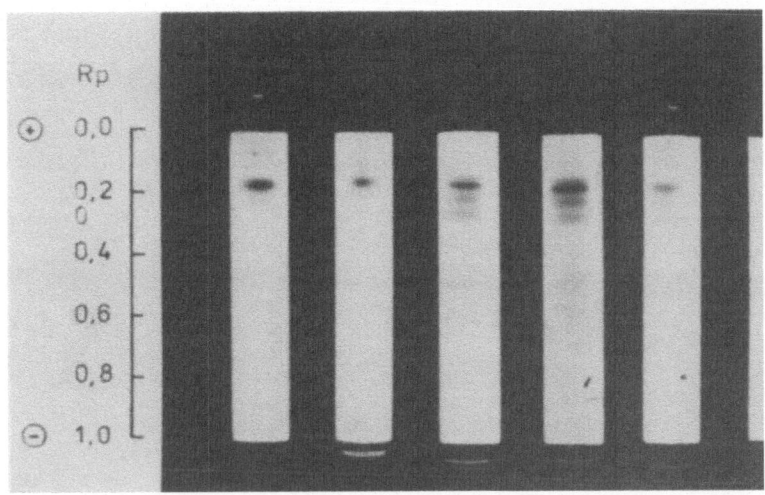

Fig. 2. Disc electrophoresis of seed legumin (acid polyacrylamide gel with 7.5% acrylamide (w/v), pH 4.3) according to Scholz et al.[18]. From left to right: *Vicia faba* L., *Pisum sativum* L., *Lupinus albus* L., *Lupinus luteus* L., *Glycine max.* (L.) Merr. (Merbach, unpublished).

Table 5. Effect of nitrogen fertilization on concentration of some amino acids in different seed proteins. Material from pot experiments represented in Fig. 1 and Table 4. Data in mg N/100 mg protein-N. ON = without N, S = +N prior to sowing, F = +N in the period of flowering. The amino acid estimations were performed by Dr. Sternkopf

Plant	Seed protein	Lysine			Methionine		
		ON	S	F	ON	S	F
Lupinus albus L.	Albumin	7.25	7.21	7.15	0.15	0.14	0.15
(cv. Kiewskij mutant)	Legumin	6.41	6.45	6.37	traces	traces	
	Vicilin	8.27	8.15	8.11	0.32	0.33	0.34
Glycine max. (L.)	Albumin	9.44	9.46	9.43	0.59	0.59	0.57
(cv. Fiskeby V)	Legumin	5.31	5.34	5.33	0.78	0.79	0.77
	Vicilin	8.29	8.30	8.35	0.37	0.38	0.36

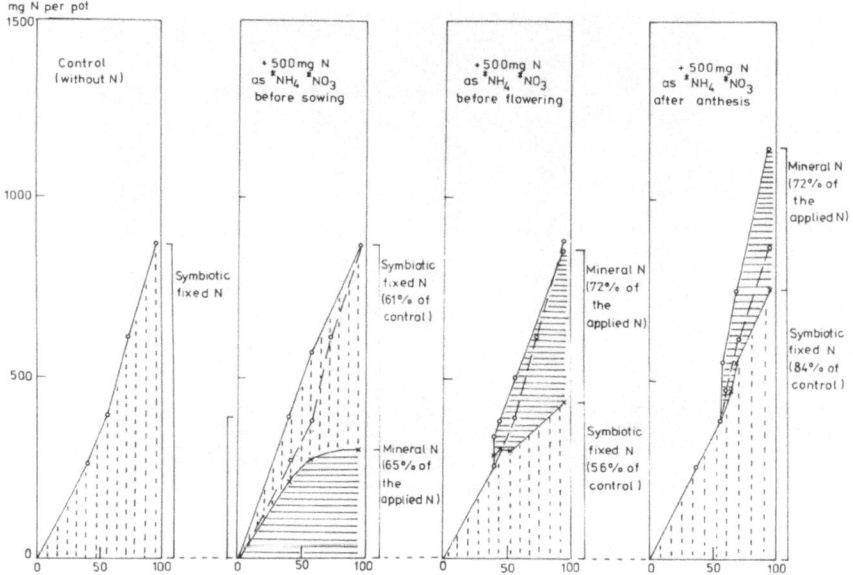

Fig 3 The influence of mineral N on the symbiotic N_2 fixation of *Pisum sativum* L. (according to Schilling et al.[17]). Nitrogen amounts include only shoot matter and not roots (the nitrogen content of roots was not estimated). The increase in fixed nitrogen after flowering was partially caused by translocation of nitrogen from roots to shoots. This conclusion is based upon comparisons with similar experiments that included root investigations.

demonstrate that the additional nitrogen could be found preferentially in reserve proteins of seeds rich in lysine.

The question is open why one group of leguminous plants is able to transform mineral N into additional seed proteins and the other one is not. Was the nitrogen fixing system of the one group sensitive to mineral N and that of the other one not? Fig. 3 shows the behaviour of the "non-sensitive" species *Pisum sativum* L. in case of mineral N application in different developing stages. Mineral N impaired N_2 fixation strongly also in this case when it was applied before flowering. But the amount of mineral N taken up in a short time was higher than the amount of symbiotically fixed N_2 in plants not being fertilized with mineral nitrogen. Therefore N fertilized plants contained for a longer time higher nitrogen amounts than the ones which were not fertilized. If nitrogen was applied after flowering we found no significant inhibition of N_2 fixation because this process had been completed. It was very interesting to notice that all species being "non-sensitive" to mineral N behaved like peas (Table 6). These species finished the N_2 fixation after the period of flowering. In contrast to this behaviour, "sensitive" species fixed N_2 until the stage of ripeness (Table 6) and they did not form additional proteins when they were fertilized at any stage. The different cultivars did not vary in this respect. That suggests that the symbiotic N_2

Table 6. Dependence of nitrogen fixation upon developing stage of leguminous plants. Pot experiments. Inoculation of seeds with suitable Rhizobium preparations prior to sowing. Data are given in mg/pot. The origin of N in plants could be determined because of ^{15}N-labelling of soil nitrogen. Nitrogen of the seed-grain was substracted from the total nitrogen in plants. No nitrogen fertilization. The data were obtained in co-operation with Dr. Merbach. *, **, *** = significant at the 5%, 1%, 0.1% level

Plant, substrate	Time of analysis	Nitrogen found in plants originating from				Nodule number per pot
		soil	air			
Lupinus albus L.						
(cv. Kiewskij mutant)	flowering	212.8	611.3	(92)		85
soil	ripeness	261.8*	666.4	(100)		86
(cv. Drushba)	flowering	453.7	138.3	(94)		100
soil	ripeness	600.0*	146.4	(100)		102
Pisum sativum L.	flowering	—	144.3	(95)		130
(cv. Auralia)	ripeness	—	152.3	(100)		124
sand						
Glycine max. (L.) Merr.	flowering	319.0	414.1	(92)		322
(cv. Fiskeby V)	ripeness	1080.7***	448.9	(100)		77**
soil						
Lupinus luteus L.	flowering	—	281.8	(34)		70
(cv. Gülzower Süße Gelbe)						
sand	ripeness	—	834.1**	(100)		68
soil	flowering	205.3	722.0	(46)		78
	ripeness	255.8**	1586.0**	(100)		84
Vicia faba L.	flowering	—	382.4	(62)		33
(cv. Fribo)						
sand	ripeness	—	619.9**	(100)		35

Table 7. Effect of different carbon supply of plants on nitrogen fixation. Pot experiments with soil (^{15}N labelled soil N), seed-grains inoculated with suitable Rhizobium preparations. Sucrose application: 2% sucrose solution ($1.85 \cdot 10^7$ Bq/g sucrose) was sprayed on leaves and stems daily for a period of 50 days beginning at the time of flowering. Plants were harvested when ripe.

CO_2 application: Whole plants were exposed to air containing 0.02 vol$\cdot -$% CO_2 or higher CO_2 concentrations at perspex boxes for 14 days beginning at the end of flowering. $^{14}CO_2$ application: One older leaf growing in the neighbourhood of the third node was exposed for 1 h to $^{14}CO_2$ in natural air. 24 h after the exposure plants were analysed (period of pod and seed growth).

The data were obtained in co-operation with Dr. Merbach and E. Freye. Biostatistical criteria as in Table 6.

Analytical characteristics	*Lupinus albus* L. (cv. Kiewskij mutant)			*Vicia faba* L. (cv. Fribo)		
	0.03% CO_2	0.1% CO_2	0.03% CO_2 sucrose	0.02% CO_2	0.14% CO_2	0.04% CO_2 ^{14}C labelled
N_2-fixation in 14 days (mg N/pot)	88.1 (100)	(377)**	–	522 (100)	(206)*	–
N_2-fixation in 50 days (mg N/pot)	108 (100)	–	(483)**	–	–	–
Nodule number per pot	10	11	–	311	367	
Increase of dry matter in g/pot (14 days)	3.7 (100)	(399)**	–	27.9 (100)	(169)*	g/plant in 5 d 0.67
of that pods incl. seeds	0.96 (100)	(438)**		12.1 (100)	(115)	0.59
roots	0.47 (100)	(419)**		3.4 (100)	(241)*	0
^{14}C-distribution in percent pods incl. seeds						66.6
roots						7.6

Table 8. Energy cost of N_2 fixation (references were taken from Phillips[20])

Theoretical energy cost of the basic reactions	$N_2 + 16$ ATP $+ 8$ $e^- + 10$ $H^+ \to 2$ $NH_4^+ + H_2 + 16$ ADP $+ 16$ Pi. With a P/O ratio of 2 for oxidative phosphorylation the nitrogen reduction requires 1 mole of glucose/N_2 or 2.57 g C/g N		Refs.[20]
Whole plant energy cost in nodulated legumes (determined experimentally)	*Glycine max.* (L.) Merr. *Pisum sativum* L. *Vicia faba* L. *Lupinus albus* L.	5.0–10.4 g C/gN 6.8 g C/gN 6.7 g C/gN 4.0– 6.5 g C/gN	Allam 1931 Mahon 1977 Mahon 1979 Pate and Herridge 1978
	Lupinus albus L. (cv. Kiewskij mutant)	7.0 g C/gN	unpublished results

fixation of "sensitive" species is more effective than that appropriate process of "non-sensitive" ones. Further investigations showed (Table 7) that application of sucrose and exposure of plants to air with higher CO_2 concentration reactivated N_2 fixation in white lupins after flowering. Therefore, the cause of early termination of N_2 fixation must be a lack of assimilates in root nodules. Obviously, the pods were very strong sinks in this period and they "directed" the assimilate stream quantitatively to themselves in "non-sensitive" species (see also refs.[8,19]) and root nodules "starved". In case of "sensitive" species bacteroids received assimilates during the period of pod growth for N_2 fixation.

These effects are understandable because of the high C requirement for N_2 fixation (6–8 g C/g N, see Table 8). This carbon requirement is about 3.4 g C/g N higher than nitrate nutrition of plants (*Lupinus albus* L., unpublished). Unfortunately, it is not possible to discuss the problems of assessing such data here. The practically assayed carbon requirement was also higher than the calculated value because energy is necessary for the formation and maintenance of nodules including bacteroids. Therefore, intimate physiological relationships exist between host plants and Rhizobia and it is clear that different strains of Rhizobium (Table 9) in combination with the same cultivar vary with regard to their efficiency. The tendency in Table 9 demonstrating the same efficiency order of strains in all cultivars need not be generally applicable.

The great importance of assimilate supply for the nutrition of bacteroids is clear and additionally estimations of net-CO_2-assimilation at nodulated and non-nodulated plants demonstrated 12–25 percent higher net-assimilation-rates of the nodulated ones (*Lupinus albus* L., unpublished). That suggests that root nodules may regulate the CO_2 assimilation of the host plant. But the extent of this

Table 9. Effect of different Rhizobium lupini preparations on nitrogen fixation of *Lupinus albus* L. and *Lupinus luteus* L. (cv. Kiewskij mutant and Gülzower Süße Gelbe in sand culture, cv. Drushba in soil culture). Soil effects were eliminated (^{15}N labelled soil N, subtraction of N_2 fixation caused by spontaneous infections). The experiments were performed by Dr. W. Merbach and R. M. Fraustein

	Lupinus albus L.						*Lupinus luteus* L.		
	cv. Kiewskij mutant			cv. Drushba			cv. Gülzower Süße Gelbe		
Treatment	Dry matter g/pot	fixed N mg/pot	Nodule number per pot	Dry matter g/pot	fixed N mg/pot		Dry matter g/pot	fixed N mg/pot	Nodule number per pot
1 Without inoculation	9.8(100)	0	–	32.8(100)	–		6.2(100)	0	–
2 + Azotogen	10.2(104)	0	–	33.1(100)	–		6.5(104)	0	–
3 + Rhizotorfin	13.1(134)	91(100)	18	45.8(140)	56(100)		26.9(434)	638(100)	60
4 + USSR strain 367a	15.4(157)	188(207)	18	–	–		–	–	–
5 + Polish strains Ts, Cz, 271	16.1(164)	217(238)	20	–	–		–	–	–
6 + Combination 4 + 5	16.4(167)	302(332)	19	40.3(123)	67(120)		32.9(531)	835(131)	68
LSD$_{0.05}$ Tukey	1.7	45.4	16	7.1	30		7.8	94	10

effect seems to vary between the different species. In *Vicia faba* L. the additional carbon requirement for N_2 fixation was supplied by the host plant because this did not produce higher yields in the presence of mineral N. The "saved" energy could not be transformed into additional dry matter. In the "non-sensitive" group of plants we obtained opposite results. Hence it follows that the reaction to mineral nitrogen can be a criterion of efficiency within a symbiosis. When the amounts of dry matter and proteins are the same in the presence and absence of mineral N the symbiosis will be very effective. Therefore, such a test can be a new method for discovering more effective Rhizobium-plant combinations. Nevertheless, it would be very important to ascertain the mechanism that regulates the photosynthesis of host plants. Probably, it will not be a simple source/sink relationship. In Freye's experiments[5] change of the sink capacity did not affect the net-CO_2-assimilation rate of *Vicia faba* L.

The great importance of physiological relationships between Rhizobia and host plants is not only valid for a true symbiosis. It can also be important for associations between non-legumes and N_2 fixing bacteria according to Hess[7]. Here it would be very useful to find bacteria mutants that have lost the mechanism of nitrogenase repression by mineral $N^{6,9}$. Only in such cases a "mixed" nitrogen nutrition from mineral N and atmospheric N_2 would be possible and useful.

References

1 Bothe H and Trebst A Eds. 1981 Biology of Inorganic Nitrogen and Sulphur. Springer Verlag, Berlin.
2 Bremner J M 1965 Inorganic forms of nitrogen. *In* Methods of Soil Analysis. Ed. C A Black. pp 1179–1232. American Society of Agronomy, Inc. Madison.
3 Faust H 1959 Untersuchungen über die Mineralstoffabgabe einjähriger Pflanzen. Ph. Thesis. Friedrich Schiller Univ. Jena (DDR).
4 Faust H, Bornhack H, Hirschberg K, Jung K, Junghans P and Krumbiegel P 1981 ^{15}N-Anwendung in der Biochemie, Landwirtschaft und Medizin. Schriftenreihe Anwendung von Isotopen und Kernstrahlungen in Wissenschaft und Technik 5. Isocommerz GmbH, Berlin.
5 Freye E 1981 Photosyntheseleistung und Assimilatverteilung bei Ackerbohnen (*Vicia faba* L.) als Grundlage zur Erklärung von Samenertragsschwankungen. Ph. Thesis. Martin Luther Univ. Halle-Wittenberg (DDR).
6 Hennecke H 1981 Regulation of nitrogenase biosynthesis in free-living and symbiotic N_2-fixing bacteria: a comparison. *In* Biology of Inorganic Nitrogen and Sulfur. Eds. H Bothe and A Trebst. pp 309–316. Springer Verlag, Berlin.
7 Hess D 1981 *In vitro* associations between non-legumes and rhizobium. *In* Biology of inorganic nitrogen and sulfur. Eds. H Bothe and A Trebst. pp 287–298. Springer Verlag, Berlin.
8 Kamata E 1957 Morphological and physiological study on nodule formation in soybeans. II. Relations between the foliar application of carbohydrates and nodule formation. Proc. Crop Sci. Soc. Japan 26, 58–60.
9 Kleiner D, Phillips S and Fitzke E 1981 Pathways and regulatory aspects of N_2 and NH_4^+

assimilation in N_2-fixing bacteria. *In* Biology of inorganic nitrogen and sulfur. Eds. H Bothe and A Trebst. pp 131–140. Springer Verlag, Berlin.

10 Merbach W 1978 Influence of N-fertilizing in the period of flowering on N-fractions of *Lupinus albus* L. seeds. Abh. AdW Berlin, Abt. Math, Naturwiss, Technik 4, 103–106.

11 Merbach W and Schilling G 1980 Wirksamkeit der symbiontischen N_2-Fixierung der Körnerleguminosen in Abhängigkeit von Rhizobienimpflung, Substrat, N-Düngung und ^{14}C-Saccharoselieferung. Zbl. Bakt. II. Abt. 136, 98–118.

12 Reissmann R 1969 Untersuchungen über die stoffliche Verarbeitung zusätzlicher Mineralstickstoffmengen durch die vegetativen Teile von *Sinapsis alba* L., *Helianthus annuus* L. und *Trifolium pratense* L. Ph. Thesis. Friedrich Schiller Univ. Jena (DDR).

13 Schalldach I 1965 Untersuchungen über Stickstoffumsatz und-verwertung bei Leguminosen, durchgeführt mit ^{15}N-markierten Verbindungen bei *Pisum sativum* L. Ph. Thesis. Friedrich Schiller Univ. Jena (DDR).

14 Schilling G 1971 Untersuchungen über die Ertragsbildung bei einjährigen Samenpflanzen. Mitt. Dt. Akad. Naturforscher Leopoldina. Reihe 3, 17, 99–114.

15 Schilling G 1980 Zur Ernährung der Pflanzen mit Luftstickstoff. Wiss. Fortschritt 30, 467–471.

16 Schilling G and Schalldach I 1966 Untersuchungen über Transport, Einbau and Verwertung spät gedüngten Mineralstickstoffs bei *Pisum sativum* L. Albr. Thaer. Arch. 10, 895–907.

17 Schilling G, Schalldach I and Polz S 1967 Neue Ergebnisse über die N-Düngung zu Leguminosen, erzielt auf der Grundlage von Versuchen mit ^{15}N-markiertem NH_4NO_3. Wiss Z. Friedrich Schiller Univ. Jena, Math. Nat. Reihe 16, 385–389.

18 Scholz G, Richter J and Manteuffel R 1974 Studies of seed globulins from legumes. I. Separation and purification of legumin and vicilin from *Vicia faba* L. by zone precipitation. Biochem. Physiol. Pflanzen 166, 163–172.

19 Wong P P 1980 Nitrate and carbohydrate effects on nodulation and nitrogen fixation (acetylene reduction) activity of lentil (*Lens esculenta* Moench). Plant Physiol. 66, 78–81.

20 Phillips D A 1980 Efficiency of symbiotic nitrogen fixation in legumes. Annu. Rev. Plant Physiol. 31, 29–49.

Changes in contents of N, P and K in maize hybrids at different nutrient regimes

P. ANDONOVA and T. KUDREV
M. Popov Institute of Plant Physiology, Sofia, Bulgaria

Key words Maize hybrids Nitrogen Phosphorus Potassium accumulation Productivity

Introduction

A comparative study of Bulgarian maize hybrids showed their unequal productivity at the same nutrient regime, as well as differences in effectiveness of nitrogen and phosphorus, introduced additionally in the soil by means of mineral fertilization.[5,7] In these same hybrids the quantity of protein and oil in the grain were investigated and also the ratios between the different higher fatty acids and it was demonstrated that they depend both on the nutrient regime and the peculiarities of the hybrids[2,8]. The accumulation of NPK in grain and forage maize was followed under field conditions.

Materials and methods

Maize hybrids for grain Knezha 2L 602 and Knezha 8 were grown on typical chernozem in the region of the city of Knezha. Forage hybrids Knezha 3L 621 and Knezha 2L 611 were grown on chernozem-smolnitza in the region of the city of Yambol, up to milk-wax ripeness. The plants were irrigated, maintaining the soil moisture over 75% in maize for grain and over 80% of the field water holding capacity, in maize for forage. The fertilizer doses during the different years were not the same and were calculated on the basis of soil content of nutrient substances and the requirements of plants of basic elements for the formation of a definite yield[6]. For three years on average in maize growing for grain, the following fertilizer doses were applied: T_1-N_{14} P_6 K_5; T_2-N_{24} P_{14} K_{14}; T_3-N_{31} P_{23} K_{15}, kg/da. In maize used for forage we applied in average for two years: T_1-N_{15} P_{12} K_{10}; T_2-N_{24} P_{20} K_{14}; T_3-N_{33} P_{28} K_{247} kg/da. The controls were unfertilized plants (T_0). The sum of the effective temperatures during the trial years has been sufficient for the ripening of the selected hybrids. The insufficient quantity of precipitation was compensated by watering during the vegetative period. At the end of the vegetative period in maize for grain and in the milk-wax phase in maize for forage, samples were taken from the separate organs of the maize plants. In the air dry material the contents of nitrogen was determined by Kjeldahl phosphorus—colorimetrically as phosphorus-molybdenum compound, potassium—on the flame photometer.

Results

The percentage content of nitrogen in the leaves, the stems, the grain and the corn cob of the two hybrids for grain is not the same (Fig. 1). In these organs

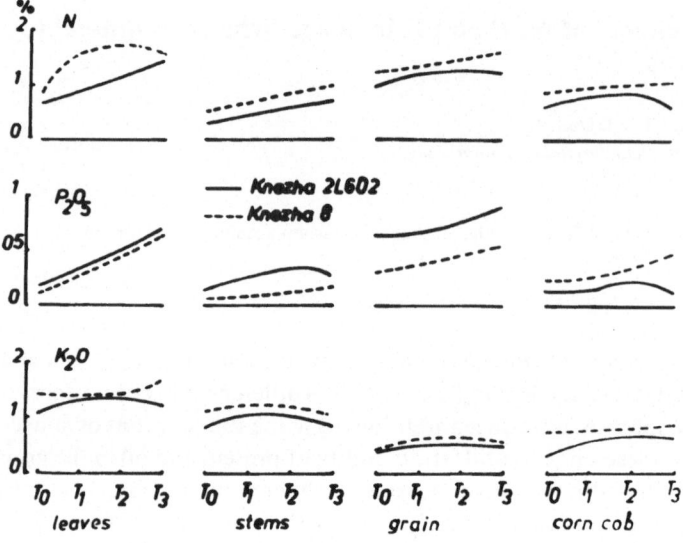

Fig. 1. Influence of the nutrient regime on the NPK percentage content in maize hybrids for grain.

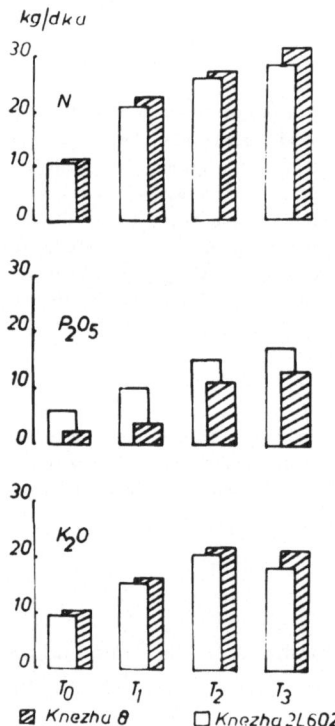

Fig. 2. Influence of the nutrient regime on the NPK content in maize hybrids for grain.

higher nitrogen content is found in Knezha 8. The phosphorus content in leaves, stems and grain is higher in Knezha 2L 602 in comparison with Knezha 8. In Knezha 8 the phosphorus content of the corn cob is greater. Higher potassium is found in the leaves, stems, grains and the cob of Knezha 8 in comparison with that of the plants of hybrid Knezha 2L 602.

The concentration of nitrogen, phosphorus and potassium in the soil exerted a considerable influence on the percentage content of nitrogen and phosphorus in the separate organs and the effect increases along with the increase in the concentration. The maximum percentage content however does not always coincide with the highest concentration of the nutrient medium. Similar results have also been obtained in another investigation[1].

The nitrogen quantity accumulated in the overground part of the plants is greater in hybrid Knezha 8 (Fig. 2). The plants of Knezha 2L 602 contain more P than Knezha 8 and the plants of Knezha 8 accumulate more potassium than those of Knezha 2L 602.

Fig. 3. shows that Knezha 3L 621 and Knezha 2L 611 differ considerably in percentage content of nitrogen in the separate organs, and the differences are clearer with higher concentration of NPK in the soil. The leaves and the spadix of Knezha 2L 611 are richer in nitrogen, compared with Knezha 3L 621, while in the stems and the tassel the opposite is found. In all organs one may observe higher phosphorus and potassium contents in Knezha 2L 611 in comparison with Knezha 3L 621, but these differences are more clearly expressed for P only in the leaves. A connection between the concentrations of NPK in the soil and their

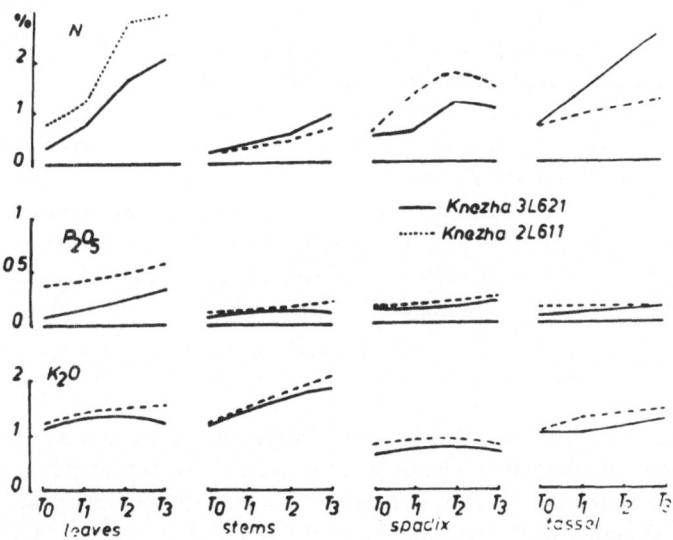

Fig. 3. Influence of the nutrient regime on the NPK percentage content in maize hybrids for forage.

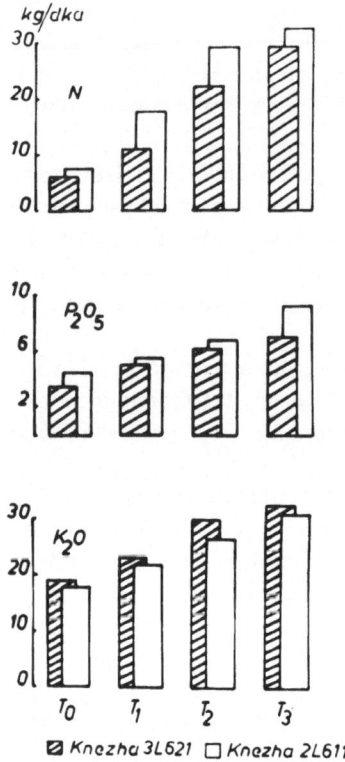

Fig. 4. Influence of the nutrient regime on the NPK quantity in maize hybrids for forage.

content in the plants, is seen for nitrogen, which increases in the leaves, the stems, the spadix, the tassel, along with the increase of the concentration.

The quantity of the accumulated nitrogen at the levels of the main nutrient elements in the soil, is higher in Knezha 2L 611 in comparison with Knezha 3L 621, and this is true for phosphorus (Fig. 4).

The productivity of the tested hybrids for grain and forage rises when the concentrations of the nitrogen, phosphorus and potassium in the soil increases. In maize for grain greater reaction in certain conditions are observed for Knezha 2L 602 and in the forage maize for Knezha 3L 621 (Table 1).

Discussion

The results obtained show that considerable differences in the uptake of nitrogen, phosphorus and potassium and their accumulation in different organs. In unequal concentrations of the main nutrient elements these differences are seen more clearly. Together with the unequal productivity, these are reflected in the total quantities, accumulated in the plants. When NPK levels are increased in

Table 1. Influence of the different nutrient regime on the yield in maize hybrids for grain and forage

Hybrid	Variants			
	T_0	T_1	T_2	T_3
Maize for grain				
Knezha 2 L 602				
Grain	663	1144	1205	1243
Vegetative biomass	868	1189	1356	1473
Total biomass	1531	2333	2561	2716
Knezha 8				
Grain	610	1077	1082	1088
Vegetative biomass	766	1096	1493	1515
Total biomass	1376	2173	2575	2603
Maize for forage				
Knezha 3L 621				
Spadix	652	915	927	1011
Vegetative biomass	1150	1364	1419	1573
Total biomass	1802	2279	2346	2584
Knezha 2 L 611				
Spadix	743	777	917	1095
Vegetative biomass	952	1025	1224	1266
Total biomass	1695	1802	2141	2361

most cases one observes an increase of the nitrogen, phosphorus and potassium content in the plants, expressed in a different degree for the organs and the elements. It was established that in equal contents of NPK in the soil, some of the tested hybrids for grain or forage, accumulate more nitrogen, others—phosphorus or potassium. This obviously leads to changes in the proportion between the nitrogen, phosphorus and potassium contents in the plants. Differences are connected probably with the genetic ability of the hybrids to take up and utilize nitrogen, phosphorus and potassium.

This inference is in accordance with the view that the proportion between the nutrient elements in the plants is constant and is a hereditary characteristic[3].

References

1   Andonova P 1980 Izmeneniya v s'd'rzhanieto na azota, phosphota, kaliya, kalciya i magneziya v carevichni rasteniya pod vliyania na koncentraciyata n hranitelnata sreda. Phiziologia na rastaniyata VI 3, 65–75.
2   Andonova P, Kudrev T, Katzarov A and Dencheva S I 1979 Alterations in the protein complex

of grain in maize hybrids under the influence of mineral nutrition depending on their variety peculiarities. Mineral Nutrition of Plants 2, 120–126.
3   Lavrichenko V M and Zhurbickiy Z N 1976 Sootnosheniya elementov pitaniya v rasteniyah kak vidovoe genotipicheskoe yavlenie. Agrokhimia, 9.
4   K'drev T 1977 Hranene na pshenicata i carevicata. Izd. na BAN, Sofia.
5   K'drev T, Andonova P, Kacarov A and Dencheva S 1978 Izuchavane produktivnostta na carevichni hibiridi posredstvom optimizirane na hraneneto im. Phiziologia na rasteniyata IV, III, 608–619.
6   K'drev T and Andonova P 1979 Ustanovyavane na optimalni s'otnosheniya i koncentraciya na makroelementi pri otglezhdaneto na carevica. Phiziologia na rasteniyata V, 2, 71–80.
7   Stoyanov Ir 1983 Influence of NO_3^-, NH_4^+ and temperature on the growth and uptake of minerale elements by young maize plants. Plant Physiol. 2 (*In press*).
8   Tyankova L, Stoyanova Z, Andonova P, Dencheva S, Katzarov A and Kudrev T 1981 Influence of mineral nutrition on maize-grain fatty acids measured by gas chromatography. II-nd International Youth Symposium, Varna, 23–27. X.

A triaxial ratio diagram and its use in comparative plant nutrition

W. G. BRAAKHEKKE
Department of Soil Science and Plant Nutrition, Agricultural University, De Dreijen 3, 6703 BC Wageningen, The Netherlands

Key words Calcium Cation selectivity Magnesium Ratio diagram Triaxial

Introduction

Most research in nutrient uptake and metabolism is concerned with single nutrients and nutrients other than the one investigated are usually left out of consideration, keeping their supply constant at a non-limiting level. Interactions between nutrients are mostly dealt with as adventitious phenomena in single nutrient studies.

From the point of view of a plant physiologist it may be sensible to study the uptake, transport and metabolism of different nutrients separately, but this approach will not suffice when we try to understand the response of the whole plant or a population of plants to the nutritional status of the substrate, as in ecophysiological or agronomical research. The subordinate position of nutrient interactions as a research subject is certainly not in accordance with their importance for the growing plant, as the following points may demonstrate:

It has been recognized for some time now that ratios of cations are superior to single cation concentrations as a means of characterizing the cation supply of soils. Ion uptake by plants is governed by ionic ratios in the substrate rather than ion concentrations. Differences in selectivity between plant species are characterized better by element ratios in their tissues rather than element concentrations. Notably, plants differ greatly in the extent to which their uptake of one nutrient is influenced by the supply of other nutrients.

The biomass a plant is able to produce with increasing amounts of one nutrient is largely dependent on the supply of other nutrients. A balanced supply of nutrients is necessary for optimum production and this balance may differ between species.

Evidently in the whole chain of relations between soil composition and plant production interactions between nutrients are important, especially since in many field situations several nutrients may act as limiting factors simultaneously.

A lack of interest in nutrient interactions is, apart from the habit of

M.R. Sarić and B.C. Loughman (eds.), Genetic Aspects of Plant Nutrition.
ISBN-13: 978-94-009-6838-7
©*1983 Martinus Nijhoff/Dr W. Junk Publishers, The Hague/Boston/Lancaster.*

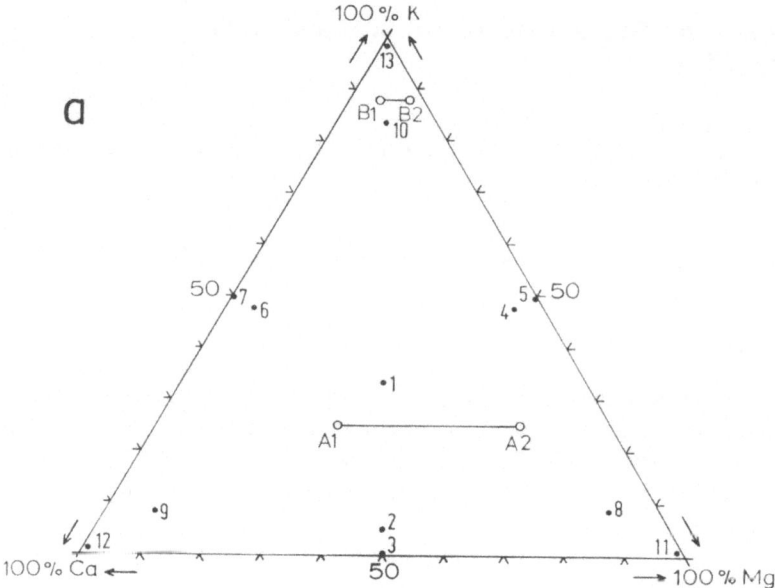

one-factor-thinking, mostly due to the complication of the subject. Results of experiments in which the supply of several nutrients is varied simultaneously are difficult to grasp. The use of suitable graphical methods of presentation may help a great deal in this respect. Especially in comparative studies there is a need for a convenient presentation of ion uptake patterns of species, taking nutrient interactions into account.

The triaxial ratio diagram presented here has been developed for the purpose just mentioned. In this diagram the relative importance of three components is represented by a single point. The usual way to do this is in a triangular diagram, but this has some drawbacks that the triaxial ratio diagram has not. I will first explain the difference between these diagrams and then show one of the many applications of the triaxial ratio diagram, namely in obtaining a comprehensive view of the relation between the cation composition of the substrate and the cation uptake pattern of different plants growing on it.

Triangular diagram versus triaxial ratio-diagram

Commonly the relative importance of three components is presented in a triangular diagram in which the percentages of the components are plotted along three axes that are perpendicular to the sides of an equilateral triangle (Fig. 1a). This diagram serves reasonably well when the components are of approximately equal importance. When the relative composition is less even, however, differences in composition are displayed less clearly.

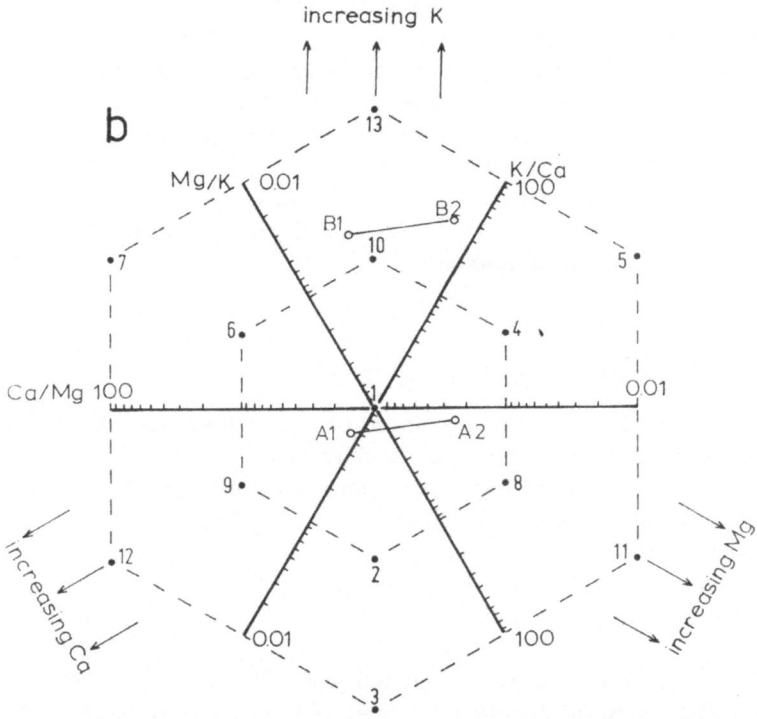

Fig. 1. Comparison of the position of some points in a triangular diagram (a) and in a triaxial ratio diagram (b). The points A1, A2, B1 and B2 represent the arbitrary nutrient compositions listed in Table 1. The points 1 to 13 ($\cdot$ – – $\cdot$) represent the nutrient solutions listed in Table 2. The arrows indicate the direction in which a cation increases. Further explanation in text.

This is seen by comparing the distance between the points A1 and A2 with the distance between B1 and B2 in Figure 1a. These points represent the relative importance of K, Ca and Mg in two imaginary plant species (A and B), differing in preference for K, grown on two substrates (1 and 2), differing in Ca/Mg ratio. The difference in Ca and Mg concentrations in plants on substrate 1 and 2 is

Table 1. Four arbitrary cation compositions plotted in Figure 1

	Concentrations			Percentages			Ratios		
	K	Ca	Mg	K	Ca	Mg	K/Ca	Mg/K	Ca/Mg
A1	167	300	200	25	45	30	0.55	1.20	1.50
A2	167	100	400	25	15	60	1.67	2.40	0.25
B1	3500	300	200	87.5	7.5	5	11.67	0.06	1.50
B2	3500	100	400	87.5	2.5	10	35.00	0.11	0.25

chosen the same for both species (see Table 1). Yet the distances A1—A2 and B1—B2 in the triangular diagram suggest otherwise. This may lead to a misinterpretation of the variability in two of the components (Ca and Mg) at different values of the third component (K).

The above problem does not exist in a triaxial ratio diagram (Fig. 1b), which, in addition, goes along with the convention of using ionic ratios in selectivity studies. A triaxial ratio diagram makes use of the principle that two of the three possible ratios between three components (not counting reciprocal ratios) determine the remaining third ratio:

$$K/Ca \times Ca/Mg \times Mg/K = 1$$

or

$$\log (K/Ca) + \log (Ca/Mg) + \log (Mg/K) = 0$$

This matches the principle that the coordinates on two of three axes laying in the same plane determine the coordinate on the third axis.

Figure 1b shows how the three ratios are plotted on logarithmic scales along three axes intersecting at angles of 60°. In the triaxial ratio diagram a difference in relative composition is represented by the same distance everywhere in the diagram. It is seen that the distance A1—A2 is equal to B1—B2. This is essential for a comparison of plant uptake responses to changes in the substrate composition.

The difference in K concentration between pair A1—A2 and pair B1—B2 results in a shift in "vertical" direction. In general, when two ratios change to the same extent, while the third ratio remains constant, a point shifts parallel to the bisector of two axes, which is perpendicular to the third axis. Therefore the directions perpendicular to the axes indicate changes in the proportion of a single component.

The triaxial ratio diagram is apt to present the relative importance of three components in widely divergent fields of science, ranging from studies in ion exchange equilibria to species compositions in vegetations. I will give here one application in the field of plant nutrition.

Application in a study of cation selectivity

The results of over a century of plant-analytical research have convincingly shown the existence of genetically determined differences in the cation uptake patterns of plants. These differences are interesting for scientific as well as practical reasons (ecology, chemotaxonomy, crop husbandry, animal nutrition). Yet our knowledge of the variation and distribution of uptake patterns in the plant kingdom is very incomplete and disordered. The main reason is that most of the analytical information is incompatible, because the plants analyzed were grown under different conditions.

Nutrient availability in particular has a strong influence on the chemical

Table 2. Composition of nutrient solutions in the experiment of Braakhekke[1] in meq/l.

Solution	K	Ca	Mg
1	2	2	2
2	0.2	2	2
3	0.02	2	2
4	2	0.2	2
5	2	0.02	2
6	2	2	0.2
7	2	2	0.02
8	0.2	0.2	2
9	0.2	2	0.2
10	2	0.2	0.2
11	0.02	0.02	2
12	0.02	2	0.02
13	2	0.02	0.02

In all solutions: Na: 2 meq. l^{-1}; $NO_3:H_2PO_4:SO_4 = 4:1:1$ (as equivalents), 5 Fe, 0.5 B, 0.5 Mn, 0.05 Zn, 0.02 Cu, 0.01 Mo ($mg \cdot l^{-1}$).

composition of a plant. To free the determination of uptake patterns from environmental noise, plants of different species have to be grown on the same substrate. Because species differ in the extent that their uptake pattern varies under the influence of variation in substrate, it is also important to grow plants of each species on different substrates. To avoid the accumulation of indigestable information, a deliberate choice has to be made of a few substrates, attuned to the kind of information wanted and the method of data processing and presentation chosen.

The triaxial ratio diagram has shown its merits in a comparative study of cation uptake patterns following the approach outlined above[1]. Plants of six species were grown simultaneously from germination to harvest on thirteen nutrient solutions with different cation concentrations. These concentrations were maintained by daily additions of increasing amounts of salt solutions, adjusted after analysis of the culture solution twice a week. The concentrations of K, Ca and Mg (Table 2) were chosen so as to give a regular pattern in a triaxial ratio diagram, consisting of two nested regular hexagons, one twice the size of the other, with one solution in the centre (Fig. 1b).

The cation compositions of two species grown on these solutions are plotted in Figures 2a and b, together with the compositions of the substrates. The pattern of each species forms an image of the hexagonal pattern of the substrate, might it be transposed, reduced and deformed. Some points are missing as the plants on some solutions did not yield enough for analysis. The differences in size, place and shape of the species hexagons demonstrate differences in degree of

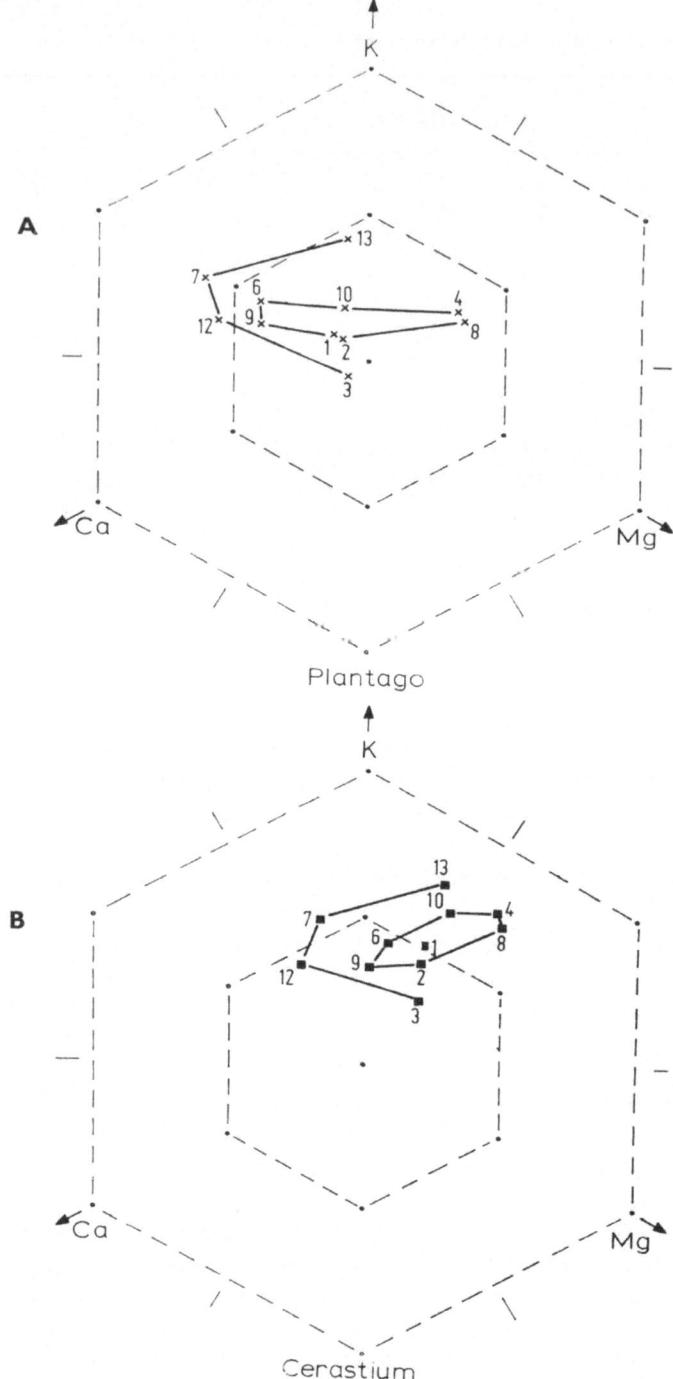

Fig. 2. The cation composition of *Plantago lanceolata* and *Cerastium holosteoides* grown in nutrient solutions 1 to 13 (· – – ·) listed in Table 2, plotted in a triaxial ratio diagram. The axes (left out) are the same as in Figure 1b. Further explanation in text.

selectivity, in preference for a specific cation and in interactions between the three cations during the uptake process, respectively. In both species the ability of selective uptake has reduced the differences in cation composition in the substrate to much smaller differences inside the plant. From the position of its hexagon it follows that *Cerastium holosteoides* has a strong preference for K and disfavours Ca, except when very high K/Ca ratios are present in the solution.

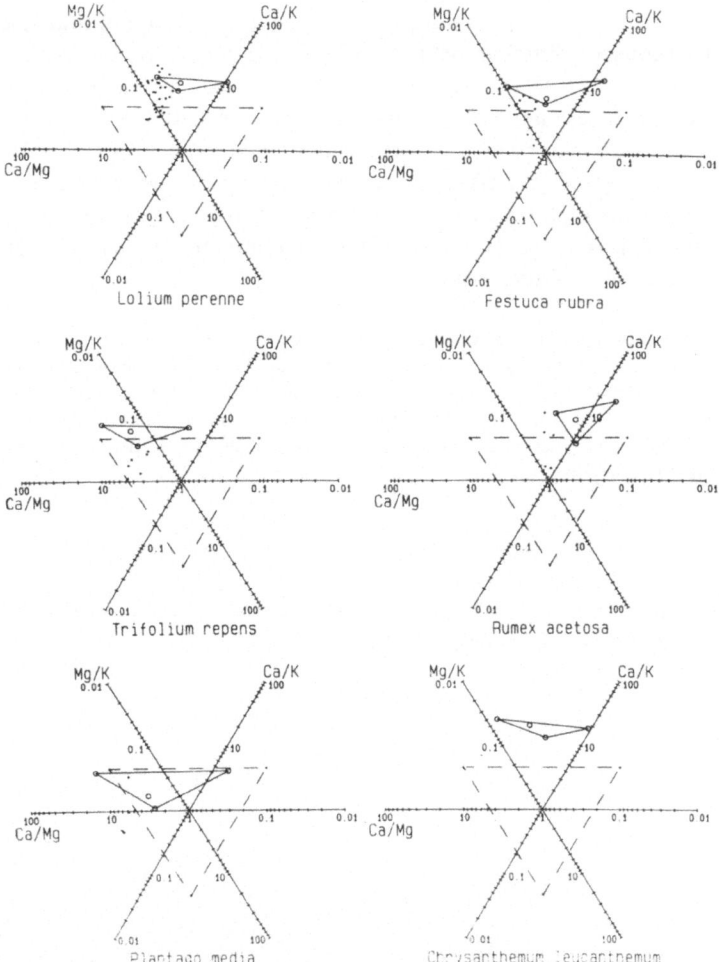

Fig. 3. The cation composition of 6 plant species grown on watercultures and in the field, plotted in triaxial ratio diagrams. The composition of the four water cultures (indicated by · – – ·) was the same as nr 1, 2, 4 and 6 of Table 2. The composition of plants grown on watercultures is indicated by o – – o. The field data (indicated by dots) are unpublished results of Dirven and De Vries (1957–1964), obtained from a wide variety of Dutch grasslands.

Plantago lanceolata is characterized by a small preference for K and a wide variation in Ca/Mg ratio, indicating a lack of discrimination between these two bivalent ions.

In another experiment following the same approach less nutrient solutions and more species were involved. Figure 3 shows some results. It seems worthwhile to investigate a large number of species in this way to provide a conspectus of cation uptake patterns, with which other analytical data may be compared. One may for instance compare the composition of plants grown in the field with the uptake pattern found on the nutrient solutions to obtain information about the availability of cations in different field situations. Field data of the species investigated so far show that the cation ratios of plants of the same species growing on different places, may differ as much as factor 10 to 100 (Fig. 3). This shows how careful one has to be in drawing conclusions about cation selectivity from analysis of field grown plants. It appeared that most of the field grown plants contain relatively high amounts of Ca compared to plants grown on the nutrient solutions. This is not surprising, as Ca is the dominant ion on the cation exchange complex of most Dutch soils.

Acknowledgements Part of this work was carried out at the Centre for Agrobiological Research (C.A.B.O.), Wageningen, supported by the Foundation for Fundamental Biological Research (B.I.O.N.), which is subsidized by the Netherlands Organization for the Advancement of Pure Research (Z.W.O.). I thank Professor ir. J. G. R. Dirven for making his field data available to me, Miss M. A. Hendrikx for carrying out the experiment of Fig. 3 and Dr. W. H. van Riemsdijk for his help in programming the computer plots.

Reference

1　Braakhekke W G 1980 On coexistence: a causal approach to diversity and stability in grassland vegetation. Agr. Res. Rep. 902. Pudoc, Wageningen.

Tolerance of barley and other cereals to manganese-deficient calcareous soils of South Australia

R. D. GRAHAM, W. J. DAVIES, D. H. B. SPARROW and J. S. ASCHER
Department of Agronomy, Waite Agricultural Research Institute, Glen Osmond, 5064 South Australia and Deparment of Agriculture, Port Lincoln, 5606, South Australia

Key words Barley Chlorosis Genotype Manganese concentration yield

Introduction

Manganese deficiency is still a widespread problem in southern Australia, though the magnitude of crop loss from it often depends on seasonal conditions. On the calcareous sands, however, manganese deficiency is both chronic and severe, yields without manganese fertilizer ranging from 40 to 75% of those with manganese[9]. Further since the pH of these soils is high (pH 8-9) soil-applied manganese fertilizer quickly reverts in the soil to unavailable forms, and one or two foliar sprays of manganese during growth are commonly required for satisfactory cereal production[9].

Barley is traditionally the cereal grown on the calcareous sands. This is partly a matter of climate and partly a belief that wheat is more susceptible to manganese deficiency. Varietal differences among cereals in susceptibility to manganese deficiency have been reported, (for example Batey[1], Munns *et al.*[7]) and since a new barley was about to be released and trends were appearing in the use of existing cultivars, a study of varietal performance on these soils was undertaken.

Materials and methods

Sites

The experimental sites, Port Lincoln (5 km west of the town) and Wangary (80 km north west) are located on Eyre Peninsula, South Australia, in a Mediterranean climate with an annual rainfall of approximately 500 mm (winter incidence). Both sites have calcareous sandy soil derived from shells of marine origin. These soils contain 80-90% $CaCO_3$, have a pH (in water) of 8-9, bulk densities of only 0.6 g cm^{-3} and have, for sands, extremely high water holding capacity.

Experimental design

The 1980 experiment at Port Lincoln compared the performance of seven genotypes with and without both soil and foliar applications of manganese fertilizer in a split-plot design with four replications.

Genotypes
Rye (*Secale cereale* cv. S.A. Commercial)
Wheat (*Triticum aestivum* cv. Halberd)
Wheat (*Triticum aestivum* cv. Oxley)

M.R. Sarić and B.C. Loughman (eds.), Genetic Aspects of Plant Nutrition.
ISBN-13: 978-94-009-6838-7
©1983 Martinus Nijhoff/Dr W. Junk Publishers, The Hague/Boston/Lancaster.

Triticale (*Triticosecale* × cv. Coorong)
Barley (*Hordeum vulgare* cv. Weeah)
Barley (*Hordeum vulgare* cv. Clipper)
Barley (*Hordeum vulgare* cv. Galleon)

Soil manganese treatments
0 kg ha^{-1} Mn
50 kg ha^{-1} Mn

Foliar manganese treatments
0 kg ha^{-1} Mn
1 kg ha^{-1} Mn (as MnSO$_4$ at tillering and stem extension)

Sub-plot size was 1.6 × 20 m. Seed of rye, wheat, barley and triticale were sown at 50, 65, 80 and 90 kg ha^{-1} respectively with basal fertilizer of 200 kg ha^{-1} superphosphate containing copper, zinc, cobalt and molybdenum together with 75 kg ha^{-1} (NH$_4$)$_2$SO$_4$. Soil-applied manganese was drilled in a compound fertilizer with the superphosphate.

During tillering, the youngest leaf with a ligule (youngest expanded blade or YEB) was sampled for analysis and grain yield was obtained by small-plot harvester.

The 1981 experiment at Wangary compared 72 barley genotypes from a world collection sown in plant breeders' plots of 4 rows × 5 m. Entries occurred at random in each of two parallel blocks. A feature of the design was the repetition of a check genotype (Weeah, the manganese-efficient genotype in 1980) every fourth plot throughout the array. Thus, each entry was compared for yield (and appearance) to its nearest Weeah neighbours, from which a relative yield was computed. The design enabled a detailed assessment of soil variability which was indeed very high in one direction (down slope).

Plots were sown on July 17, 1981 by precision drill with basal fertilizer as for 1980, but no manganese was used. During the season, visual assessment of plots was made at tillering and stem extension by two people. A scale of 1–5 was used (1 for the least vigorous, most chlorotic plot; 5 for maximum greenness and vigour). YEBs were collected at tillering for analysis of manganese by atomic absorption spectrophotometry following nitric-perchloric digestion. Grain yield was estimated by hand harvest (1m of row from the middle rows of each plot).

Results

Manganese deficiency was so severe at Port Lincoln in 1980 that Galleon barley and Coorong triticale, the most sensitive genotypes, appeared mid-season to be dying in −Mn plots. However a warm dry September followed by prolonged wet conditions in October apparently released native soil manganese to such an extent that these plots produced 1.8 and 1.5 t ha^{-1} respectively. Despite the recovery, these yields represented only 56% and 62% relative grain yield. Rye and Weeah barley appeared tolerant of the soil deficiency and produced 100% relative grain yields, that is, did not respond to manganese fertilizer. The wheats and Clipper barley were intermediate (Table 1). We term Weeah barley and the rye manganese–efficient genotypes and Coorong triticale and Galleon barley manganese–inefficient genotypes on the basis of their relative yields. This is an agronomic definition.

The concentration of manganese in the YEB at tillering reflected the relative

Table 1. Response of cereal genotypes to manganese fertilizer on a deficient calcareous sand in South Australia. Manganese was applied by both soil and foliar pathways. Means of four replications

Genotype	Grain yield, t ha^{-1}		Mn concentration*, μg g^{-1}	
	−Mn	+Mn	−Mn	+Mn
Tritcale cv. Coorong	1.5	2.4	9	11
Wheat cv. Halberd	2.0	2.8	13	15
Wheat cv. Oxley	1.7	2.4	13	15
Rye cv. S.A. Commercial	1.8	1.8	18	22
Barley cv. Clipper	2.8	3.3	11	13
Barley cv. Weeah	3.3	3.3	12	13
Barley cv. Galleon	1.8	3.2	8	10
LSD ($P<0.05$) gen. x Mn: 0.38				

* Concentration of Mn in youngest expanded blade at tillering.

yields of the genotypes in spite of the effect of the late rains. The wheats may be slightly anomalous in having higher Mn concentrations than their relative yields would suggest: this implies a higher internal requirement for manganese in wheats than in barleys.

The grain yields of the 1981 experiment were surprising for the extremes of response observed; it was possible to divide the collection into three primary classes with some assurance despite high variability of the soil with slope (Table 2). The manganese deficiency at this site was so severe that many genotypes died in midseason. Among the latter (group C, Table 2) was an entire family, 16 in all, of barleys all derived from Cl 3576, an introduction from Egypt. This group includes Galleon, a newly-released, high-yielding feed barley which was extremely sensitive to manganese deficiency in 1980 as well.

A second group (Table 2, group B) which included the established malting barley, Clipper, showed a small degree of tolerance while the 10 barley lines in group A all showed a high degree of tolerance. This group had excellent visual ratings for vigour and greenness during tillering (except 76T006-1 which improved its appearance and growth relative to the others as it matured). Some of the lower relative yields in this group (Duckbill, Resibee, Research) may be attributed in part to late maturation, that is, poor adaptation to the Mediterranean environment.

The concentrations of manganese in the YEB, though perhaps slightly higher in group A, are generally extremely low throughout all groups. However, because of much greater vigour in group A genotypes than in group C, uptake of manganese into the shoot was probably greater in group A (compared to C) by a

Table 2. Relative grain yield of some of the barleys from a world collection grown in manganese–deficient calcareous soil at Wangary, South Australia. Yields are calculated yields relative to the nearest check plots of Weeah barley sown in every fourth plot. Also shown are the visual rating for chlorosis [scale of 1 (chlorotic) to 5 (green)] assessed at tillering stage, and the concentration of manganese in youngest expanded blades at stem extension for selected genotypes. Means of duplicates

Genotype	Relative Yield %	Chlorosis at tillering	Mn Concentration $\mu g\ g^{-1}$
Group A			
Golden Promise	93	5	6.4
WI 2598	86	5	4.4
Maythorpe	62	4	4.0
Duckbill	54	5	5.4
Maltworthy	71	5	–
Research	56	5	–
Resibee	53	5	–
Weeah	97	5	4.1
Dampier	69	5	5.1
76T006-1	178	3	4.2
Group B			
WI 2477	16	3	3.2
WI 2468	19	3	3.7
Forrest	18	2	–
Clipper	15	3	3.6
Arivat	25	2	–
Prior	52	3	–
Suifu	49	3	–
Noyep	37	4	4.3
Athenais	12	3	–
Hiproly	9	3	–
Marocaine	7	4	3.6
Group C			
Proctor	7	3	—
Ketch	3	2	—
Corvette	3	2	—
Riso 9265	0	2	—
La Mesita	0	1	4.0
Nudinka	4	2	—
Akka	8	1	—
Indian dwarf	3	1	—
Galleon	8	1	3.8
CI 3576	0	1	—
WI 2591	3	1	3.6
LSD (P<0.05)	39		

Table 3. Group A barleys tolerant of manganese deficiency

1. *Plumage—Archer line*

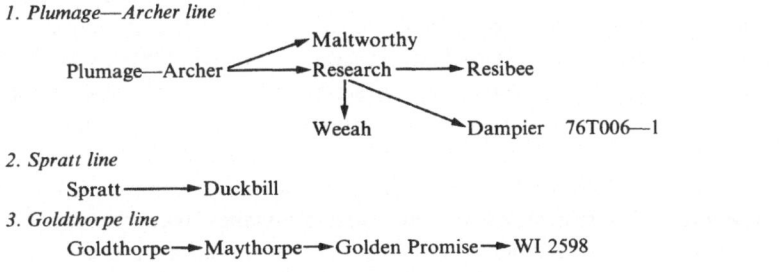

2. *Spratt line*

   Spratt ⟶ Duckbill

3. *Goldthorpe line*

   Goldthorpe ⟶ Maythorpe ⟶ Golden Promise ⟶ WI 2598

factor of 10 or more on average. Of necessity, however, the plots were not sampled for vegetative yields.

Discussion

Geneticists predict that triticale should generally be intermediate in character between its parental types, rye and wheat, as indeed was the case for tolerance to deficiencies of copper[3] and zinc[5]. In this study the triticale was more sensitive to manganese–deficient soil than either the wheats or rye; however, this triticale was not genetically related to the wheat or rye cultivars in the experiment. Other triticales have proved to be much more tolerant than Coorong (unpublished data). Rye is particularly interesting: although this cultivar has a low yield potential owing to self–incompatibility and poor seed set, it is nevertheless well adapted to many low-yielding, South Australian sandy soils, being as it is, copper-, zinc- and manganese-efficient and also tolerant of both acid and alkaline conditions generally[6].

The critical level of manganese in the YEB of cereals is thought to be in the range 15–20 μg g^{-1} (see Graham and Loneragan[4] for wheat). Thus the analyses reporting in this paper are indicative of the low manganese status of the plants in both experiments, especially the second. Values of 4–6 μg g^{-1} Mn are exceptionally low. Differences among these genotypes in critical concentrations of manganese are unknown but may be small, and if so, manganese uptake may be a better indicator of manganese efficiency.

As it happened, the three barley genotypes grown in 1980 fell into the three different classes defined in the 1981 study. Thus, the less precise data from 1981 are supported by the more quantitative data obtained in 1980 (large plots, four replications) under less severe manganese deficiency.

Currently, we define the manganese efficiency of a genotype as the relative grain yield (ratio of $-$Mn/$+$Mn) obtained from field plots on manganese deficient soil. This is an agronomic definition and does not imply mechanisms or processes. In the strictest sense, a value applies to a genotype only for the type of soil on which it was obtained. Ideally, at least one genotype should approach 100% relative yield at the site and a sufficiently large and wide selection of genetic material used. This technique ought then to be adequate to identify major, or single gene control of efficiency (or inefficiency) among the genotypes as well as effective combinations of minor genes. The possibility of different ranking under milder deficiency (for example, see Boken[2]) is ignored, at least in the first stages of the investigation, in favour of studying a wider range of genetic material. In the 1981 field study, a further simplification of the above was used: instead of +Mn plots for all genotypes, performance without added manganese was compared to that of a known manganese–efficient genotype grown in check plots arrayed throughout the experimental area. This type of design is easily sown with plant breeders' equipment and is suitable for preliminary investigation of a large collection of genetic material but differences in potential yield may affect interpretation.

The interest in screening the world barley collection lies in the genetic associations within groups. All progeny of Cl 3576 (16 in all) were manganese–inefficient and fell into group C despite a wide range of other parental material involved in the crosses; this suggests that manganese inefficiency in Cl 3576 is dominant and closely linked to traits seen to be desirable by the plant breeder. Cl 3576, an introduction from near Alexandria, Egypt, has conferred considerable yield advantage and disease resistance on its progeny under more highly fertile conditions than exist at Wangary. It is of some significance in South Australia that the new cultivar, Galleon, is so sensitive: at present Galleon is not recommended for calcareous soils, pending further agronomic work on its manganese fertilizer requirements. We note in passing that good yields were obtained with Galleon in another trial with exceptionally high rates of manganese fertilizer (both soil and foliar applied).

Group A genotypes fell into three genetically related groups, each having a pedigree going back to traditional English landraces or their derivatives, as shown in Table 3.

The derivation of tolerance from old English lines is most interesting since the south–eastern half of England is formed on Jurassic and Cretaceous marine limestones on which manganese deficiency is well known; manganese deficiency is also widespread on organic soils in this area (Batey[1]). This then appears to be the origin of the considerable tolerance to manganese deficiency which we have demonstrated in the Australian barley cultivars, Weeah and Dampier, two barleys which have become more popular with farmers cropping the calcareous sands despite the superior yield potential of newer cultivars.

The barleys in group B showed a moderate amount of chlorosis midseason or poor grain yield or both. Clipper, a well established Australian cultivar, and Proctor, a parent of Clipper fell into this group despite having Plumage–Archer in their pedigrees. These two barleys have not retained the high manganese efficiency shown by Weeah and Dampier. Batey[1] recorded that Proctor was less sensitive than some other barleys. Prior is an old well–adapted farmer's selection and its moderate grain yield may be due to its adaptation more than to any tolerance to manganese deficiency. On the other hand, Suifu, a late Japanese dwarf barley may have some useful tolerance.

Golden Promise and its progeny have proved to be remarkably well adapted to the South Australian environment. They have good yield potential, quality and disease resistance, and being tolerant to manganese deficiency, may provide a valuable cultivar for the calcareous sands in the near future.

Acknowledgements We acknowledge assistance of Mr S. Cornish with the field work. We thank Mr. D. Ritchie and Mr. D. Ottens who provided and prepared the land on which these experiments were carried out.

References

1 Batey T 1971 Manganese and boron deficiency. Technical Bulletin 21, 137–149. *In* Trace Elements in Soils and Crops. Ministry Agric. Fish. Food, London.
2 Boken E 1966 Studies on Methods for Determining Varietal Utilization of Nutrients. Thesis. Royal Veterinary and Agricultural College, Copenhagen.
3 Graham R D 1978 Tolerance of triticale, wheat and rye to copper deficiency. Nature 271, 542–543.
4 Graham R D and Loneragan J F 1981 The critical level of manganese for wheat. Proc Nat Workshop on Plant Anal. pp 95–96. Feb 15–18, 1981. Goolwa, S A.
5 Harry S P and Graham R D 1981 Copper-zinc interactions in wheat, rye and triticale growing on soils of contrasting pH. p 361 *In* Copper in Soils and Plants. Eds. J F Loneragan, A D Robson and R D Graham. Academic Press.
6 Harry S P and Graham R D 1981 Tolerance of triticale wheat and rye to copper deficiency and low and high pH. J. Plant Nutr. 3, 721–730.
7 Munns D N, Johnson C M and Jacobson L 1963 Uptake and distribution of manganese in oat plants. I. Varietal variation. Plant and Soil 19, 115–126.
8 Reuter D J, Heard T G and Alston A M 1973 Correction of manganese deficiency in barley crops on calcareous soils. 1. Manganous sulphate applied at sowing and as foliar sprays. Aust. J. Exp. Agric. Anim. Husb. 13, 434–439.
9 Reuter D J, Heard T G and Alston A M 1973 Correction of manganese deficiency in barley crops on calcareous soils. 2. Comparison of mixed and compound fertilizers. Aust. J. Exp. Agric. Anim. Husb. 13, 440–445.

Bacterial disease and the iron status of plants

BRUCE C. HEMMING* and GARY A. STROBEL
Department of Plant Pathology, Montana State University, Bozeman, MT 59717, USA

Key words Bacterial disease Chelates Iron (Fe^{+++}) *Pseudomonas* spp. Phytotoxin Siderophore Syringomycin

Abstract An antibiotic from a *Pseudomonas syringae* van Hall (1902) strain chelates iron and is similar to microbial high-affinity Fe(III) transport compounds termed siderophores. The previously unidentified basic amino acid residues of syringomycin, a hexapeptide phytotoxin from *P. syringae*, were identified as δ-N-hydroxyornithyl residues. This amino acid is common to hydroxyamate siderophores. These antibiotics inhibit the growth of the eucaryotic organisms, *Geotrichum candidum*, *Ceratocystis ulmi* and a *Rhodotorula* species. Of the 176 bacterial strains examined for antibiosis, inhibition by many strains increased on media containing Fe(III). The nature of the antibiotic activities of these isolates is discussed in relation to the possible commonality to known antibiotics, siderophores and phytotoxins produced by fluorescent pseudomonad species. The enhancement by iron of phytotoxin production supports the contention that the iron status of plants and the ability of a pathogen to acquire iron may significantly affect the progression of bacterial plant disease.

Introduction

Within the agriculturally important group of fluorescent bacteria of the genus *Pseudomonas* are found saprophytic soil and epiphytic strains; furthermore, the fluorescent pseudomonads are one of the largest groups of plant pathogenic bacteria. Some of these bacteria aggressively colonize the roots of plants and may be considered beneficial, deleterious or neutral to the plant[18,19]. A number of beneficial strains have been classified in the *Pseudomonas fluorescens-putida* group and used as seed inoculants on crop plants[10]. Some phytopathogenic strains are classified as within the more nutritionally fastidious *P. syringae* group[17]. Syringomycin, a non-host specific phytotoxin which also exhibits a wide spectrum antibiotic activity, is produced by *Pseudomonas syringae* pv. *springae* and is considered essential for pathogenicity[6]. Such toxins are among the likely agents of the deleterious effects on plant growth. The beneficial activity of certain isolates is considered in part as the result of the production of extracellular siderophores which complex the iron in the root environment, making it less available to competing microflora, thus alleviating the stress on plants by certain soil-borne pathogens[10].

The crystal and molecular structure of ferric pseudobactin, a siderophore from

* Present address: Monsanto Corporate Research Laboratories, 800 N. Lindbergh Blvd., St. Louis, MO 63167.

M.R. Sarić and B.C. Loughman (eds.), *Genetic Aspects of Plant Nutrition.*
ISBN-13: 978-94-009-6838-7
©1983 *Martinus Nijhoff/Dr W. Junk Publishers, The Hague/Boston/Lancaster.*

a plant growth-promoting fluorescent pseudomonad strain, has been determined and shown to consist of a linear hexapeptide of both L and D-amino acids attached to a fluorophoric quinoline derivative[25]. The singular characteristic of both beneficial and deleterious fluorescent pseudomonads is the production of such water-soluble fluorescent pigments. The siderophores or related substances produced by fluorescent pseudomonads include, in addition to pseudobactin, pyoverdine[14], ferribactin[12] and compound S[13]. Only partial structures of these other compounds are known.

The results of a systematic analysis of 176 fluorescent pseudomonad isolates from several host plant species and a concomitant study of the effects of iron on the procaryotic and eucaryotic microbial inhibitory activities of these bacteria allowed us to examine the strain specificity of such interactions[9]. Along an analogous line of logic as to the importance of iron in influencing bacterial virulence of animal disease agents[27,28], we were prompted to speculate that the iron status of plants and the ability of a pathogen to acquire iron may significantly affect the progression of microbial plant diseases[9]. We now present the partial isolation and characterization of an antimycotic with iron-chelating activity from a selected *P. syringae* van Hall isolate. This isolate has been applied as a bacterial treatment against Dutch Elm disease[15,23]. The antimycotic activity is compared to the hexapeptide phytotoxin, syringomycin, whose previously unidentified basic residues[5] are now shown to consist of δ-N-hydroxyornithyl residues. These residues are characteristic of hydroxamate siderophores. The results are presented as further evidence of the probable integral role of iron in plant disease interactions of particular phytopathogenic fluorescent pseudomonad isolates.

Abbreviations HPLC, High-performance liquid chromatography; NMR, nuclear magnetic resonance spectroscopy; SR, syringomycin; TLC, thin-layer chromatography.

Materials and methods

The source of all organisms and growth procedures used in this study follow those given previously[9] with the exception of the source of *Pseudomonas syringae* B3A. This isolate and an authentic sample of syringomycin were provided by Dr. D. F. Myers and Dr. J. E. DeVay, respectively (Dept. of Plant Pathology, University of California, Davis, CA). Rhodotorulic acid was a gift of Dr. Thomas Emery, Department of Chemistry and Biochemistry, Utah State University, Logan, UT. A sample of aerobactin was kindly supplied by Dr. J. B. Neilands, Department of Biochemistry, University of California, Berkeley, CA. The full details of all procedures have been provided elsewhere[8], but pertinent information is now outlined. The basic medium for all antibiotic work consisted of a modified mineral-salts medium[2], in which 1% glucose replaced 1% glycerol as the carbon source in the medium. In cases of iron supplementation, the medium contained 10 ppm $FeCl_3$ which was added after autoclaving from a filter-sterilized, fresh 0.1% aqueous solution. This medium referred to as Dye's medium was used in broth culture or as plates of 1.2% Oxoid purified agar obtained from K. C. Biological Inc., Lenexa, KY. Care was taken with the media to minimize traces of iron.

Cell-free supernatant preparation was obtained by the addition of cold acetone (2:1 v/v) to 4 to 5-day-old litter broth cultures followed by centrifugation (15,000 $\times g$, 10 min) and concentration

(20 ×) by flash evaporation (30°C). Such preparations served as chromatography samples or were further extracted (3 ×) with n-Butanol (1:1 v/v- and again concentrated (30 ×) preparatory to chromatographic analysis.

Conventional column chromatography was conducted using Bio-Gel P-2 (200-400 mesh) packed to provide a bed 90 cm × 1.4 cm (i.d.) in size. Water served as the eluting solvent by gravity flow. The effluent was collected in 4.0-ml fractions following spectrophotometric monitoring (Gilson Holochrome spectrophotometer) at a wavelength of 280 nm. Fractions were concentrated (4 ×) by flash evaporation prior to bioassay of 50-μl aliquots and iron determination by atomic absorption spectroscopy.

High pressure liquid chromatography was utilized in the exclusion mode. The use of two prototype I-60 columns in series (on consignment from Waters Assoc., Inc.) allowed size separation of crude fractions and of biologically active P-2 column fractions on an analytical scale. Further chromatographic conditions are presented with the resulting separation (Fig. 3). Instrumentation included the Waters Associates 6000A and M-45 pump modules, a 660-gradient programming module, a U6K universal injector and 440-absorbance detector fitted with a filter for 280-nm wavelength detection.

Thin-layer chromatography was performed using silica gel, 250 μm hard-surfaced analytical layer plates (5 cm × 20 cm) with 254-nm fluorescent indicator (R. T. Baker Co.). Development was completed in 2.5 hours in a pre-equilibrated chamber lined with 3MM paper and soaked in the solvent. The solvent was a mixture of n-butanol:acetic acid:pyridine:deionized distilled water (15:3:10:12). Total sample volume applied was generally 50 μl. Observation was completed under long and short UV, light but also included the spray reagents, 0.25% ninhydrin in butanol, Folin-Ciocalteu reagent for hydroxamate siderophore detection[24], and a reagent for specific colorimetric detection of o-diphenols, and 3,4-dihydroxyphenylalanine-containing peptides[26]. Detection of iron-binding activity on thin-layer plates was achieved by the addition of 40 μl of a 0.1% solution of $FeCl_3$ to 100-μl aliquot of a concentrated butanol extract prior to chromatography. Application of a 50-μl sample to the TLC plate, followed by development and detection using $2N$ HCl and 1% aqueous potassium ferrocyanide, allowed the observation of Fe(III)-binding activity as a light blue spot or band[4]. An autoradiogram of the TLC plate was made in cases where Fe^{59} in the form of $FeCl_3$ (0.2 mCi/m) served as the iron addition to the extract as described above. This was done by exposing XR5 Kodak X-Omat RP x-ray film to the TLC plate for 2 weeks while stored at 4°C. Film development was by an automated rapid x-ray processing system.

Amino acid analysis of chromatographic fractions and standards was executed on a Beckman model 120C amino acid analyzer equipped with both an intermediate-length column (23 cm) containing type AA-15 custom research resin and a short column (5.5 cm) containing type AA-27 resin. Samples (1–3 mg) were hydrolyzed in vacuo with 1.0 ml constant boiling HCl at 110°C from 0.5 to 15 hours. The hydrolysate was concentrated by flash evaporation to dryness after which 2.0 ml of sample buffer was added. The buffer, sodium citrate (pH 2.20 0.2 N Na^+), of commercial preparation (Pierce Chemical Co.) served as the sample buffer. For particularly labile fractions, the concentration procedure following the hydrolysis was amended to include the following steps. To the 1.0 ml hydrolysate was added 1.0 ml deionized distilled water which was then concentrated by flash evaporation at 40°C to a volume of 1.0 ml. A second 1.0-ml volume of water was added followed by concentration to a volume of between 100 and 200 microliters. To the final concentrate was added 1.8 to 1.9 ml of sample buffer.

Paper electrophoresis was conducted on acid hydrolysates and single amino acid standards using an LKB Multiphor 2117 electrophoresis apparatus and Brinkman regulated power supply. The technique was adapted from that of described by Emery[3]. The electrophoresis buffer consisted of the mixture, pyridine:glacial acetic acid:water (14:10:930) at pH 5.0. Development was carried out on Whatman no. 1 paper at 200 volts for a 45-minute duration. Samples (25 μl) were applied from hydrolysates or standards (1 mg/ml). Visualization of spots was by means of ninhydrin or tetrazolium chloride spray. Hydrolysis of the deferri form of the siderophores, aerobactin and rhodotorulic acid, provided the standards, N-OH-lysine and N-OH-ornithine, respectively. These standards were used for electrophoresis and for the amino acid analysis experiments.

Results

Antibiotic activity of *Pseudomonas* M27m was exhibited against all three eucaryotic organisms, *Geotrichum candidum*, *Ceratocystis ulmi* and a *Rhodotorula* species, in Dye's medium supplemented with 15 mM histidine or 1% (w/v) casamino acids. Supplementation of Dye's medium containing casamino acids with a series of two-fold concentrations of $FeCl_3$ ranging from 0.02 mM to 0.32 mM showed a corresponding increase in the diameter of the inhibition zones with increasing Fe(III) concentration, when strain M27m was oversprayed with *Geotrichum candidum*. Table 1 makes a comparison of eucaryotic inhibition in the presence and absence of $FeCl_3$ from 54 antimycotic producing isolates of the 176 isolates previously studied[9]. In contrast to Pseudomonas strain M27m, *Pseudomonas syringae* strain B3A, a syringomycin producing isolate, exhibited the same size of inhibition zone against *Geotrichum* with or without the addition of 10 µg/ml $FeCl_3$ to the medium. Some isolates (Table 1) did not respond equivalently in changes of relative zone size for all three indicators. The number of such isolates are listed under 'remainder' of Table 1.

Fractionation by Bio-gel P-2 column chromatography of acetone-treated and subsequently concentrated cell-free supernatant preparations produced from cultures of strain M27m grown in the presence and absence of iron exhibited quantitative differences, but demonstrated no major qualitative change in the

Table 1. Comparison of eucaryotic inhibition in the presence and absence of ferric chloride

Indicators	No. of producers	No. of producer isolates which show the zone size relationships indicated			
		B<C	B~C	B>C	Remainder
G R C	26	1	1	14	10
G R	7			5	2
G C	2			2	0
R C	3			2	1
G	1				0
R	11			6	0
C	4		1	3	0
Total	54				

Geotrichum (G)
Rhodotorula (R)
Ceratocystis (C)

Media:
B dyes+casamino acids (0.1%)+$FeCl_3$ (10 ppm)
C dyes+casamino acids (0.1%)

elution profiles obtained at 280 nm (Fig. 1). From the column fractions collected, two regions of greater total iron concentration were noted for both samples. The first region (Fig. 1) was located at the void volume of the column (Fractions 14 and 15), while the second region included fractions 24 thru 27. Fraction 26 contained the principle amount of the yellow-green fluorescent pigment. This fraction closely correlated with the elution position of cyanocobalamin (1357 M.W.), a standard reference compound. Bioassay tests of the individual fractions revealed that the antibiotic activity appeared in the void volume suggesting an apparent molecular weight equal to or greater than 1800 Daltons. Fraction 26, containing the fluorescent pigment, exhibited a fungistatic activity against *Geotrichum candidum*.

A qualitative amino acid composition analysis of active material obtained from the exclusion chromatography revealed the presence of an unidentified basic, ninhydrin-reactive substance in the hydrolysate. This residue had a retention time of 30 minutes on the short column of the analyzer, eluting prior to histadine (33 min) and ammonia (39.5 min) as determined from standard reference samples. The retention time of this basic residue did not match the time of either ornithine (26 min) or lysine (27 min). A second unidentified material produced an incompletely resolved peak at 37 minutes as a shoulder to a larger peak from ammonia.

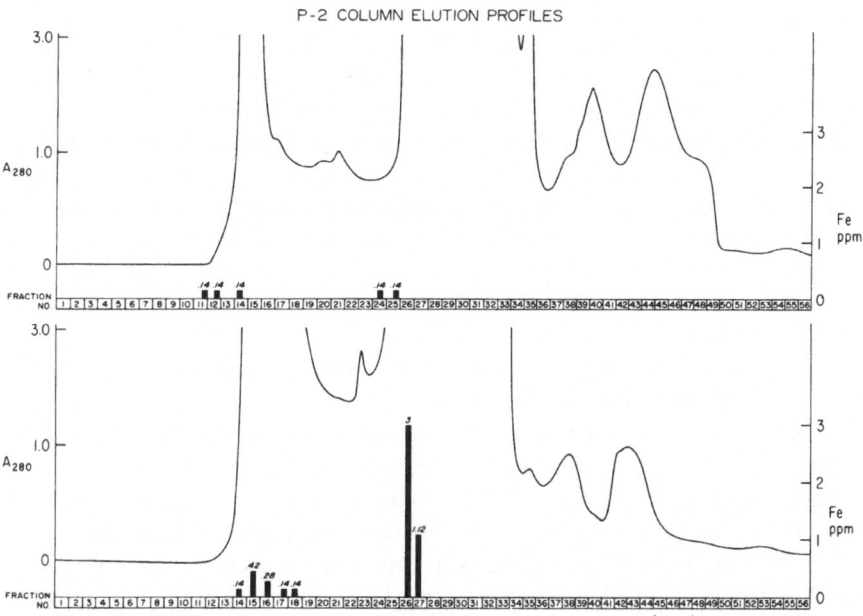

Fig. 1. Elution profiles of absorbance (280 nm) and iron concentration (atomic absorption spectrometry) from Bio-Gel P-2 column chromatography of supernatant preparations produced from cultures grown in the absence (*top*) and presence (*bottom*) of iron.

A quantitative amino acid analysis of the hydrolysate (11 h at 110°C, in vacuo, 6 N HCl) of 0.5 mg of syringomycin substantiated the previously reported composition of this hexapeptide phytotoxin as serine, phenylalanine, an unknown basic amino acid and arginine in a molar ratio of 2:1:2:1, respectively[5]. The unknown basic amino acid of this peptide was reported to elute 1.6 minutes ahead of lysine with the use of a single-column amino acid analyzer. In the system employed here, a 30-minute retention time was determined for this unknown residue in correspondence with the unknown material from the M27m antibiotic analysis.

Increased resolution for amino acid analyses was attained by the use of the intermediate length column of the Beckman 120C analyzer. Retention times of the following reference amino acids were determined in search of a possible tentative identification by retention time correspondence to the unknown residues: ornithine (62 min), anserine (63.5 min), lysine (64 min), histidine (77.2 min), L-3-methylhistidine (78.7 min), L-1-methylhistidine (81 min), ethanolamine (92 min), ammonium ion (98.2 min) and arginine (146.2 min). The matching unknown residues of M27m antibiotic and SR eluted at 74.8 minutes. The second peak arising from an unidentified compound of the M27m antibiotic material exhibited the same retention time as ethanolamine. The increased column length resolved the ethanolamine and ammonia peaks. The short column did not provide baseline resolution for these two peaks. The siderophore from the red yeast, *Rhodotorula pilimanae* Hendrick at Burke[1], provided a standard of N-hydroxyornithine[3]. A 2.5 hour hydrolysis yielded a retention time of 30 minutes, otherwise apparent degradative products were obtained from longer hydrolysis times or the presence of trace iron contamination.

Paper electrophoresis of acid hydrolysates of syringomycin, aerobactin, rhodotorulic acid, and M27m antibiotic, with detection by a tetrazolium spray[22], confirmed the presence of omega-N-hydroxy amino acids in these samples.

The migration distances from the origin toward the cathode for the reference compounds, ornithine and lysine, were 4.8 cm and 4.6 cm, respectively. The hydrolysates of rhodotorulic acid and aerobactin showed that δ-N-hydroxyornithine migrated 3.7 cm and δ-N-hydroxylysine migrated 3.5–3.6 cm, respectively. SR and M27m antibiotic hydrolysates each produced a single spot at 3.7 cm which reduced the tetrazolium reagent. The M27m antibiotic hydrolysate produced at least 3 distinct spots when examined with ninhydrin overspray at distances of 4.6 cm, 3.7 cm and 2.6 cm from the origin.

Thin-layer chromatography (TLC) of aliquots of antibiotically active butanol extracts obtained following the acetone treatment of supernatant preparations of M27m cultures disclosed several ninhydrin-reactive substances as separate bands on silica gel plates. In addition, development of the plates using Folin-Ciocalteu reagent revealed three bands. This reagent is known to react with hydroxamate siderophores in an uncharacterized manner[24]. These three Folin reagent-positive and ninhydrin-reactive bands gave approximate Rf values of 0.64–0.66 (band A),

0.55 (band B) and 0.39 (band C). Band A was bright red-orange in colour when sprayed with ninhydrin and possessed antibiotic activity. The test for o-phenols and 3,4-dihydroxyphenylalanine residues in peptides also disclosed three yellow (positive test) bands with the same Rf values.

TLC separation of the fluorescent fraction, fraction 26 of the P-2 column experiment, demonstrated the composition to be predominantly of band C material. This fraction was not fungicidal. The material is concluded to be the candidate compound for the fluorescent siderophore of this strain.

Addition of $FeCl_3$ to a butanol extract, separation by TLC and detection with potassium ferrocyanide reagent, produced two light blue bands with approximate Rf values of 0.52 and 0.60. This reagent detects the presence of Fe(III).

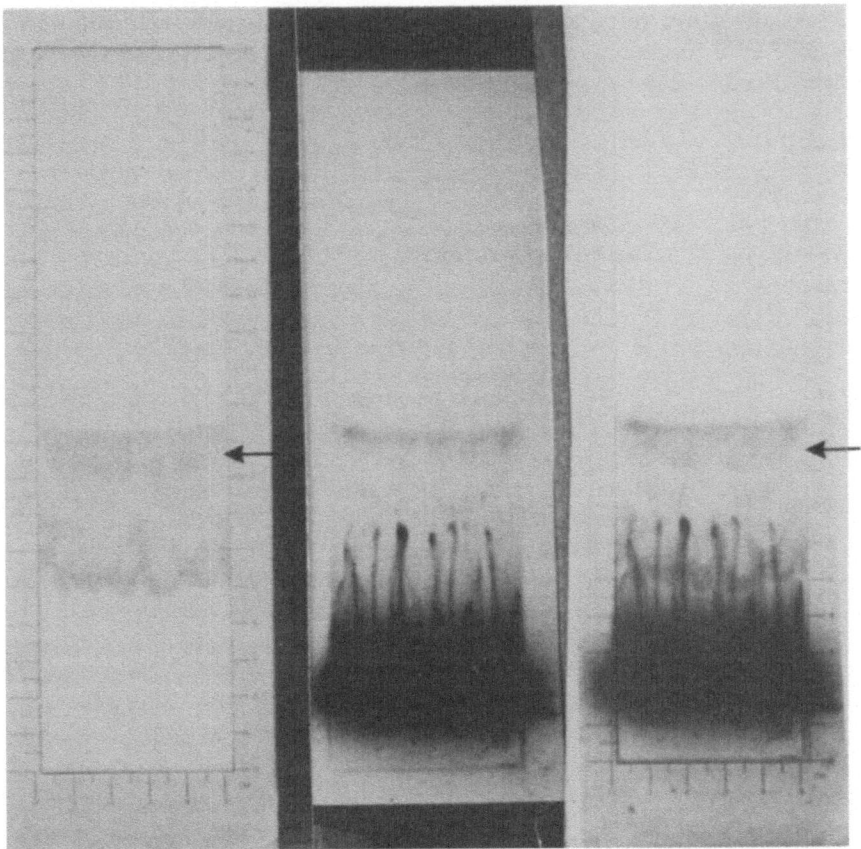

Fig. 2. Autoradiographic determination of iron binding of separated components in the TLC separation of butanol extract. On the *left* is shown a TLC plate previously sprayed with ninhydrin. Exposed x-ray film, the autoradiogram, is shown over white paper in the *center* of the figure. On the *right* the film is placed atop the TLC plate. The *arrow* indicates the antibiotically active band which is shown to bind Fe^{59}.

Autoradiography of TLC plates developed for separation of Fe^{59} containing extracts shows three iron-containing bands (Fig. 2). Unbound $FeCl_3$ is responsible for the exposure of the film at the origin. Fe(III) does not migrate on silica gel unless bound to some ligand. The antibiotically active band contains iron.

NMR analysis of 3 mg of band A material collected from the supernatant liquor of an iron-free broth culture substantiated the presence of a paramagnetic substance since no spectrum could be obtained.

The fungicidal activity of band-A material exhibited characteristics of a relatively small molecular weight compound, which neither correlated with the apparent molecular weight of the fungicidal activity which eluted in the void volume of the Bio-gel P-2 column nor that expected of SR. HPLC of the biologically active void volume fraction from the conventional column under disassociative conditions, as opposed to water elution, dispersed an associated low molecular weight compound (Fig. 3).

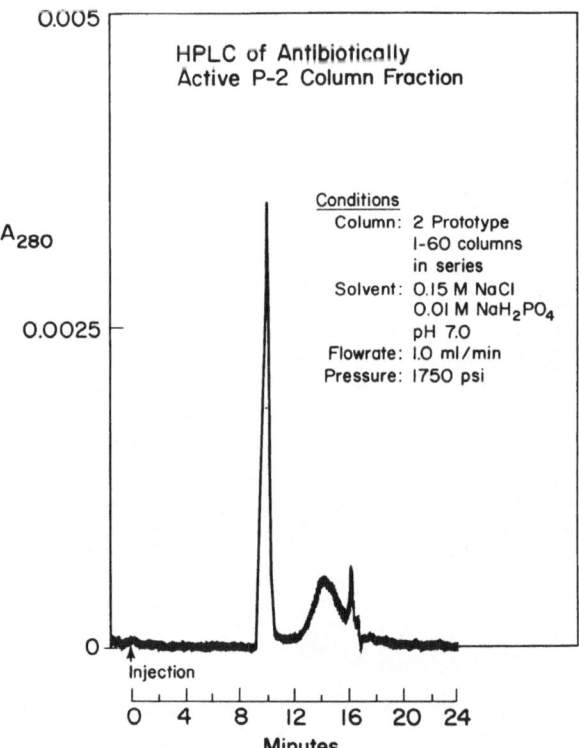

Fig. 3. Exclusion chromatography in a buffer by HPLC of an active preparation from a conventional P-2 column separation where only water had served as the eluting solvent.

Discussion

The results of the amino acid analyses, electrophoretic and chromatographic studies are interpreted as confirming the identity of the residue in syringomycin and M27 in antibiotic which gives rise to the 30-minute peak (short column) or the 74.8-minute peak (intermediate length column) of the analysis as δ-N-hydroxyornithine. The autoradiographic experiments of M27m supernatant preparations indicate the presence of at least three iron-complexing compounds. One possessed a fungicidal activity (syringomycin-like) and another exhibits characteristics similar to those of pseudobactin (*i.e.* yellow-green fluorescence quenched by iron addition, approximate size, and presense of fungistatic activity). A third produces reactions to indicator reagents on TLC plates similar to the other two compounds. It has not been excluded that these compounds are interrelated (*i.e.* represent the instability of the parent molecule). Further characterization of these compounds is required for the unequivocal resolution of the question of identity. The presence of the hydroxamate group in syringomycin denotes a possible relationship of this peptide to the recently characterized fluorescent pseudomonad siderophores. The molecular instability of pyoverdine, a fluorescent siderophore containing the hydroxamate residue in a heptapeptide has been described by Meyer and Abdallah[14].

The number of different N-hydroxylated amino acid-containing peptide products of the fluorescent pseudomonads suggest the testable hypothesis that siderophores or similar compounds from various species or ecotypes may exhibit heterogeneity in the peptide moiety of such complexes. It is plausible that ferribactin, a decapeptide, syringomycin, a hexapeptide and syringotoxin, a pentapeptide; are derived from or are precursors of strain or species specific siderophores which differ in the peptide portion of the molecules. The study of fluorescent products from *P. aeruginosa* has implicated that pigments of clinical strains apparently contain an identical fluorophore bound to larger molecules that differ with respect to their chromatographic properties[21]. These differences provide a unique fingerprint for the selective identification and characterization of certain species of pseudomonads and could be a useful approach to selection of particular strains of plant-associated pseudomonads of agricultural importance.

Interest has focussed also on the chemical characterization of siderophores from other plant-associated bacteria. The structure of agrobactin, a siderophore from the crown gall disease incitant, *Agrobacterium tumefaciens*, has been determined[16]. Siderophore synthesis has been reported to be a common attribute of bacterial pathogens[7]; however, in the case of agrobactin, its synthesis may be a survival mechanism of the pathogen *ex planta*[11]. Similarly, siderophore synthesis may assist pathogens to prosper *in planta mortua* following disease related death[7]. We emphasize that the relationship of syringomycin to the hydroxamate siderophores produced by the fluorescent pseudomonads is as yet unclear;

however, the presence of δ-N-hydroxyornithine in a phytotoxic compound may be of evolutionary significance in the development of bacterial pathogens. In addition, the stimulation by iron of the production of syringomycin or similar compounds underscores the probable integral role of iron in plant disease and the importance of the iron status of the plant in host–pathogen interactions.

Acknowledgements We thank Dr. R. Olsen for providing $Fe^{59}Cl_3$ used in this study. Dr. T. Emery and Dr. S. Rogers provided critical suggestions during the chemical analysis. We express gratitude to Patsy Guenther for typing the manuscript. This work was supported in part from grants from the Chevron Chemical Company (Ortho division) and US-AID contract No. DSAN-C-0024.

References

1 Atkins C L, Neilands J B and Phaff H J 1970 Rhodotorulic acid from species of *Leucosporidium, Rhodosporidium, Rhodotorula, Sporidiobolus*, and *Sporobolomyces*, and a new alanine-containing ferrichrome from *Crytococcus melibiosum*. J. Bacteriol. 103, 722–733.
2 Dye D W 1963 The taxonomic position of *Xanthomonas stewartii* (ERW Smith 1914) Dowson 1939. New Zealand J. Sci. 6, 495–506.
3 Emery T F 1966 Initial steps in the biosynthesis of ferrichrome. Incorporation of δ-N-acetyl-δ-N-hydroxyornithine. Biochemistry 5, 3694–3701.
4 Feigl F and Anger V 1972 Spot test for iron *In* Spot Tests in Inorganic Analysis. 6th Edition Elsevier Publ. Co., NY.
5 Gross D C and DeVay J E 1977 Production and purification of syringomycin a phytotoxin produced by *Pseudomonas syringae*. Physiol. Plant Pathol. 11, 13–28.
6 Gross D C, DeVay J E and Stadtman F H 1977 Chemical properties of syringomycin and syringotoxin; toxigenic peptides produced by *Pseudomonas syringae*. J. Appl. Bacteriol. 43, 453–463.
7 Harrington G J & Neilands J B 1982 Isolation and characterization of dimerum acid from *Verticillium dahliae*. J. Plant Nutrition 5(4–7), 675–682.
8 Hemming B C 1982 Plant-associated fluorescent pseudomonads: their systematic analysis, microbial antagonism and iron interaction. PhD dissertation. Montana State University, Bozeman, Montana.
9 Hemming B C, Orser C, Jacobs D L, Sands D C and Strobel G A 1982 The effects of iron on microbial antagonism by fluorescent pseudomonads. J. Plant Nutrition 5(4–7), 683–702.
10 Kloepper J W, Leong J, Teintze M and Schroth M N 1980 Enhanced plant growth by siderophores produced by plant growth-promoting rhizobacteria. Nature 286, 885–886.
11 Leong S A and Neiland J B 1981 Relationship of siderophore mediated iron assimilation to virulence in crown gall disease. J. Bacteriol. 147, 482–491.
12 Mauer B, Muller A, Keller-Schierlein W and Zahner H 1968 Ferribactin ein Siderochrom aus *Pseudomonas fluorescens* Migula. Arch. Microbial. 60, 326–339.
13 McCracken A R and Swinburne T R 1979 Siderophores produced by saprophytic bacteria as stimulants of germination of conidia of *Collectotrichum musae*. Physiol. Plant Pathol. 15, 331–340.
14 Meyer J M and Abdallah M A 1978 The fluorescent pigment of *Pseudomonas fluorescens*: Biosynthesis, purification and physiochemical properties. J. Gen. Microbiol. 107, 319–328.
15 Myers D F and Strobel G A 1982 *Pseudomonas syringae* as a microbial antagonist against Ceratocystis ulmi in the apoplast of American Elm. Transact. British Mycol. Society (*In press*).

16 Ong S A, Peterson T and Neilands J B 1979 Agrobactin a siderophore from *Agrobacterium tumefaciens*. J Biol Chem 154(6), 1860–1865.
17 Palleroni N J 1978 The Pseudomonas Group. Meadowfield Press Ltd, Durham, England
18 Schroth M N and Hancock J G 1981 Selected topics in biological control. Ann. Rev. Microbiol. 35, 453–476.
19 Schroth M N & Hancock J G 1982 Disease-suppressive soil and root-colonizing bacteria. Science 216, 1376–1381.
20 Shelly D C, Quarles J M and Warner I M 1980 Identification of fluorescent *Pseudomonas* species. Clin. Chem 26, 1127–1132.
21 Shelly D C, Warner I M and Quarles J M 1980 Multiparameter approach to the "Fingerprinting" of fluorescent pseudomonads. Clin. Chem. 26, 1419–1424.
22 Snow G A 1954 Mycobactin a growth factor for *Mycobacterium johneri*. Part II. Degradation and identification of fragments. J. Chem. Soc. 1954, 2588–2596.
23 Strobel G A and Lanier G N 1981 Dutch Elm disease. Sci. American 145(2), 56–66.
24 Subramanian K N, Padmanaban G and Sarma P S 1965 Folin-Ciocalteu reagent for the estimation of siderochromes. Anal. Biochem 12, 106–112.
25 Teintze M, Hossain M B, Barnes C L, Leong J and van der Helm D 1981 Structure of ferric pseudobactin, a siderophore from a plant growth promoting *Pseudomonas*. Biochemistry 20, 6446–6457.
26 Waite J H and Tanzer M L 1981 Specific colormetric detection of o-diphenols and 3,4-dihydroxyphenylalanine containing peptides. Anal. Biochem. 111, 131–136.
27 Weinberg E D 1978 Iron and infection. Microb. Rev. 42, 45–66.
28 Weinberg E D 1982 Iron and neoplasia. Biol. Tracer Element Research 3, 55–80.

Fertilizers and nutrient relations in some genotypes of cereals

B. LÁSZTITY
Research Institute for Soil Science and Agricultural Chemistry of the Hungarian Academy of Sciences, Budapest, Hungary

Key words Calcium Genotypes Nitrogen Nutrient content Phosphorus Potassium Sodium Triticale Winter rye

Introduction

The nutrient status of plants is jointly determined by internal genetic factors and external environmental factors. Plant nutrient contents are specifically characteristic of the stage of development, the species, variety or hybrid, the plant organ, and various parts or tissues of the organs. Among the external, abiotic factors, mineral nutrition or application of fertilizers has the greatest effect on the nutrient concentration in the majority of cases, within certain limits. Plant nutrient status is also significantly influenced by genetic factors, such as the capacity and dynamics of nutrient uptake, and the utilisation and distribution of assimilates; the majority of these properties are genetically controlled. In recent years many authors have concentrated on research into the nutritional peculiarities of various phenotypes[1,2].

In the present paper the nutrient status and dry matter accumulation of two genetically similar cereals—winter rye and triticale have been studied under intensive fertilizing conditions in the tillering phenophase.

Materials and methods

A field experiment was set up on a carbonate, weakly humous sandy soil in four replications in 1980. The two variables were fertilization and genotype. The soil on the experimental area was poorly supplied with nutrients. In various treatments (Table 1) 200 kg of nitrogen, 500–1000 kg of P_2O_5 and K_2O fertilizer were applied per hectare. The fertilizer was applied in the form of calcium ammonium nitrate, superphosphate and potassium chloride. The experimental varieties were Kecskeméti H. winter rye and KT—77 triticale. The samples consisted of the complete above ground plant material from 4 metres per plot and were taken during the tillering phenophase (Feekes 3.), when previous experiments and the literature showed the plants to be most responsive to nutritional status[3].

The plant material was weighed, and samples were taken for the determination of the N, P, K, Ca and Na contents. The dry matter and nutrient data were evaluated with a desktop computer using variance analysis.

M.R. Sarić and B.C. Loughman (eds.), Genetic Aspects of Plant Nutrition.
ISBN-13: 978-94-009-6838-7
©1983 Martinus Nijhoff/Dr W. Junk Publishers, The Hague/Boston/Lancaster.

Results and discussion

The accumulation of plant dry matter is shown in Table 1. The data indicate that rye accumulates dry matter more rapidly both over the average of the fertilization doses and in the individual treatments. Triticale also accumulates more dry matter as a function of fertilizer. The results show that under identical nutritional conditions rye develops more intensively.

The nutrient data for plant samples taken at the tillering stage are shown in Table 2. The nutrient concentration increased significantly due to the effect of nitrogen in both plants and in all the relevant treatments. Over the average of the treatments the dry matter content of rye proved to be significantly higher. In treatments without nitrogen no significant differences were found between the two plants.

The P content increased simultaneously with the rising fertilizer dose in both plants. Comparing the P content of the two plants, that of rye was greater in all cases, probably due to the stronger root system. The K concentration increased significantly in both genotypes as the result of K application, and also over the average of the two plants. The K content of rye was also significantly greater than that of triticale over the average of the fertilizer treatments. However, this better response of rye was not evident in all the individual fertilizer treatments.

The Ca content significantly increased after fertilizer application in rye and over the average of the two plants, but the increase was not significant for triticale. Over the average of the fertilizer treatments higher contents were recorded in rye in the case of Ca, too. In treatments without nitrogen the difference between the two plants was not significant.

Sodium, as a conditional nutrient, did not differ significantly either over the fertilizer treatments (with one exception) or between the plants.

Table 1. Effect of the fertilization on the dry matter accumulation at tillering stage (DM kg/ha)

Treatment	R	T	LSD$_{5\%}$	Mean
1, Ø	559	383		471
2, N$_{200}$	778	494		636
3, P$_{500}$K$_{500}$	516	335	146	425
4, N$_{200}$P$_{500}$	912	538		757
5, N$_{200}$K$_{500}$	976	456		716
6, N$_{200}$P$_{500}$K$_{500}$	808	585		697
7, N$_{200}$P$_{1000}$K$_{1000}$	863	529		696
LSD$_{5\%}$		146		55
Mean	744	474	103	

R = rye T = triticale P = P$_2$O$_5$ kg/ha K = K$_2$O kg/ha N = N kg/ha

Table 2. Effect of the fertilization on the nutrient content of plants at tillering stage

Treatment	N %				P %				K %			
	R	T	LSD$_{5\%}$	Mean	R	T	LSD$_{5\%}$	Mean	R	T	LSD$_{5\%}$	Mean
1	3.34	3.19		3.26	0.46	0.38		0.42	2.79	2.59		2.69
2	4.76	3.96		4.36	0.51	0.37		0.44	3.21	2.74		2.97
3	2.87	2.97		2.92	0.52	0.44		0.48	3.24	3.01		3.12
4	4.97	4.17	0.4	4.57	0.64	0.49	0.05	0.56	3.12	2.71	0.40	2.92
5	4.97	3.86		4.42	0.46	0.38		0.42	3.35	3.21		3.28
6	4.98	4.02		4.50	0.65	0.44		0.55	3.66	3.21		3.43
7	4.74	4.08		4.41	0.61	0.51		0.56	3.69	3.56		3.63
LSD$_{5\%}$		0.4		0.15		0.05		0.01		0.40		0.15
Mean	4.38	3.75	0.28		0.55	0.43	0.01		3.29	3.00	0.29	

Treatment	Ca %				Na %			
	R	T	LSD$_{5\%}$	Mean	R	T	LSD$_{5\%}$	Mean
1	0.54	0.43		0.48	0.10	0.10		0.10
2	0.75	0.50		0.62	0.12	0.10		0.11
3	0.48	0.39		0.43	0.09	0.08		0.08
4	0.74	0.48	0.14	0.61	0.13	0.10	0.04	0.11
5	0.77	0.45		0.61	0.12	0.09		0.10
6	0.72	0.46		0.59	0.12	0.09		0.11
7	0.74	0.46		0.60	0.13	0.10		0.11
LSD$_{5\%}$		0.14		0.05		0.04		0.02
Mean	0.68	0.45	0.1		0.11	0.09	0.03	

R = rye T = triticale

The two cereals were also examined with respect to the correlation between nutrient status and dry matter production. A significant correlation was found between the N content at tillering and the amount of dry matter accumulated in the course of the vegetation period. This correlation was closer in the case of triticale.

The difference between the two genotypes is clearly visible (Fig. 1.), since the optimum N content proved to be higher for rye.

Summary and conclusions

The effect of intensive fertilizer application on the nutrient content and

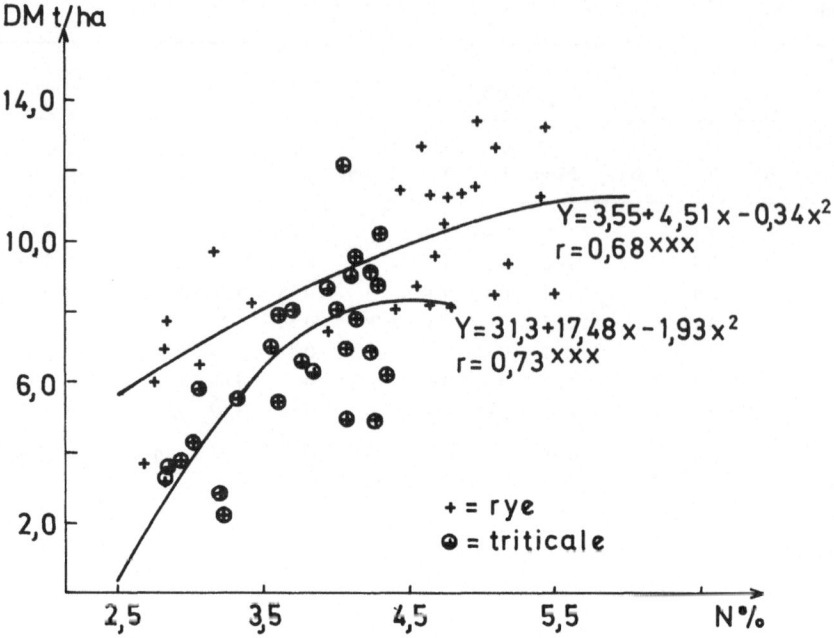

Fig. 1. Relationship between N–content of the fresh plant at tillering and the dry matter production of the plant parts grown above ground.

development of winter rye and triticale was studied in the tillering phenophase under field conditions.

The results lead to the following conclusions:

(1) The accumulation of dry matter increased significantly in both cereals due to N, NP and NPK fertilization. The increase was significantly greater for winter rye. The concentration of the macro-elements (NPK) significantly increased over the fertilizer treatments in the plants examined. For all three elements the greater increase was found in winter rye, which showed a significant surplus in all cases over the average of the fertilizer treatments.
(2) The Ca content also increased due to the effect of fertilizer and to a significantly greater degree for rye.
(3) No significant difference could be demonstrated in the Na concentration either between the two cereals or, with one exception, between the fertilizer treatments.

Under identical environmental conditions the genetic difference between the two cereals as regards to mineral nutrition was reliably demonstrated.

References

1 Gashaw L and Hugwira L M 1981 Ammonium-N and nitrate-N. Effects on the growth and mineral compositions of triticale, wheat and rye. Agron. J. 73, 47–51.

2 Kastori R 1981 Sadržaj i raspodela biogenih elemenata u pšenici. Separate. ed. DXXXVI. of the Serb. Acad. of Sci., Dep. of Sci., Vol 53, 79–101.

3 Lásztity B and Kádáv 1978 Az öszi buza szárazanvyag telhalmozódásának, valamint tápanyagfelvételének tanulmányozása szabadföldi kisérletben (Study of dry matter accumulation and nutrient uptake in winter wheat under field conditions). Agrokémia és Talajtan 27, 429–444.

Nitrogen fixation in soybean depending on variety and *R. japonicum* strain

Z. SARIĆ and Ali H. FAWZIA
Faculty of Agriculture, University of Novi Sad, Yugoslavia and Research Station of Plant Protection, Basrah, Iraq

Key words Nitrogen fixation *Rhizobium japonicum* Soybean varieties

Introduction

The extent of nitrogen fixation depends on numerous factors. A number of ecological and agrotechnical factors affect the symbiotic relationship between the host plant and nodular bacteria. The population of nodular bacteria in the soil under soybean usually includes strains of different efficiency[9]. Therefore, the characters of the *R. japonicum* strain which penetrated the root causing nodule formation are important for the establishment of the symbiotic relationship. *R. japonicum* strains differ among themselves not only in nitrogen fixation but also in the capacity to synthesize different growth regulators[8], which in turn affects the efficiency of nitrogen fixation of the strains in question[10]. Specific characters of the host plant affect also the extent of nitrogen fixation and soybean varieties which take up increased amounts of nitrogen when grown without their natural symbiont are also superior in nitrogen fixation when inoculated[11].

The objective of this study was to examine the effect of different *R. japonicum* strains on nitrogen fixation in certain soybean varieties.

Materials and methods

We examined the following ten soybean varieties: Braungelbe, Dieckmans Grüngelbe, Hodgson, Evans, Traverse, Yoshicuataya, Iregi, Udabinska, Kirovogradska, and NSMM.

The five *R. japonicum* strains used for inoculation (24, 33, 511, 512, and 518) were produced at the Department of Microbiology in Novi Sad and the Institute of Soil Science in Belgrade.

Seed was disinfected in 75% ethyl alcohol for several minutes, rinsed in distilled water and left in trays to germinate. Seven days later, the seedlings were placed in pots filled with sterilized sand (ten replications). Each variety was planted with and without inoculation. Reid-York's nutritive solution without nitrogen was added to the sand after 15 and 22 days. The experimental plants were grown for 32 days, uprooted, and the roots separated from the nodules and the shoots. The plants were dried and weighed prior to nitrogen content determination after the method of Kjeldahl. Nitrogen content was calculated per plant and the amount of fixed nitrogen was calculated from the difference in nitrogen contents between inoculated and uninoculated plants. The obtained results are shown in Figures 1, 2, 3, 4 and 5.

M.R. Sarić and B.C. Loughman (eds.), *Genetic Aspects of Plant Nutrition.*
ISBN-13: 978-94-009-6838-7
©1983 *Martinus Nijhoff/Dr W. Junk Publishers, The Hague/Boston/Lancaster.*

Results and discussion

The results obtained show that a symbiotic relationship was established between the majority of the varieties examined and *R. japonicum* strains. The variety Yoshicuataya was an exception because it formed a symbiotic relationship only with strain 511. Also, the varieties Korovogradska and Traverse failed to establish the relationship with strains 24 and 33, respectively. All *R. japonicum* strains formed nodules on the roots of the examined varieties but the nodules varied in number and weight. The nodules were most numerous with the varieties Evans and NSMM, least numerous with Yoshicuataya, Udabinska, and Traverse. The number of nodules with individual varieties varied largely in dependence of *R. japonicum* strain used. Smallest differences in the number of nodules among the strains were found with the varieties Kirovogradska, Hodgson and Dieckmans Grüngelbe. The differences were larger with the other varieties (Fig. 1).

Large differences in the amount of fixed nitrogen which occurred between the varieties as well as within the same variety depended on the soybean variety examined and *R. japonicum* strain used. Smallest differences among the strains were found in the varieties Dieckmans Grüngelbe and Hodgson. With the other varieties, the differences in the amount of fixed nitrogen among the strains were large. Strain 518 fixed the largest amount of nitrogen in the varieties NSMM and Braungelbe, strain 511 in Evans and Udabinska, strain 512 in Hodgson, strain 24 in Diekmans Grüngelbe, and strain 33 in Iregi (Fig. 2).

The majority of the strains were inefficient in the variety Yoshicuataya, while single strains displayed inefficiency in the varieties Traverse and Kirovogradska.

The nodules formed on the roots of the varieties also differed in nitrogen

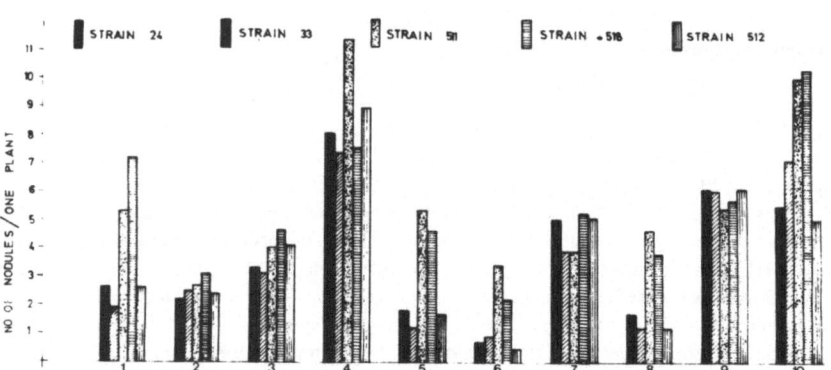

Fig. 1. Number of nodules in some soybean varieties inoculated with different strains of *Rhizobium japonicum*.
1. Braungelbe, 2. Dieckmans Grungelbe, 3. Hodgson, 4. Evans, 5. Traverse, 6. Yoshicuataya, 7. Iregi, 8. Udabinska, 9. Kirovogradska, 10. NSMM.

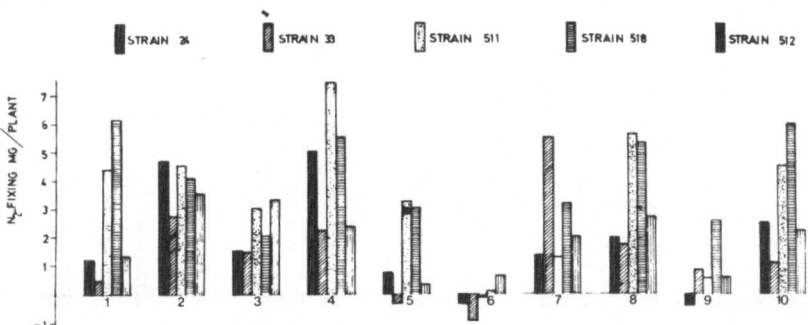

Fig. 2. N_2-fixing in whole plant of some soybean varieties inoculated with different strains of *Rhizobium japonicum*.

content (Fig. 3). Nitrogen content in the nodules was not always correlated with the number of nodules, as exemplified by the varieties NSMM, Udabinska, and Iregi.

In the majority of the varieties the roots of the inoculated plants had decreased nitrogen contents. It was especially noticeable with the varieties NSMM and Dieckmans Grüngelbe, which had reduced nitrogen contents in the roots in combination with all *R. japonicum* strains. In the other varieties, nitrogen content in the roots of inoculated plants depended on the strain used. Conversely, the varieties Evans and Iregi had increased nitrogen contents in the roots in combination with all *R. japonicum* strains (Fig. 4).

Differences were found in nitrogen content in the shoots both among the soybean varieties and among *R. japonicum* strains. Nevertheless, the majority of the varieties had higher nitrogen contents in the shoots of the inoculated plants. With the varieties NSMM and Udabinska, all *R. japonicum* strains brought increases in nitrogen content in the above-ground parts. With the other varieties, nitrogen content varied in dependence of *R. japonicum* strain used. In contrast to

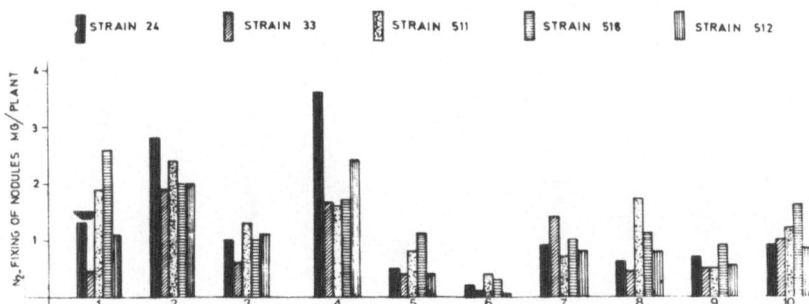

Fig. 3. N_2-fixing in nodules of some soybean varieties inoculated with different strains of *Rhizobium japonicum*.

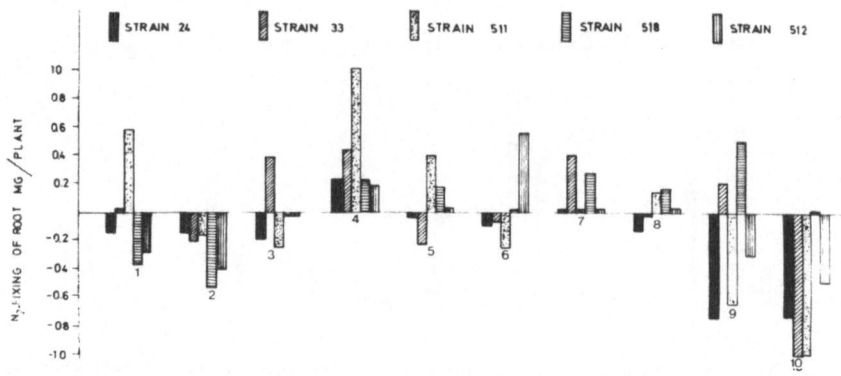

Fig. 4. N$_2$-fixing in roots of some soybean varieties inoculated with different strains of *Rhizobium japonicum*.

all other varieties, Yoshicuataya reduced the nitrogen content in the above-ground parts, *i.e.* nitrogen fixation did not occur during the period of plant development. It means that the plants did not establish symbiotic relationships with nodular bacteria (Fig. 5).

Strain 512, the only strain that displayed a certain degree of efficiency in the variety Yoshicuataya, caused a reduction of nitrogen content in the above-ground parts of the variety Evans. Furthermore, strain 511 caused a reduction of nitrogen content in the above-ground parts of the variety Iregi although the strain was capable of nitrogen fixation (Fig. 2). It may be assumed that *R. japonicum* strains affect not only the amount of fixed nitrogen but also the rate of transportation of fixed nitrogen from roots to above-ground parts.

The results obtained enabled us to conclude that nitrogen fixation in soybeans

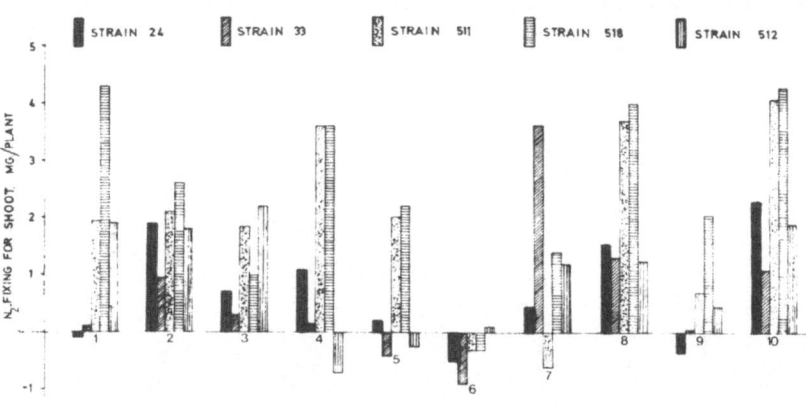

Fig. 5. N$_2$-fixing in shoots of some soybean varieties inocoulated with different strains of *Rhizobium japonicum*.

depends on specific characters of the *R. japonicum* strain used. In other words, the dynamics of establishment of symbiotic relationship between soybean and *R. japonicum* is determined by the characters of variety and strain used[6], *i.e.*, some symbioses are more efficient than others. Although our results showed that some varieties and strains did not establish symbiotic relationship during the stage of development examined, we could not say that new relationships would not be established in later vegetative growth, especially if we take into consideration the fact that the varieties examined differed in vegetative period. With the soybean varieties, usually only a single strain had the maximum nitrogen fixation, i.e. the variety had the most successful symbiotic relationship with that strain. The symbiotic relation between *R. japonicum* and soybean is a variable complex[7,12]. Soybean plant genotype and *R. japonicum* genotype are of equal importance in the establishment of their symbiotic relationship.

At early stages of development, inoculated plants used larger amounts of reserve matter, fall back in growth and development, develop a smaller vegetative area, and have lower photosynthetic activity than uninoculated plants[7]. A portion of reserve matter is used for the multiplication of nodular bacteria, their growth, and other changes necessary for the establishment of symbiotic relationship.

The results obtained in this study as well as those of other authors show that efficiency of *R. japonicum* strains varies in different soybean varieties[1,2,3]. It is therefore necessary to select *R. japonicum* strains according to soybean genotypes. There are a number of criteria which may be used in selection for the efficiency of strains: dry matter production[4], acetylene reduction, growth stimulator production[9], nitrogen concentration and content, etc. When selecting *R. japonicum* strains, one should not overlook the limiting ecological–climatic factors which determine the production of grain and proteins in soybean[5].

References

1 Abel G H and Erdman L W 1964 Response of Lee soybeans to different strains of *Rhizobium japonicum*. Agric J. 56, 423–424.
2 Balasundari V R, Iswaran V and Sundara Rao W V B 1972 Interactions between soybean (*Glycine max* L Merr.) genotypes and different isolates of *Rhizobium japonicum*. Indian J. Agr. Sci. 42, 387–389.
3 Caldwell B E and Vest G 1970 Effects of *Rhizobium japonicum* strains on soybean yields. Crop. Sci. 10, 19–21.
4 Chhonkar P K and Negi P S 1971 Response of soybean to rhizobial inoculaton with different strains of *Rhizobium japonicum*. Indian J. Agr. Sci. 41, 741–744.
5 Lagacherie B 1982 Selection de souches de *Rhizobium japonicum* sur leur efficacite a fixer l'azote en symbiose. CETOM Informations techniques. N° 78, 3–14.
6 Sarić Z 1953 Efektivnost nitraginizacije i neki faktori koji utiču na njezino povećanje. (The effect of nitrogenization on the yield of soya). Zemljište i biljka 1, 157–167.
7 Sarić Z and Sarić M 1959 Utilcaj nitraginizacije na rast i razvićesoje (The influence of

nitrogenization on the growth and development of soybean). Savremena poljoprivreda 10, 819–835.
8 Sarić Z, Milić V and Hazem T 1978 Sinteza materija rastenja tipa indola, gibereblina i fenola od strane nekih sojeva *R. japonicum* različite efektivnosti. (Biosynthesis of plant growth substances by some *Rhizobium japonicum* strains of various efficiency). Arhiv za poljoprivredne nauke 115, 29–41.
9 Sarić Z and Milić V 1981 Uticaj plodoreda na nastajanje populacije *Rhizobium japonicum* u černozemu (The formation of *Rhizobium japonicum* population in chernozem soil). Arhivza za poljoprivredne nauke 146, 293–296.
10 Sarić Z and Milić V 1981 Sinteza materija rastenja nekih soyeva *Rhizobium japonicum* izolovanih iz soje gajene u plodoredu. (Biosynthesis of plant growth regulators by some *Rhizobium japonicum* strains isolated from the soil under several-crop rotation). Arhiv za polyjoprivredne nauke 148, 487–500.
11 Sarić M and Krstić B 1982 Genetisch bedingte Unterschiede im Stickstoffgehalt verschiedener Sojasorten. Arch. Acker-und Pflazenbau Bodenk. No 12, 755–761.
12 Sloger Ch. 1969 Symbiotic effectiveness and N_2 fixation in nodulated soybean. Plant Physiol. 44, 1666–1688.

Characteristics of sugar beet varieties on chernozem semigley in Baranja

B. TODORČIĆ,* B. BERTIĆ,* M. ŠEPUT,** V. VUKADINOVIĆ,* I. FOLIVARSKI** and V. LAKTIĆ***
*Agricultural Faculty, Osijek, Yugoslavia and **Development Service, PIK "Belje", Darda, Yugoslavia and *** Factory of Sugar, Beli Manastir, Yugoslavia*

Key words Concentration Macronutrients Micronutrients Sugar beet Root Yield

Introduction

The best choice of sugar beet varieties for particular agroecological conditions requires field testing, because the characteristics of varieties depend on many biotical and abiotical factors.

The capacity of varieties to produce the largest quantity of biological sugar per hectare is their most important characteristic[4].

Ability of varieties to utilize the effective soil fertility is shown by the content and distribution of nutrient ions in parts of sugar beet and this factor immediately influences the synthesis and accumulaton of sucrose[5,6].

The varieties were arranged on the basis of the yield of biological sugar. The analysis of concentrations of essential elements and sodium in plant parts served to establish varietal specificity in relation to their requirements for nutrients.

Materials and methods

Twenty-nine different varieties of sugar beet were grown in the field experiment in Baranja, on chernozem–semigley soil (humus = 2.0%, P_2O_2 = 11.43 mg/100 g, K_2O = 17.86 mg/100 g, pH_{H_2O} = 7.2 and pH_{KCl} = 6.4), in 1981. The field experiment was conducted in a randomized block system, each variety in four replicates on experimental plots of 114 m^2.

All varieties had the same level of soil fertility (138 kg N/ha, 308 kg K_2O/ha and 140 kg P_2O_5/ha). Nitrogen and potassium were given in the autumn with basic tillage, but phosphorus was given with the preceding crop (wheat, 1979).

Seeds were sown on 4 April 1981 and plants harvested on 28 September. The root yield was determined by weighing roots of 10 m^2 (four replicates) of the experimental parcels.

Elemental composition was determined after wet-ashing[1] for macroelements and dry-ashing in muffle furnace at 500°C for 3 h[3], for microelements. Nitrogen was determined by micro-Kjeldahl, phosphorus by molybdatevanadate method and all other elements by atomic absorption spectrophotometry[7].

M.R. Sarić and B.C. Loughman (eds.), Genetic Aspects of Plant Nutrition.
ISBN-13: 978-94-009-6838-7
©1983 Martinus Nijhoff/Dr W. Junk Publishers, The Hague/Boston/Lancaster.

Table 1. The yield of root and biological sugar yield and polarization of investigated sugar beet varieties

Group	Variety	Biological sugar yield dt/ha	Root yield dt/ha	Root yield $x - x_i$ D	Polarization %	Polarization $x - x_i$ D
I	Regine	112.62	701	94.83	16.06	0.45
	Gemo 3967 M	109.53	690	83.83	15.87	0.26
	Maribo	107.80	688	81.83	15.73	0.12
	Kawemaya	104.40	648	41.83	16.07	0.46
	Nova Dima	103.74	691	84.83	15.05	−0.56
	Arigomono	101.33	678	71.83	14.93	−0.68
	H−H Mona	100.68	648	41.83	15.54	−0.07
	Novo Gemo	100.01	651	44.83	15.36	−0.25
II	Jasica	99.66	644	37.83	15.46	−0.15
	Prima	99.54	620	13.83	16.04	0.43
	Kawekatja	98.50	633	26.83	15.53	−0.08
	Kawepura	97.64	597	−9.17	16.36	0.75
	Orisant	97.38	635	28.83	15.31	−0.30
	Monopur 79	97.33	659	52.83	14.74	−0.87
	Maribo M CR	96.62	673	66.83	14.36	−1.25
	7709	96.21	596	−10.17	16.10	0.49
	Cermona	95.61	624	17.83	15.32	−0.29
	Monofort org.	91.85	582	−24.17	15.76	0.15
	KWS 863	86.11	571	−35.17	15.60	−0.01
	7918	88.41	562	−44.17	15.73	0.12
	Dima	87.84	571	−35.17	15.40	−0.21
	Maribo M-2	87.47	526	−80.17	16.62	1.01
III	Carpo	84.64	521	85.17	16.27	0.66
	Monofort (hom.)	84.55	534	72.17	15.85	0.24
	KWS 051	83.19	518	88.17	16.20	0.59
	Kawetija	83.40	551	55.17	15.20	−0.41
	Buram	81.52	550	−56.17	14.94	−0.67
	AL Mona	80.35	498	−108.17	16.10	0.49
	Kaweduca	79.86	521	85.17	15.31	−0.30
	x̄	94.54	606		15.61	
	LSD 5%	9.12	58.1		1.47	
	LSD 1%	12.10	77.0		1.95	

Results and discussion

The most important characteristic of different sugar beet varieties is the biological sugar yield per hectare. The varieties were arranged in three groups:

I group with 100–115 dt/ha sugar (8 varieties)
II group with 85–99 dt/ha sugar (14 varieties)
III group with 70–84 dt/ha sugar (7 varieties)

Along with the sugar yield, the yield of root and polarization, and their deviations from the average values (Table 1) are shown. The varieties which had the higher than average root-yield belong to the 1st group with the highest sugar yield. Polarization had the same influence on the biological sugar yield.

The correlation coefficients were:

root yield:sugar yield $r = 0.941^{++}$ $(y = 10.967 + 0.138x)$
polarization:sugar yield $r = -0.50$ $(y = 107.97 - 0.860x)$
root yield:polarization $r = -0.382^{+}$ $(y = 17.574 - 0.0032x)$

The partial influence of root yield and polarization on the sugar yield was established by multiple regression ($\hat{y} = -101.09 + 0.061 x_1 + 6.276 x_2$), and was shown in Figure 1.

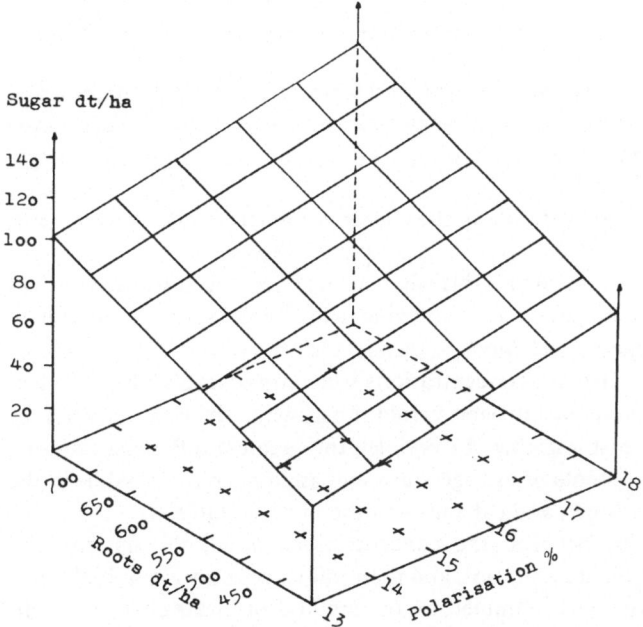

Fig. 1. Dependence of biological sugar yield on root production and polarisation.

Table 2. Average concentrations of the macro- and microelements in dry matter of the leaf blade, petiole and root of sugar beet

	N	P	K	Ca	Mg	Na	Fe	Mn	Zn	Cu
	% in dry matter						ppm in dry matter			
Leaf blade										
x	3.6	1.6	4.0	0.94	0.76	1.7	199.0	124.8	57.8	6.9
k_v %	17.8	11.4	7.7	17.10	3.63	10.6	21.5	28.8	48.5	9.6
LSD 5%	0.24	0.1	0.12	0.16	0.008	0.07	16.3	13.7	10.7	0.2
1%	0.33	0.1	0.16	0.08	0.011	0.09	22.0	18.4	14.4	0.3
Petiole										
x	1.6	1.1	3.5	0.51	0.68	1.7	104.0	35.8	19.6	4.5
k_v %	24.4	18.9	12.2	19.78	5.82	15.8	40.7	45.0	40.7	25.6
LSD 5%	0.15	0.1	0.2	0.04	0.014	0.10	16.1	6.1	3.0	0.4
1%	0.20	0.1	0.2	0.05	0.019	0.14	21.7	8.3	4.1	0.6
Root										
x	0.98	0.64	0.89	0.09	0.27	0.22	377.2	35.8	15.9	2.3
k_v %	35.81	25.2	14.3	31.9	5.00	27.40	17.5	36.6	73.2	51.6
LSD 5%	0.13	0.06	0.15	0.01	0.005	0.02	16.31	13.66	10.7	0.25
1%	0.18	0.08	0.07	0.01	0.007	0.03	22.01	18.43	14.4	0.34

The varietal specificity was established, and concentrations of the macro- and microelements in various parts of the sugar beet plants were investigated in the terminal period of vegetation. The average values obtained in these investigations are shown in Table 2.

Differences between the varieties in the content of macro- and microelements in plant parts were most obvious in the microelements. In leaf blades the concentration of zinc was most variable whereas in petiole the most variable was the concentration of manganese, iron and zinc, while in the root the concentrations of zinc and copper varied most.

Variations of macroelement concentrations were considerably lower. In leaf blades the concentrations of nitrogen varied most, while the concentration of manganese was the least variable. In petioles the greatest differences among varieties were in the contents of nitrogen, calcium and phosphorus, while in the root the content of nitrogen and calcium was the most variable.

Mutual relationships between the concentrations of all the macro- and microelements in leaf blades and root, and their influences on yield of biological sugar and polarization, was examined by the method of linear correlation. In Table 3, those correlation coefficients which are statistically significant are shown.

Table 3. The correlation coefficients

Part of plant	Macroelements	r	Microelements	r
leaf blade	Ca:Mg	0.6589++	Mn:Fe	0.4325+
	Ca:Na	0.8641++	Zn:Cu	−0.3725+
	Mg:Na	0.6985++		
root	Na:polariz.	−0.3828+	Fe:Zn	0.4544+
	K:Na−	0.3987+		
	Ca:Mg	0.5106++		
	P:sugar yield	0.6675++		

Conclusions

In a 1-year investigation of twenty-nine sugar beet varieties, differences in biological sugar yield, root yield, percent of polarization and concentrations of elements in leaf blade, petiole and root of sugar beet were established.

Varieties showed significant differences in the biological sugar yield (79.86–112.62 dt/ha), root yield (498–701 dt/ha) and in the percentage of polarization (14.36–16.62%).

Significant relations between the root yield and the biological sugar yield ($r = 0.941^{++}$) were established but not between the percentage of polarization and biological sugar yield. Differences in the microelement of varieties were greater than those of the macroelements. Variability was more marked in the root than in the petiole, and least in the leaf blade. The concentration of every element increased in the same order.

References

1 Faller N 1971 Razaranje biljnog materijala kombinirano vodikovim peroksidom i perhlornom kiselinom. Agrohemija 13, 225–300.
2 Faller N and Vukadinović V 1981 Uloga makro i mikroelemenata u fiziološko-biohemijskim procesima šećerne repe. Fiziologija šećerne repe, Srpska akademija nauka i umetnosti Beograd, 73–94.
3 Isaac R A and Johnson C W 1975 Collaborative study of wet and dry ashing techniques for elemental analysis of plant tissue by atomic absorption spectrophotometry. AOAC 58 (3), 436–440.
4 Sarić M and Kovačević V 1981 Sortna specifičnost mineralne ishrane šećeren rape. Fiziologija šećerne repe, Srpska akademija nauka i umetnosti Beograd, 57–73.

5 Todorčić B, Crnogorac S and Blaženka Bertić 1981 Uticaj makroelemenata na prinos i kvalitet šećerne repe. Srpska akademija nauka umetnosti, Beograd, 97–118.
6 Todorčić B, Crnogorac S and Blaženka Bertić 1981 Uticaj mikroelemenata na prinos i kvalitet šećerne repe. Srpska akademija nauka i umetnosti, Beograd, 119–133.
7 Welz B 1975 Atom Absorptions Spectroscopie 2. (Völlig neu bearbeitete Auflage). Verlag Chemie GMBH, Weienheim, BR Deutschland.

SECTION V
GENETICAL INVESTIGATIONS CONCERNED WITH SELECTION OF GENOTYPES FOR A MORE EFFECTIVE USE OF MINERAL ELEMENTS

The search for and interpretation of genetic controls that enhance plant growth under deficiency levels of a macronutrient*

W. H. GABELMAN and G. C. GERLOFF
Department of Horticulture, University of Wisconsin, Madison, WI 53706, USA

Key words Calcium Nutrient stress Phosphorus Potassium Root growth

Summary Techniques developed to measure growth of tomatoes and beans under limiting amounts of either P, K or Ca in solution culture reveal differences among strains. Genetic analysis permits estimates of gene action for control of efficiency and the isolation of improved segregants. The genetic isolates have value in studying mechanisms contributing to uptake, transport and utilization processes.

Introduction

Papers published in the 1930's provided preliminary insights into genetic controls of mineral nutrition of plants[6,7,10], but surprisingly little was published in the next three decades that might bring breeders, geneticists, plant physiologists and soil scientists closer in research on the problem. Why? Perhaps because developed agricultural countries found fertilizer to be a relatively inexpensive production input and research emphasized maximum production per unit of added fertilizer. Developing countries, where strains efficient in response to threshold levels of nutrients would have been very important, lacked research capabilities to exploit the research of Harvey[6], Lyness[7] and Smith[10].

During the 1940's soybeans unable to utilize Fe normally provided plant physiologists with opportunities to compare "efficient" and "inefficient" utilization of that element[11]. The concept of using genetic mutants to elucidate physiological mechanisms was applied to studies on Fe utilization. But Weiss' studies and research by others were limited to comparisons of mutants incapable of effective plant growth when compared to adapted agricultural cultivars.

Materials and methods

Our approach to research on efficiency of nutrient utilization is based on the following logic:
1. Plant species adapted to unique mineral nutrient soil conditions express genetic systems which exist, in part, in many economic species.

* This study has been supported in part by the National Science Foundation (USA) on Grant DAR-7912261 and by the College of Agriculture and Life Sciences, University of Wisconsin, Madison, Wisconsin, USA.

M.R. Sarić and B.C. Loughman (eds.), Genetic Aspects of Plant Nutrition.
ISBN-13: 978-94-009-6838-7
©1983 Martinus Nijhoff/Dr W. Junk Publishers, The Hague/Boston/Lancaster.

2. An appropriate test should identify plants able to grow more efficiently in response to nutrient stress. These isolated plants would then serve as efficient (E) parents for intraspecies hybridization and genetic analysis.

The response of *Lycopersicon esculentum* Mill (tomato) and *Phaseolus vulgaris* L. (common bean) to low P, K, N and Ca has been emphasized. Recently responses in *Medicago sativa* L. (alfalfa) to low K has been measured.

Nearly all studies have been limited to the vegetative stage of growth using liquid nutrient cultures. A limiting amount of the element to be studied is provided at the onset of an experiment and is not supplemented (Table 1). Thus total growth is limited by the element studied. Care must be exercised to avoid element substitution.

Differences in chlorosis due to lack of K may be used to discriminate strain differences in beans[9]. In alfalfa, however, because more rapidly growing plants exhibit K-deficiency sooner than slower growing plants, symptom differences may be misleading.

Preference is given to the use of total dry weight produced per unit of nutrient in the plant as an estimate of efficiency. The efficiency ratios are expressed as KER, PER and CaER for K, P, and Ca, respectively. All plant parts are easily removed from single plants grown in solution cultures. Roots and above ground portions of the plants are readily separated, dried, and weighed. Plants identified as efficient may be weighed (fresh weights only), transferred to a complete Hoagland's solution, and grown to maturity for purposes of seed production.

As soon as lines are confirmed as efficient (E) and inefficient (I), series of controlled reciprocal matings are made as follows:

efficient × efficient
efficient × inefficient
inefficient × inefficient

The F_1 progeny are each backcrossed to their respective parents and simultaneously allowed to self pollinate. Thus, by retaining remnant seed of parents and F_1 hybrids, adequate seed can be accumulated for simultaneous comparative tests of the following:

1. Parents
   P_1 self pollinated
   P_2 self pollinated
3. Backcross progeny
   $(P_1 \times P_2) \times P_1$
   $(P_2 \times P_1) \times P_2$
   $(P_1 \times P_2) \times P_2$
   $(P_2 \times P_1) \times P_2$

2. Reciprocal F_1 hybrids
   $P_1 \times P_2$
   $P_2 \times P_1$
4. F_2 progeny
   $(P_1 \times P_2)$
   self pollinated

The simultaneous test of all these progenies in a single environment minimizes the possibility of unknown environmental interactions interfering with comparisons based either on fresh weight of plant or on deficiency symptoms. The method of selecting efficiency must correlate with dry matter production[2,3]. The use of reciprocal F_1 hybrids provides an estimate of maternal and/or cytoplasmic effects.

The simultaneous test of the parents, hybrids, backcross and F_2 progeny of the hybrids provides an opportunity to estimate and compare population means. Each of the means can be expressed in terms of the F_2 mean (m); and additive (a), dominance (d), additive × additive (aa), additive × dominance (ad) and dominance × dominance (dd) gene effects. The 6 parameters are estimated from the population means using the relationships described by Gamble[4].

The sensitivity of tests as measured by differences in dry matter accumulation per g of stress element (K, P, or Ca) is supported by co-efficients of variation that are uniformly smaller for P_1, P_2 and F_1 and larger for BC and F_2 progeny. The sensitivity of the tests also allows us to identify

Table 1. Selected nutrient levels used to identify differential strain response to deficiency stress levels

Crop	Element (mg/plant)^z			
	K	P	N	Ca
Bean	5	2^y		
Tomato	5	3	35	10
Alfalfa	2.8			

^{z} Total amount of element in solution culture at start of experiment. The stressed element is not replenished during the growth period.
^{y} Bean seeds may contain more than 2 mg P necessitating the use of 3 mg/plant in some studies.

transgressive segregation among progeny whether the parents differ or are similar in nutrient efficiency. Lastly, the isolation of plants differing in efficiency in F_2 and BC populations provides an opportunity to produce F_3 and BC F_2 seed respectively for studies on heritability[2].

Results and discussion

Differences in response to potassium stress in beans

a) Genetic information Progeny derived from hybrids between 2 efficient and 2 inefficient strains segregated in a manner which can be assumed to be a single gene, *ke*, with efficiency of K utilization as homozygous recessive[9]. The F_1 hybrid between the two efficient parents is also efficient indicating that Pl 177,760 from Italy and Pl 180 761 from Germany carry the same gene for this response to low K, and are *ke ke*.

b) Information derived from the genetic studies Prior to the development of chlorosis the lower leaves reflex and an abscission layer starts forming which presumably could trap K in the older leaves. However K was transported from the older leaves to the growing tip prior to the development of the abscission layer. In contrast, there was a greater retention of K in the older leaves of the efficient plants, the leaves did not reflex, the abscission layer did not form and chlorosis did not develop. A higher K level was also found in the roots of K-efficient plants suggesting that differential K mobility was associated with the genetic differences in efficiency and inefficiency observed in these studies.

There is no evidence of Na substitution for K in beans.

Differences in response to potassium stress in tomatoes

a) General information Strains were isolated that varied widely in dry matter

Table 2. Frequency distribution of total plant dry weight (in grams) for the populations derived from the cross of parental lines 94(I) and 98(E). Plants grown at 5 mg of K per plant in nutrient solution for 30 days

Pedigree	Number of plants per class												N	Mean	SE	C.V. %
	0.53– 0.67	0.68– 0.82	0.83– 0.97	0.98– 1.12	1.13– 1.27	1.28– 1.42	1.43– 1.57	1.58– 1.72	1.73– 1.87	1.88– 2.02	2.03– 2.17	2.18– 2.32				
P$_1$ (94)		2	3	5									10	0.94	0.04	12.15
P$_2$ (98)								1	3	2		3	9	1.96	0.08	12.88
F$_1$					2	7	10	1					20	1.44	0.02	6.92
BCP$_1$			1	10	9	8	2						30	1.20	0.03	12.06
BCP$_2$			1	2	1	3	5	9	7	3			30	1.63	0.05	15.26
F$_2$	1	1	4	7	15	11	7	11	1	2			60	1.33	0.04	20.81

production. The K-efficient strains produced nearly 80% more dry weight at 5 mg K/plant than the inefficient strains but when grown at 200 mg K there were no differences in dry weight production among the 2 efficient and 2 inefficient lines used in the genetic studies. The origin of the 4 selected lines was as follows:

Line 94(I) = PL 289286 from Hungary
Line 139(I) = PL 117563 from Brazil
Line 98(E) = PL 126452 from Peru
Line 42(E) = Cultivar 'Oshogbo' from Nigeria

b) Genetic information Differences in yields of efficient and inefficient lines were heritable. There was no evidence of maternal effects. Because the results obtained from 4 crosses between efficient and inefficient lines were similar, segregation data are presented (Table 2) only for one family 94(I) × 98(E). Population distributions and means for this inefficient × efficient family suggested additive genes effects. The F_1 was intermediate compared to the 2 parents, the BCP_1 and BCP_2 distributions were skewed toward the P_1 and P_2 parents respectively. The F_2 means also were intermediate to the parents with individual plants distributed across and over the range of both parents. Narrow sense heritability was estimated to be in excess of 60%. Heterotic responses were not observed in the inefficient × inefficient and efficient × efficient families, although there was evidence of recombination in F_2 and BC populations of both families.

The means of all populations, analyzed by the methods of Gamble (4), provided estimates of various gene effects (Table 3). The major contribution to variation by additive and dominance gene effects is indicated by the relative magnitude of parameter a and d to parameter m. Additive effects made the major contribution in the efficient × efficient families, while dominance effects were

Table 3. Mean estimates of the 6 gene effects based on total dry wt for the 6 tomato families[z]

Crosses[y]	m	a	d	aa	ad	dd
94(I) × 98(E)	1.325**	−0.430**	0.361	0.368*	0.083	0.246
94(I) × 42(E)	1.153**	−0.432**	0.349*	0.240	−0.084	−0.145
139(I) × 98(E)	0.895**	−0.353**	0.073	0.158	0.024	−0.125
139(I) × 42(E)	0.868**	−0.248**	0.690**	0.624**	0.083**	−0.927**
94(I) × 139(I)	0.540**	0.034*	0.386**	0.384**	0.010	−0.523**
98(E) × 42(E)	1.303**	0.106**	0.408**	0.260*	0.049	−0.393*

[z] The signs of parameter a depend upon the parents being considered as P_1 or P_2; because the inefficient lines had been used as P_1, most of the estimates of parameter a are negative.
[y] = Inefficient, E = Efficient.
*, ** Significant at 5% (*) and 1% (**) probability levels by use of t test.

important only in 94(I) × 42(E) and 139(I) × 42(E). Epistatic gene effects were also important in 139(I) × 42(E).

c) Information obtained from genetic studies Response to sodium Cultivated tomatoes are not generally known to use sodium as a substitute for potassium. However, comparison of the 4 parents used in this study clearly showed that both efficient parents responded to Na when K was limited (Table 4). Line 42 appeared to be responsive to Na at all levels of K although the 4 replications used in this study were inadequate to provide clear support at the 50 and 200 mg K levels (Table 5). This study was repeated by Chen Hang, a Chinese Scholar visiting our laboratories, with the same results. Our data suggest that Na may be an essential element for Line 42. KER values were increased by Na even with 25 mg K available. The general similarity of Makmur *et al.* data[2] and the more recent data by Chen Hang suggest that this response to Na may be significant at 50 and 200 mg K.

Differences in response to potassium stress in alfalfa

Alfalfa was selected as an economic crop to which large amounts of K are applied in the USA and which is produced primarily for its vegetative growth. Even though a very heterozygous tetraploid, we felt alfalfa a good choice, among economic crops, for applying our techniques to a crop improvement program. The program, funded by the National Science Foundation, Washington, D. C., is too new to supply firm data. However, we have preliminary evidence that

Table 4. Analysis of variance of KER differences for four tomato lines grown at two levels of Na (high vs. low) at four levels of K

Line		K (mg)			
		5	25	50	200
98(E)	M.S.	4189.19	20.72	1.70	36.01
	F	10.13**	ns	ns	ns
42(E)	M.S.	15343.95	3017.28	443.24	235.25
	F	37.09**	7.29**	ns	ns
94(I)	M.S.	300.74	65.41	305.45	19.73
	F	ns	ns	ns	ns
139(I)	M.S.	2282.08	34.56	718.76	366.29
		5.52*	ns	ns	ns

* Significant at 5% level.
** Significant at 1% level.

Table 5. The effect of K-Na interaction on yield, Na concn, and KER of 4 tomato lines grown at 4 levels of K in high and low Na nutrient solutions

K (mg)	Variable[z]	98(E)[y] Low Na	98(E)[y] High Na	42(E) Low Na	42(E) High Na	94(I) Low Na	94(I) High Na	139(I)[y] Low Na	139(I)[y] High Na
5	Dry wt (g)	1.31	1.64	0.93	1.44	0.92	1.04	1.18	1.07
	KER	245	298	172	273	154	168	222	175
	Na%	0.22	1.36	0.21	1.31	0.16	1.14	0.13	0.76
25	Dry wt (g)	3.45	3.80	2.37	3.97	2.55	2.80	3.21	3.31
	KER	157	151	112	167	105	112	136	131
	Na%	0.12	1.40	0.11	1.26	0.12	1.07	0.10	0.69
50	Dry wt (g)	5.05	5.63	3.57	4.77	3.55	4.32	4.61	5.33
	KER	115	114	87	104	77	91	128	109
	Na%	0.09	1.47	0.09	1.36	0.09	1.01	0.08	0.69
200	Dry wt (g)	6.14	6.05	4.62	6.72	4.73	5.50	5.40	8.52
	KER	34	39	26	39	34	30	29	45
	Na%	0.08	1.01	0.07	1.20	0.06	0.68	0.05	0.92

[z] LSD of KER values: P, 5% = 37.25; P, 1% = 49.55. LSD of dry wt: P, 5% = 0.25; at 5 mg K: P, 1% = 0.35.
[y] E = Efficient, I = Inefficient.

selection of K-efficient alfalfa plants is possible and that matings among K-efficient plants provide progeny more K-efficient than plants obtained from intermated plants that are somewhat K inefficient.

Differences in response to phosphorous stress in beans

a) General information Bean lines were screened for efficiency of P utilization in nutrient culture at 2 mg P per plant (seed P + added P). The P was removed rapidly from nutrient solution and nearly all the P stored in cotyledons was exported before the cotyledons abscissed. From 54 lines screened, 6 were selected to represent extremes in response to P stress and were classified as efficient, moderately inefficient, and inefficient based on dry weight production and PER.

The most efficient lines produced approximately 100% more dry wt per plant than the most inefficient (Table 6). These differences reflect the ability of the plant to convert a finite amount of P into dry matter. The large root dry wt. of Line 11 was a surprise, apparently reflecting P retention in the roots of this plant since the P concentration in the roots of Line 6, 11, and 12 were very similar. Higher mobility of P was detected in the inefficient and moderately inefficient lines than in the efficient lines. Net photosynthesis by the lower leaves of plants growing under P stress reflected P concentration in the leaves[12]. Since the P is remobilized more rapidly from the leaves of the inefficient strains, net photosynthesis on any date was greater for the efficient line because of its greater leaf P concentration.

b) Genetic information Progeny tests suggested a wide range of genetic controls for P utilization. Families derived from the parent lines 11 and 3 (Table 7) indicate dominance for efficiency in the F_1 hybrid[12]. The opportunity to select more efficient segregates in the F_2 and BC populations is supported although the

Table 6. Dry wt and P concn in 6 selected bean lines grown at 2 mg P per plant

Efficiency class	Line	Dry wt (mg)			P concn (%)		
		Top[z]	Root	Total	Tops	Roots[y]	Total
Inefficient	3 (W 185)	484 c	106	590	0.26±.05	0.30	0.26[y]
Moderate	1 (PI 180751)	748 b	126	874	0.17±.06	0.24	0.18
	6 (PI 180755)	777 b	124	901	0.20±.03	0.18	0.20
	9 ('Slimgreen')	719 b	103	822	0.19±.03	0.26	0.20
Efficiency	11 (PI 206002)	1141 a	356	1497	0.13±.02	0.17	0.15
	12 ('Sprite')	1093 a	236	1329	0.15±.03	0.18	0.15

[z] Mean separation between lines by Duncan's multiple range test.
[y] A weighted average calculated from the means for % P and dry wt of the tops and roots of each line.

Table 7. Mean, standard deviation, and distribution of plants for PER (tops) of the family of line 3 × line 11. Plants were cultured under phosphorus stress at 3 mg phosphorus per plant

Pedigree	Mean	Number of plants per class midpoint											Total plants
		350	400	450	500	550	600	650	700	750	800	850	900	
	(mg)													
3	467 ± 76	2	3	2	3	2	1							13
11	734 ± 72							2	4	4	2	3		15
F_1	779 ± 54								2	5	5	2	1	15
3 × F_1	702 ± 163			1	1	1	2	2	6	4	5	2		24
F_2	711 ± 126				2	4	2	1	1	6	3	4	1	24

Table 8. Distribution of PER (tops) for the family line 1 × line 9 when cultured under P stress at 3 mg P per plant

Pedigree	Generation	Mean	Number of plants per midpoint																	Total plants
			300	350	400	450	500	550	600	650	700	750	800	850	900	950	1000	1050	1100	
1	P_1	617±25							8	7										15
9	P_2	625±42						1	7	5	2									15
1×9	F_1	776±61							1		2	6	4			2				15
9×1	F_1	763±75						2				7	4	1		1				15
(1×9)×1	BC	824±65									3	4	5	12	6					30
(9×1)×1	BC	787±112				1					5	8	10	2		2	1	1		30
(1×9)×9	BC	780±123				2			1		2	8	5	9		1	1	1		30
(9×1)×9	BC	852±127					1	1		1	1	3	4	3	4	9	2	1		30
(1×9)–1	F_2	707±124		1	4	2	4	1		7	20	19	12	10	2		1			83
(9×1)–1	F_2	693±97			3	2	2	1	5	15	20	22	8	2		1				81

potential gain may not be great. The families derived from the 2 moderately inefficient lines 1 and 9 (Table 8) suggest overdominance in the F_1 and segregates in the F_2 and BC population that far exceeded either parent. Among the F_2 and BC populations of family 9 × 1 the PER values of 32 of 284 plants were greater than 900 and over 50% higher than either parent. Estimates of the heritability of the family performances were made subsequently by Fawole et al. Broad sense heritability estimates are high. Narrow sense heritability estimates varied between 0.68 and 1.13 for BC and F_2 segregates indicating that performance of the F_2 and BC individuals reflected genotype and that selection for efficiency by plant breeders would be effective using these techniques[1].

Fawole et al.[3] estimates of gene effects support the suggestion of Whiteaker et al.[12] that different genetic systems are operating in this material. Additive, dominance and the additive × additive and dominance × dominance epistatic gene effects are generally significant, while the additive × dominance epistasis is of no real importance.

c) Root growth of Line 11 under P stress The extensive root growth by Line 11 under P stress has interesting implications. In soils phosphorus diffusion is limited; acquisition of P from a soil very low in available P therefore could reflect the extensiveness of the root system. This response could be important in the adaptation of plants to the acid soils high in Al and Fe in those parts of the world where P is so tightly bound that P amendments have little beneficial effect on plant growth.

The extensive root growth of Line 11 is heritable[2]. Total root weight measured in the various bean families suggested different genetic systems. The progenies obtained in the family from Lines 11 and 1 indicate overdominance for increased root growth in the F_1 and the opportunity to select plants with larger root systems than the more efficient parent (Line 11). Two isolates with about 35% more root than Line 11 were identified (Table 9). The broad sense heritability in the families involving Line 11 are between 69–90%. Selection for the large root character under conditions of P stress would be feasible in a breeding program.

Differences in response to calcium stress in tomatoes

a) General information Giordano initiated a study to determine the inherent variability among tomato strains to low Ca because of problems in Ca nutrition of that crop in Brazil[5]. Strains of tomato were found with large differences in dry matter (Table 9) when grown in a solution providing 10 mg Ca/plant (2L sol'n at 5 ppm Ca). These differences were less pronounced at 20 mg/plant and non-existent at 400 mg/plant. Unlike studies at stress levels (separately) for K, P, and N in which nearly all the stressed element was removed from the nutrient solution, there is a strain difference in the amount of Ca removed from the solution culture (Table 10). Because each plant had its own separate medium

Table 9. Frequency distribution of root dry weight (g) for the populations derived from the cross of lines 11 and 1; plants grown at 3 mg P per plant

Pedigree	Number of plants per class midpoint										N	$\bar{X}$	SE	CV %
	0.35	0.45	0.55	0.65	0.75	0.85	0.95	1.05	1.15	1.25				
11(E)[z]						5	9	1			15	0.930	0.051	5.55
1(MI)			3	8	4						15	0.767	0.074	9.65
F_1						17	6	6	1		30	1.017	0.085	8.36
$11 \times F_1$				4	4	3	5	2	1	1	20	0.871	0.171	19.51
$1 \times F_1$		8	2	1	2		3	2	2		20	0.706	0.278	39.66
F_2	4	10	7	9	8	9	4	2	1	1	55	0.689	0.218	31.93

[z] E = efficient, MI = moderately inefficient.

Table 10. Calcium concentration (%) and total Ca uptake per plant in 6 tomato lines grown in solutions with initial Ca levels of 10 and 400 mg/plant.

Added Ca per plant (mg)	Line	Ca concn total plant (%)	Total Ca per plant (mg)
10	35(E)[z]	0.23 ab[y]	6.98 bcd[y]
	39(E)	0.22 a	8.07 d
	113(E)	0.21 a	6.66 bcd
	67(I)	0.24 abc	5.02 ab
	118(I)	0.28 c	5.64 bc
	139(I)	0.26 bc	3.55 a
400	35(E)	2.98 b	186.35 a
	39(E)	2.94 b	227.45 a
	113(E)	2.83 b	222.87 a
	67(I)	2.98 b	204.94 a
	118(I)	2.65 ab	170.61 a
	139(I)	2.39 a	152.59 a

[z] E = efficient; I = inefficient
[y] Mean separation between lines by Duncan's multiple range test separately for each level of Ca supplied to plants.

Table 11. Frequency distribution of total plant dry weight (grams) for the populations derived from the cross of parental lines 39(E) and 139(I).

Generation	Number of plants per class										N	x̄	s	C.V. %
	1.0–1.5	1.6–2.1	2.2–2.7	2.8–3.3	3.4–3.9	4.0–4.5	4.6–5.1	5.2–5.7	5.8–6.3	6.4–6.9				
P_1 (39)						3	5	2			10	4.86	0.469	9.65
BCP_1			1	6	9	10	4				30	4.47	0.613	13.71
F_1					4	10	5	1			20	4.96	0.380	7.67
F_2		2	2	4	8	16	7	1			40	4.55	0.790	17.37
BCP_2			4	6	10	9					29	4.16	0.537	12.90
P_2 (139)		2	2	6							10	3.30	0.518	15.70

these strain differences in Ca removal must be mediated by the plant and, therefore, reflect genetic differences. Additional research indicated that these differences were not pH mediated.

b) Genetic information The progenies of the family formed by mating 39(E) and 139(I) are of special interest (Table 11). The parents removed 81.7% and 35.5% of the Ca respectively from the 10 mg of solution Ca and produced corresponding mean values of 4.86 and 3.30 grams of dry weight. The F_1 exhibited dominance for dry weight production; the segregating populations suggest that genetic variants were easily identifiable. The CaER's of this family show no difference among mean values for the two parents, the F_1 hybrid and all three segregating progenies (Table 12). Thus the genetic control of Ca uptake accounts for the ultimate differences in dry weight. Both the inefficient and efficient parents were equally capable of utilizing the Ca removed from the solution. In the other three families formed from hybrids of E and I parents, yield (dry weight and CaER) reflected both differences in uptake and differences in efficiency of Ca utilization following uptake. Studies on the nature of the gene effects of Ca nutrition imply that several models may account for the major differences. This would suggest more than one mechanism. In the E × I families a simple additive-dominance model was adequate to explain the CaER data. Broad sense heritability was high and selection among segregating progeny would be effective.

Comparative evaluation of K, P, and Ca studies

Parents varying widely in dry matter production were isolated in every study in our program. Mechanisms responsible for the dry weight differences appear to involve a capacity to function effectively at relatively low tissue concentrations of N, P, and Ca, substitution of Na for K, and greater uptake from Ca of a limited supply.

Efficient strains for P and K tended to retain a larger amount of the stressed element in roots and older leaves suggesting differences in remobilization. However, strains of Ca-efficient plants removed more Ca from the solution than did inefficient plants which suggest that more active remobilization of Ca in tomatoes may contribute to efficiency.

In every study heritability estimates were high and selection for "efficient" segregates would be effective. In some progeny genetic recombinants much more "efficient" than either parent indicate that plant materials much more efficient than the most efficient parents can be synthesized.

The unique responses of bean line 11 (extensive root growth under low P) and tomato line 42 (response to Na in studies on K stress) were unanticipated useful finds among germplasm and were probably discovered because the research was carried out in a liquid culture medium. It is recognized that screening in solution cultures may not effectively detect root features critical in nutrient acquisition in

Table 12. Frequency distribution of total plant CaER for the populations derived from the cross of parental lines 39(E) and 139(I).

Generation	Number of plants per class										N	x̄	s	C.V. %
	400–430	431–461	462–492	493–523	524–554	555–585	586–616	617–647	648–678	679–709				
P₁ (39)			1	0	1	2	4	2			10	584.07	44.663	7.65
BCP₁				4	10	7	6	3			30	565.47	38.278	6.77
F₁				2	2	7	6	3			20	579.71	36.525	6.30
F₂	1	0	0	6	9	10	9	3	2		40	567.35	47.089	8.30
BCP₂			2	4	9	6	5	2	0	1	29	559.21	45.481	8.13
P₂ (139)				1	4	1	4				10	563.15	34.135	6.06

soils. However, the reproducibility and convenience of solution cultures have been largely advantageous in our studies.

References

1. Fawole I 1979 Heritability of efficiency in phosphorus utilization in beans (*Phaseolus vulgaris* L.) grown under phosphorus stress. Ph.D. Thesis. University of Wisconsin, Madison. 97 p.
2. Fawole I, Gabelman W H and Gerloff G C 1982 Genetic control of root development in beans (*Phaseolus vulgaris* L.) grown under phosphorus stress. J. Am. Soc. Hortic. Sci. 107, 98–100.
3. Fawole I, Gabelman W H, Gerloff G C and Nordheim E V 1982 Heritability of efficiency in phosphorus utilization in beans (*Phaseolus vulgaris* L.) J. Am. Soc. Hortic. Sci. 107, 94–97.
4. Gamble E 1962 Gene effects in corn (*Zea mays* L.). I Separation and relative importance of gene effects for yield. Can. J. Plant Sci. 6, 582–586.
5. Giordano L de B, Gabelman W H and Gerloff G C 1982 Inheritance of differences in calcium utilization by tomatoes under low-calcium stress. J. Am. Soc. Hortic. Sci. 107, 664–669.
6. Harvey P H 1939 Hereditary variation in plant nutrition. Genetics 24, 437–461.
7. Lyness A S 1936 Varietal differences in the phosphorus feeding capacity of plants. Plant Physiol. 11, 665–688.
8. Makmur A, Gerloff G C and Gabelman W H 1978 Physiology and inheritance of efficiency in potassium utilization in tomatoes grown under potassium stress. J. Am. Soc. Hortic. Sci. 103, 545–549.
9. Shea P F, Gabelman W H and Gerloff G C 1967 The inheritance of efficiency in potassium utilization in strains of snapbean, *Phaseolus vulgaris* L. Proc. Am. Soc. Hortic. Sci. 9, 286–293.
10. Smith S N 1934 Response of inbred lines and crosses in maize to variations of nitrogen and phosphorus supplied as nutrients. J. Am Soc. Agron. 26, 775–780.
11. Weiss M G 1943 Inheritance and physiology of efficiency in iron utilization in soybeans. Genet. 28, 253–268.
12. Whiteaker G, Gerloff G C, Gabelman W H and Lindgren D 1976 Intraspecific differences in growth of beans at stress levels of phosphorus. J. Am. Soc. Hortic. Sci. 101, 472–475.

Rationale of selection for specific nutritional characters in crop improvement with *Phaseolus vulgaris* L. as a case study*

P. B. VOSE
Centro de Energia Nuclear na Agricultura, C. P. 96, 13400 Piracicaba, S. P., Brasil

Key words Acid soils Cation exchange capacity Fertilizer use Nitrogen *Phaseolus vulgaris* Phosphorus Selection Soil stresses

Abstract Genetic effects are obtainable for any aspect of transport, accumulation, and efficiency of nutrient use by plants, and for virtually any element. Some of the important characters are: tolerance to acid soils (18% of soils or 2.4 billion ha), tolerance to high pH induced Fe-chlorosis, and tolerance to salinity (about 1,000 m ha). Genotypes which made better use of N and P would be the means of saving fertilizers, especially important to developing countries. A 10% economy of fertilizer use represents a minimum world saving of US$6 billion annually.

Phaseolus vulgaris is taken as a model to show that although we know quite a lot about the extent of its nutritional variation, e.g. adaptability to acid soils, and the crop's utilization of N and P, we are handicapped in exploiting this because of lack of genetic information. This in turn depends on knowledge of specific mechanisms, and investigating these must be a priority.

Scope of nutritional variation

I can recall that when I was writing an article 20 years ago[48] attempting to develop a coherent philosophy and approach towards varietal differences in plant nutrition it was possible to attempt a complete review of literature published up to that time. It is a gratifying measure of the increased research in this area that this is now scarcely practical. Similarly, in 1966[23] it was quite hard to organize as part of a larger meeting, a special section on genetical aspects, but now we are meeting here for a whole week.

It is apparent from recent and earlier review publications (*e.g.*[38,52,57]) that one can almost be certain of obtaining genetic effects for any aspect of transport, accumulation and efficiency of use for virtually any element. The possibility of using mutation techniques to increase the range of variation and to induce sought for nutritional response characteristics in otherwise desirable varieties should also not be overlooked[50]. The challenge now is to apply such variation through improved crop plants.

Although a lot of recent work has concentrated on the identification and selection of cultivars able to withstand mineral stresses in problem soils, *e.g.* differential resistance to toxicities, such as due to Al, Mn, heavy metals and NaCl, it is necessary to appreciate that cultivars which are efficient in their use of the major nutrients N and P, and of certain microelements, could be at least of

* IAEA Project BRA/5/009

M.R. Sarić and B.C. Loughman (eds.), Genetic Aspects of Plant Nutrition.
ISBN-13: 978-94-009-6838-7
©*1983 Martinus Nijhoff/Dr W. Junk Publishers, The Hague/Boston/Lancaster.*

equal value at the very highest levels of soil fertility and fertilization, through the more efficient use of fertilizers.

We still don't understand varietal effects relating to phosphorus. It appears we need to distinguish between genotypes which are adapted to growing with low P, from genotypes which apparently have superior ability to gather P from low-P soils, probably through having a better developed or more branched root system.

To use fertilizer-N more efficiently must be a major objective. We now realize that some cultivars can produce more dry matter per unit of nitrogen supplied, but they are not always the highest economic yielders. Some genotypes make better use of the nitrogen they take up by using it to produce economic yield, for example grain as opposed to straw.

As a group I believe we have not made sufficient effort to emphasize the economic importance of our research, though a recent report[14] has done something to redress this. In this paper I look at the economic relevance, and take an overview of nutritional variation in a 'model' crop; beans, *Phaseolus vulgaris*, with the objective of both evaluating the type of information we have and indicating the major gaps.

Economic background

During this century, soil science and agronomy have concentrated on changing soils to make them suitable for crops, rather than attempting to breed crops better adapted to specific soil conditions. This approach, coupled with greater use of fertilizer, has in general been successful, *e.g.* average yields of grain in Europe have more than doubled in 30 years since the Second World War. At the same time plant breeders have most often tried to get the maximum potential adaptability into their breeding material and have then selected under widely differing climatic and soil conditions. This was the technique adopted with the largely successful international plant breeding programmes for wheat[5] and rice[6].

Poor soils vs *more food*: The approach has been successful because it has concentrated on developing varieties for what we may call 'non-extreme' soils. However, we are now facing a situation where most of the land still remaining in the World which can be developed for improved agriculture is on soils which have inherent problems which cannot easily be ameliorated, either because of the extent or cost, or both. About 25 per cent of the World's area of cultivable soils has relatively acute problems, such as extreme acidity or alkalinity, with too much manganese or aluminium, or too little iron or zinc, or excess salinity. Many developing countries will need to bring new, less fertile and problem soils into cultivation, and require specially adapted varieties[51].

It has been estimated for 1981–82 that 68 low income countries need to import 35 million tons of cereals and other foods, but reputedly there are likely to be only 27 million tons available[46]. It is clear that developing countries must grow more of their own food, not only because most of them lack the foreign exchange necessary to purchase on the international market, but also because food from outside is now becoming unavailable. In part, the food production problems of developing countries are socio-political, but there are many inherent plant-soil problems which need to be solved too.

For example, in Africa at the present time more and more people are sharing the same or less food, and yet a recent study[21] has suggested that there is sufficient land not only to meet present needs but the requirements up to the year 2000, even assuming the present rate of population increase. Thus there are an estimated 800 million hectares of potential arable soils in Africa of which only 160 million hectares are at present used[37]. Specific adaptability of crop varieties to soil and climate seems especially important in Africa, where the modern broadly adapted varieties have made little impact due to wide variation in edaphic, climatic and cultural factors. In Asia it has been suggested[24] that increased rice production could come from 48 million hectares.

Fertilizer use: The great increase in the cost of oil has been reflected in the price of fertilizer, and we can no longer make plans for crop improvement based on inexhaustible supply of fertilizer input. Developing countries at the present time have 75 per cent of the World's population but at present use only about 15 per cent of the output of fertilizer. It would clearly be desirable for them to use more, but poverty, labour-intensive farming, and lack of foreign exchange make it unrealistic to expect very large increases in fertilizer use in the short term. In 1976 it was still possible to anticipate[43] that by the year 2000, 40% of the demand for N and P fertilizers would be from the developing countries. One could wish that this was likely, but it no longer seems possible, bearing in mind the present costs of fertilizer and the general economic situation. For example, in Brasil it is believed that fertilizer sales in 1981 were 30% down. There is some evidence too that projected increases in fertilizer use in agriculturally highly developed countries are not as great as would be expected, due to higher cost. It is apparent that there are changed circumstances and that we must seek greater efficiency of fertilizer utilization, both through scientific application and by using more efficient plant varieties.

It is possibly not too meaningful to try and put an exact financial value on say, a modest 10% improved efficiency of N, P and K fertilizer use by more effective cultivars, but it is worth establishing the scale of the basic numbers. Extrapolating from existing studies[4,45] we can anticipate that World fertilizer consumption this year will be of the order of 120 + million tons, of which about half will be N. The earlier[45] forecast for 2000 suggested that consumption might increase to as much as 300 million tons, but this has been revised down to somewhat over 240 million tons[4]. Taking the 1981 figure, if we consider a saving of 10% on 120 million tons *i.e.* 12 million tons, costing say 50 cents/kilo, this would represent and *annual* saving to the World farm industry of a minimum of US$6 billion. In reality, fertilizer prices in most countries are already higher than 50 cents/kilo. At the end of 1981, U. S. A. prices were on average 58 cents/kilo N as urea and 82 cents/kilo N as ammonium sulphate[47] while Brazilian prices for the corresponding grades were the equivalent of 68 US cents and US$1.00 respectively[3]. There are potential capital savings too in scaled-down requirements for fertilizer plant. Stangel[43] calculated that 10% improved use of fertilizer N between then and 2000 would result in a capital saving on plant of US$11.2 billion at 1977 prices, or probably with inflation about US$18 billion in 1982. Taken together, these potential savings are not small.

Extent of major soil problems

Consider the extent of the major soil problems of acidity, Fe deficiency and salinity:

Acid soils: The typical acid soil has high levels of available Al and Mn, combined frequently with a lack of Ca, and with high capacity for P immobilization. The major soils of the tropics are Ferralsols and Acrisols, and together with fluvisols comprise 18 per cent of total World soil area over 2.4 billion hectares[11]. Acid soils stretch from Korea, Japan, China, Vietnam, Thailand to Burma and Bangladesh, an down towards Malaysia, Philippines, Indonesia and Sri Lanka. Similar acid soils extend on a broad land right across equatorial Africa to Senegal, Gambia and Sierra Leone (FAO, 1976). In low lying coastal areas very low pH acid sulphate soils, formed from marine sediments with both high aluminium and salinity can be especially toxic.

In South America most soils are acid, often extremely so, except West of the Andes *e.g.* Chile and south of latitude 30°. Acidity is by no means solely a problem in the sub-tropics. In the temperate zones of the Northern Hemisphere there is a belt of acid Podzols, with potentially high available Mn and Al, which stretch across Europe and N. America. On these soils, acidity may not be an acute problem, but maintaining a sufficiently high pH is a continuing management problem and it is already apparent that specific breeding for Al tolerance *e.g.* in wheat in N. America is worthwhile.

Iron deficiency: Iron is seldom deficient *per se* but deficiency problems in plants are caused by lack of availability due to high pH, free calcium carbonate or bicarbonate in irrigation water often coupled with water stress. Some of the worst cases of Fe-deficiency occur on long established orchard soils which have high levels of Ca and Cu from fungicide sprays. The areas affected by Fe-deficiency were considered at length recently[53] and need only be summarised here.

Over 25% of the World land surface is calcareous and potentially susceptible to Fe-deficiency problems on such soils as Chernozems, Rendzinas and Xerosols.

The main areas affected are large areas of mid-USA, parts of Southern Africa, and Southern Russia north of Baltic and Caspian areas, and the Eastern Mediterranean and Near East and many areas in India. In the USA 5% of the cultivated area, amounting to more than 4.8 million ha of the 22 states west of the Mississipi is deficient in available-Fe for susceptible crops such as soybeans and sorghum[27]. In W. Europe the areas are relatively small and the deficiency often the result of bad soil management.

In the USA plant breeding programmes for soybeans and sorghum are beginning to take into account selection for Fe-deficiency tolerance and we can anticipate this development being extended to other countries, especially with avocados, citrus and vines.

Salinity: Although saline soils are often conceived as primarily suffering from excess NaCl, they usually also have excess Ca salts. This general alkalinity also makes them coincidentally liable to Fe-deficiency problems. It is difficult to be precise about the worldwide extend of saline soils, but it is generally considered to be growing. It has been suggested that there is more land going out of irrigation due to salinity, than there is new land coming into irrigation. Salinity problems are primarily restricted to arid and semi-arid regions, often under irrigation, but are not limited to irrigation agriculture.

Worldwide, there are about 1000 million ha affected by salinity[25]. About a third of irrigated soils, amounting to almost 77 million ha are said to be sufficiently affected by salinity as to adversely affect crop growth[12]. There are more extensive areas of saline soils in Europe than generally supposed. Szabolcs[44] reported approximately 30 million ha of salt affected soils, two-thirds of which were Solonetz soils, in E. Europe.

In the Far-East the Indo-Gangetic plain has many salinity affected areas, while Ponnamperuma[33] noted that salinity is one of the major factors reducing rice yields on possibly as much as 40 million ha of flat land in deltas, estuaries and coastal fringes of the humid tropics. It is now recognized that selection for salinity tolerance in rice must often be a major objective for these difficult soils.

None of the data for problem soils presented here is new, but I make no apologies for this. We have to keep on presenting these facts until they are more widely known and better appreciated by those responsible for supporting research. To attempt to compute meaningful data relating to potentially improved yields and monetary return is not possible, but it is obvious that we are not talking of insignificant amounts. We must keep emphasizing this.

A model crop case study (*Phaseolus vulgaris* L.)

In an earlier paper[49], it was noted that it would be interesting to carry out a complete integrated study of the factors governing nutritional variation in one species, preferably using the same varieties. Such a study still remains utopic, but what is now possible is the assembly of a composite picture of the nutritional variability that is available for exploitation in a model crop. Additionally such a review may highlight where critical information is lacking. Beans, *Phaseolus vulgaris* L., has been chosen here, but similar treatment could now be given to wheat, rice, maize and sorghum.

Nitrogen nutrition: The study of the efficiency of N-utilization in Phaseolus and other legumes is a particular challenge, as not only is it necessary to consider N-utilization per se but also the efficiency of N_2-fixation. It would be of comparatively little use if we had a highly efficient utilizer of nitrogen which was a poor N_2-fixer.

A study by Amaral[1] of over a hundred varieties of Phaseolus in solution culture showed differences in the efficiency of N-conversion into grain and leaves. Table 1 shows the range of per cent N in leaves and grain, which is clearly very large. In general, a high content of N in leaves indicates that less is available for the economically important grain, and is a rough measure of efficiency.

This study did not include the effect of nitrogen fixation and was confined to solution culture. We[36] have taken such studies a little further, in a field experiment which compared five cultivars (including three used by Amaral[1]) and evaluated N_2-fixation by ^{15}N isotope dilution technique. Table 2 shows that there are quite big differences in both the ratio kg seed/kg N_2-fixed and also kg seed/kg

Table 1. The range of nitrogen concentration (per cent) found in the grain and leaves of 104 cultivars of *Phaseolus vulgaris* L. grown with high and low N-nutrition(**)

Selected cultivars	N in grain (%)		N in leaves (%)	
	High N nutrition	Low N nutrition	High N nutrition	Low N nutrition
Rosinha	5.50*	2.33	3.66*	1.34
Mexico	4.12	3.90*	1.79	1.40
Goiano precoce	4.67	3.35	3.42	2.10*
Carioca	4.37	1.45	2.47	1.45
Costa Rica	3.88	2.25	2.03	1.29
Mulatinho	3.29+	2.28	1.73	1.40
Mexico 163	4.54	1.15+	1.94	1.15
Mulatinho 1208	3.74	2.34	1.37+	1.37
Costa Rica 1031	4.84	3.41	3.31	1.11+

Tukey (1%) = 0.95

*,+ represent highs and lows of range

** derived from Amaral[1]

Table 2. Parameters of dinitrogen fixation and utilization efficiency of Phaseolus cultivars grown in the field with 20 kg N/ha given as a "starter"**

Varieties	Seed Yield kg/ha	N₂-Fixed Total kg/ha	N₂-Fixed As % in crop	Ratio kg seed/kg N N₂-Fixed	Ratio kg seed/kg N all sources	Harvest Index* per cent Yield	Harvest Index* per cent Nitrogen
Carioca precoce	545	46	63	12.2	7.4	19.0	26.9
Goiano precoce	271	25	38	13.0	4.1	11.3	22.8
Costa Rica	593	58	65	11.0	6.7	11.6	24.2
Carioca	917	65	65	15.0	9.1	18.3	31.9
Moruna	343	37	61	9.1	5.7	12.5	13.1
LSD (<0.05)	149	29.2	n.d.	6.5	4.2	n.d.	n.d.

* Harvest Index = $\frac{\text{Seed}}{\text{Total}} \times 100$

** derived from Ruschel et al.[36]

n.d. = not determined

N derived from all sources (*i.e.* from soil, fertilizer and N_2-fixation). It is noteworthy that such variation can occur in highly selected cultivars.

Our data suggest that efficiency of nitrogen utilization i.e. as measured by seed production per unit of plant nitrogen is not directly linked to efficiency of dinitrogen fixation. Goiano precoce for example, a '60-day' short season variety, has only 40% of plant nitrogen due to fixation, and while having a low yield of grain showed a quite high Harvest Index for N, as well as yield of seed per kg N_2-fixed. Carioca which had the greatest seed yield showed good nitrogen fixation and good conversion of nitrogen, however measured.

The yield of Moruna was also not related to N_2-fixation. This is a recently developed variety[31] yet showed up rather badly except for N_2-fixation. There is some evidence that it was selected and tested under high levels of N-fertilization[32] which suggests that a selection process in which prime importance was given to obtaining disease resistance, can inadvertently select away from highly N-efficient types.

In Phaseolus we are still not able to ascribe variations in N-utilization to differences in nitrate reductase activity, nitrate absorption, or the ability to break down and retranslocate stored nitrogen from the foliage to the grain during seed filling, as has been found for wheat varieties Anza and UC44—11 by Huffaker and Rains[22] and others[19,34]. Also working with wheat, Mikesel and Paulsen[29] showed that protein in the grains was strongly influenced by translocation efficiency from lower leaves. It is not possible to say yet how important N-redistribution is in Phaseolus for N-use efficiency but we have found (unpubl.) that a short pulse of ^{15}N given as early as the first leaf stage can be detected in the grain at maturity, so clearly it has some influence.

Phosphorus requirement: Apart from nitrogen, phosphorus is the single most important nutrient in Phaseolus production, and here some quite large genotype differences in response have been recorded, primarily by the Wisconsin group. Whiteaker *et al.*[56] screened over fifty lines for efficiency of P utilization at very low levels of P. Lines were recognized as representing extremes in response to phosphorus stress, and classified as efficient or inefficient based on dry weight production per unit P. Values ranged from 380 to 671 mg dry wt/mg P. It was found that at higher levels of nutrient P some lines showed growth response while others didn't, and that lines efficient at stress P levels were not necessarily responders. Net photosynthesis per unit of P was higher in an efficient than in an inefficient line, under P stress.

These results show clearly that in Phaseolus differences in P utilization are related to the participation of P in metabolic processes, and are not only due to absorption. That is not to say that differences in root habit *e.g.* depth or fine branching might not also be influential in P-acquisition on low P soils. Unfortunately, published studies on P-efficiency in Phaseolus, carried out under controlled conditions, have been with relatively high levels of combined-N and have yet to take into account the effect of P on nodulation and nitrogen fixation. The P-content of nodules is the highest of any plant tissue and may be as great as 0.5%[20]. Moreover, Cassman et al.[8] found that the P-requirement of soybeans may be increased as much as 50% when they are required to depend entirely on N_2-fixation for their nitrogen supply. This suggests that it is essential for future studies of P-efficiency in legumes to consider the effect on nitrogen fixation. Of course, to a certain extent that has been automatically taken into account in selection of superior cultivars for acid soils, considered in the next section.

Tolerance to acid, high-Al soils: For a Phaseolus variety to grow successfully on the acid soils of Brasil and other similar acid Ferralsols and Acrisols it has to be tolerant to high levels of Al and able to grow with relatively low P. Differential tolerance to Al toxicity is well recognized in beans (*e.g.* ref.[16,26,41]). There is in general a sharp division between black beans, usually much more acid tolerant, and non-black beans. With high-Al, Venezuela 350 (black) had greater nodule number and greater nitrogenase activity than Carioca (white)[18]. There also seems to be a major division between beans developed in the N. Hemisphere compared with the much more Al-tolerant S. American cultivars.

It is not too surprising that the latter are Al-tolerant, considering the generations of natural selection that are behind every landrace. Even so, CIAT is finding it desirable to screen for improved tolerance[9,42]. At a practical level it may be clear that there are big differences in Phaseolus tolerance to acid soils and high-Al, but it still doesn't seem entirely clear what is the real basis of tolerance. Is it tolerance to high-Al, low-P, low-Ca, or all three; are these factors interlinked; which is more important; has a branched root system an important rôle in P-acquisition?

We need answers to these points for a fuller understanding of cultivar tolerance to soil acidity. For example, in a CIAT trial it was thought that cv. Carioca, a widely grown Brazilian landrace variety was very effective in using a low soil phosphorus but was not really tolerant to high Al, (Thung, pers. comm., 1981). Yet, it is difficult to believe that Carioca is not tolerant to quite high levels of Al, as how else did a landrace variety achieve success on Brazilian low pH soils? This emphasizes the problem of evaluating the acid soils tolerance mechanism in Phaseolus. In wheat, which is much less dependent on P and Ca for good yields, then tolerance to Al is the prime concern.

Cation exchange capacity (CEC) has been shown to be an indicator of adaptability to soil types, with species adapted to soils of low pH having low CEC[10,28]. It was found with ryegrass, *Lolium perenne* L., that an experimental line selected for high tolerance to Al had also been simultaneously selected towards low CEC[54]. Low CEC has also been found to be associated with Al-tolerance in wheat and barley[17]. We carried out a study[55] in which we screened 23 Phaseolus cultivars, thought to be reasonably adapted to acid soils, for CEC. Partial results

Table 3. The range of cation exchange capacity (CEC) found in 23 genotypes of *Phaseolus vulgaris* L.*
(low CEC genotypes are putatively acid tolerant)

Range		CEC mequ/100 g dwt	Number of genotypes
High		26–30	3
Median		22–26	8
Low		18–22	12
	Low: Carioca	21.7	
	Rosinha G-2	21.4	
	CIAT P382-A	19.0	
	CIAT P277-A	18.3	

LSD ($P=0.05$)	1.29

* derived from Vose and Tulmann Neto[35]

are shown in Table 3, indicating a quite exceptionally wide species range of 19.0–29.9 mequ/100 g dwt.

Divided into high, low and median portions of the range, 12 lines fell into the 18–22 mequ Group, 8 lines into the median 22–28 mequ, and only 3 lines into the top 26–30 mequ, Group. Thus 12 of the 23 lines fell into the 19–22 mequ/100 g dwt group which is very low for Leguminosae, overlapping with the range for Graminease. Carioca and Rosinha, the well established landraces, fell into this group. The two lowest recorded genotypes are CIAT experimental lines. It appears that CEC provides another approach to determining adaptability of cultivars to acid soils, without any pre-judgement as to whether they are Al-tolerant or capable of growing with low P.

Other nutritional characters: Shea et al.[39,40] study of potassium utilization at low and high K level was able to classify genotypes into efficient and inefficient users of K, with results suggesting that control is by a single pair of alleles possibly with modifier genes, and that efficiency is recessive. Differential response to low K nutrition appeared to be associated with efficiency in K utilization at the metabolic level, and variation was not associated with root size nor with higher levels of K in efficient plants.

Of the minor elements, zinc is most often a problem in Phaseolus, usually linked with excess P[7]. Ellis[13] and Ambler and Brown[2] noted differential susceptibility to Zn-deficiency, the former finding that the variety Sanilac was much more susceptible to deficiency than Saginaw. Polson and Adams[30] later found that with excessive Zn levels Saginaw tolerated higher levels than Sanilac. It seems unusual for a variety to be susceptible to both high and low levels of an element. One suspects that response to Zn effects is closely related to phosphorus metabolism and that variation in Phaseolus sensitivity to high P-levels is also connected. Thus Rice[35] found that tolerance of Phaseolus to high P nutrition was inherited maternally, with efficiency associated with large-seeded varieties. Maybe large seeded varieties carry larger reserves of seed-Zn for subsequent plant utilization, even with high P.

Conclusions

The first part of this paper showed that the selection and breeding for crop varieties specially adapted to problem soils could be the means of increasing production on very many thousand hectares of soil. The past policy of developing widely adapted crop varieties has been basically successful, but the areas of potential arable soils left for farming, but having stress problems, is large enough to warrant specially selected varieties. Indeed there is probably a large area of soil, particularly in Africa, S. America and Asia which can probably never be satisfactorily farmed without nutritionally fitted cultivars.

Even on fertile soils, free of abnormal problems, the identification and breeding of varieties more efficient in their use of N and P could result in substantial fertilizer and monetary savings, of the order of billions of dollars.

Considering Phaseolus as a typical crop case study, it is apparent that there is considerable variation in nutritional characters available for exploitation for almost every requirement. A major limitation is our knowledge of the genetics. Only for potassium do we have any very good genetical information. The reason that we don't have good data for N and P efficiency and for soil acidity tolerance is not hard to seek. Probably, unlike potassium we are concerned with very complex multi-component systems, and until we do research on specific mechanisms and interactions it will remain difficult to define recognizable characters for genetic studies. It is noticeable that where there is a very clearly defined situation such as Al-tolerance in wheat, or Fe-deficiency induced chlorosis in soybeans, cation content in maize, *etc.* then we do have good genetic information.

Investigating specific mechanisms of nutritional variation must therefore remain a priority. That is why many basic studies presented at this meeting are of particular importance.

References

1 Amaral F A L 1975 Eficiência de utilização de nitrogênio, fóstoro e potassio de 104 variedades de feijoeiro (*Phaseolus vulgaris* L.). Tese Doutoramento ESALQ, Piracicaba, S. P., Brasil.
2 Ambler J E and Brown J C 1969 Cause of differential susceptibility to zinc deficiency in two varieties of navy beans, *Phaseolus vulgaris* L. Agron. J. 61, 41–43.
3 ANDA 1981 Preços de fertilizantes. Associação Nacional para Difusao de Adubos. Tabela, de zembro 1981 (São Paulo).

4   ANON 1980 The Global 2000 Report to the President. Vols 1 and 2. Rept. prepared by U S Council on Environmental Quality and the Department of State, U.S. Govt. Printing Office (Washington, DC).
5   Borlaug N E 1973 Building a protein revolution on grain legumes. *In* Nutritional Improvement of Food Legumes by Breeding, pp 7–11. Protein Advisory Group Symposium, Rome 3–5 July 1972, UN, New York.
6   Brady N C 1975 Rice responds to science. *In* Crop Productivity, Research Imperatives. Eds. A W A Brown *et al.* pp 62–96. Kettering Foundation, Yellow Springs, Ohio.
7   Burleson C A, Dacus A D and Gerard C J 1961 The effect of phosphorus fertilization on the zinc nutrition of several irrigated crops. Soil Sci. Soc. Am Proc. 25, 365–368.
8   Cassman K G, Whitney A S and Fox R L 1981 Phosphorus requirements of cowpea and soybean as affected by mode of N nutrition. Agron. J. 73, 17–22.
9   CIAT 1977 Centro Internacional de Agricultura Tropical. Annu. Rept. 1976, CIAT (Cali).
10 Drake M, Vergris I and Colby W G 1951 Cation-exchange capacity of plant roots. Soil Sci. 72, 139–147.
11 Dudal R 1977 Inventory of the major soils of the World with special reference to mineral stress hazards. *In* Proc. Workshop Beltsville, Ed. M J Wright. Cornell Univ. Agric. Exp. Sta. Special Pub. Ithaca (New York) 3–13.
12 Eckholm E P 1975 Salting the earth. Environ. 17, 9–15.
13 Ellis B G 1975 Response and susceptibility. *In* Zinc deficiency, a Symposium. Crop Soils 18, 10–13.
14 Epstein E 1980 Impact of applied genetics on agriculturally important plants: mineral metabolisms. Report to the Office of Technology Assessment, U. S. Congress. Mimeo pp. 42.
15 Food and Agriculture Organization of the UN (FAO) 1976 FAO-UNESCO Soil Map of the World, Vol. VI, Africa. UNESCO (Paris).
16 Foy C D, Armiger W H, Fleming A L and Zaumeyer W J 1967 Differential tolerance of drybean, snapbean and lima bean varieties to an acid soil high in exchangeable aluminium. Agron. J. 59, 561–563.
17 Foy C D 1974 Effects of aluminium on plant growth. The Plant Root and its Environment. Ed. E W Carson. pp 601–642. Univ. Press Va. (Charlottesville, Va.).
18 Franco A A 1981 Acidity factors limiting nodulation, nitrogen fixation and growth of *Phaseolus vulgaris*. Ph.D. Thesis. Univ. Calif. (Davis).
19 Gallagher L W 1976 Genetic variation in nitrate reductase activity in two wheat genotypes (*Triticum aestivum*) and its relation to nitrogen assimilation. Ph.D. Thesis. Univ. Calif. (Davis).
20 Graham P H and Rosas J C 1979 Phosphorus fertilization and symbiotic nitrogen fixation in common beans (*Phaseolus vulgaris* L.). Agron. J. 71, 925–926.
21 Higgins G M, Kassan A A, Naiken L and Shah M M 1981 Africa's agricultural potential. Ceres 14, 13–21.
22 Huffaker R C and Rains D W 1978 Factors influencing nitrate acquisition by plants assimilation and fate of reduced nitrogen. *In* Nitrogen in the Environment. Vol 2. Eds. D R Nielsen and J G MacDonald pp 01–43. Academic Press (New York).
23 IAEA 1967 Proc Symp Isotopes in Plant Nutrition and Physiology IAEA (Vienna).
24 Laidlaw K 1978 Report of the Trilateral Commission scheme to double rice productivity in Asia. The Times (London) 23 August.
25 Massoud F I 1974 Salinity and alkalinity as soil degradation hazards. Rept. FAO/UNEP Expert Committee on Soil Degradation. FAO (Rome).

26 Miranda L N and Lobato E 1978 Tolerância de variedades de feijão e de trigo ao aluminio e à baixa disponibilidade de fósforo no solo. R. Bras. Ci. Solo 2, 44–50.
27 Mortvedt J J 1975 Iron chlorosis. Crop Soils 27, 10–12.
28 McLean E O, Adams D and Franklin R S 1956 Cation-exchange capacity of plant roots as related to nitrogen contents. Soil Sci. Soc. Am. Proc. 20, 345–347.
29 Mikesell M E and Paulsen G M 1971 Nitrogen translocation and the role of individual leaves in protein accumulation in wheat grain. Crop Sci. 11, 919–922.
30 Polson D E and Adams M W 1970 Differential response of navy beans *Phaseolus vulgaris* to zinc. 1. Differential growth and elemental composition at excessive zinc levels. Agron. J. 62, 557–560.
31 Pompeu A S 1978 Aroana e Moruna, cultivars de feijoeiro (*Phaseolus vulgaris* L.) para o Estado de São Paulo. Bragantia 37, LXXIII–LXXVI.
32 Pompeu A S, de Almeida L. d'Artagnan, Schmidt N C and Loberto L C 1978 Comportamento de linhagens e cultivares de feijoeiro (*Phaseolus vulgaris* L.) no Vale do Paraíba, S. P. Bragantia 37, 93–101.
33 Ponnamperuma F N 1977 Screening rice for tolerance to mineral stresses. *In* Proc Workshop, Beltsville. Ed. M J Wright. Cornell Univ. Agric Exp. Sta. Special Pub., Ithaca (New York) 341–353.
34 Rao K P, Rains D W, Qualset C O and Huffaker R C 1978 Crop Sci. (cited in Huffaker and Rains 1978).
35 Rice R 1974 Physiology and inheritance of differential growth response under high phosphorus levels among different lines of beans, *Phaseolus vulgaris* L. Ph.D. Thesis. Univ. Wisconsin, Madison. Diss. Abstr. 35, 522 O–B.
36 Ruschel A P, Vose P B, Matsui E, Victoria R L and Tsai Saito S M 1982 Field evaluation of N_2 – fixation and nitrogen utilization by Phaseolus bean varieties determined by ^{15}N isotope dilution. Plant and Soil 65, 397–407.
37 Saouma E 1981 A new food order for Africa. Ceres 14, 22–26.
38 Saric M R 1981 Genetic specificity in relation to plant mineral nutrition. J Plant Nutrition 3, 743.
39 Shea P F, Gabelman W H and Gerloff G C 1967 The inheritance of efficiency in potassium utilization in strains of snap beans, *Phaseolus vulgaris* L. Am. Soc. Hortic. Sci. Proc. 91, 286–293.
40 Shea P F, Gerloff G C and Gabelman W H 1968 Differing efficiencies of potassium utilization in strains of snap beans, *Phaseolus vulgaris* L. Plant and Soil 28, 337–346.
41 Spain J M, Francis C A, Howeler R H and Calvo F 1975 Differential species and varietal tolerance to soil acidity in tropical crop and pastures. *In* Soil Management in Tropical America. Eds. E Bornemiza and A Alvarado. pp 308–329.
42 Spain J M 1977 Field studies on tolerance of plant species and cultivars to acid soil conditions in Colombia. *In* Proc Workshop, Beltsville, 22–23 November 1976, pp 213–222. Cornell Univ. Spec. Pub. (Ithaca, N. Y.)
43 Stangel P J 1977 World fertilizer in relation to future demand. *In* Proc Workshop Beltsville. Ed M J Wright. Cornell Univ. Agric. Exp. Sta. Special Pub., Ithaca (New York) 31–46.
44 Szabolcs I 1971 European Solonetz soils and their reclamation. Akademia Kiado (Budapest).
45 UNIDO 1977 World-wide study of fertilizer industry, 1975–2000. United Nations Industrial Development Organization (Vienna).
46 USDA 1981 Report, FAO in action. Ceres 14, 11.
47 USDA 1981 Agricultural Prices. Statistics Reporting Service, Pub. Crop Reporting Bd. 12, 30.
48 Vose P B 1963 Varietal differences in plant nutrition. Herbage Abstracts 33, 1–13.

49 Vose P B 1967 The concept application and investigation of nutritional variation within crop species. *In* Proc Symp Isotopes in Plant Nutrition and Physiology 1967 IAEA (Vienna).
50 Vose P B 1981 Potential use of induced mutants in crop plant physiology studies. *In* Proc Symp Induced Mutations a Tool in Plant Research. IAEA (Vienna), pp 159–181.
51 Vose P B 1981 Crops for all conditions. New Scientist 89, 688–690.
52 Vose P B 1982 Chapter 4; Effects of genetic factors on nutritional requirements of plants. *In* Contemporary Bases for Crop Breeding, Eds. P B Vose and Stig Blixt. Pergamon Press (Oxford and New York) *In press*.
53 Vose P B 1982 Iron nutrition in plants: a World overview. *In* Proc Int Symp Iron Nutrition and Interaction in Plants. Provo, Utah, 12–14 August 1981, Eds. S D Nelson *et al In press*.
54 Vose P B and Randall D J 1962 Resistance to aluminium and manganese in plants related to variety and cation-exchange capacity. Nature London 196, 85–86.
55 Vose P B and Tulmann Neto A 1982 Varietal differences in cation exchange capacity of *Phaseolus vulgaris. In preparation*.
56 Whiteaker G, Gerloff G C, Gabelman W H and Lindgren D 1976 Intraspecific differences in growth of beans at stress levels of phosphorus. J. Am. Soc Hortic. Sci. 101, 472–475.
57 Wright M J Ed. 1977 Plant adaption to mineral stress in problem soils. Proc. Workshop, Beltsville, Maryland 22–23 November 1976. Cornell Univ. Agric. Exp. Sta. Special Pub., Ithaca (New York).

Low input varieties: definition, ecological requirements and selection

M. DAMBROTH and N. El BASSAM
Intitut für Pflanzenbau und Pflanzenzüchtung der Bundesforschungsanstalt für Landwirtschaft Braunschweig—Völkenrode (FAL), Bundesallee 50, D-3300 Braunschweig, FRG

Key words Breeding Fertilizer Soil Varieties Yield

Summary The growth and yield of plants is determined by their genetic potential and the availability of ecological growth factors (solar energy, water, nutrients, CO_2 etc.). The realization of this potential is closely related to the dominating environmental factors. Appropriate genotypes for a given ecological area have higher yield stability.

The concept of plant breeding for low input varieties (= efficient varieties) considers the ecological constraints and not the gene pools. Local management practices and other facilities should also be taken into consideration. Our aim is not the highest possible yield, regardless of the level of inputs, but suitable genotypes for a given ecological habitat.

The contribution discusses the interactions between physiological – with emphasis on photosynthesis, function and efficiency of plant root systems – environmental and genetic aspects of plant growth. Means to reduce inputs and consequences for plant breeding are also discussed.

Introduction

Primitive prehistoric man could live from the abundance of nature surrounding him, present-day man, with his limited resources, cannot exist without employing his intelligence and directing nature's evolution. Plant science, therefore is not a matter of man's curiosity in trying to understand nature, but an essential requirement for his existence and for the existence of future generations. Genes are ultimately responsible for all physiological, developmental and morphological characteristics of plants and plant breeding sciences are fundamental tools to increase and ensure food production by means of genetic improvement of the respective cultivated crops.

Two decades ago, the introduction of so-called "high yielding varieties" in developing countries inspired big hopes. These varieties are high responsive varieties due to their critical inputs such as fertilizers and water control; supposed to form the basis for what is popularly known as the "green revolution".

About two decades later, we now realize the need for a new orientation in plant breeding. Ecological conditions local management practices and other facilities must be taken into consideration as well as the gene pools. The highest possible yield regardless of the level of inputs is not the aim but appropriate genotypes for a given ecological habitat with higher yield stability must be produced.

Definition

The development of new genotypes or variety requires 6–15 years and therefore, a clear formulation of breeding objectives is essential.

Low input or appropriate varieties may be characterized as improved varieties which might be developed by means of natural selection or breeding. They have been or are adapted for specific ecological and growth conditions. Adaptation means an effective use of growth promoting factors of the local habitat such as light, natural fertility and compatibility with growth inhibiting factors such as drought, salinity or oxygen shortage in soils *etc.*

Adapted varieties and appropriate crop management practices contribute to realize the production potential of any agro-ecological region. They should also help to improve and maintain soil fertility and production stability. The physiological and morphological factors (root system, leaves, height of plants, number of leaves and position, number of stomata, osmotic conditions of drought and disease resistance) responsible for the adaptation are genotypically and environmentally controlled. Land races of plants possess the best requirement for adaptation. They are to be considered as "couple × complex evolutionary populations", including the indigenous landraces as well as the exotic provenances[13].

Ecological requirements

Climatic components

The efficiency of cultivars depends on their genetical yield potential and the realization of this potential is closely related to the dominating environmental factors[2,4]. Rarely is it the only factor limiting agricultural production or is the complete answer to overcoming food supply problems. Fig. 1 summarizes the

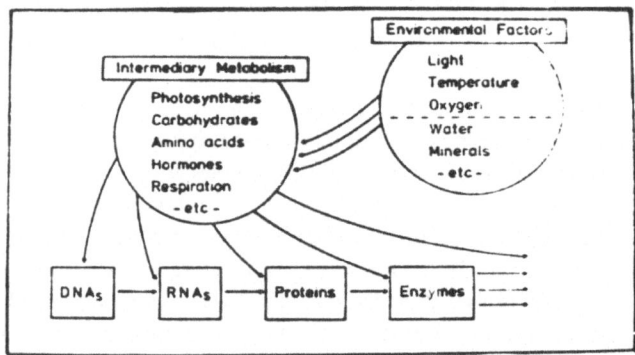

Fig. 1. The influence of environmental factors and intermediary metabolism on the molecular biological pathway.

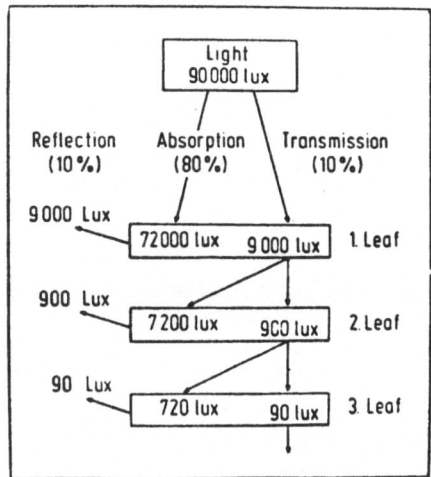

Fig. 2. Utilization of light energy, horizontal leaf position (ref.[17]).

influence of the ecological components (solar energy, oxygen, soil *etc.*) and intermediary metabolism on the molecular biological pathway.

The achievement of the potential photosynthetic efficiency depends on efficient utilization of light energy. Fig. 2 demonstrates that the third leaf in a horizontal leaf-orientation can absorb only 1% of the light energy compared with the first leaf.

Rice varieties with shorter and more erect leaves are more efficient than the long leaves of the tall traditional varieties. This erect-leaf geometry allows light to penetrate deeply into the leaf (Fig. 3) canopy so that even the lower leaves receive sufficient light to carry on photosynthesis and produce the carbohydrates necessary for growth. The winter season in the tropics typically has a higher level of solar radiation than the summer because there are fewer cloudy days. Varieties which can take advantage of this high level of solar radiation are more responsive

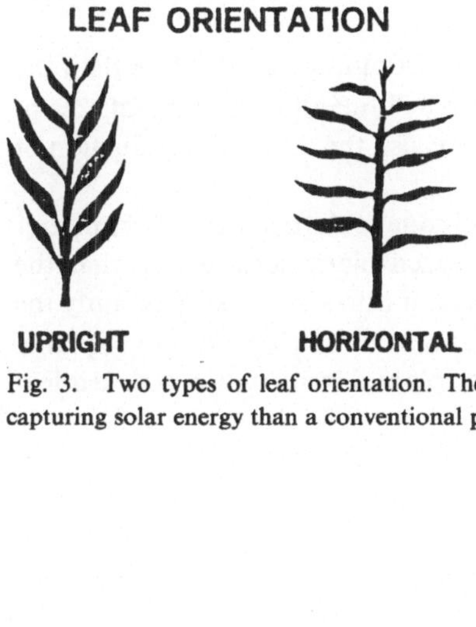

LEAF ORIENTATION

UPRIGHT   **HORIZONTAL**

Fig. 3. Two types of leaf orientation. The upright leaf type plant should be more efficient in capturing solar energy than a conventional plant with dense drooping foliage.

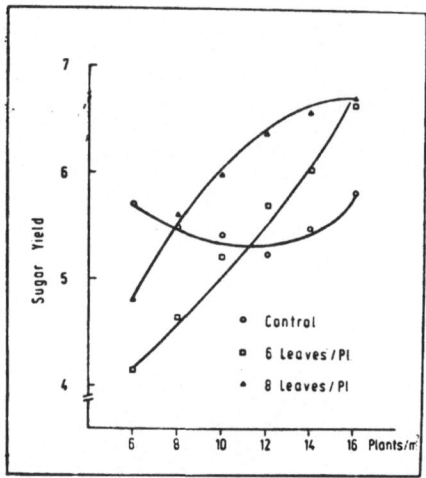

Fig. 4. Sugar yield (t/ha) in relation to crop density and number of leaves.

to other inputs. Even within a given season, date of planting has a sizable effect on yield stability. Early planting allows full advantage to be taken of favourable conditions in order to achieve a complete development of the vegetative cover before entering the grain formation phase. This is an important effective consequence of early planting because full vegetative development means that the plant develops maximum photosynthetic capacity relative to the other production factors. This photosynthetic capacity of the plants is one of the most important elements determining final grain yield[4,5,15].

The relative stagnation of sugar beet yields has probably been caused by the high leaf formation potential of sugar beet. About 70% of the developed leaf apparatus does not achieve its potential photosynthetic efficiency because of mutual shadowing in the crop. Greenhouse and field experiments have proved that sugar beet is able to produce higher sugar yields per hect. with a mechanically reduced leaf apparatus (8 leaves/plant) with the reduction of distance between the rows to 30 cm and with increased crop densities of 100 000 plants per hectare[6].

Sugar beet with 6 and 8 leaves produce more root mass per hectare than the control plants with plant densities of more than 8 pl./m^2 although the leaf area of the single control plants is greater and therefore also the total leaf production of the control plants is greater (Fig. 4).

The root development of the control plants is due to the decrease of dry matter production per m^2 leaf area caused by increased plant densities, so that the assimilates are not translocated into the roots but into the leaves. Certainly the total dry matter production of the control plants is always higher than of plants with a reduced number of leaves but a higher plant density with a reduced leaf number leads to a higher sugar yield.

These results show that a maximum yield achieved in field experiments does not represent a physiological limitation of presently available cultivars but demonstrates only that portion of the genetic potential which is realisable by the optimal utilization of the present means of cultivation.

The high increase of leaf yields did not lead to yield losses of sugar because the number of plants per unit of area has been reduced simultaneously and the stress of competition remained constant. Adaptation of the crop density by reducing the number of plants per ha by about 10 000 each decade is no solution for effective yield increase. An increase can only be achieved by raising the crop density. But this is impossible with the present genotypes.

This dilemma can be solved by an alteration of the appearance of the plants. When only 30% of the leaf area attains the full photosynthetic productivity in a plant stand it is useful to reduce the leaf apparatus to approximately this value in such a manner that two plants share the place of one without increasing the stress of competition in the growing leaf cover. These results show that an increase of the economic yield can be achieved by reducing the yield limiting leaf material of individual plants.

Breeders should select genotypes with few leaf buds which by an adequate plant arrangement per area will quickly build a closed crop cover photosynthetically active for a long time.

Soil and fertilizer

Crop yield, as mentioned above, is determined by (1) the genetic potential of the plant being cultivated; (2) the relative availability of inorganic nutrients, water, CO_2, and light energy during the growing cycle; and (3) the degree of interference from living organisms and physical factors in the production systems. The soil contributes to crop growth by providing the mechanical support for plants and by acting as the medium through which roots obtain mineral nutrients, oxygen and water.

Figure 5 illustrates that, as yields increase, the ability of the soil to supply adequate amounts of nutrients decreases and fertilizer use rises. Observations which lie well above the confidence limits for the curve represent areas of very high native soil fertility (United States, Argentina), very favourable input: output price ratios (Yugoslavia, Japan), or very favourable environmental factors such as solar radiation and irrigation (Egypt). Observations lying below the curve represent cases where the necessary complement of technical-managerial inputs is not adequately supplied (El Salvador) or a single major input (such as irrigation water) is in short supply (Portugal, Israel). However, it should be noted that the fertilizer application rates shown for each country in Figure 5 are obtained by dividing the total use of $N + P_2O_5 + K_2O$ in that country by the total estimated arable hectares in the country. The resulting application rate obtained may differ considerably from the application rate on grain crop land. In some countries, cash or nonfood crops receive heavier fertilizer applications than grain

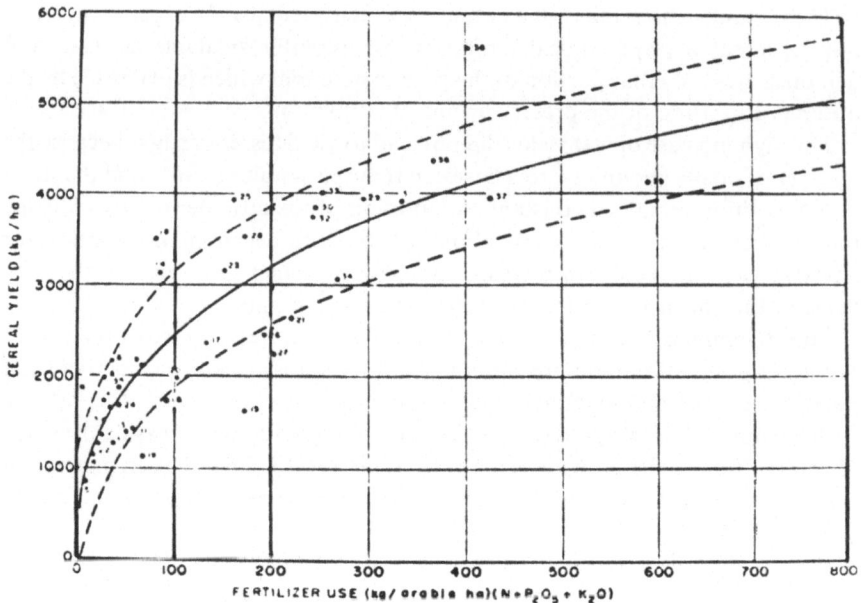

Fig. 3. National use of fertilizer and grain yields, 40 countries, 1972–76 (UN 1981)[15].

1 Argentina	11 Brazil	21 Poland	31 Korea
2 Pakistan	12 Chile	22 Australia	32 Denmark
3 Turkey	13 Southadeica	23 Italy	33 United Kingdom
4 India	14 Yugoslavia	24 Peru	34 Norway
5 Syria	15 El Salvador	25 Egypt	35 China
6 Indonesia	16 Spain	26 Finland	36 Switzerland
7 Philippines	17 Greece	27 Israel	37 Fed. Rep. Germany
8 Canada	18 United States	28 Sweden	38 Japan
9 Colombia	19 Portugal	29 France	39 Belgium/Lux.
10 Mexico	20 Sri Lanka	30 Austria	40 Netherlands

crops. In some countries much or even most of the fertilizer used is not applied to arable cropland but to grassland (pasture or rangeland). In such cases, a deceptively high figure is obtained by dividing the fertilizer use by arable hectares (e.g. Netherlands and Australia)[15].

The highest yield of rye at Timmerlah with high natural productivity was achieved without N-fertilization, and the splitting of application of nitrogen fertilizers did not increase the yield. At Völkenrode it was necessary to add 120 kg N (in 3 rates) in order to obtain the same yield (Table 1), and similar results were obtained with a land race of wheat in Turkey[1] (Fig. 6).

Characteristics, function and efficiency of plant root system

About 30% of the plant genome is involved in growth and development of roots[16]. The potential of the primary production of plant communities cannot be

Table 1. Influence of the location and N-fertilization on grain yield of rye

Time of application kg N/ha					Grain yield t/ha			
					Völkenrode		Timmerlah	
F/G	H	J	M	O	1971	1972	1971	1972
—	—	—	—	—	2.72	2.86	4.69	4.68
40	—	—	—	—	3.68	5.21	4.42	4.44
80	—	—	—	—	4.09	4.79	3.57	3.90
120	—	—	—	—	4.07	4.05	3.22	3.47
160	—	—	—	—	3.90	3.55	3.26	3.18
200	—	—	—	—	3.77	3.49	3.02	2.88
40	—	—	40	—	4.53	5.27	4.30	4.62
40	—	40	40	—	5.02	5.47	4.66	4.23
40	40	40	40	—	4.60	4.90	3.31	4.00
40	40	40	40	40	4.84	4.60	3.25	4.03

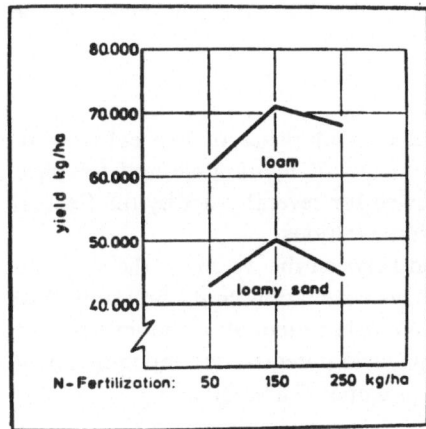

Fig. 6. Influence of soil type and N-fertilization on sugar beet yield, average of 3 years.

fully realized without records of the function, structure and efficiency of the root system under different environmental conditions.

Investigations on genetic variation and control of root development have suffered from distinct paucity of experimental tools due to the natural habitat of the roots.

Efficiency of plant roots

The efficiency of roots mainly depends on their absorptive capacity and

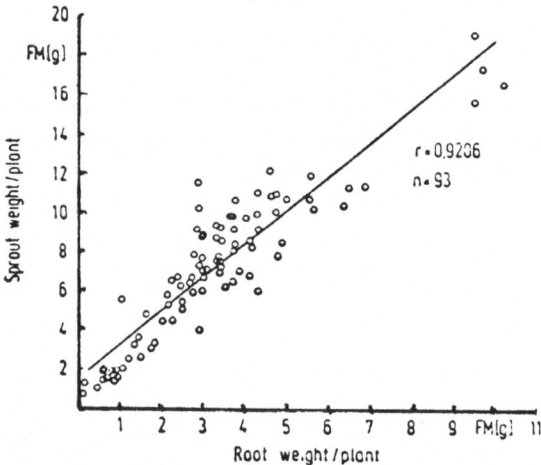

Fig. 7. Correlation between the weight of the root and shoot systems of barley plants of 5 cultivars.

extension and both features are genetically and environmentally dependent. However, several environmental growth factors and cultivation measures may mask or abolish the genetical variability. Specific root efficiency can be determined under controlled conditions.

Extension of roots

Some morphological and phenological characteristics of five cultivars of spring barley have been investigated by means of hydroponic and the yield potential of these cultivars has been followed for several years by the Federal Office for Cultivar Registration under field conditions.

A very close correlation has been found between the weight of the root and shoot systems of all plants (Fig. 7), but no correlation could be established between the root weights, root lengths and yield potential of the investigated cultivars. Plants of cultivars carrying a high yield potential develop more nodal axes than those cultivars of a lower yield potential (Table 2)[8].

Table 2. Yield potential of five varieties of barley and number of nodal axes in a constant environment after four weeks from germination in solution culture

Variety	Yield potential	Number of the nodal roots
Aramir	high–very high	10.2 ± 3.8
Aura	medium–high	5.9 ± 1.9
Union	low	4.7 ± 2.1
Stankas F. G.	very low	4.3 ± 1.1
Hauters S. G.	old variety	2.7 ± 1.9

Water permeability of roots

The routes which nutrient and water follow from the soil to shoots, across the cortex to the vascular stele, thence upwards in the xylem are separately controlled and the ratio in which nutrients and water enter plants can differ greatly from that in the solution external to their roots. In this contribution water permeability will be understood and measured as water absorption by the root system.

A method (Fig. 8) has been developed for *in situ* determination of genotype root efficiency under controlled conditions, *i.e.* soil density, soil water potential and nutrient supply. The roots can develop in soil columns (total length 120 cm) of 4 compartments in plexiglass tubes of 30 cm each. An adjustable constant soil water potential can be maintained during the whole vegetation period by means of ceramic cells which are inserted in the soil of a well defined density. The cells are connected to an automized system for water supply at a given soil water potential.

Effects of soil compaction, nutrient, salt concentration and soil water potential can be examined at different root zones. The rational carrying out of a big number of treatments enables the investigation of several varieties of genotypes.

By means of the above method, the root development and efficiency of 3 varieties of spring barley have been examined. The figures 9–11 demonstrate the daily and cumulative water uptake of the different root zones at different water supply regimes.

The results showed that at lower soil water potential (-0.1 bar) the main

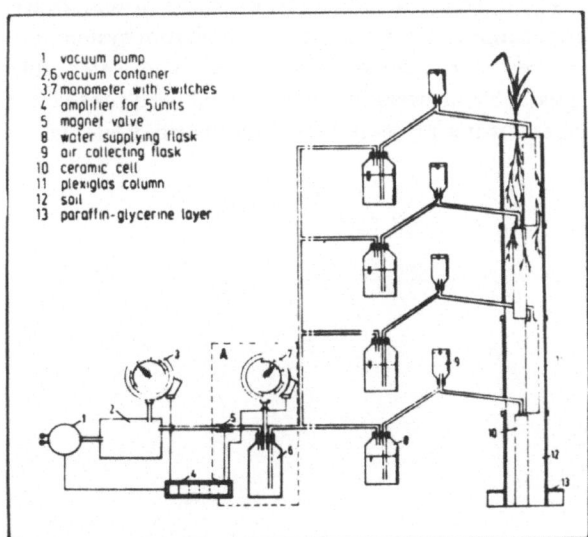

Fig. 8. Installation for root studies.

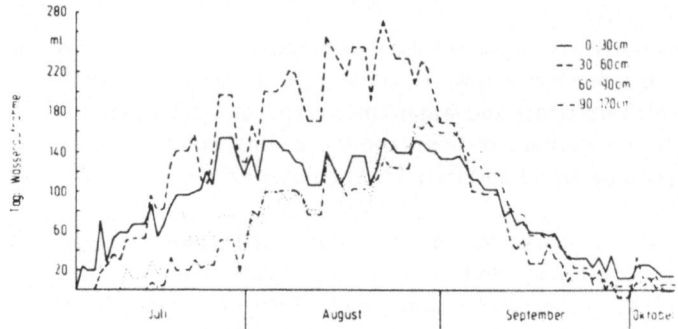

Fig. 9. Daily water uptake (ml) of the 4 root zones of spring barley (variety: Abessinien), soil water potential: −0.1 bar.

active zones of water uptake were the zones 0–30 and 30–60 cm (Fig. 10). At higher levels of potential (−0.35 bar) the most active part was the zone 90–120 cm (Fig. 11). The dry matter of roots of the zone 90–120 cm weighs only half that of the zone 0–30 cm (Fig. 12), El Bassam. The described techniques meet the following requirements:

a—examination of an adequate root length with possibility or differentiation
b—adjustable constant soil water potential
c—adjustable soil density
d—rational carrying out of a big number of treatments which enable the investigations of several varieties of genotypes.

The experimental set-up is suitable for investigation of effects of temporary changes in environmental conditions in a part or on the whole root system, *i.e.* temporary wet conditions in a certain root zone, due to high ground-water table and the set-up may provide valuable answers in models which need root water uptake terms as functions of soil water pressure head/root distribution, *etc.*

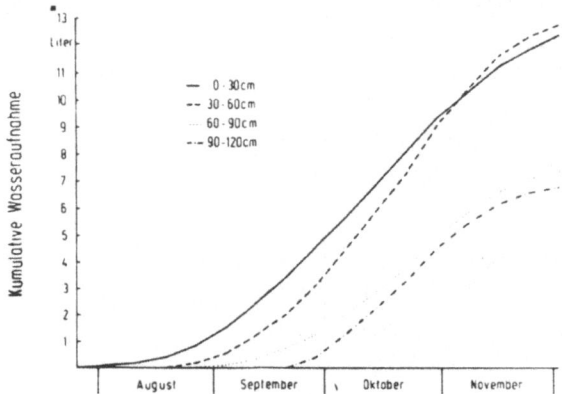

Fig. 10. Cumulative water uptake (litre), variety Aura, soil water potential −0.1 bar.

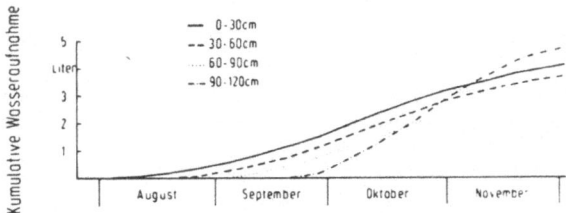

Fig. 11. Cumulative water uptake (litre), variety Aura, soil water potential −0.35 bar.

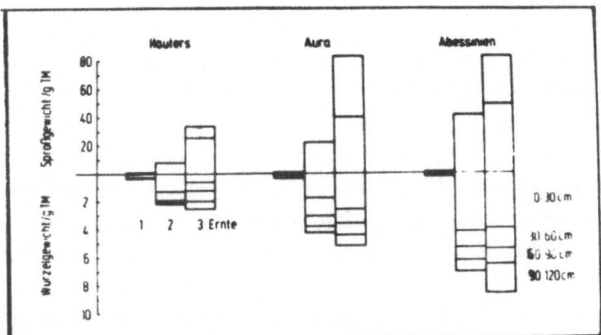

Fig. 12. Weights (g dry matter) of the sprouts (straw and grains) and of the 4 root zones at 3 growth stages.

Fig. 13 demonstrates the phosphorus status in soils after harvesting. The extraction of phosphorus was higher through the old variety Hauters Sommergerste and the land variety from the arid part of Ethiopia than through the "modern" variety Aura.

These are the first results of the experimental approach used to investigate some of the physiological aspects of crops and the interaction between genetic factors, environmental conditions and production.

Consequences for plant breeding and crop management

Overemphasis on a single production input such as high fertilizer rates detracts from the complex and interrelated nature of the agricultural process. The key role of providing adequate inputs which complement fertilizer use could be characterized as the "package approach" or appropriate inputs.

Increased rates of return to traditional agriculture as a whole are obtainable only through the provision of a new or improved package of production technologies including development of crop varieties with the genetic capability to exhibit large responses to natural features of the location. The term location or local conditions does not comprise only soil, climate but also man, technical facilities, infrastructure and relation between prices and costs[2].

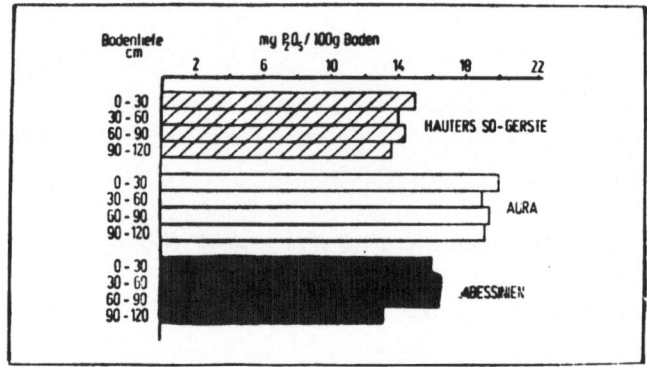

Fig. 13. Phosphorus contents of the soils (lactate soluble mg P_2O_5/100 g soil) in the 4 depths after harvesting of 3 barley varieties.

Breeding programmes should, therefore, consider the ecological and environmental conditions in order to find the appropriate genotype.

Appropriate input or the "package approach" must include the following aspects:

Development of varieties with high transformation efficiency of the local growth conditions (solar energy, nutrients, water, disease resistance),

Conservation of soils and its natural productivity to prevent erosion and salinization,

The optimum utilization of air and soil borne nutrient elements and low inputs of renewable energy.

References

1 Atanasiu N 1977 Der Tropenlandwirt, 71, 12–19.
2 Binsack D 1981 "Low input technology" ein Ansatz? Seminar Low Input Technology, GTZ Eschborn.
3 Bommer D and Dambroth M 1970 Zur Physiologie der Ertragsbildung von Kulturpflanzen der gemäßigten Zone. Potassium symposium 1970. Institut International de la potasse, Berne Suisse 95–106.
4 Dambroth M 1981 Neue Wege im Pflanzenbau beschreiten! DLG-Mitteilungen 18, 978–980.
5 Dambroth M 1979 Konsequenzen "moderner" Sorten und Fruchtfolgen für die Düngung. Kali-Briefe (Büntehof) 14, 447–457.
6 Dambroth M and Bramm A 1980 Consequences of physiological yield aspects concerning the leaf information of sugar beets for breeding. Compte rendu 43e congrès d'hiver, Institut International de Recherches Betteravières. 267–276.
7 Dambroth M and Hondelmann W 1981 Some notes on the collection of land-races of cultivated plants indigenous to the territory of the Federal Republic of Germany. Kulturpflanze XXIX, 41–45.
8 El Bassam N 1981 Genetical variation in efficiency of plant root systems. *In* Structure and

Function of Plant Roots. Eds. R Brouwer et al. Martinus Nijhoff/Dr W Junk Publishers. The Hague/Boston/London. pp 295–299.
9. El Bassam N and Jawad K S 1975 Influence of different N-forms on growth and quality of Mexipak wheat under semi-arid conditions. Landwirtsch. Forsch. 31/I. Sonderheft, 122–126.
10. El Bassam N and Sommer C 1980 Development and efficiency of plant roots at different regimes of water supply. Seminaires sur l'irrigation localisée. III. Influence de l'irrigation localisée sur la morphologie et al physiologie des racines, Sorrento, Italie. pp 127–137.
11. El Bassam N and Sommer C 1980 Ein Methode zur in situ-Ermittlung der Leistungsfähigkeit des Wurzelnetzes von Genotypen. Z. Acker- und Pflanzenbau 149, 391–397.
12. Kramer P J 1969 Plant and Water Relationships. New York: McGraw Hill Book Co.
13. Rohrmoser K 1979 Saatguterzeugung in der Grundbedürfnis-Strategie. Entwicklung+Ländlicher Raum 3/79, 14–16.
14. Russel R S 1977 Plant Root Systems: Their Function and Interaction with the Soil. McGraw-Hill Book Company (UK) Limited.
15. United Nations 1980 Fertilizer Manual, Development and Transfer of Technology. Series No 13. UN, New York 18–31.

Biomass and harvest index as indicators of nitrogen uptake and translocation to the grain in sorghum genotypes

G. ALAGARSWAMY and N. SEETHARAMA
International Crops Research Institute for the Semi-Arid Tropics (ICRISAT), ICRISAT Center, Patancheru, A.P., 502 324, India

Key words Biomass Breeding Genetic variation Grain yield Nitrogen uptake *Sorghum bicolor* Translocation index

Introduction

Nitrogen is usually the most limiting nutrient for crop production and the poor recovery of applied fertilizer nitrogen by crops is of world wide concern. The differential response of sorghum (*Sorghum bicolor* L. Moench) genotypes to applied nitrogen suggests that differences in nitrogen uptake, translocation and accumulation in the grain exist[1]. This paper deals with (i) the extent of variation in sorghum for the above characters, (ii) correlations of these traits with agronomic traits such as days to flower, biomass and grain yield and, (iii) the implications of (i) and (ii) in breeding and crop management.

Materials and methods

Two trials were conducted at the ICRISAT Center on deep Vertisols during the post rainy (winter) season under high fertility (100 kg N, 60 kg P_2O_5/ha) and intensive plant protection with four replications. Trial I contained 40 elite breeding entries and Trial II contained 48 selected germplasm lines. Trial I received a basal application of fertilizer (100 kg N, 60 kg P_2O_5) and no irrigation. Trial II recieved 60 kg N and 60 kg P_2O_5 in the seed bed and 40 kg N 28 days after planting. Irrigation was applied on the 22nd, 42nd, and 63rd day. Each plot consisted of 4 rows of 5 m length; the plant population was 12.1/m^2.

At maturity, 10 plants were randomly harvested for growth analysis. Dry weights of plant parts (culm, leaf, grain and chaff) were determined after drying the samples at 80°C for 24 hours. N concentration was determined on dry samples (ground in a Willey mill) using the standard micro-Kjeldhal method.

Tissue N percentages were multiplied by their dry weights to determine total content per plant part. GPP was computed by multiplying grain N percentage by 6.25. Harvest index (HI: grain wt./total above ground dry wt.) and nitrogen translocaton index (NTI: grain N/total N of the above-ground parts) were expressed as percentages.

Abbreviations GPP, grain protein percentage; HI, harvest index; N, Nitrogen; NTI, nitrogen translocation index; P, probability

Submitted as C.P. No. 71 by the International Crops Research Institute for the Semi-Arid Tropics (ICRISAT).

M.R. Sarić and B.C. Loughman (eds.), Genetic Aspects of Plant Nutrition.
ISBN-13: 978-94-009-6838-7
©*1983 Martinus Nijhoff/Dr W. Junk Publishers, The Hague/Boston/Lancaster.*

Results and discussion

Variations in agronomic and nitrogen physiology traits (Table 1)

There was significant ($P < 0.05$) genotypic variation for all the traits listed in Table 1. The variability was generally larger in germplasm lines (Trial II) than in elite breeding material adapted to the postrainy season (Trial I).

Plants in Trial II were generally late in flowering. This was partly due to a week's delay in sowing during the cooler winter season and partly due to variations in genotypes. Irrigation further delayed maturity and the biomass was substantially higher in Trial II. In spite of lower HI, the grain yield in Trial II was also higher than in Trial I. Both plant and grain N contents were higher in Trial II, but the efficiency of transfer of dry matter or N to the grain (HI and NTI, repectively) was higher in Trial I.

Correlations between traits (Table 2)

The biomass was positively correlated with days to flower and grain yield in both the trials. Biomass was also found to be significantly and negatively correlated with HI in Trial I; this was probably due to water stress during the grain filling stage.

GPP was negatively correlated with HI, indicating that a proportionally greater transfer of dry matter occurs than N, resulting in its dilution. Kramer[4] has shown that the measured genetic variability for GPP is less after adjusting for HI.

Both total plant N and the N in the grain were strongly correlated with total biomass and grain yield. They were usually correlated with all those traits with

Table 1. Phenotypic variability in agronomic and nitrogen physiology traits in sorghum

	Trial I				Trial II			
	Mean	Standard deviation	Range	Check CSH1	Mean	Standard deviation	Range	Check CSH1
Grain yield (g/plant)	30.2	10.9	54.3–9.0	38.2	43.3	16.30	95.1–21.6	62.3
Biomass (g/plant)	68.0	22.5	132.0–31.2	69.2	119.2	53.7	278.7–50.5	118.7
HI (%)	44.7	9.2	60.7–20.9	55.2	38.5	9.1	53.9–13.7	52.2
Days to flowering	69.0	6.5	95.0–57.0	68.0	75.3	10.1	99.0–61.0	73.0
Protein % in grain	9.86	0.86	11.631–7.988	8.463	10.44	1.63	15.63–6.25	10.56
Total N in plant (g)	0.628	0.203	1.140–0.224	0.639	1.13	0.39	2.3–0.6	1.45
Total N in grain (g)	0.471	0.164	0.826–0.159	0.517	0.71	0.26	1.56–0.36	1.00
NTI (%)	74.6	7.9	85.7–54.8	80.9	63.1	8.4	79.5–34.0	69.4

Table 2. Phenotypic correlation coefficients between agronomic and nitrogen physiology traits in sorghum. (The upper row values refer to Trial I; values for Trial II are found in parentheses)

	Grain yield	Biomass	HI	Days to flowering	Grain protein %	Total N/ plant	Total N/ grain
Biomass	0.755 ***† (0.701)***‡						
HI	0.520 *** (0.202)	−0.109 (−0.504)***					
Days to flowering	0.048 (0.339)*	0.475 ** (0.743)***	−0.414 * (0.656)***				
Grain protein %	−0.327 * (−0.237)	−0.096 (0.147)	−0.355 * (−0.377)**	0.112 (−0.026)			
Total N/plant	0.889 *** (0.812)***	0.870 *** (0.929)***	0.229 (−0.269)	0.245 (0.596)***	0.031 (0.229)		
Total N/grain	0.969*** (0.917)***	0.781 *** (0.761)***	0.452 ** (0.066)	0.082 (0.325)*	−0.104 (0.159)	0.956 *** (0.916)***	
NTI	0.510 ** (0.340)*	0.007 (0.237)	0.837 *** (0.762)***	−0.303 (−0.597)***	−0.425 ** (−0.018)	0.155 (−0.023)	0.422 ** (0.159)

† Corresponds to Trial I (n = 40)
‡ Corresponds to Trial II (n = 48)

* $P < 0.05$
** $P < 0.01$
*** $P < 0.001$

which biomass and grain yields were correlated. This shows the greater importance of increasing dry weight than of increasing N concentration itself, for harvesting more nitrogen in the grain. Thus, for any further increased grain protein content, after maintaining high HI, one should increase the biomass productivity, which will be influenced by crop growth duration. Maintenance of green-leaf area for a longer duration and the continued uptake of N during grain filling[5] should be given due consideration for increasing GPP at high yields.

Nitrogen uptake and metabolism are energy dependent and hence greater photosynthetic activity is required to support higher uptake[2]. Plants flowering late have larger leaf area[6] and possibly deeper roots. Such plants are usually large in size and have greater capacity for photosynthesis, N uptake and assimilation. Preliminary results show that in general the genotypes capable of promoting nitrogen fixation are late maturing and large in size (S. P. Wani, ICRISAT, personal communication). In such plants the correlation between GPP and grain yield is likely to be insignificant (Trial II). Late maturing plants give more stover (fodder) with same grain yield than early genotypes[6]. Tucker and Bennett[7] have reported that short-season sorghum hybrids have a shorter critical period (6–7th leaf to boot) for N uptake than long season hybrids, and hence demand a higher level of N during this period. They are also likely to lose much N to weeds since their ability to compete with the latter is less (as the canopy is less competitive) than long-season cultivars. However, where irrigation is not available or the season is short, selection for early flowering and high HI is the best strategy for harvesting maximum grain and protein.

Breeding for nitrogen uptake and translocation

In both the trials the total N in the plant and NTI were strongly and positively correlated with total biomass and HI. This suggested that selection for high biomass and HI is sufficient to ensure high N uptake and translocation. Similarly, F_3 progenies selected for high biomass and HI also have high N content and NTI[3].

Exceptions to the general trends do exist however. From re-evaluation of the

Table 3 Variability in biomass, grain yield, nitrogen uptake, and translocation in selected sorghum genotypes

	P-721	IS 858	IS 2223	Diallel 642	IS 6380
Biomass (g/plant)	60.6	79.1	53.6	61.5	70.2
Grain yield (g/plant)	18.3	27.2	21.8	24.1	15.1
HI (%)	39.6	36.6	41.5	33.7	26.7
Total N (g/plant)	0.78	0.77	0.48	0.56	0.67
NTI (%)	49.2	52.1	57.4	40.3	34.6

original study it was possible to identify genotypes with approximately the same biomass and HI which varied considerably in nutrient uptake and transfer to the grain. For example, P-721 and Diallel 642 both have similar dry weights (60.6 and 61.5 g/plant respectively, Table 3), but P-721 takes up 39 per cent more N than Diallel 642 (0.78 and 0.56 g N/plant, respectively) and has nine per cent higher NTI. Such individual differences are usually masked when the data for the whole set of heterogenous genotypes are analysed.

In order to determine whether such differences are significant in a practical breeding program, we made crosses between contrasting parents and selected F_2 plants for a range of dry weights per plant and HIs. In F_3 progenies, estimates of dry weight, grain yield, and N in the grain and whole plant were made[3]. The biomass and total nutrient taken up by F_3 progenies in each family were again strongly correlated [*e.g.*, in progenies of IS 2223 × IS 6380 cross the correlation coefficient (r) between dry weight and total nitrogen was 0.91, $P<0.01$]. We observed similar relationships between HI and NTI (r = 0.88, $P<0.01$). Therefore, we conclude that selection for biomass and HI also effectively includes selection for traits concerned with nutrient efficiency.

Hence the routine screening for these traits is unnecessary; only the parents used in the crosses or the material in the final stages of breeding program need to be analyzed for N to confirm that they have more than average N uptake and translocation efficiency.

Acknowledgements We are grateful to H. Doggett for guidance, K. L. Sharawat for N analysis and our colleagues for comments on the manuscript.

References

1 Alagarswamy G 1977 Nitrogen uptake and transfer to the grain. *In* International Sorghum Workshop, Hyderabad. Proc. 6–13 March 1977, ICRISAT, Patancheru, A.P., 502 324, India.
2 Bhatia C R and Rabson R 1976 Bioenergetic consideration in cereal breeding for protein improvement. Science 194, 1418–1421.
3 ICRISAT (International Crops Research Institue for the Semi-Arid Tropics), 1981. Annual Report 1979–80. pp 28–29. Patancheru, A.P., 502 324, India.
4 Kramer T H 1979 Environmental and genetic variation for protein content in winter wheat (*Triticum aestivum* L.). Euphytica 28, 209–218.
5 Rao N G P and Venkateswarlu J 1974 Genetic analysis of some exotic × Indian crosses in sorghum. III. Heterosis in relation to dry matter production and nutrient uptake. Indian J. Genetics and Pl. Breed. 31, 156–176.
6 Seetharama N 1977 Variability in sorghum growth stages. *In* International Sorghum Workshop, Hyderabad. Proc. 6–13 March 1977. ICRISAT, Patancheru, A.P., 502 324, India.
7 Tucker B B and Bennett W F 1968 Fertilizer use on grain sorghum. *In* Changing Pattern in Fertilizer Use. Eds. L B Nelson, M H McVickar, R D Munson, L F Seatz, S L Tisdah and W C White. pp 189–220. Soil Sci. Soc. of America Inc. Madison USA.

Genetic basis of mineral nutrition in *Triticum aestivum*
II. Effect of the cytoplasm on the absorption of nutrient elements

B. BOCHEV, E. NEIKOVA-BOCHEVA, G. GANEVA and T. FROLOZHKI
Institute of Genetics, BAS—Sofia and Institute of Soil Science and Yield Programming "N. Poushkarov"—Sofia, Bulgaria

Key words   Aneuploid   Cytoplasm   Genotype   Nutrition   *Triticum aestivum*

Introduction

This study represents only a part of a large scale programme concerning the genetic basis of the mineral nutrition of wheat. The complete series of different aneuploid conditions developed recently along each chromsome of cv. Chinese spring[6], as well as a large set of alloplasmatic lines of the same cultivar with the participation of cytoplasms from the basic Aegilops and Triticum species[7,8] are an important pre-condition for a principally new approach in the investigations concerning the genetic basis of mineral nutrition in *T. aestivum*.

The first goal of our investigations into that problem was to assess the effect of individual chromsomes on the assimilability of the basic nutrient elements in *T. aestivum*[1].

The aim of the present study was to establish the effect of the different Aegilops and Triticum cytoplasms in alloplasmatic lines of cv. Chinese spring (*T. aestivum*) on the assimilability of the basic mineral nutrient elements and their participation in the formation of reproductive and vegetative organs in view of their practical use in wheat genetics and breeding.

Materials and methods

A pot experiment was carried out including alloplasmatic lines of Chinese spring kindly presented to us by Prof. Dr. Tsunewaki with donor cytoplasm from various Aegilops and Triticum species (Table 1), which belongs to the D, S, S', G and M* plasma types. The soil used in the experiment was leached chernozem, the level of phosphorus was optimal corresponding to 0.3 mg/l P in 0.01 M CaCl$_2$ and with 300 mg of N and K$_2$O per kg of soil, applied as monocalcium phosphate, ammonium nitrate and potassium sulfate respectively.

After the plants had been harvested the total quantity of dry matter was determined, as well as the content of nitrogen after wet ashing of phosphorus by dry ashing and colorimetrically by the vanadat method, and of K, Ca, Mg, Zn, Mn, Cu and Fe by atomic absorption. The analytical data of the assimilated quantities of the individual elements in the grain and the straw of the experimental plants

* The cytoplasm type is designated after Maan[4] and Tsunewaki *et al.*[8].

Table 1. Plant material

N	Plasma type	Donor cytoplasm	Genome	N	Plasma type	Donor cytoplasm	Genome
C-04	D	Ae. squarrosa	D	C-22	S	T. dicoccum var. Vernal	AB
C-28	D	Ae. cylindrica	CD	C-33	S	Ae. kotschyi	C^uS^v
C-36	D	Ae. ventricosa	DM^v	C-34	S	Ae. variabilis	C^uS^v
C-08	S,G	Ae. speltoides	S=B	C-10	S'	Ae. sharonensis	S'
C-25	G	T. timopheevi	AG	C-05	M	Ae. comosa	M
C-21	S	T. dicoccoides v. spontanem	AB	C-52	S	T. aestivum var. Chinese spring	ABD

were recalculated in scale units, corresponding to one-fifth of the total variation of all alloplasmatic lines and the controls. The effect shown by the action of the individual cytoplasms on the assimilation of nutrient element in the grain and straw was determined by scale differences from the control (c) cv. Chinese spring (Cs) in its own cytoplasm.

Results and discussion

Data concerning the percentage content of separate mineral nutrient elements in the grain and in the straw, expressed in scale differences from the control, are presented in Figs. 1 to 8.

It is evident that the cytoplasms of different alloplasmatic Cs lines have a specific effect of separate nutrient elements in the grain and straw. Data in the figures show that positive changes produced by the effect of various cytoplasms are more pronounced in the grain than in the straw. The cytoplasms of the *Aegilops* species *squarrosa* and *speltoides* contribute to the increase of the percentage content of more nutrient elements in the grain. In this respect the cytoplasms of *Ae. cylindrica*, *Ae. kotschyi* and *Ae. ventricosa* have a lower effect. Nitrogen content in the grain increases under the influence of *T. dicoccoides*, *T. dicoccum*, *T. timopheevi*, *Ae. kotschyi* and *Ae. variabilis* cytoplasms. The cytoplasms of *Ae. squarrosa*, *Ae. speltoides* and *Ae. sharonensis* have a positive effect on the phosphorus content in the grain, while potassium is positively influenced not only by these cytoplasms, but also by the cytoplasms of *Ae. comosa* and *Ae. variabilis*. Potassium in the grain increases only under the effect of *T. dicoccoides* cytoplasm. *Ae. squarrosa* and *Ae. speltoides* cytoplasms contribute to the increase of Mg and Zn in the grain. A similar effect, but less expressed in respect to Mg is also produced by the cytoplasms of *Ae. sharonensis* and *T. dicoccum*.

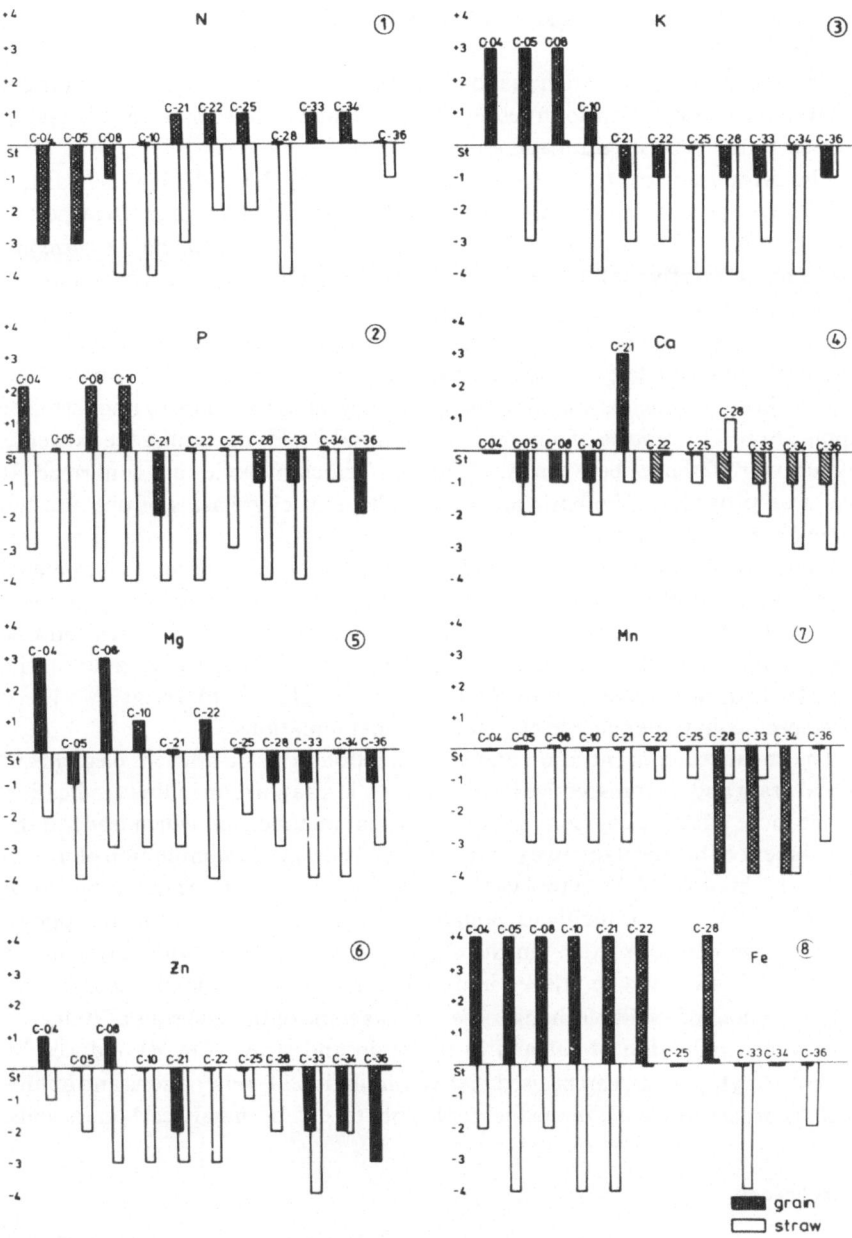

Figs. 1–8. Relationships between the different cytoplasms and nutrients.

Cytoplasms have the highest positive effect on the concentration of iron (Fig. 8). Most of them exert a negative influence on the content of various elements in the straw.

The specific effect of individual cytoplasms on the utilization of the nutrient elements is manifested also in respect to the uptake of nutrient elements and in respect to the yield of grain and straw.

The cytoplasms of *Ae. ventricosa* and *Ae. kotschyi* have the highest positive effect on the accumulation of three basic macroelements in the grain. In respect to nitrogen and potassium uptake the cytoplasms of *Ae. cylindrica* and *Ae. variabilis* also exert a positive influence, as do the cytoplasms of *Ae. squarrosa* and *Ae. comosa* in respect to phosphorus and potassium.

The results obtained prove the specific effect of various cytoplasms on the concentration and total uptake of nutrient elements and are of pronounced interest both to genetics and plant breeding, and to agrochemistry and fertilizer application. The reports of a number of authors[2,3,4,5,7,8] concerning the existence of genetic differences between Aegilops and Triticum species are confirmed. It has been proved, on the basis of these results, that cytoplasms of one and the same plasma type, originating from various Aegilops and Triticum species, exert different influences on the uptake and participation of different nutrient elements in the formation of wheat reproductive and vegetative organs. A modifying effect of the cytoplasm on the genes and the gene complexes is assessed, which controls the assimilation of the nutrient elements. This fact is of great significance for the transfer of desired genetic information from its relatives to wheat by way of interspecific and intergeneric chromosome manipulations.

The enhanced concentration and accumulation of basic nutrient elements in wheat grain and their respective lower quantity in straw are an indirect indication of the more effective course of nutrient element metabolism, which leads to the enrichment of the reproductive organs. The considerable accumulation of iron in the grain, in most of the cytoplasms tested, can be of use in the development of suitable genotypes for highly carbonate soils, where conditions for the appearance of iron chlorosis exist. On the other side, the reduced Mn content in the grain and straw could be utilized in developing genotypes tolerant to soils with higher content of available manganese, such as some of the acid soils of Bulgaria. The typical reduction of calcium in the grain and straw (Fig. 4), induced by almost all cytoplasms, can be used in the assimilation of some cations and anions such as potassium, zinc, manganese, phosphate, etc. on highly carbonate soils.

Conclusions

The results of these investigations show that the cytoplasms of the different Aegilops and Triticum cytoplasms in the alloplasmatic lines of cv. Chinese spring have a modifying influence on the summed up effect of the genes controlling the uptake of basic nutrient elements.

The highest positive effect on the assimilability of nutrient elements was exerted by cytoplasms originating from *T. dicoccum, Ae. kotschyi, Ae. ventricosa* and *Ae. cylindrica*. These cytoplasms should be given priority in interspecific and intergeneric chromosome manipulations, aimed at development and reconstruction of desired genotypes of the existing wheat cultivars.

References

1 Bochev B, Neykova E, Ganeva G, Mincheva M and Dimitrova F 1980 Genetic bases of mineral nutrition of *T. aestivum* L. Plant Physiol. t.V. 17–27.
2 Fukasawa H 1959 Nucleus substitution and restoration by means of successive backcrosses in wheat and its related genus Aegilops. Japan. J. Bot. 17, 55–91.
3 Kihara H 1951 Substitution of nucleus and its effects on genome manifestations. Cytologia 16, 177–193.
4 Maan S 1973 Cytoplasmic and cytogenetic relationships among tetraploid species. Euphytica 22, 287–300.
5 Panajotov I and Gotsov K 1975 Results of nucleus substitution in Aegilops and Triticum species by means of successive backcrosses with common wheat. Wheat Inf. Serv. 40, 20–22.
6 Sears E 1954 The aneuploids of common wheat. Missouri Agric. Exp. Sta. Res. Bull. 572, 59.
7 Tsunewaki K 1973 Genetic differentiation of the cytoplasms in wheat and Aegilops (Abstr.). Genetics 74, 280–281.
8 Tsunewaki K, Mukai T, Endo T Ryo, Tsuji S and Murata M 1976 Genetic diversity of the cytoplasm in Triticum and Aegilops. V. Classification of 23 cytoplasm into eight plama types. Japan. J. Genet. 51 (3), 175–191.

Genetic variability of the efficiency of nutrient utilization by maize (*Zea mais* L.)

G. CACCO, G. FERRARI and M. SACCOMANI
Institute of Agricultural Chemistry, University of Padua, Italy

Key words Efficiency Hybrids Lines Maize Nitrate Sulphate

Introduction

Mineral nutrients are transformed by crops into biomass through several sequences of chemical reactions starting from each nutrient ion. The coordination of the transformation rates within each sequence and among the various sequences has been recognized to be a main factor of crop productivity[5]. Therefore the genetic control of this coordination is related to that of the efficiency of nutrient utilization by plants. Breeding for more efficient use of nutrients has been carried out mainly on the basis of a direct evaluation of the yields of different genotypes in conditions of equal nutrient availability. However nutrient utilization can be split into an uptake and an assimilaton step: both have been recognized to be metabolic steps mediated by carrier and enzyme proteins respectively. Since proteins are involved in both steps, these are under genetic control and the problem arises whether the regulation of the uptake is coordinated with that of the assimilation step. The two steps can be considered as consecutive reactions, and the total efficiency is controlled by the lower rate step. The objectives of our study were to answer the following questions: (I) what is the range of genetic variability for the efficiency of uptake and for that of the assimilation step?; (II) in the maize genotypes obtained by current breeding practices is the uptake efficiency equivalent to that of assimilation?; (III) do the modifications of the genome (crossing, ploidy changes, mutations) affect the two steps independently or coordinately?; (IV) do the uptake and the assimilation steps give different responses to changes in environmental conditions, namely in nutrient availability?

Materials and methods

The seeds of inbred and hybrid genotypes were obtained from the Maize Experimental Station, Bergamo. Multiplication of the inbreds and assay of performance of diallelic crosses were carried out partly at the same Bergamo Station and partly at the De Kalb Center, Chiarano. The procedures used to obtain excised roots and to evaluate sulfate uptake by excised or by whole plant roots have been described[8]. In the case of NO_3^- the absorption rate was evaluated by a depletion procedure[6].

Results

Ranges of genetic variability of uptake and assimilation activities

In the case of sulfate the comparison between the uptake efficiency of excised roots and the activity of the key-enzyme of sulfate assimilation ATP-sulfurylase (Table 1) demonstrated that the range of variability within a large number of maize genotypes was wider for the uptake (0.174 to 0.372) than for the enzyme activity (132 to 160). Also for nitrate the uptake step showed a wider range (0.192 to 0.862) of variability with respect to the nitrate reductase activity (0.35 to 0.79) (Table 2).

Comparative efficiency of the uptake and assimilation steps

Sulfate uptake was tested in a group of maize hybrids, using both excised and whole plant roots (Table 3). In excised roots the sulfate cumulative uptake varied with the genotype and was positively correlated with grain yield (Cacco et al.[4].) The uptake by whole plant roots was positively correlated with that of excised roots, but the variability was less and the correlation with grain yield showed a higher level of significance in the former. The uptake step was affected by the assimilation step to different extents within maize hybrids, demonstrating that the relative efficiency of the two steps was under genetic control. ATP-sulfurylase activity and the root to shoot translocation rate were not correlated with grain yield. Therefore the genetic control of the regulatory assessment between the assimilation and uptake steps of sulfate should be improved, in order to overcome the inadequacy of the uptake efficiency.

Effects of genome modifications on the relative efficiency of the uptake and assimilation steps

In the case of crossing within a diallel set of F_1 hybrids a close relation

Table 1. Efficiency of sulfate uptake and ATP-sulfurylase in maize genotypes

Hybrids		SO_4^{2-} pool size in excised roots μeq $\cdot$ mg^{-1} protein	ATP sulfurylase nmol Pi $\cdot$ min^{-1} $\cdot$ mg^{-1} protein
XL 22	a	0.252±0.0390	160± 15.9
	b	0.174±0.0266	144± 16.1
	c	0.186±0.0245	139± 9.3
XL 41	a	0.252±0.0277	156± 4.1
	b	0.240±0.0432	148± 5.2
	c	0.222±0.0328	136± 10.2
XL 48	a	0.300±0.0381	146± 11.1
	b	0.318±0.0311	148± 4.7
	c	0.372±0.0624	132± 1.5

Table 2. Efficiency of nitrate uptake and nitrate reductase in maize genotypes

Maize line	NO_3^- uptake	Nitrate reductase activity (μmol $\cdot$ h^{-1} $\cdot$ mg^{-1} protein)
33–16	0.192±0.023	0.63±0.05
L 1058	0.317±0.024	0.41±0.04
WF 9	0.315±0.015	0.47±0.04
B 14	0.415±0.025	0.35±0.08
33–16 × L 1058	0.851±0.027	0.41±0.09
33–16 × WF 9	0.502±0.012	0.79±0.06
33–16 × B 14	0.412±0.022	0.35±0.03
L 1058 × WF	0.862±0.020	0.44±0.07
L 1058 × B 14	0.755±0.028	0.42±0.07
WF 9 × B 14	0.380±0.032	0.37±0.09

appeared between the uptake and the assimilation steps of sulfate. The level of heterosis for sulfate uptake varied concordantly with that of ATP-sulfurylase[2]. This was not valid for nitrate uptake and nitrate reductase[7]. Within the set of hybrids obtained by breeding during the last 50 years changes in the sulfate uptake capacity were accompanied by parallel changes in the ATP-sulfurylase activity with few exceptions. Within another set of hybrids the sulfate uptake efficiency was not correlated with the level of ATP-sulfurylase activity. When modifications in the ploidy level were considered, the response of sulfate and potassium uptake was characterized by enhancement of its efficiency at

Table 3. Sulfate uptake by excised and by whole plant roots of different maize genotypes

Hybrids		Grain yield 100 kg $\cdot$ ha^{-1}	SO_4^{2-} cumulative uptake in excised roots μeq $\cdot$ mg^{-1} protein	SO_4^{2-} uptake rate by roots of intact plants μeq $\cdot$ min^{-1} $\cdot$ mg^{-1} protein
XL 22	a	109.5±5.25	0.252±0.0390	0.83±0.012
	b	109.0±6.32	0.174±0.0266	0.83±0.020
	c	100.8±2.92	0.186±0.0245	0.71±0.068
XL 41	a	110.6±6.85	0.252±0.0277	0.88±0.037
	b	102.0±3.16	0.240±0.0432	0.75±0.056
	c	103.0±2.59	0.222±0.0328	0.78±0.074
XL 48	a	120.0±4.92	0.300±0.0381	1.09±0.053
	b	117.1±3.04	0.318±0.0311	1.03±0.089
	c	116.5±4.54	0.372±0.0624	0.86±0.094

Table 4. Changes in the efficiency of sulfate uptake (μeq · min^{-1} · mg^{-1} protein) and assimilation on steps in response to SO_4^{2-} + NO_3^- shortage. Enzyme activities in nmol · min^{-1} · mg^{-1} protein.

Nutrient condition	SO_4^{2-} uptake rate		ATP-sulfurylase		α-Acetylserine-sulfhydrylase		Nitrate reductase	
	XL1	XL2	XL1	XL2	XL1	XL2	XL1	XL2
Control	4.2	2.9	207	175	57.3	53.2	5.43	8.15
SO_4^{2-} + NO_3^- shortage	4.7	2.7	178	158	115	80.8	1.42	2.63

increasing ploidy while that of the metabolic steps was not correlated with the ploidy level[1]. However, maize is not actually involved in this kind of modification of the genome. Mutations demonstrated different effects on the relative efficiency of the uptake and assimilation. In the case of the opaque-2 maize mutant, the two activities were concordantly modified with respect of the wild genotype[4]. On the contrary the Ht mutation (*Helminthosporium turcicum* resistance) affected the sulfate uptake, while the two assimilation steps mediated by ATP-sulfurylase and α-acetylserinesulfhydrylase were unchanged[3]. We must conclude that genome modifications can affect the relative efficiency of the uptake and assimilation steps in a different way, depending on the kind of changes of the genome and on the genetic background of the material subjected to the genome modifications.

Responses of the uptake and assimilation steps of sulfate to nutrient shortage

In two maize genotypes (XL 1, XL 2), submitted to SO_4^{2-} and NO_3^- shortage, the responses of the uptake and assimilation steps were as follows (Table 4). Both hybrids showed no significant change in the uptake efficiency and in ATP sulfurylase activity while α-acetylserinesulfhydrylase was enhanced and nitrate reductase was depressed. Therefore the responses of the uptake and assimilation steps to nutrient shortage were different. The genetic control of these responses should be taken into account in breeding practices.

References

1 Cacco G, Ferrari G and Lucci G C 1976 Uptake efficiency of roots in plants at different ploidy levels. J. Agric. Sci. Camb. 87, 585–589.
2 Cacco G, Ferrari G and Saccomani M 1978 Variability and inheritance of sulfate uptake efficiency and ATP-sulfurylase activity in maize. Crop Sci. 17, 503–505.
3 Cacco G, Ferrari G and Saccomani M 1979 Coordination of sulfate uptake, ATP-sulfurylase and OASS activities in maize roots as affected by Ht mutation. Maydica 24, 247–254.
4 Cacco G, Saccomani M and Ferrari G 1977 Development of sulfate uptake capacity and ATP-sulfurylase activity during root elongation in maize. Plant Physiol. 60, 582–584.

5 Hageman R H, Lang E R and Dudley J E 1967 A biochemical approach to corn breeding. Advances in Agronomy 19, 45–86.
6 Neyra C A and Hageman R H 1975 Nitrate uptake and induction of nitratereductase in excised corn roots. Plant Physiol. 56, 692–695.
7 Saccomani M and Cacco G 1980 Genetic regulation of the nitraterductase and nitrate uptake activities in maize. Riv. di Agron. 14, 228–230.
8 Saccomani M, Cacco G and Ferrari G 1981 Efficiency of the first steps of sulfate utilization by maize hybrids in relation to their productivity. Physiol. Plantarum 53, 101–107.

Genetic differences in phosphorus absorption among white clover populations

J. R. CARADUS*
Agricultural Botany Department, Reading University, Reading RG6 2AS, UK

Key words Clover Phosphorus Populations Root Shoot *Trifolium repens* White clover

Summary Eight semi-natural white clover populations and two cultivars were grown in culture solutions containing 10 ppm and 0.01 ppm phosphorus (P). The rate of P uptake by the intact plants was then measured in solutions containing 10 ppm P.

Phosphorus uptake per unit root length was twice as great by plants previously grown at 0.01 ppm P than those grown at 10 ppm P. Large differences in total P uptake were found among populations regardless of the pretreatment; most of this variation was accounted for by differences in root length. Only small differences were found between populations for P uptake per unit root length, and then only after pretreatment with 10 ppm P; this variation was largely accounted for by relative growth rate and shoot %P.

Introduction

White clover is an important component of pastures in temperate regions, because of its ability to fix nitrogen and produce a high quality feed. However, the content of clover in swards is very dependent upon an adequate phosphate content in the soil. In many regions this can only be achieved by frequent application of superphosphate. The high cost of this treatment has therefore led to attempts to maintain clover at lower phosphorus (P) contents in the soil. One attempt has been to breed white clovers more tolerant to low P soils, and yet still maintain productivity[2,3]. This approach has met with limited success; populations with small leaves and dense stolon production, which partition a large proportion of P into stolon and root production[4], have performed better under intense grazing on low fertility hill country in New Zealand, than larger leaved forms, though the latter are more productive on higher fertility lowland soils[13,14].

This study was designed to determine whether natural populations of white clover, collected from a wide range of soils, differ in their ability to absorb P, to determine what morphological and physiological characteristics were associated with such differences, and whether such differences were related to the ecological origins of the populations.

* Permanent address: Grasslands Division, DSIR, Palmerston North, New Zealand.

M.R. Sarić and B.C. Loughman (eds.), Genetic Aspects of Plant Nutrition.
ISBN-13: 978-94-009-6838-7
©1983 Martinus Nijhoff/Dr W. Junk Publishers, The Hague/Boston/Lancaster.

Methods

Eight semi-natural populations of white clover and two cultivars (Table 1) were selected for study. Pre-rooted stolon tips from eight genotypes of each population were grown in a growth cabinet with 14 h days (25°C) and 10 h nights (15°C); the light intensity was 220 micro-einsteins $m^{-2} sec^{-1}$.

Four plants were grown in each pot (175 ml). Two P concentrations (0.01 and 10 ppm P) were applied. All other nutrients were added in adequate amounts, and pH adjusted to 5.5. Nutrient solutions were renewed weekly.

After 24 days, the plants were transferred to nutrient solution containing 10 ppm P, and the P content of the solution determined at 3 h intervals for 15 h, using the method of Eibel and Lands[6]. Plants were then harvested and measurements made of root length[7], root and shoot dry weight, and volume of nutrient solution remaining. Root and shoot P contents (%) were also determined.

Results

The populations differed in relative growth rate (RGR) at both high and low P levels ($p<0.001$ and $p<0.05$ respectively) (Fig 1). There were significant ($p<0.001$) differences among populations for root length and root/shoot ratio after both pretreatments (Table 1). Populations differed in shoot %P and root %P ($p<0.001$ and $p<0.05$ respectively) only after the low P pretreatment (Table 2).

The populations differed in response to P concentration ($p<0.01$) when RGR was considered (Fig. 1). However there was no significant relationship ($r = -0.56$ ns) between the response of populations to P and the P status of their soil origin.

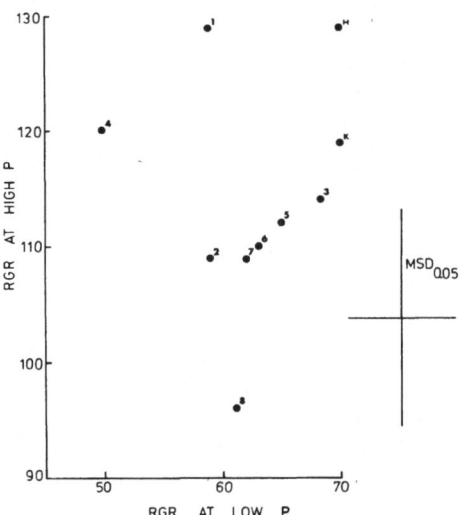

Fig. 1. The relative growth rate (RGR) ($mg\ g^{-1}\ FW\ day^{-1}$) of eight semi-natural populations and two cultivars of white clover when grown at high (10 ppm) and low (0.01 ppm) concentrations of phosphorus.

Two populations from low-P soils (1 and 4) showed the greatest response to P, while a population from high-P soil (8) showed the least response (Figure 1).

Populations differed significantly ($p<0.001$) in total uptake of P (Table 2), regardless of whether they had been pretreated with 10 ppm P or 0.01 ppm P. Multiple regression analysis showed that root length accounted for 98% of variation between populations in total P uptake by low P pretreated plants and 87% for high P pretreated plants.

There were much smaller differences among populations in P uptake per unit root length (Table 2) and only then differences following growth at 10 ppm P were significant. Populations collected from low-P soils (1, 2, 3 and 4) generally had lower rates ($p<0.05$) of P uptake per unit root length (0.066 mg P m^{-1} root h^{-1}) than populations collected from high-P soils (0.074 mg P m^{-1} root h^{-1}) after low P pretreatment, but not after high P pretreatment. Only 54% of variation in P uptake per unit root length, after low-P pretreatment, could be accounted for by multiple regression; root/shoot ratio and water influx rate were two most significant variables. After high-P pretreatment, 92% of the variation between populations in P uptake per unit root length was accounted for by RGR, water influx rate and shoot %P.

Table 1. The root length and root/shoot ratio of eight semi-nutrial populations and two cultivars of white clover, after being grown in 0.01 ppm and 10 ppm P.

Population number	Site details			Root length (cm)		Root/shoot ratio	
	Name	Soil pH	Soil P (ppm)	0.01	10	0.01	10
1	Finedon	8.1	2	243	896	0.29	0.21
2	Halfmoon Common	4.7	<1	243	658	0.27	0.19
3	New Forest	4.7	20	223	484	0.26	0.17
4	Rothamsted	5.2	<1	252	705	0.28	0.17
5	Rothamsted	7.1	2	244	704	0.19	0.15
6	Hillfield	7.2	111	201	551	0.24	0.20
7	Rothamsted	4.8	144	249	687	0.26	0.20
8	Rothamsted	6.7	117	237	577	0.24	0.18
H	'Huia'	—	—	578	1546	0.23	0.33
K	'Kent Wild White'	—	—	315	803	0.22	0.35
	Mean	—	—	278	761	0.27	0.19
P	—	—	—	***	***	***	***
MSD$_{0.05}$	—	—	—	197	515	0.09	0.05

Table 2. The per cent P shoot and root of white clover populations after being grown at 0.01 ppm P; and the total P uptake (mg P h^{-1}) and P uptake per unit root length (mg P m^{-1} h^{-1}) from a 10 ppm P solution after being grown in 0.01 ppm or 10 ppm P nutrient solution. Details of the populations are given in Table 1.

Population	% P shoot	% P root	P uptake per pot		P uptake per root length	
	0.01	0.01	0.01	10	0.01	10
1	0.54	0.98	0.018	0.036	0.073	0.040
2	0.63	0.93	0.017	0.019	0.073	0.028
3	0.59	0.94	0.014	0.019	0.059	0.042
4	0.70	0.97	0.017	0.020	0.066	0.029
5	0.52	0.86	0.015	0.018	0.061	0.024
6	0.61	1.03	0.014	0.024	0.072	0.045
7	0.57	0.90	0.018	0.017	0.074	0.025
8	0.58	0.92	0.018	0.012	0.075	0.021
H	0.60	0.70	0.042	0.059	0.073	0.038
K	0.90	1.11	0.024	0.024	0.077	0.031
Mean	0.62	0.93	0.020	0.025	0.070	0.032
P	***	*	***	***	ns	*
MSD$_{0.05}$	0.23	0.32	0.017	0.025	—	0.022

Discussion

The rapid rate of P uptake by rapidly growing plants, such as Huia (Table 2), may result from: (a) greater supplies of metabolites to the roots and hence greater nutrient uptake, or (b) increased nutrient demand by the shoot, or (c) synthesis of more carriers[5]. Both Williams[12] and White[11] found that, when not grossly deficient, the rate of P uptake was regulated more by internal factors, *i.e.* demand, than by external factors, *i.e.* supply. In the present study RGR and shoot %P were among the most important factors accounting for variation among populations in P uptake per unit root length, when plants were previously grown at high P. Root length was also found to be of great importance in understanding the differences in absolute P uptake among populations. Root pruning experiments may determine the actual effect of root length on P uptake by separating the size of the absorbing surface from total plant size and RGR.

There were significant differences between populations in P uptake per unit root length following growth at high-P but not at low-P. Only a few such previous comparisons have been made using intact plants grown in nutrient solution; small, though significant, differences have been found among maize genotypes[1,8].

In this study, populations from low-P soils had lower rates of P uptake per unit root length than those from high-P soils (Table 2). Chapin[5] has also noted equivalent differences among species and populations adapted to nutrient deficient soils. It has been suggested that even a low root absorption capacity is adequate to absorb nutrients, such as P, whose availability is limited by diffusion at low levels of supply[9,10]. Selections for genotypes with large root systems, without detrimental effects on shoot production, may improve performance of white clover on low-P soils.

Acknowledgement Thanks to Dr Roy Snaydon for helpful discussion.

References

1　Baligar V C and Barber S A 1979 Genotypic differences of corn for ion uptake. Agron. J. 71, 870–872.
2　Caradus J R and Dunlop J 1978 Screening white clover plants for efficient phosphorus use. *In* 'Plant Nutrition 1978'. Proceedings of 8th International Colloquium on Plant Analysis and Fertilizer Problems. N Z Auckland, A R Ferguson, R L Bieleski and I B Ferguson. N.Z. DSIR Information Series No. 134, 75–82.
3　Caradus J R, Dunlop J and Williams W M 1980 Screening white clover (*Trifolium repens* L.) plants for different responses to phosphate. N.Z.J. Agric. Res. 23, 211–217.
4　Caradus J R and Williams W M 1981 Breeding for improved white clover production in New Zealand hill country. *In* 'Plant Physiology and Herbage Production'. Proceedings of British Grassland Society Symposium, April 1981. Nottingham pp 163–169.
5　Chapin E S 1980 The mineral nutrition of wild plants. Annu. Rev. Ecol. Syst. 11, 233–260.
6　Eibel H and Lands W E M 1969 A new, sensitive determination of phosphate. Anal. Biochem. 30, 51–57.
7　Marsh B a'B 1971 Measurement of length in random arrangement of lines. J. Appl. Ecol. 8, 265–267.
8　Nielsen N E and Barber S A 1978 Differences among genotypes of corn in the kinetics of phosphorus intake. Argon J. 70, 695–698.
9　Nye P H 1977 The rate-limiting step in plant nutrient absorption from soil. Soil Sci. 123, 292–297.
10　Nye P H and Tinker P B 1977 'Solute movement in the soil-root system'. Studies in Ecology. Vol. 4. Blackwell Sci. Publ.
11　White R E 1972 Studies on mineral ion absorption by plants. I. The absorption and utilization of phsophate by *Stylosanthes humilis, Phaseolus atropurpurea* and *Desmodium intortum*. Plant and Soil 36, 427–447.
12　Williams R F 1948 The effect of phosphorus supply on the rates of intake of phosphorus and nitrogen and upon certain aspects of phosphorus metabolism in gramineous plants. Aust. J. Sci. Res. B1, 333–361.
13　Williams W M and Caradus J R 1979 Performance of white clover lines on New Zealand hill country. Proc. N. Z. Grassland Assoc. 40, 162–169.
14　Williams W M, Lambert M G and Caradus J R 1982 Performance of a hill country white clover selection relative to other clover cultivars in hill country. Proc. N. Z. Grassland Assoc. 42, 188–195.

Genetic variation in nitrogen nutrition of grasses and cereals: possibilities of selection

P. J. GOODMAN
Welsh Plant Breeding Station, Plas Gogerddan, Aberystwyth SY23 3EB, UK

Key words Barley Grain Mutants Nitrate Nitrogen Redistribution Reductase Ryegrass Selection Uptake

Introduction

Nitrogen nutrition is a complex process with many stages. In the present work, three stages have been chosen for detailed genetic studies: ion uptake, nitrate reduction and nitrogen redistribution. The first stage in nitrogen nutrition, ion uptake, is proportional to the size of the root system and to the velocity of ion uptake per unit root size[3]. The ion absorbed is usually nitrate, so that nitrate reductase is needed as the second stage in forming organic nitrogen. Several other stages are then needed to produce protein. Protein nitrogen formed in the leaves is part of the economic product of grasses, but in cereals a further stage is needed, the redistribution of protein from leaves to form grain protein. These stages, with others, determine the efficiency of nitrogen nutrition in grasses and cereals (Fig. 1).

Nitrogen uptake rate and root size in grasses

Plant material

The grasses used were *Lolium perenne* vars S23, Melle and Norlea, *L. multiflorum* vars RvP and S22; *L. perenne* × *L. multiflorum* 'Hybrid 5' selected for increased nitrogen uptake and yield[2,3] and its 'Backcross 25', the result of reselecting Hybrid 5 × *L. perenne* for yield.

Methods

There were three experiments, the first with 14-week-old seedlings, the second with 8-week-old tillers, and the third with 1-year-old sward plants. The seedlings were grown in 'complete' nutrient solutions with 0.05, 0.1, 0.25 or 1 mM NaNO$_3$[1]; the tillers with 0.1 or 1 mM NaNO$_3$, and nitrate depletion was measured by 'Orion' specific ion electrode in stirred cultures. The sward plants were sampled by harvesting the foliage, then taking soil cores and washing the roots free of soil. Shoot and root fresh and dry weights were recorded. Nitrogen concentrations in tops were determined by Kjeldahl analysis.

M.R. Sarić and B.C. Loughman (eds.), Genetic Aspects of Plant Nutrition.
ISBN-13: 978-94-009-6838-7
©1983 Martinus Nijhoff/Dr W. Junk Publishers, The Hague/Boston/Lancaster.

Results and discussion

Comparisons between varieties, of root fresh weight and nitrate uptake velocity per unit fresh weight of root, showed significant differences (Table 1). The ranking of varieties also differed with maturity.

Hybrid 5 seedlings absorbed rapidly and indeed this was one character selected in Hybrid 5, but relative uptake was less in sward (Table 1). RvP differed in having slow uptake in seedlings, but large roots, giving moderate total uptake per plant, but this uptake actually increased at maturity. This type of behaviour with relative uptake increasing in the mature plants, should ensure that the plants compete strongly, and so persist, in sward. Other varieties in the experiment showed characters intermediate between Hybrid 5 and RvP.

Estimates of Michaelis constant, the nitrate concentration required for half maximum uptake rate, ranged from 50 μMN in Norlea to 230 μMN in Hybrid 5. These values are larger than those reported by other workers using fast flowing nutrient solutions[10]. Further tests are therefore needed, with faster flowing, undepleted solutions to confirm our results.

Huffaker and Rains[4] have published preliminary results on nitrate uptake in cereals. As with grasses, comparisons are needed in cereals using fast flowing undepleted nutrient solutions to determine the nitrogen requirements of different varieties.

Nitrate reductase activity in grasses

Materials and methods

The same varieties of ryegrass were used as in the previous experiment. 13-week-old tillers were used to measure *in vivo* nitrate reductase activity by the method of Tokarev[8].

Table 1. Nitrogen uptake rate ($\mu g \cdot h^{-1} \, g^{-1}$ root fresh weight) root fresh weight (g plant^{-1} or $g \cdot m^{-2} 10^{-2}$) and total uptake ($\mu g \cdot h^{-1} \cdot pl^{-1}$ or $\mu g h^{-1} \, m^{-2} 10^{-2}$) of *Lolium* varieties

Variety	14-week-old seedlings			8-week-old tillers			1-year-old sward		
	Uptake rate	Root wt.	Total uptake	Uptake rate	Root wt	Total uptake	Uptake rate	Root wt	Total uptake
Hybrid 5	33	3.7	122	21	5.0	105	102	14.6	1489
Backcross 25	–	–	–	24	4.7	113	79	18.0	1422
Melle	25	3.5	88	27	1.7	46	63	21.8	1373
Norlea	17	2.5	43	19	2.5	48	121	12.0	1452
RvP	0.7	13.2	9	24	3.3	79	159	13.0	2067
S23	–	–	–	18	3.0	54	64	24.0	1536
S.E.	–	–	–	1.7	0.7	–	23	5.8	–

Table 2. Nitrate reductase activity ($\mu g\ h^{-1}\ g^{-1}$ dw leaf or $\mu g\ h^{-1}\ plant^{-1}$) and dry matter production (g plant^{-1}) of 13-week-old rooted tillers. (Adapted from Goodman[3])

Variety	Leaf nitrate reductase	Shoot dry weight	Total plant nitrate reductase
Hybrid 5	195	1.67	339
Backcross 25	147	1.71	266
Melle	144	0.64	97
Norlea	173	0.60	112
RvP	179	1.23	230
S23	163	0.99	170
S.E.	1.4	0.21	–

Results and discussion

Including weight differences, nitrate reduction per plant in Hybrid 5 was three-fold that of Melle (Table 2). Shoot dry weight was twice as much in Hybrid 5 as in Melle or Norlea. The nitrate reductase activities found in this experiment greatly exceeded the rates of nitrate uptake, which shows that nitrate reductase activity does not always limit plant growth. It seems more likely that limitation occurs at certain times of day or season of the year[1,3]. Timing and duration of

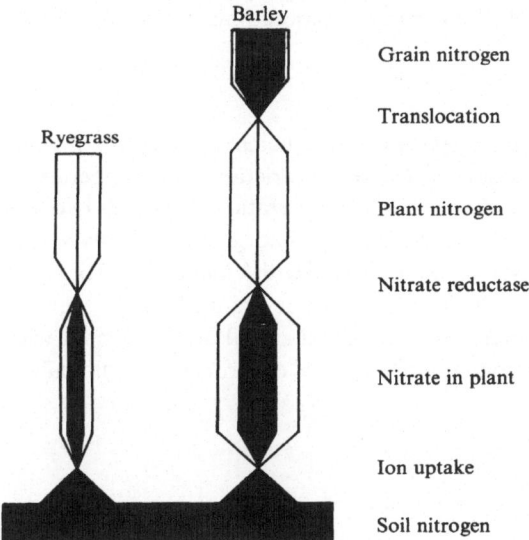

Fig. 1. Genetic range of successive stages of nitrogen flux in ryegrass and barley, represented as width of the upright columns. Minimum values (*dark*), maximum values (*light*). Log scale $\mu mN \cdot g^{-1}$ (fw)$\cdot h^{-1}$. Recalculated from Bowerman et al.[1], Goodman[3], Huffaker et al.[4], Rahman[6] and Tokarev et al.[8].

Table 3. Nitrate reductase activity (μM h^{-1} g^{-1} fw leaf) in barley mutants at two temperatures of growth

Variety	10°C	20°C	
Viner	4.0	3.8	
Chlor 29	2.8	1.4	S.E. = 0.25
Chlor 18	0.3	1.5	
Az 12	–	1.0	
Az 13	–	0.9	

nitrate reductase activity are therefore important. They may depend on energy supply, which links photosynthesis, growth and nitrogen nutrition[10]. Breeders may need to consider these processes together.

Nitrate reductase mutants in cereals

Materials and methods

In barley, mutations have been induced in var. Viner: 'Chlor 18' and 'Chlor 29', on treatment with ethyl methane sulphonate by the Novosibirsk, USSR workers[8]; and in var Steptoe: 'Az 12' and 'Az 13' on treatment with azide by the Washington State, USA workers[9]. These varieties have now been compared at different temperatures in controlled environment conditions, for *in vivo* nitrate reductase activity.

Results and discussion

All the mutants had significantly less nitrate reductase activity than Viner (Table 3). Interactions also occurred between varieties and temperature. Although Viner activity scarcely changed with temperature, Chlor 29 had less activity at the higher temperature, while Chlor 18 had greater activity at the higher temperature. The weighted means (compared with Viner control) of Az 12 and Az 13 did not differ significantly from Chlor 18 or Chlor 29. Unfortunately the interesting environmental interactions in this material are associated with much variability in seed batches, so that more detailed experiments are difficult to perform[5].

Grain nitrogen in cereals

Materials and methods

A field experiment with twenty varieties of barley was analysed for grain yield, N% and total grain nitrogen. In a second, diallel, experiment to determine heritability, seven of the twenty varieties were grown in the glasshouse and similarly analysed[6].

Table 4. Yield and nitrogen concentration in field-grown barley varieites (after Rahman[6])

	Grain N yield (mg·pl^{-1})	Grain (N%)	Grain Yield (g·pl^{-1})	Straw N yield (mg·pl^{-1})
Universe	120	1.44	8.23	16.12
Mazurka	111	1.68	6.59	19.46
Risø 1508	107	1.77	6.03	16.66
Bussell	149	2.40	6.21	18.72
Betzes	94	1.93	5.29	11.36
Gondar	121	2.11	5.72	21.32
Abyssinian 16	80	2.01	3.98	15.86
S.E.	8	0.05	0.36	1.70

Results and discussion

Nitrogen concentrations, grain yields, total grain and straw nitrogen differed significantly in seven varieties of the twenty, tested in the field (Table 4). Although nitrogen concentrations and grain yield were inversely related, grain nitrogen yield was significantly heritable in the diallel. Nitrogen redistribution, as measured by the ratio Grain N Yield:Straw N yield was 60% greater in Bussell and Betzes than in Abyssinian 16. Efficient nitrogen redistribution is essential to modern varieties.

Conclusions

In grasses, selections have been successful in varying root size and rate of uptake of nitrogen, but these characters vary with time. Seedling nitrogen efficiency may favour plant establishment, but sward nitrogen efficiency is likely to confer persistency.

Nitrate reductase activity has interactions with temperature in cereals and possibly with time in grasses. Each of these variations in nitrogen nutrition may, in turn, be related to growth, so that nutrition needs studying in a wider context of plant development[10].

In cereals, not only do uptake and nitrate reductase need study, but nitrogen redistribution is also very important. Negative interactions are found between nitrogen concentration and grain yield which suggest again that growth and development are closely associated with nitrogen nutrition of different varieties.

References

1   Bowerman A and Goodman P J 1971 Variation in nitrate reductase activity in *Lolium*. Ann. Bot. 35, 353–366.

2 Goodman P J 1977 Selection for nitrogen responses in *Lolium*. Ann Bot 41, 243–256.
3 Goodman P J 1981 Genetic control of nitrogen response in *Lolium* species. *In* Plant Physiology and Herbage Production. Ed. C E Wright. Occasional Sympsoium No. 13, British Grassland Society, pp 131–136.
4 Huffaker R C and Rains D W 1978 Nitrate acquisition by plants. *In* Nitrogen in the Environment. Eds. D R Nielsen and J G MacDonald 2, 1–43.
5 Oh J Y, Warner R L and Kleinhofs A 1980 Effect of nitrate reductase deficiency upon growth, yield and protein in barley. Crop Sci. 20, 487–490.
6 Rahman M A 1978 Genetic variation in nitrogen utilization in barley. Ph.D. Thesis, University of Wales.
7 Tokarev B I 1977 Methods for the determination of the level of nitrate reductase activity in wheat and barley. *In* Effects of Chemical Agriculture in Siberia. Siberian Department, Academy of Sciences of U.S.S.R. (*In Russian*), pp 58–65.
8 Tokarev B I and Shumny V K 1977 Detection of barley mutants with a low level of nitrate reductase activity after seed treatment with ethyl methane sulphonate. Genetica (Moscow) 13, 2098–2103 (*In Russian*).
9 Warner R L, Lim C J and Kleinhofs A 1977 Nitrate reductase deficient mutants in barley. Nature, London 269, 406–407.
10 Wild A and Breeze V G 1981 Nutrient uptake in relation to growth. *In* Physiological Processes Limiting Plant Productivity. Ed. C B Johnson. Butterworth, London, pp 331–344.

Variability and genetic control of aluminium tolerance in sorghum genotypes*

P. R. FURLANI**, R. B. CLARK, W. M. ROSS and J. W. MARANVILLE
Department of Agronomy and U.S. Department of Agriculture, Agricultural Research Service, University of Nebraska, Lincoln, NE 68589, USA

Introduction

Plants grown on acid soils often have limited growth and productivity. One of the major problems in acid soils is aluminium (Al) toxicity on plants[12]. Aluminium toxicity is particularly serious in acid subsoils where it is difficult to lime and amend the condition. Hence, root growth in acid soils is usually greatly inhibited with subsequent decreases in nutrient and water uptake by the plants.

A promising method of enhancing plant growth in acid soils is to develop plants that tolerate excess levels of Al. Plant genotypes tolerant to Al could not only improve plant growth in acid soils but also could decrease lime and fertilizer inputs needed to assure good plant growth. Differential tolerances to Al among genotypes have been reported in many crop species[6,13,17,19,31]. Differences among genotypes of many crop species for Al tolerance under laboratory, greenhouse, and growth chamber (controlled environment) conditions have correlated well with field dry matter and grain yields[14,16,20,25]. Controlled environment screening methods for Al tolerance in plants should enhance the identification of superior germplasm for plant breeding programs. Controlled environment screening methods are usually more rapid and less expensive than field screening. In addition, controlled environment screening methods also assess more germplasm in a given period of time, reduce the amount of germplasm to be tested in the field, and allow for year-round testing.

Differences among genotypes of many plant species for Al tolerance have been demonstrated to be under genetic control[8,11]. Most studies have shown that inheritance of Al tolerance is quantitative with high heritabilities and a predominance of additive gene action[19]. Although differential responses of Al tolerance have been noted among sorghum [*Sorghum bicolor* (L.) Moench] genotypes[1,2,4,9,10,15,18,21,22,23,30], information on the genetic control of Al tolerance in sorghum is limited.

The objectives of this study were to determine (1) variability of sorghum genotypes to Al tolerance, and (2) inheritance characteristics of Al tolerance in sorghum.

* Published as Paper No. 6894, Journal Series, Nebraska Agricultural Experiment Station.
** Present address: Instituto Agronomico, C.P. 28,13,100 Campinas, S.O., Brazil.

M.R. Sarić and B.C. Loughman (eds.), Genetic Aspects of Plant Nutrition.
ISBN-13: 978-94-009-6838-7
©1983 Martinus Nijhoff/Dr W. Junk Publishers, The Hague/Boston/Lancaster.

Materials and methods

Screening genotypes for Al tolerance

Two hundred S_1 progenies from the NP16BR sorghum population and 94 inbred lines were screened for variability in Al tolerance in nutrient solutions using the techniques of Furlani and Clark[15]. In brief, this method consisted of placing 8-day-old seedlings in 0 and 148 μmol Al (64 μmol P) where plants were allowed to grow in treatment solutions 10 days before experiments were terminated. At this time, visual Al toxicity symptoms on roots (ALTS) were assessed and rated from 0 = no Al toxicity to 4 = severe Al toxicity in whole number increments. Symptoms for Al toxicity on sorghum plants and the composition of nutrient solutions have been described[15]. Standard or control genotypes were grown with each experiment for comparative purposes: TX415 was an Al-susceptible genotype and SC369-3-1JB and NB9040 were used as Al-tolerant genotypes based on previous results[15]. Plant (root and top) dry weights and seminal root lengths were measured to assess genotypes for Al tolerance. Relative seminal root lengths (RRL) is defined as the length of seminal roots of plants grown with Al divided by length of seminal roots of plants grown without Al.

Two sets (three plants per set) of 21 genotypes were placed on each plate for growth in each container of solution (126 plants/container). Each genotype set was randomized for position on the plate. Treatment solutions were replicated three or four times in a split-plot design. Controlled environment chambers were used to grow the plants with 17 hours light at $28 \pm 1°C$ and 7 hours dark at $23 \pm 1C°$. Lamps for the light were incandescent plus metal halide (Sylvania metalarc BU-HOR 40*) giving 350 μE m^{-2} sec^{-1} at plant height (60 cm below the lamps).

Studying inheritance of Al tolerance

Single-cross hybrids were made as follows and assessed for tolerance to Al toxicity: toxicity: (1) TX414, SC33-9-8-E4, NP9040, and SC369-3-1JB males were crossed onto CK60 and KS35 females

Table 1. Ratings of visual Al toxicity symptom on roots (ALTS) of sorghum parental lines (in parentheses) and single-cross hybrids

Male		Female		
		CK60 (3.0)*b*	KS35 (3.0)*b*	Mean
		------ Hybrid ------		
TX415	(4.0) *a*	3.2 a†	3.2 a	3.2 *a*
SC33-9-8-E4	(3.1) *b*	2.8 b	2.8 b	2.8 *b*
NB9040	(1.2) *c*	1.5 c	1.0 c	1.2 *c*
SC369-3-1JB	(1.2) *c*	1.0 c	1.0 c	1.0 *c*
Mean		2.1 a	2.0 a	

† Mean followed by a common letter are not significantly different at $P = 0.05$ according to Duncan's multiple range test.

* Mention of trademark, proprietary product, or vendor does not constitute a guarantee or warranty of the product by the U.S. Department of Agriculture and does not imply its approval to the exclusion of other products or vendors that may also be suitable.

Table 2. Ratings of visual Al toxicity symptoms on roots (ALTS) of sorghum genotypes with different cytoplasms

Parents	
TX415	4.0 *a*†
CK60A	3.0 *b*
CK60B	3.0 *b*
Hybrids	
CK60 × TX414	3.2 ab
KS34 × TX414	3.0 ab
KS35 × TX414	3.2 ab
KS36 × TX414	3.0 ab
KS37 × TX414	2.8 b
KS38 × TX414	3.5 a
KS39 × TX414	3.0 ab

† Means followed by a common letter are not significantly different at $P = 0.05$ according to Duncan's multiple range test

to produce eight hybrids (Table 1); (2) TX414 was crossed to seven females (CK60, KS34, KS35, KS36, KS37, KS38, and KS39) with different cytoplasms to produce hybrids (Table 2); and (3) SC500-6-1, NB9040, TX415, and Plainsman males were crossed to Wheatland, Martin, and Redlan females to produce 12 hybrids (Table 3). These parental lines and their hybrids were assessed for Al tolerance using ALTS and RRL. In the case of the hybrids and parents in Table 3, these genotypes were also assessed for Al tolerance using relative longest root length (RARL). RARL is the length of the longest adventitious root of plants grown with Al divided by the length of the longest adventitious root of plants grown without Al. These plants were grown in nutrient solutions similar to the experiments described except that fewer plants were maintained in the containers (60 plants/container). Seedlings were suspended in the nutrient solutions from the container lid by wrapping plants with small strips of sponge rubber. Plants were in Al treatment solutions 10 days. The cytoplasms are described elsewhere[27,29].

Results and discussion

Effects of Al on plants

Sorghum seedlings grew well and developed normally when grown in nutrient solutions without added Al. Plants were in the 5- to 7-leaf stage when experiments were terminated. No unusual symptoms (element deficiency or excess) were noted in the plants grown without Al. With Al added to the solutions, many genotypes developed symptoms characteristic of Al toxicity; that is, small, stubby, brittle, coralloid, less-branched, and dark-coloured roots, and reddish (characteristic of P deficiency) or interveinally chlorotic (similar to Fe deficiency) leaves. More details of Al toxicity symptoms on sorghum have been described[5,7,15]. Relatively wide differences were noted among sorghum genotypes for Al toxicity according to ALTS (Table 4), which was useful to assess

Table 3. Ratings of visual Al toxicity symptom on roots (ALTS) and relative longest adventitious root length (RARL) of sorghum parental lines (in parameters) and single-cross hybrids

Male		Female			
		Wheatland	Martin	Redlan	Mean
ALTS rating					
		(1.8) *c* Hybrid	(3.0) *b*	(3.0) *b*	
NB9040	(1.0) *d*	2.0 de†	1.8 ef	1.6 f	1.8 *C*
SC500-6-1	(1.9) *c*	2.2 de	1.6 t	2.5 cd	2.1 *BC*
Plainsman	(3.6) *a*	1.8 ef	2.8 bc	2.2 bc	2.2 *B*
TX415	(4.0) *a*	3.2 b	3.8 a	4.0 a	3.7 *A*
Mean		2.3 A	2.5 A	2.6 A	
RARL					
		(0.73) *bc* Hybrid	(0.63) *c*	(0.36) *d*	
NB9040	(0.93) *b*	1.19 a	0.93 b	0.90 b	1.00 *A*
SC500-6-1	(1.18) *a*	0.87 b	1.17 a	0.82 b	0.95 *A*
Plainsman	(0.60) *c*	0.58 c	0.47 c	0.49 c	0.51 *B*
TX415	(0.60) *c*	0.58 c	0.52 c	0.47 c	0.52 *B*
Mean		0.81 A	0.77 A	0.67 B	

† Means followed by a common letter within experiments are not significantly different at $P=0.05$ according to Duncan's new multiple range test

genotype differences to Al tolerance. ALTS ratings classified fewer genotypes into the Al tolerant category compared to RRL (Tables 4, 5).

Fewer differences were noted among sorghum genotypes for either root or top dry matter weights of juvenile plants grown with or without Al (data not presented). In many cases, dry matter weights were higher for plants grown with low levels of Al compared to plants grown without Al. Even roots of plants that were shorter in length and had relatively severe Al toxicity symptoms often had higher dry weights than roots of plants without Al toxicity symptoms. Beneficial effects of low levels of Al have been noted for many plant species[13]. Dry matter yield differences between plants of this age grown with and without Al did not appear useful to assess sorghum genotype differences in Al tolerance.

Seminal root lengths of sorghum plants grown with Al are generally shorter than for plants grown without Al[15]. This was generally noted for the genotypes in this study, and the correlations between RRL values and ALTS values were

Table 4. Frequency distribution of sorghum lines (I) and S_1 progenies from the NP16BR sorghum population (II) for Al tolerance according to relative seminal roots lengths (RRL) (average of 6 sets of plants)

Increments of distribution†	Sorghum group	
	I	II
1.11–1.20	2	3
1.01–1.10	5	34
0.91–1.00	20	54
0.81–0.90	24	50
0.71–0.80	21	46
0.61–0.70	19	17
0.51–0.60	2	2

† $RRL = \dfrac{\text{Root length of plants grown with Al}}{\text{Root length of plants grown without Al}}$ in increments of 0.10.

Table 5. Frequency distribution of sorghum lines (I) and S_1 progenies from the NP16BR sorghum population (II) for Al tolerance according to visual Al toxicity symptoms of roots (ALTS)

Increment of distribution†	Sorghum group	
	I	II
< 1.0	1	0
1.0–1.3	3	0
1.4–1.6	0	1
1.7–1.9	1	7
2.0–2.2	1	4
2.3–2.5	4	9
2.6–2.8	11	11
2.9–3.1	5	30
3.2–3.4	14	36
3.5–3.7	29	37
3.8–4.0	25	65

† Ratings for ALTS were 0 = no Al toxicity symptoms to 4 = severe Al toxicity symptoms in increments of 0.3.

statistically significant but were relatively low (<0.40) and appear to have little predictive value. In many cases, Al caused RRL values to be larger than 1.0. Even though RRL was significantly correlated with ALTS, the use of RRL to assess sorghum genotypes for Al tolerance may not be practical because of the large number of genotypes showing only small changes of RRL. For example, more genotypes fell into the tolerant range when assessed by RRL than when assessed by ALTS. Only 45% of the sorghum lines had seminal roots inhibited by more than 20% (Table 4) compared to nearly 80% of the sorghum lines having relatively intolerant ALTS values (Table 5).

Genotype variability to Al tolerance

Sorghum genotypes showed relatively wide ranges of Al tolerance when assessed according to RRL (Table 4) and ALTS (Table 5). At the level of Al used to screen genotype differences in Al tolerance, all genotypes showed at least some Al toxicity symptomology. A rating of 1.0 was given to genotypes even if trace symptoms of Al toxicity were noted. Thus, the rating system from 1.0 to 4.0 would be more representative of the range of symptoms noted, rather than 0 to 4.0 ratings. This is pointed out by the differences noted among the standard genotypes used for comparisons (Table 6). Assessment of Al tolerance by ALTS classified most of the sorghum genotypes as intolerant to Al and assessment by RRL gave a more even distribution of genotypes over the Al toxicity range. The ALTS system appeared to be more practical for assessing Al tolerance because it was more definitive and allowed more accurate classification of genotypes. Even though most genotypes were intolerant to Al, some Al tolerant genotypes were noted. The range of Al tolerance within the population was about the same as that noted among the lines. This indicates that progress should be made for Al tolerance through either population improvement methods or from pedigree breeding.

The sorghum lines consisted of numerous commonly used materials in the Nebraska breeding program, which included yellow endosperm, greenbug resistant, conversion-line, hybrid, and some forage materials. The NP16BR sorghum population has been described by Ross and Kofoid[28].

Table 6. Aluminium tolerance of standard sorghum genotypes used to compare other genotypes (average of 17 experiments)

Genotype	ALTS†	RRL†
TX415	4.0±0.2	0.81±0.16
NB9040	1.4±0.4	1.06±0.20
SC369-3-1JB	1.3±0.4	1.01±0.19

† ALTS = visual Al toxicity symptoms on roots and RRL = relative seminal root lengths.

Genetic studies of Al tolerance

For the male parents in the experiment of Table 1, TX415 was very susceptible, SC33-9-8-E4 was moderately susceptible, and NB9040 and SC369-3-1JB were tolerant to Al. The female parents, CK60 and KS35, were relatively susceptible to Al (Table 1). The F_1 hybrids with TX415 and SC33-9-8-E4 were moderately susceptible to Al and TX415 derived hybrids tended to be more susceptible than SC33-9-8-E4 derived hybrids. The hybrids derived from NB9040 and SC369-3-1JB were tolerant to Al. The male parents appeared to have genes dominant for Al tolerance as expressed in the single-cross hybrids. The female parents appeared to carry recessive genes.

The KS34 to KS39 cytoplasms, like milo cytoplasm, produce male sterility with CK60[27,29] and they gave Al responses similar to the A and B lines of CK60 (Table 2). When TX414 (a sister line to TX415 selected from the same cross) was crossed to the cytoplasms, differences for Al tolerance were not apparent in F_1 hybrids, and the different cytoplasm sources did not appear to differ for Al tolerance. KS37 × TX414 had the lowest ALTS rating but it was not statistically significant from CK60 × TX414. Identification of lines differing in cytoplasmic effects for Al tolerance requires further study. Reciprocal crosses involving Al-tolerant and Al-intolerant parental lines would be an appropriate approach but would have to involve many lines. No differences for cytoplasmic effects on Al tolerance have been reported for barley[24] and maize[26], but maternal effects for tolerance to Mn toxicity were noted in soybeans[3].

Wheatland (Al-tolerant) and TX415 (Al-susceptible) produced Al-susceptible F_1 hybrids unlike other crosses where tolerance was dominant (Table 3). TX415 needs to be studied further to determine whether genes for susceptibility in this hybrid are, in fact, dominant. In this same study, RARL was used to assess Al toxicity. RARL was used because Al in solution has pronounced effects on adventitious root development (Furlani and Clark, unpublished). Since adventitious roots usually do not appear on plants until after seedlings have been placed into Al treatment solutions, the effect of Al on adventitious root growth can be assessed without consideration for initial root growth as is the case for seminal root growth. The Al susceptible male parents, TX415 and Plainsman, had less adventitious root growth and thus had lower RARL values compared to the Al tolerant male parents, NB9040 and SC500-6-1. Aluminium tolerance was expressed in the F_1 hybrids of the latter. The conribution of the female parents to RARL values was relatively small, even though the trends of the hybrids for RARL followed the trend of the female parents.

Variability in Al tolerance was found in the sorghum genotypes tested, and Al tolerance in F_1 hybrids appeared to be expressed by dominant genes, but an exception was noted with TX415 in that susceptibility appeared to be dominant. Our results indicate that improvement for Al tolerance can be achieved in sorghum plant breeding programs using either recurrent selection or pedigree breeding methods.

References

1. Bastos C R and Gourley L M 1982 Rapid screening of sorghum seedlings for tolerance to low pH and aluminium. 742 p. *In* L R House, L K Mughogho and J M Peacock (Eds.) Sorghum in the Eighties. Vol. 2. Int. Crops Res. Inst. Semi-Arid Trop., Hyderabad, India.
2. Brown J C, Clark R B and Jones W E 1977 Efficient and inefficient use of phosphorus by sorghum. Soil Sci. Soc. Am. J. 41, 747–750.
3. Brown J C and Devine T E 1980 Inheritance of tolerance or resistance to manganese toxicity in soybeans. Agron. J. 72, 898–904.
4. Brown J C and Jones W E 1977 Fitting plants nutritionally to soils. III. Sorghum. Agron. J. 69, 410–414.
5. Clarke R B 1982 Plant response to mineral element toxicity and deficiency. pp 71–142. *In* M N Christiansen and C F Lewis. (Eds) Breeding Plants for less favorable Environments. John Wiley & Sons, New York, NY.
6. Clark R B and Brown J C 1980 Role of the plant in mineral nutrition as related to breeding and genetics. pp 45–70. *In* L S Murphy, E C Doll and L F Welch (Eds.) Moving up the yield Curve: Advances and Obstacles. Am. Soc. Agron., Madison, WI.
7. Clark R B, Pier P A, Knudsen D and Maranville J W 1981 Effect of trace element deficiencies and excesses on mineral nutrients in sorghum. J. Plant Nutr. 3, 357–374.
8. Devine T E 1982 Genetic fitting of crops to problem soils. pp 143–173. *In* M N Christiansen and C F Lewis (Eds.) Breeding plants for less favorable environments. John Wiley & Sons, New York, NY.
9. Duncan R R 1981 Variability among sorghum genotypes for uptake of elements under acid soil field conditions. J. Plant Nutr. 4, 21–32.
10. Duncan R R, Dobson J W Jr and Fisher C D 1980 Leaf element concentrations and grain yield of sorghum grown on an acid soil. Commun. Soil Sci. Plant Anal. 11, 699–707.
11. Duvick D N, Kleese R A and Frey N M 1981 Breeding for tolerance of nutrient imbalances and constraints to growth in acid, alkaline and saline soils. J. Plant Nutr. 4, 111–129.
12. Foy C D 1974 Effects of aluminium on plant growth. pp 601–642. *In* E W Carson (Ed.) The plant root and its environment. Univ. Press Virginia, Charlottesville, VA.
13. Foy C D, Chaney R L and White M C 1978 The physiology of metal toxicity in plants. Annu. Rev. Plant Physiol. 29, 511–566.
14. Foy C D, Lafever H N, Schwartz J W and Fleming A L 1974 Aluminium tolerance of wheat cultivars related to region of origin. Agron. J. 66, 751–758.
15. Furlani P R and Clark R B 1981 Screening sorghum for aluminium tolerance in nutrient solutions. Agron. J. 73, 587–594.
16. Howeler R H and Cadavid L F 1976 Screening of rice cultivars for tolerance to Al-toxicity in nutrient solutions as compared with a field screening method. Agron. J. 68, 551–555.
17. Jung G A (Ed.) 1978 Crop tolerance to suboptimal land conditions. Am. Soc. Agron. Madison, WI.
18. Konzak C F, Polle E and Kittrick J A 1976 Screening several crops for aluminium tolerance. pp 311–327. *In* M J Wright (Ed.) Plant Adaptation to Mineral Stress in Problem Soils. Cornell Univ. Agric. Exp. Stn., Ithaca, NY.
19. Lafever H N 1981 Genetic differences in plant response to soil nutrient stress. J. Plant Nutr. 4, 89–109.

20 Lafever H N, Campbell L G and Foy C D 1977 Differential response of wheat cultivars to Al. Agron. J. 69, 563–568.
21 Malavolta E, Nogueira F D, Oliveira I P, Nakayama L and Eimori I 1981 Aluminium tolerance in sorghum and bean-methods and results. J. Plant Nutr. 3, 687–694.
22 Pitta G V E, Schaffert R E and Borgonovi R A 1979 Evaluation of sorghum parents and hybrids to high acidity soil conditions. pp 127. *In* Proc. 12th Ann. Brazil Maize Sorghum Rev., Goiania, Brazil.
23 Pitta G V E, Trevisan W L, Schaffert R E, deFranca, G E and Bahia A F C 1976 Evaluation of sorghum lines under high acidity conditions—preliminary note. pp 553–557. *In*: Proc. 11th Ann. Brazil Maize Sorghum Rev., Piracicaba, Brazil.
24 Reid D A 1969 Genetic control of reaction to aluminium in winter barley. pp 409–413. *In*: R A Nilan (Ed.) Barley Genetics II. Proc. 2nd Int. Barley Genet. Sympt., Washington State Univ. Press, Pullman, WA.
25 Reid D A, Jones G D, Armiger W H, Foy C D, Koch E J and Starling T M 1969 Differential aluminium tolerance of winter barley varieties and selections in associated greenhouse and field experiments. Agron. J. 61, 218–222.
26 Rhue R D 1979 Differential aluminium tolerance in crop plants. pp 61–80. *In* H Mussell and R C Staples (Eds.) Stress physiology in crop plants. John Wiley & Sons, New York, NY.
27 Ross W M and Hackerott H L 1972 Registration of seven isocytoplasmic sorghum germplasm lines. Crop Sci. 12, 720–721.
28 Ross W M and Kofoid K D 1976 Sorghum breeding and genetics. pp 166–188. *In*: Development of improved high-yielding sorghum cultivars. Annu. Report No. 9., Coop. Univ. Nebraska, U.S. Dept. Agric., Rockerfeller Found., and U.S. Agric. Int. Devel., Lincoln, NE.
29 Ross W M and Kofoid D C 1979 Effect of non-milo cytoplasms on the agronomic performance of sorghum. Crop Sci. 19, 267–270.
30 Schaffert R E, McCrate A J, Trevisan W L, Bueno A, Meira J L and Rhykerd C L 1975 Genetic variation in *Sorghum bicolor* (L.) Moench for tolerance to high levels of exchangeable aluminium in acid soils of Brazil. pp 151–160. *In* Int. Sorghum Workshop. Univ. Puerto Rico, Mayaguez, PR.
31 Wright M J (Ed.) 1976 Plant adaptation to mineral stress in problem soils. Cornell Univ. Agric. Exp. Stn., Ithaca, NY.

Genetic basis of N-utilization in wheat

E. KISS, A. BÁLINT, K. DEBRECZENI and J. SUTKA
University of Agricultural Sciences, Gödöllö and Research Institute of Hungarian Scientific Academy, Martonvásár (J.S.), Hungary

Key words Genotypes Grain Hybrids Lines Nitrogen Yield

Introduction

Effective nutrient utilization is becoming a more important economic question. One of the basic problems of modern plant production is the selection and breeding of genotypes and varieties which have efficient nutrient absorption and translocation systems[2] and the efficiency of nitrogen assimilation is a fundamental aspect of this problem. According to Vose[6] there are great differences between genotypes in the ability for absorbing NH_4^+ or NO_3^-, in the level and activity of nitrate reductase, in the size of storage pools of nitrate, and in the capacity to mobilise and translocate nitrogen from leaves and other parts to developing grain.

Materials and methods

In 1978/79 the nitrogen utilization of hybrid wheat ($F_1 = P_1 \times P_2$) and the parents (P_1, P_2) was studied in a pot experiment. The pots were filled with 5.0 kg brown forest soil and the following treatments were used: Untreated control ϕ, 600 mg N/pot (I) and 900 mg N/pot (II); the N:P:K ratio was 1.2:0.8:1.0. ^{15}N-labelled compound NH_4NO_3 and urea were applied as nitrogen fertilizers, which were added in three equal portions to the soil: before sowing, before stem extension and in the heading stage. 4 replicates of 15 plants each were used.

In 1980 the Chinese Spring/Cheyenne chromosome substitution lines were studied for nitrogen utilization by applying $^{15}NH_4$ $^{15}NO_3$, on two levels of nitrogen: Untreated control and 600 mg N/pot. The other parameters of the experiment are the same as that of the former one.

After harvesting the total dry matter, grain yields were measured and the samples prepared for nitrogen determination and ^{15}N isotopic analysis by emission spectometry (Isonitromat 5201 automatic ^{15}N analyzer). The grain, chaff, leaves and stems were separately tested.

Results and discussion

The main results of the first experiment can be seen in Table 1. There is a significant fertilizer effect on total dry matter and grain yield with P_1 and F_1 on both level and type of N. The crop yield of P_2 surpassed the control only in the $NH_4 NO_3$ II treatment and the grain nitrogen percentage of the three genotypes

Table 1. Nitrogen utilization of hybrid wheat F_1 and P_1, P_2

Treatment		Total dry matter g/pot	Grain yield g/pot	N% in grain	N-utilization % in grain
Control	P_1	23.3	5.6	1.72	–
	P_2	18.6	4.9	2.06	–
	F_1	16.3	5.0	1.81	–
I NH_4NO_3	P_1	49.3	15.6	2.20	38.0
	P_2	40.8	7.3	3.04	24.5
	F_1	42.2	13.4	2.42	35.4
II NH_4NO_3	P_1	53.9	14.7	2.84	34.9
	P_2	47.7	8.3	3.20	22.0
	F_1	54.6	16.3	2.54	34.8
I $CO(NH_2)_2$	P_1	45.8	15.4	2.10	32.8
	P_2	35.7	5.9	2.89	16.3
	F_1	36.7	11.3	2.49	27.1
II $CO(NH_2)_2$	P_1	55.3	16.6	2.45	29.2
	P_2	47.4	5.9	3.45	14.9
	F_1	50.3	13.1	2.55	26.2
$LSD_{0.05}$ between any two combinations		5.3 g	2.7 g	0.27 %	6.9 %

increased significantly compared with the control. The nitrogen utilization percentage values calculated from the data of isotopic analysis are nearly the same with P_1 and F_1 but they are higher than that of P_2. The N utilization of F_1 differs significantly from the P_2 parent only. As for the effect of N source on N utilization percentage the values on NH_4NO_3 are higher than urea, but the differences are significant only in P_2 and F_1.

The amounts of nitrogen derived from the fertilizer are illustrated in Fig. 1. The uptake of N from urea is lower than from NH_4NO_3 and if the treatment combinations are compared this is seen only on the higher N level with F_1 and P_2. It seems to be possible that the P_2 parental form is more responsive to urea and this effect appears in F_1 too, though the influence of maternal form in nitrogen utilization is more significant.

The results of experiments conducted with substitution lines are included in Table 2. The NPK significantly increased the straw and grain yield in all of the lines and the same is true of the nitrogen percentage of the grain. Since the experiment aimed at determining the effect of individual chromosomes of

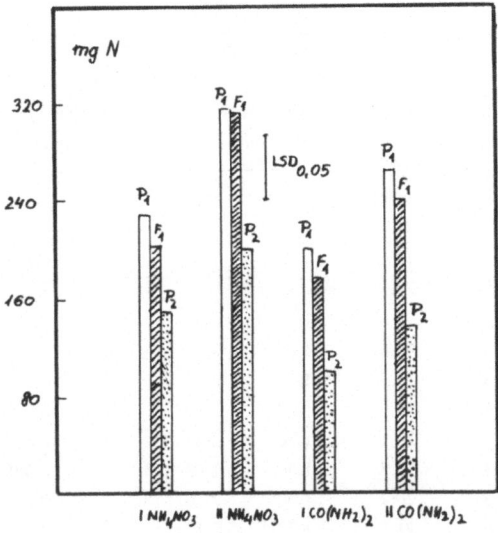

Fig. 1. Amount of nitrogen (mg/pot) derived from the fertilizer in the hybrid wheat combination (P_1, P_2, F_1) as affected by level and source of nitrogen

Cheyenne on the traits responsible for nitrogen utilization of Chinese Spring, the results are discussed in comparison with this later variety.

There were no significant differences between the lines and Chinese Spring in grain yield and %N in the control treatment. Chinese Spring and Cheyenne differ from each other on this level of nutrient supplement only in N concentration. On the effect of fertilizing the difference between Chinese Spring and Cheyenne becomes significant in grain yield, because the yield of Cheyenne is very low, but the lower N content of Cheyenne is not significant. Similar results were obtained with the two varieties by Morris et al.[4] in a greenhouse experiment conducted in order to study the effect of 1D chromosome of Cheyenne on the flour quality. Comparing the lines with the recipient variety the grain yields of lines 4B, 7B, 5D are significantly lower then that of Chinese Spring. The %N in the grain of lines 7B and 5D surpasses that of Chinese Spring but the cause cannot be satisfactorily explained on the basis of our experimental data. Sasaki et al.[5] obtained a significant increase in flour protein content over the parental cultivars for a Cheyenne 5D substitution in Chinese Spring.

The relative enrichments, which show what portion of the nitrogen derives from the fertilizer, in the lines 1A, 2A, 3A, 6A, 4B, 7B, 5D and in Cheyenne are lower than that of Chinese Spring.

The quantity of nitrogen taken up from the tagged compound can be seen in Fig. 2. The lines differing from Chinese Spring in amount of nitrogen absorbed by the total plant are underlined (2A, 3A, 5A, 6A.) The amount of nitrogen

Table 2. Fertilizer response with Chinese Spring/Cheyenne substitution lines (ϕ = control)

Variety, line	Mass of the straw g/pot		Grain yield g/pot		N percentage in grain		^{15}N relative enrichment	
	ϕ	NPK	ϕ	NPK	ϕ	NPK	ϕ	NPK
Cheyenne	16.7	40.4	6.1	10.9	1.47	2.32	–	0.613
Chinese spring	16.9	39.6	8.5	21.0	1.70	2.42	–	0.635
1A	17.1	38.3	7.8	18.8	1.69	2.49	–	0.618
2A	18.5	37.8	7.8	19.7	1.74	2.43	–	0.605
3A	18.9	38.6	8.5	20.6	1.68	2.29	–	0.601
4A	17.4	39.0	7.5	21.1	1.76	2.35	–	0.627
5A	18.0	43.9	7.3	20.7	1.80	2.40	–	0.608
6A	19.1	39.8	9.3	20.7	1.60	2.30	–	0.603
7A	21.0	46.3	8.0	21.3	1.72	2.49	–	0.604
1B	18.8	38.9	8.1	19.8	1.63	2.43	–	0.623
3B	20.1	40.0	6.8	20.3	1.71	2.46	–	0.618
4B	20.2	40.9	8.5	17.8	1.63	2.39		0.635
5B	20.5	40.8	6.6	18.3	1.70	2.59	–	0.632
6B	19.9	37.1	9.0	18.9	1.58	2.60	–	0.619
7B	17.9	37.5	8.1	14.1	1.50	3.03	–	0.642
1D	21.4	37.8	8.6	19.9	1.67	2.45	–	0.633
2D	19.6	40.4	8.3	20.1	1.75	2.51	–	0.619
3D	20.3	37.1	8.5	20.3	1.68	2.45	–	0.622
4D	22.1	41.1	8.4	19.3	1.65	2.48	–	0.630
5D	21.5	43.8	8.0	15.6	1.75	2.81	–	0.624
6D	23.1	42.8	8.9	18.9	1.62	2.58	–	0.632
7D	25.1	44.7	8.8	20.7	1.61	2.50	–	0.639
Average	19.73	40.30	8.06	18.9	1.67	2.49	–	0.622
LSD 0.05 between any two combinations	3.5		2.6		0.20			0.020

deposited in the kernels differ in Cheyenne and 1A, 2A, 3A, 6A, 4B, 7B, 5D lines as indicated by the X.

The nitrogen utilization percentage for the total aerial plant material and grain can be seen in Table 3. The numbers of the third column of this table are characteristic for the distribution of nitrogen taken up from the fertilizer and

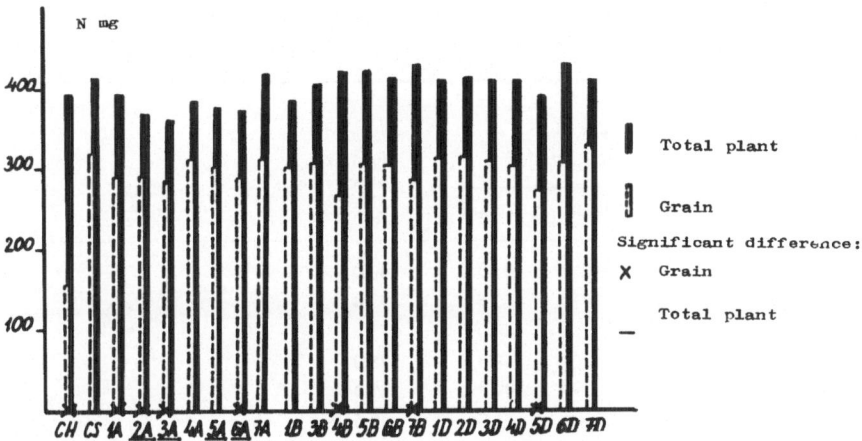

Fig. 2. Amount of nitrogen (mg/pot) taken up from the fertilizer in total plant material and the grain of Chinese/Cheyenne substitution lines

they give the ratio of fertilizer N in grain and whole plant, *i.e.* the so called nitrogen harvest index in per cent[1].

The nitrogen utilization percentage calculated for the total aerial plant material gave significant differences from Chinese Spring in the following lines: 2A, 3A, 5A, 6A, but the Chinese Spring Cheyenne difference did not reach the level of significance. At the same time the nitrogen utilization percentages in the grain are significantly lower in Cheyenne and 1A, 2A, 3A, 6A, 4B, 7B, 5D lines. Chinese Spring surpasses not only the lines which gave much lower yield (4B, 7B, 5D) but also other lines, which differ in any other traits determining nitrogen utilization: for example lines 2A, 3A (relative enrichment). It is interesting that line 1A compared with Chinese Spring differs also in nitrogen utilization, though it does not differ significantly from it in the other features. However the multiplication of these small factors can give a large difference in the final result.

As for the nitrogen harvest index the following differences can be observed: in addition to Cheyenne, which has an extremely low value, the lines 4B, 5B, 7B, and 5D gave a lower nitrogen harvest index than Chinese Spring. This may suggest, that under our experimental conditions the translocation of N into the grain of Cheyenne was smaller and the 4B, 5B, 7B, 5D chromosomes of Cheyenne may carry genes rensponsible for N transport. The higher N content in leaves of Cheyenne and 4B, 7B, lines seems to support this supposition. The ears of line 4B are awned similarly to Cheyenne and the time of flowering and maturity of 5D was near to that of Cheyenne. There appear to be several chromosomes responsible for nitrogen utilization, since it is a complex physiological trait including yield, N content, ability for N uptake and translocation *etc.*, which are separately determined by polygenes[3].

Table 3. Nitrogen utilization % and nitrogen harvest index % of the Chinese Spring/Cheyenne chromosome substitution lines (1980)

Variety or line	Percentage of nitrogen utilization		Fertilizer N harvest index %
	Total plant	Grain	
Cheyenne	64.9	25.9	40.0
Chinese spring	69.3	53.5	77.4
1A	65.0	48.0	73.9
2A	61.9	48.3	78.0
3A	59.7	47.4	79.4
4A	64.3	51.7	80.5
5A	63.2	50.4	79.8
6A	62.6	47.8	76.4
7A	69.8	51.8	74.2
1B	64.2	50.0	78.0
3B	67.8	51.1	75.3
4B	70.8	44.2	62.4
5B	70.7	49.1	69.4
6B	68.9	50.5	73.4
7B	72.0	47.6	66.7
1D	68.0	51.7	75.9
2D	69.2	51.8	74.9
3D	68.6	51.2	74.7
4D	69.4	50.2	72.3
5D	64.7	45.1	69.7
6D	71.8	52.7	73.4
7D	68.4	53.9	78.8
LSD$_{0.05}$	5.7	4.4 %	5.9 %

Conclusion

A wheat hybrid and two parents were studied in our first experiment to determine the genetic influence on nitrogen utilization. The great differences observed between the two parents and between the male form and hybrid suggest that the ability for nutrient utilization is genetically determined. The response to applied nitrogen of the three chosen varieties depended on both the dose of nitrogen and the source.

The data and statistical evaluation of the fertilizer experiment conducted with substitution lines also support the relation between mineral nutrition and genetic background. The substitution of the following chromosomes of Cheyenne into

Chinese Spring affected the nitrogen utilization of total plant and kernel: 2A, 3A, 5A, 6A and 1A, 2A, 3A, 6A, 4B, 7B, 5D, respectively.

Acknowledgement We thank Dr. Zoltán Kertész, Cereal Research Institute Szeged, Hungary, for providing the hybrid wheat and its parents.

References

1 Dubois J B and Fossati A 1981 Influence of nitrogen uptake and nitrogen partitioning efficiency on grain yield and grain protein concentration of twelve winter wheat genotypes *Triticum aestivum* L. Z. Pflanzenzüchtung 86, 41–49.
2 Epstein E 1972 Mineral Nutrition of Plants. Principles and perspectives. pp 332–338. John Wiley and Sons Inc.
3 Konzak C F 1977 Genetic control of the content amino acid composition and processing properties of proteins in wheat. Adv. Gen. 19, 407–582.
4 Morris R, Mattern P J, Schmidt J W and Johnson V A 1978 Quality tests of Cheyenne wheat chromosome 1D substitution in Chinese Spring. Wheat Information Service Nos 45, 46, 12–17. Kihara Institute for Priological Research Yokohama Japan.
5 Sasaki M, Ogawa S, Nishiyama K and Ito T 1973 Chromosomal analysis of biochemical properties of wheat flour in relation to its baking quality. Genetics 74 (2 part 2), 239–240 (Abstr.).
6 Vose B P 1981/82 Contemporary Bases for Crop Breeding. 8–14. Ed. Pergamon.

The ear-leaf percentage of nitrogen, phosphorus and potassium in maize (*Zea mays* L.) inbred lines and their diallel progeny

V. KOVAČEVIĆ
Agricultural Institute, Osijek, Yugoslavia

Key words Diallel crossing Ear leaf Hybrids Inbred lines Maize Nitrogen Phosphorus Potassium

Introduction

When differing genotypes of corn are grown on the same soil, under identical climatic conditions, differences in the quantity of nutrients taken up may indicate differences in the efficiency of nutrient uptake[1,2,3,4,5,6,7,9,10,12] *etc.*

Most breeders select plants under high levels of fertility to insure obtaining genotypes with the greatest yield potential. This may result in an unconscious selection of some genotypes with below-average efficiency in nutrient utilization. Efficiency of nutrient utilization is defined as dry weight yield per unit of particular nutrient uptake (mg dry weight/mg nutrient).

Selection of genotypes for physiological efficiency in nutrient use can improve the crop productivity, the economy of fertilizer applications and possibly the prognosis of combining ability of the parents.

Material and methods

Eight inbred lines from diverse sources FAO groups 500 and 600 were grown under field conditions: Os 56, Os 64, and Os 2 (lines of the Osijek Agricultural Institute), ZP L161 and ZP R455 (lines of Belgrade Corn Research Institute), NS 358/II and NS 796 (lines of the Novi Sad Institute of Field and Vegetable Crops) and C 103 (line from USA). The inbred lines were used to produce 56 single-cross hybrids (diallel crossing). This seed was planted the following year on the same plot. Prior to planting, the experimental field was fertilized with 155 kg N, 130 kg P_2O_5 and 160 kg K_2O per hectar. The experiment was laid at in randomized block design with four replications. Each experimental plot of hybrids measured 14 m^2. Population density was 47,619 plants/ha. Ear-leaf was taken from each plot when the corn was in the tasseling stage. The leaf samples were prepared by standard methods for chemical analysis. Nitrogen was determined by micro-Kjeldahl, phosphorus by the molybdate-vanadate method and potassium by atomic absorption.

Results

Levels of N in the leaf tissue of corn inbred lines ranged from 2.29 to 2.94%. Lines OS 2, NS 358/II, ZP R455 and C 103 were lower, while Os 64 and NS 796 had higher N levels. Parental effects on ear–leaf N percentage of single–crosses were very noticeable. Maternal effects showed differences from 2.64 to 2.92% N

M.R. Sarić and B.C. Loughman (eds.), Genetic Aspects of Plant Nutrition.
ISBN-13: 978-94-009-6838-7
©1983 Martinus Nijhoff/Dr W. Junk Publishers, The Hague/Boston/Lancaster.

Table 1. Percentage of N, P and K (on dry weight basis) in the ear-leaf tissue (tasseling stage) of single-cross corn hybrids and their parents

Female	Parental components of single-crosses Male								Average	LSD	
	Os 56	Os 64	Os 2	C 103	NS 358/II	NS 796	ZP L161	ZP R455		5%	1%
Nitrogen (%N)											
Os 56	–	3.09	2.97	2.76	2.55	3.09	2.66	2.75	2.84		
Os 64	2.61	–	2.72	2.84	2.74	2.73	2.77	2.57	2.71		
Os 2	3.02	2.82	–	2.65	2.82	2.52	2.61	2.29	2.68		
C 103	3.02	2.89	2.51	–	2.55	3.11	2.76	2.57	2.77		
NS 358/II	2.74	2.96	2.81	2.57	–	2.82	2.68	2.86	2.78		
NS 796	2.62	3.02	3.01	3.03	2.59	–	3.17	3.06	2.92		
ZP L161	2.78	2.74	2.60	2.71	2.85	2.98	–	2.48	2.73		
ZP R455	3.01	2.47	2.69	2.80	2.39	2.84	2.28	–	2.64		
Average	2.83	2.85	2.76	2.77	2.64	2.87	2.70	2.65	2.73	0.04	0.06
Inbred lin.	2.71	2.95	2.29	2.38	2.37	2.84	2.73	2.37	2.58	0.12	0.16
LSD 5%				0.12					0.04		
1%				0.16					0.06		
Phosphorus (%P)											
Os 56	–	0.28	0.29	0.28	0.34	0.32	0.38	0.30	0.30		
OS 64	0.33	–	0.34	0.40	0.27	0.36	0.30	0.39	0.33		
Os 2	0.30	0.30	–	0.28	0.27	0.31	0.30	0.33	0.30		
C 103	0.28	0.28	0.28	–	0.27	0.33	0.30	0.28	0.29		
NS 358/II	0.35	0.36	0.32	0.32	–	0.35	0.36	0.35	0.35		
NS 796	0.29	0.28	0.25	0.24	0.25	–	0.30	0.29	0.27		
ZP L161	0.34	0.34	0.35	0.32	0.31	0.35	–	0.35	0.34		
ZP R455	0.34	0.36	0.38	0.32	0.32	0.39	0.33	–	0.35		
Average	0.32	0.32	0.31	0.29	0.29	0.34	0.31	0.33	0.32	0.01	0.02
Inbred lin.	0.32	0.36	0.35	0.35	0.31	0.35	0.39	0.40	0.35	0.03	0.04
LSD 5%				0.03					0.01		
1%				0.04					0.02		
Potassium (%K)											
Os 56	–	1.20	1.22	1.22	1.17	1.38	1.43	1.30	1.28		
Os 64	1.03	–	1.36	1.10	1.25	1.00	0.96	1.15	1.12		
Os 2	1.12	1.47	–	1.40	1.55	1.48	1.50	1.83	1.48		
C 103	1.40	1.12	1.50	–	1.65	1.00	1.63	1.53	1.20		
BS 358/II	1.10	1.23	1.40	1.60	–	1.75	1.63	1.35	1.50		
NS 796	1.13	0.92	1.10	1.18	1.17	–	0.92	1.23	0.09		
ZP L161	1.08	1.57	1.81	1.50	1.86	1.55	–	1.50	1.65		
ZP R455	1.53	1.23	1.56	1.43	1.72	1.38	1.45	–	1.47		
Average	1.30	1.25	1.42	1.34	1.48	1.36	1.36	1.47	1.32	0.06	0.08
Inbred lin.	0.98	0.63	1.07	1.28	0.83	0.92	1.00	1.25	1.00	0.16	0.22
LSD 5%				0.16					0.06		
1%				0.22					0.08		

and paternal effects from 2.64 to 2.87% N (average values, Table 1). The hybrids with Os 2 and ZP R455 lines (as the female component) had the lowest and with NS 796 the highest average N percentage. However, when lines NS 358/II, ZP R455 and ZP L161 were male components, then the single–crosses contained lower average N percentage, and with lines Os 64, NS 796 and Os 56 these percentages were higher.

The corn hybrids and their parents differed in the ear–leaf P percentage, too (Table 1). Lines NS 358/II and Os 56 had a low P percentage and lines ZP L161 and ZP R455 high (ranging from 0.31 to 0.40%P (paternal effects). The hybrids that included line ZP R455 as the female component had the highest P and with NS 796 the lowest P percentage (average values). However, the single–crosses that included NS 796 line as male component had the highest average P level, while that included NS 358/II had the lowest P levels.

Potassium percentage in the ear–leaf tissue of lines ranged from 0.64% to 1.28% K and of single–crosses from 0.92% to 1.86% K. Maternal effect caused differences from 1.09 to 1.65% K and paternal effect from 1.25 to 1.48% K (average values). The single–crosses that included lines ZP R455 and NS 358/II as parental component (male or female) had a high average P percentage, and with line Os 64 this percentage was low.

Discussion and conclusions

The inbred lines and their single–cross hybrids varied considerably in the ear–leaf percentage of N, P, and K. Differences between the lowest and the highest levels of N, P, and K, were 22%, 24% and 50% (inbred lines), and 28%, 38% and 51% (single–crosses) respectively. Maternal effects were more expressive (10%, 23%, and 34%) in comparison with paternal effects (8%, 15% and 15%) for N, P, and K respectively.

The comparison of the average values of N, P, and K percentages at the single–crosses which have had the same parent line (male: female component) indicated the specific differences. The hybrids of the lines Os 64, Os 2 (female parent) and NS 358/II (male parent) had the higher N percentage; the hybrids of lines NS 358/II and ZP R455 (as female parent), Os 56 (as male parent) had the higher P percentage; the hybrids of lines C 103 and ZP L161 (as female parent), Os 64 and NS 796 (as male parent) had the higher K percentage. The average values, in comparison with the values of reciprocal hybrids (with the same parent as other parental component) were highly statistically significant (LSD 1%). In the hybrids of lines NS 796 (nitrogen), Os 64, Os 2, ZP L161 (phosphorus) and Os 2 (potassium) the analogical differences were significant, too (LSD 5%). The comparison between the other parental combinations has not shown statistically significant differences.

Comparison of all 28 analogical pairs of single–crosses (correlation coefficient $r = a \cdot b : b \cdot a$) showed very complex effects of parents on the percentage of the elements investigated (r = 0.29; r = −0.07 and r = 0.44 for N, P, and K respectively). When the average values of single–crosses which had the same parent (male or female) were compared with the levels of investigated elements in this inbred line, low correlations were also found (female component: r = 0.45 r = 0.34 and r = 0.27; male component: r = 0.69 r = 0.38 and r = 0.40 for N, P, and K respectively). When the average values of single–crosses with the same parental line (male: female) were compared the correlation coefficients were: r = 0.55, r = −0.29 and r = 0.63 for N, P, and K respectively.

Thomas et al.[18] reported ranges of 12 to 104 and 33 to 107 µg Zn/g on dry weight basis for two diallel sets of corn single–cross hybrids. From the results of the diallel experiment involving nine corn inbred parent, several lines were characterized as contributing to high, intermediate or low accumulation of Zn^{16}. The Zn accumulation characteristics of corn hybrids were reviewed by Shuman et al.[17].

O'Sullivan et al.[13] investigated the concept of genetic control of the efficiency of N utilization in 146 genotypes of tomato. Under severe N stress in nutrient culture solution efficient genotypes produced as much as 45% more dry weight than inefficient genotypes. The study of 156 tomato lines under K stress[11] showed that an efficient line produced an average of 79% more dry weight than an inefficient line. Partial substitution of K by Na was important in high K efficiency. No evidence for maternal control of efficiency in K utilization was observed.

Whiteaker et al.[19] screened 54 bean lines for the efficiency of P utilization in the nutrient culture solution. Dry weight yield per mg P in the plant tissue varied from 380 to 671 mg. In some lines, growth increased as solution P increased to relatively high levels; in other lines there was no growth response.

Our results indicate the importance of heredity in the mineral composition of progeny. However, the relationships between the chemical composition of genotypes and their parents is very complex. A precise prognosis of the levels of mineral elements in progeny or the combining ability of parents on the basis of this characteristic is probably feasible.

References

1 Baker D E, Jarrell A E, Marshall L E and Thomas W I 1970 Phosphorus uptake from soil by corn hybrids selected for high and low phosphorus accumulation. Agron. J. 62, 103–106.
2 Brown J C 1967 Differential uptake of Fe and Ca by two corn genotypes. Soil Sci. 103, 331–338.
3 Brown J C and Ambler J E 1970 Further characterisation of iron uptake in two genotypes of corn. Soil Sci. Soc. Am. Proc. 34, 249–252.
4 Brown J C and Bell W D 1969 Iron uptake dependent upon genotype of corn. Soil Sci. Soc. Am. Proc. 33, 99–101.
5 Bruetsch T F and Estes G O 1976 Genotype variation in nutrient uptake efficiency in corn. Agron. J. 68, 521–523.

6   Estes G O and Bruetsch T F 1973 Physiological aspects of iron-phosphorus nutrition in two varieties of corn: I, Uptake and accumulation characteristics under greenhouse and field conditions. Soil Sci. Soc. Am. Proc. 37, 243–246.
7   Furlani P R, Hiroce R, Bataglia O C and Silva W J 1977 Acumulo de macronutrientes de silico e de materia seca por dois hibridos simples de milho. Bragantia, Sao Paulo, Campinas, 36(22) 223–229.
8   Gorsline G W, Baker D E and Thomas W I 1965 Accumulation of eleven elements by field corn. Pennsylvania Agr. Exp. Stat. Bulletin 725, 32.
9   Gorsline G W, Ragland J L and Thomas W I 1961 Evidence for inheritance of differential accumulation of Ca, Mg and K by maize. Crop Sci. 1, 155–156.
10 Kovačević V 1980 Proučavanje sepcifičnosti samooplodnih linija kukuruza u odnosu na mineralnu ishranu. Doctoral Thesis. Zbornik radova Poljoprivrednog instituta Osijek, 10(2).
11 Makmur A, Gerloff G C and Gabelman W H 1978 Physiology and inheritance of efficiency in K utilization in tomatoes grown under K stress. J. Am. Soc. Hortic. Sci. 103(4), 545–549.
12 Onochie B E 1970 Variation in absorption and assimilation of Zn by inbreds of corn (*Zea mays* L). Ph.D. Thesis. University Wisconsin, USA.
13 O'Sullivan J 1974 Variations in efficiency of N utilization in tomatoes grown under N stress. J. Am. Hort. Sci. 99(6), 543–547.
14 Pascual M R and Suarez C 1978 Ensayos comparativos de maices ricos en aminoacidos esenciales con variedades de grano comun II Composicion mineral. An. Fac. Vet. Leon Spain 24, 169–174.
15 Sarić M and Kovačević V 1978 Physiologic-genetical aspects of content of mineral nutrition elements in maize (*Zea mays* L). Proceedings of the 8th Int Colloquium on Plant Analysis and Fertilizer Problems. Auckland, N.Z., DSIR Information Series No 134, Wellington, New Zealand, 439–448.
16 Sperling D W 1968 Prediction of levels of element accumulation by Maize. M.S. Thesis, Dep. of Agronomy, The Pennsyl. State Univ.
17 Shuman L M, Baker D E and Thomas W I 1976 Zinc accumulation characteristics of corn hybrids. Agric. Exp. Station, University Park, Pennsylvania. Bulletin 811.
18 Thomas W I, Baker D E and Gorsline G W 1963 Proceedings of the Eighteenth Annual Hybrid Corr. Res. Ind. Conf. 9, pp.
19 Whitaker G, Gerlof G S, Gabelman W H and Lindgren D 1976 Interspecific differences in growth of beans and stress levels of phosphorus. J. Am. Soc. Hortic. Sci. 101, 472–475.

Genotypic variation and inheritance of mineral element content in winter wheat

I. MIHALJEV and R. KASTORI
Faculty of Agriculture, Novi Sad, Yugoslavia

Key words Inheritance Mineral content Nitrogen Phosphorus Uptake Wheat

Introduction

The increasing demand for food requires the development of new wheat varieties with increased genetic potentials for yield. Taking into consideration the rising costs of energy and the consequent rises in the prices of mineral fertilizers, emphasis should be placed on the development of varieties with improved utilization of nutrients and their rational use in the synthesis of organic matter.

Specific modes of mineral nutrition in different genotypes may be considered from various aspects. One of them is the content of mineral elements. Wheat genotypes differ significantly in their contents of mineral substances in certain plant parts and at different stages of growth and development[2,9,11]. However, knowledge of the genetics of mineral element content and the efficiency of mineral element usage by wheat is relatively limited, although such data are available for other species. Studies of the heritability of mineral element content in wheat showed that it was mostly very low. Rasmusson *et al.*[11] reported the heritability of the content of P, K, Ca, and Mg in grain to seldom exceed the value of 0.10. Austin *et al.*[2] reported the heritability of nitrogen content at the stage of stamen emergence to be 0.30 for the leaf, 0.20 for the stem with leaf sheath, 0.29 for the spike, and 0.36 for the whole plant; at the stage of full maturity, the values were 0.61 for the grain, 0.56 for the straw, and 0.48 for the whole plant. Similar results were obtained by Gorsline *et al.*[6] for the heritability of mineral element content in the leaf above the corn ear and in corn grain. The heritability ranged from 0.05 to 0.84, depending on the mineral element examined. Graham[7] reported the monogenic character of the efficiency of iron use by soybean and boron use by tomato. Shea[12] obtained similar results when studying the inheritance of the efficiency of potassium use by snap beans. Regarding the components of genetic variability, the prevalence of additive variance for the content and efficiency of the use of mineral substances was established in several instances[1,6].

It should be pointed out that several studies on various plant species indicated the existence of a positive correlation between the contents of certain elements in plants and yield level[5,10]. However, data indicating the opposite are also available.

It was the aim in this study to examine the variability and mode of inheritance of mineral element content in different wheat genotypes in order to obtain basic data which are essential in developing a wheat breeding program for these characters.

Materials and methods

This study involved seven winter wheat varieties (Sava, Bezostaia-1, NSR-2, Posavka-2, Balkan, Sremica, and NS-7000) diallelly crossed, without reciprocals, by hand in field conditions of 1979. In 1981, 21 F_1 hybrids and the parent varieties were included into a field small-plot trial established after the system of random blocks in three replications. The basic plot size was 0.4 m^2 (two rows 1 meter long, spaced apart at 0.2 m). The sowing density was 375 seeds per 1 m^2. At the stage of flowering the flag leaf was taken from randomly selected plants in all replications to make average samples for analysis. Dry matter of the flag leaf was analysed for nitrogen, phosphorus and potassium by the method of Kjeldahl, vanadate-molybdate, and flame photometry, respectively.

The experimental data obtained were processed by analysis of variance; GCA and SCA were calculated after Griffing[8] method 2, model 1. Mode of inheritance was determined by comparing mean values of F_1 hybrids against mean values of the parent varieties of the hybrid analysed.

Results and discussion

The analysis of variance for the content of N, P, and K in the flag leaf of wheat (Table 1) displayed highly significant differences ($P < 0.01$) among the genotypes, F_1 hybrids and parent varieties, enabling the application of the method of the analysis of variance of combining ability.

The content of N in the flag leaf of the parent varieties and F_1 hybrids ranged from 3.80 to 4.08% and 3.67 to 4.16%, respectively. The content of P varied from 228 to 303 mg % and 246 to 317 mg %, respectively. Finally, the content of K

Table 1. Analysis of variance of the block system for the content of N, P, and K in the flag leaf of wheat at the stage of flowering

Source of variation	d.f	Mean squares		
		N	P	K
Replications	2	0.0416	6.50	328.0
F_1 hybrids and varieties	27	0.0689**	1846.63**	11841.2**
Error	54	0.0248	16.96	219.3
Total	83			

** Significant at 1% level, based on F test

varied from 423 to 512 mg % and 422 to 674 mg %, respectively (Table 2). Large genetic variability in the content of N, P, and K in different wheat genotypes, which had been observed by other researchers too (*e.g.* refs.[2,4,11]), offers possibilities of wheat breeding for these physiological characters aimed at the development of new genotypes possessing a desired content of these elements.

The 21 F_1 hybrids display the intermediate mode of inheritance for the content of K and P in the majority of combinations (52 and 38%, respectively) (Table 3).

Table 2. Variability of the content of N, P and K in dry matter of the flag leaf of wheat at the stage of flowering

Genotype	N(%)	P(mg%)	K(mg%)
Sava × Bezostaja-1	4.05	266	597
Sava × NSR-2	4.07	271	674
Sava × Posavka-2	4.07	285	623
Sava × Balkan	3.96	254	537
Sava × Sremica	3.84	265	524
Sava × NS-7000	3.97	264	582
Bezostaja-1 × NSR-2	4.03	250	422
Bezostaja-1 × Posavka-2	4.03	299	511
Bezostaja-1 × Balkan	3.95	254	637
Bezostaja-1 × Sremica	4.14	268	521
Bezostaja-1 × NS-7000	4.16	246	472
NSR-2 × Posavka-2	4.13	246	465
NSR-2 × Balkan	3.67	289	536
NSR-2 × Sremica	3.88	277	487
NSR-2 × NS-7000	4.14	289	522
Posavka-2 × Balkan	4.05	311	502
Posavka-2 × Sremica	3.73	305	521
Posavka-2 × NS-7000	3.95	317	462
Balkan × Sremica	3.76	272	637
Balkan × NS-7000	3.95	291	523
Sremica × NS-7000	3.81	292	510
Sava	4.02	285	512
Bezostaja-1	3.90	303	460
NSR-2	3.80	296	423
Posavka-2	4.08	244	498
Balkan	3.93	235	510
Sremica	3.98	228	511
NS-7000	3.95	231	512
LSD 0.05	0.22	6.76	24
0.01	0.29	9.01	32

Table 3. Mode of inheritance of the content of N, P and K in the flag leaf of wheat at the stage of flowering, F_1 generation

Mode of inheritance	Nitrogen		Phosphorus		Potassium	
	No. of comb.	%	No. of comb.	%	No. of comb.	%
Intermediate	5	24	8	38	11	52
Dominant (−)	3	14	1	5	1	5
Dominant (+)	6	29	3	14	2	9
Negative heterosis	3	14	3	14	1	5
Positive heterosis	4	19	6	29	6	29
Total:	21	100	21	100	21	100

Regarding the inheritance of the content of N, the dominance of the better parent was expressed (29%). Positive heterosis was observed in a significant number of combinations (29%, 29%, and 19% for P, K, and N, respectively). Negative heterosis occurred in a limited number of combinations. These results show that the mode of inheritance was dependent on the combination of parent varieties. This fact should not be neglected when breeding for these characteristics.

The analysis of variance of combining ability (Table 4) showed that GCA and SCA values were highly significant for all three elements ($P < 0.01$). Furthermore, the ratio GCA/SCA was larger than 1 in all cases, indicating that the larger portion of the total genetic variability resulted from the additive gene action, although the non-additive gene action (dominance and superdominance) was present too. Similar results were obtained by Gorsline et al.[6] for the content of 10

Table 4. Analysis of variance of combining ability for the content of N, P, and K in the flag leaf of wheat at the stage of flowering

Source of variability	d.f.	Mean squares		
		N	P	K
GCA	6	2.4650**	4547**	4972**
SCA	21	0.0157**	217**	3583**
Error	54	0.0083	5.7	73
Ratio GCA/SCA		164	21	1.4

** Significant at 1% level, based on F test

elements in corn leaf and grain, Crompton et al.[3] for the content of calcium in peanut seed, and Amris et al.[1] for the use of potassium in tomato.

The major portion of genetic variability resulting from the additive gene action is of special interest in wheat breeding for a desired content of mineral elements because the additive action may be fixed in later generations. The breeding advance for these characteristics may be achieved by pedigree methods applied after crossing to obtain varieties from pure lines, which is in any case the most common method in wheat breeding.

References

1 Amris M, Gerloff G C and Gabelman H W 1978 Physiology and inheritance of efficiency in potassium utilization in tomatoes grown under potassium stress. J. Am. Soc. Hort. Sci. 103, 545–549.
2 Austin R B, Ford M A, Edrich I A and Blackwell R D 1977 The nitrogen economy of winter wheat. J. Ag. Sci. 88, 159–167.
3 Crompton C E, Wynne J C and Isleib T G 1980 Combining ability analysis of seed calcium content and adenosine phosphates in peanuts. Oléagineux 34, 71–75 (1979) P. B. A. 50, No 2 (1424).
4 De Mooy C J and Pesek J 1970 Differential effects of P, K and Ca salts on leaf composition, yield and seed size of soybean lines. Crop Sci. 10, 72–77.
5 Desai R M and Bhatia C R 1978 Nitrogen uptake and nitrogen harvest index in durum wheat cultivars varying in their grain protein concentration. Euphytica 27, 561–566.
6 Gorsline G W, Thomas W I and Baker D E 1964 Inheritance of P, K, Mg, Cu, B, Zn, Mn, Al and Fe concentrations by corn (*Zea mays* L) leaves and grain. Crop Sci. 4, 207–210.
7 Graham R D 1978 Tolerance of Triticale wheat and rye to copper deficiency. Nature. London 271, 542–543.
8 Griffing B 1956 Concept of general and specific combining ability in relation to diallel crossing systems. Austr. J. Biol. Sci. 9, 464–493.
9 Kleese R A, Rasmusson D C and Smith L H 1968 Genetic and environmental variation in mineral element accumulation in barley, wheat, and soybeans. Crop. Sci. 8, 591–593.
10 Kumbhar D D and Sonar K R 1978 Grain yield and mineral composition of rice (*Oryza sativa* L.) varieties grown under upland conditions. Intern. Rice Comm Newsletter 27(2), 7–8.
11 Rasmusson D C, Hester A J, Fick G N and Byrne I 1971 Breeding for mineral content in wheat and barley. Crop Sci. 11, 623–626.
12 Shea P E 1980 Genetic control of potassium nutrition in snap beans (*Phaseolus vulgaris* L.). Fertilizer Abstr. 13, 140.

Genetical differences in the cation accumulation capacity of mitochondria

M. G. ZAITSEVA and N. K. ZUBKOVA
K. A. Timiriazev Institute of Plant Physiology, Academy of Sciences, Moscow, USSR

Key words Calcium Cations Genotype Magnesium Mitochondria Potassium Wheat

Introduction

It is well known that animal, as well as plant mitochondria, are capable of cation accumulation. Essential differences in the cation content of mitochondria from different kinds of tissues of the same organism as well as of similar tissues of various organisms and species have also been established.

For example, the calcium content of mitochondria from nonthermogenic guinea pig tissues was as high as 84 nmoles·mg^{-1} proteins, whereas that of thermogenic brown adipose tissue was 12–15 nmoles × mg^{-1} protein[4]. The most efficient for calcium accumulation were mitochondria of lung tissues[3] while the only mitochondria capable of magnesium accumulation were those from heart muscle[3]. Finally, intraspecific inherited differences in calcium accumulation between mitochondria from the skeletal muscle of normal and distrophic hamsters were demonstrated[8].

In plant mitochondria, cation accumulation has been poorly investigated up to now and there is no evidence concerning inherited differences in cation accumulation capacity. Therefore in the present study attempts have been made to reveal a possible existence of inherited differences in cation accumulation in mitochondria isolated from roots. These data could be relevant to the problem of mineral nutrition, since it is believed that mitochondria are involved in the process of intracellular ion transfer[5]. Calcium, magnesium and sodium contents have been measured with mitochondria, isolated by standard procedures. Since cation transport and cation accumulation are involved in the regulation of mitochondrial activity[6], we determined the activity, as well as the changes in the oxygen consumption rate, related to substrate and magnesium concentration. The changes in the cation content in the presence of EDTA, as well as the movements of cations related to metabolic state transitions, induced by the addition of substrate and ADP, have also been studied.

Materials and methods

Five summer and winter cultivars of wheat (*Triticum aestivum* L.): Albidum 43, Moskovskaya 21,

Saratovskaya 38, Ilyichevka and Polesskaya 70 have been used. Plants were grown as previously described[9] for 4 to 7 days in a greenhouse on a Knop's solution.

Mitochondria were isolated at 15,000 g for 10 minutes in a medium, which contained sucrose 0.5 M, KH_2PO_4 67 mM, cysteine 0.07%, EDTA 2 mM and BSA 1 mg $\times$ ml^{-1}. Mitochondria were washed once with the same medium without cysteine. The incubation medium contained sucrose (0.48 M), KH_2PO_4 (18.5 mM) and BSA (2 mg $\times$ ml^{-1}). Cation determinations were performed after the addition of 0.5% Triton X-100[10] to the mitochondrial pellet. Potassium and sodium content were measured with a Zeiss flame spectrophotometer and calcium magnesium with a Hitachi atomic absorption spectrophotometer. Cation determinations were performed immediately after the isolation of mitochondria, as well as after 5 minutes' incubation in the presence of EDTA (3.2 mM) or 10 minutes' incubation with succinate (substrate state 4) and succinate and ADP (state 3).

The oxygen consumption rate was followed by means of a Clark type electrode. The changes in the oxygen consumption rate induced by various succinate and magnesium concentrations were used for the K_m calculations.

Protein was measured by the procedure of Lowry et al.[7] All reagents were of high purity, ADP was received from REANAL (Hungary). Statistics are given as means $\pm$ SEM.

Results and discussion

Results of experiments clearly demonstrated that mitochondria isolated from roots of several wheat cultivars substantially differ in their cation content. Statistically significant differences (t $\geqslant$ 3) were more frequent in the potassium, calcium and magnesium and rather rare in the sodium content. Moreover there were differences in the variability of the cation level of mitochondria (Table 1).

Some data indicated that the intramitochondrial cation pool level was essential for the response of mitochondria to the reagents present in the incubation medium. Mitochondria with unequal cation content differed in the optimal concentration of succinate and magnesium as evidenced by the K_m values (Table 2). The latter for succinate seemed to be dependent on the potassium content. Depending on the cation pool level, differences in the direction of cation movements, induced by the addition of EDTA, substrate and ADP were observed. For instance, in the presence of EDTA there was an influx

Table 1. Cation content of mitochondria of wheat roots

Cultivars	Cation content nmoles $\times$ mg^{-1} protein				Standard error as % of the mean value			
	Ca^{2+}	Mg^{2+}	K^+	Na^+	Ca^{2+}	Mg^{2+}	K^+	Na^+
Albidum 43	426	394	8650	550	$\pm$7.0	$\pm$7.1	$\pm$8.0	$\pm$1.6
Moskovskaya 21	224	120	3780	740	$\pm$22.3	$\pm$30.0	$\pm$11.6	$\pm$12.4
Saratovskaya 38	250	190	8480	313	$\pm$20.0	$\pm$30.5	$\pm$20.0	$\pm$38.4
Polesskaya 70	272	242	9050	232	$\pm$17.6	$\pm$9.9	$\pm$17.9	$\pm$21.1
Ilyichevka	444	374	6630	536	$\pm$9.0	$\pm$6.4	$\pm$4.5	$\pm$13.4

Table 2. K_m values for succinate and magnesium

Cultivars	K_m values (mM)	
	Succinate	Magnesium
Moskovskaya 21	5.3	0.1
Ilyichevka	8.3	—
Saratovskaya 38	14.3	—
Albidum 43	16.6	0.004

of potassium in the mitochondria of Moskovskaya 21 cultivar, in contrast to the loss of cation from the mitochondria of Albidum 43. At the same time the addition of succinate to the latter induced an influx of calcium, while there was an efflux of cation from the organelles of Moskovskaya 21. Finally, the opposite movements of magnesium were observed in the mitochondria of the above mentioned cultivars after the addition of ADP (Table 3).

The results indicate the existence of differences between cultivars in the cation accumulation capacity of mitochondria and of some relation between the intramitochondrial cation pool and the responses of the organelles to the conditions of incubation. Therefore a role of the intramitochondrial cation pool in the detection of the responses of mitochondria to environmental conditions is indicated. However, the results do not yet provide evidence as to the role of the relationship between cultivars in the detection of the cation accumulation pattern of mitochondria, because differences between rather remote cultivars such as Saratovskaya 38 and Poleslskaya 70 were lacking. Further investigations are needed in order to solve this problem.

Table 3. Accumulation (+) and losses (−) of cations (nmoles·mg^{-1} protein) as influenced by the incubation conditions of mitchondria

Incubation conditions	Cultivars					
	Albidum 43			Moskovskaja 21		
	Ca^{2+}	Mg^{2+}	K^+	Ca^{2+}	Mg^{2+}	K^+
EDTA 3.2 mM without substrate	−118	−60	−132	−100	−100	+914
EDTA 3.2 mM+Succinate 18.3 mM	+220	+52	+2195	+4	−17	+1236
EDTA 3.2 mM+Succinate 18.3 mM+ADP 0.2 mM	−25	+103	−658	+52	−92	+198

References

1. Brierley G P, Murer E, Bachman E and Green D E 1963 Studies on ion transport. II. The accumulation of inorganic phosphate and magnesium ions by heart mitocondria. J. Biol. Chem. 238, 3482–3489.
2. Chance B and Williams I R 1956 The respiratory chain and oxidative phosphorylation. *In* Advances in Enzymology (F F Nord, Ed.) 17, 65–134. Interscience Publisher INC, New York.
3. Fischer A B, Scarpa A, La Noue K F, Basset D and Williamson I R 1973 Respiration of ratlung mitochondria and the influence of Ca^{2+} on substrate utilization. Biochemistry 12, 1438–1445.
4. Flatmark T and Pedersen I I 1975 Brown adipose tissue mitochondria. Biochim. Biophys. Acta 416, 53–103.
5. Kahn J S and Hanson J B 1959 Some observations on potassium accumulaton in corn root mitochondria. Plant Physiol. 34, 621–629.
6. Lehninger A L 1964 The Mitochondrion: Molecular basis of structure and function. W. A. Benjamin Inc., New York.
7. Lowry O H, Rosebrough N J, Farr A L and Randall R J 1951 Protein measurement with the Folin–phenol reagent. J. Biol. Chem. 193, 265–275.
8. Wrogeman K, Jacobson B E and Planshaer M E 1973 On the mechanism of calcium-associated defect of oxidative phosphorylation. Arch. Biochem. Biophys. 159, 267–278.
9. Zaitseva M G, Vorobjeva E A and Titova Z V 1974 Changes of the properties of mitochondria from wheat roots related to the age of plants. Fiziol. Rastenij 21, 1154 1159.
10. Zaitseva M G and Zubkova N K 1979 Accumulation and losses of catons by mitochondria in changing metabolic states. Fisiol. Rastenij 26, 1085–1091.

Genetic variability in the systems of absorption and utilization of mineral elements

A review

V. K. SHUMNY and B. I. TOKAREV
Institute of Cytology and Genetics, Novosibirsk 630090, USSR and Siberian Institute of Soil Management and Agricultural Chemization, Krasnoobsk, Novosibirsk Region 633128, USSR

Key words Absorption systems Adaptation Barley Breeding Chlorate Genotype Mutagenesis Mutants Nitrate Productivity Review

Introduction

Vavilov[3] has been one of the first proponents of the idea that developments in the genetics of mineral nutrition are indispensable for successful introduction and breeding of plants. This is now a statement of fact, not a theoretical consideration. Indeed, it has become increasingly apparent that breeding of agronomic plants relies on the efficiency of photosynthesis and mineral nutrition. These basic processes are interrelated at all levels of organization by energy and metabolic products, as well as by means of regulated change. Thus, our present state of knowledge has allowed us to define mineral nutrition as one of the major regulated factors in plant breeding[28].

Rationality in utilization of mineral elements from fertilizers and soil is determined by the extent to which their supply conforms to the genetically determined requirements of plants, by the avalability of nutritional elements to plants of definite genotype, as well by energy resources used for uptake of these elements (Fig. 1).

To develop adapted varieties, better knowledge of how plants use mineral elements under particular soil and climatic conditions is needed. This does not call so much for accumulation of observations, but understanding of the genetic control of the key reactions providing absorption and utilization of mineral ions during biosynthesis[6,19,27,28].

The genetics of mineral nutrition provides a theoretical and practical basis for the development of crop varieties adapted to optimum utilization of mineral elements under definite soil and climatic conditions. The genetics of mineral nutrition is concerned with natural and induced variability in characters of mineral nutrition, *i.e.* it deals with all the genetic diversity of the natural adaptation of plants to a wide range of environments.

Herein is presented an attempt to analyse naturally occurring and experimentally induced variability in characters of mineral nutrition. The analysis was

M.R. Sarić and B.C. Loughman (eds.), Genetic Aspects of Plant Nutrition.
ISBN-13: 978-94-009-6838-7
©*1983 Martinus Nijhoff/Dr W. Junk Publishers, The Hague/Boston/Lancaster.*

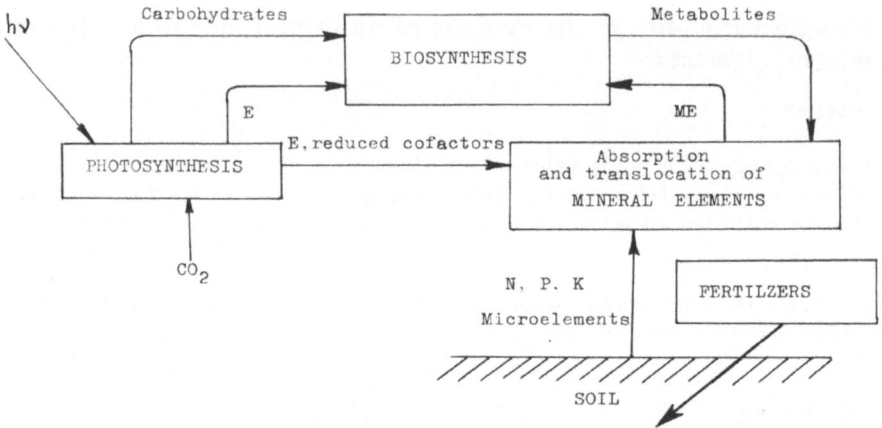

Fig. 1. A schematic presentation of the interrelation of the absorption and utilization of mineral elements by the metabolic system of plants. E = energy supply, ME = mineral elements.

performed with the aim of evaluating models of this polymorphism in terms of the genetic control of absorption and utilization of mineral elements, and to determine how experimentally induced variability helps to gain insight into this genetic control.

Genetic variation for characters of mineral nutrition as a result of selection during natural adaptation of plants to soil conditions

Variation for characters of mineral nutrition, like any other inherited variation, has been established during natural adaptation of plants to environmental conditions. Competition for mineral elements between plants in natural habitats and crop stands can be strong enough to enhance advantages arising as a consequence of genetic variation. Particular genotypes are favoured and persist in populations. As a result, diverging plant species can give rise to nutritional ecotypes[28].

The range of natural genetic variation in the characters of mineral nutrition can be larger among edaphic ecotypes than within a species. Epstein[6] has reviewed the relevant data. A brief mention of the illustrative cases from the literature will be made.

Ammonium and nitrate can be equally important components of nitrogen nutrition of higher plants. Among cranberrys, however, there occur ecotypes whose sole essential nutrient is ammonium[16]. Edaphic variability may be latent, and its overt manifestation is then brought about by nutritional imbalance. Uniformity of plants with respect to efficient use of nutrients in some populations may be more apparent than real. Large plant variability in susceptibility to potassium deficiency was observed in a population of *Arctotheca calendula* L.

When grown in nutrient solutions cultures at a low potassium concentration, the susceptible (or 'inefficient') individuals died from severe deficiency, whereas the efficient ones made good use of potassium[1]. Variation in latent edaphic characters in populations can be exploited. These latent characters can be made overt against an appropriately provocative background. This strategy has enabled us to identify plants resistant to toxic soil factors and mineral deficiencies.

Equation of the genetic variations in characters of mineral nutrition to variations in specific control of nutritional mechanisms would be an oversimplification. An adaptive character in nutrition is not the result of the action of a selective factor; it is the cumulative product of the interplay of modifier genes, of the concominant variations in correlated characters, as well as of arrays of unspecific population effects[28].

Estimates of the contribution of any factor to specific control of absorption and utilization are *a priori* biased because unspecific population effects cannot be excluded even when many sets of nutritional ecotypes are considered. This has been concluded from a survey of the data in the literature and our observations (Tokarev and Shumny, *in press*). Consequently, models based on natural edaphic diversity engender complexities giving unambiguous solutions of the problems of the genetics of mineral nutrition.

Genetic variation for characters of mineral nutrition as a result of breeding for crop productivity

The intervarietal differences in response to conditions of mineral nutrition are often large compared to the average for interspecific differences[19]. This "multidimensional" diversity is the consequence of breeding for productivity in markedly different conditions of soil and climate. The data on intervarietal variations in characters of mineral nutrition and their exploitation for better understanding of the genetic control of absorption and utilization of mineral elements have been summarized[6,20,28].

It would be instructive to recall the salient features of some of the reviewed papers.

Breeding for productivity and adaptation to acid soils has resulted in large variations in aluminium tolerance of wheat and cotton varieties[11,12].

Increased nitrogen response is justly considered to be due to breeding for grain yield of dwarf lines of wheat grown on soils to which much nitrogen was applied. The dwarf genes do not affect directly uptake and assimilation judging by the results obtained with wheat lines near isogenic for plant height[4].

Characters essential for mineral nutrition remain concealed, awaiting a provocative stimulus for manifestation, in lines bred for agronomic characteristics on nutrient-rich soils[3,34]. In fact, Epstein[6,7,8] has developed lines outstanding in tolerance to sea water from the World Barley Collection. This was achieved by use of sea water as a provocative background for screening but specific genetic combinations from the gene pool. Furthemore, Gableman et al.[9,13] have obtained evidence for large differences in responses to nutritional stress in lines of tomatoes and beans. The examples are instructive in that they demonstrate that breeding for improved yield has drastically affected the utilization of mineral elements.

Diversity has been inevitably repatterned by the adaptation of plants to conditions of mineral nutrition during their breeding. The spreading with human cultivation of plants to areas presenting novel soil and climatic conditions has promoted wide divergence of nutritional characters. Efforts to improve crops by increasing the input of fertilizers without due regard to their output are often associated with disturbance of old habitats. The challenge for plants is through adaptation to withstand mineral stresses and to reorganize accordingly their potentialities for genetic variations in nutritional characteristics.

However, improvement in characters of mineral nutrition has not been the goal of either natural, or artificial selection. For this reason, varieties or lines under selection may rely on very different mechanisms of adaptation to modified soil conditions. This may impose constraints to use of variations in characters of mineral nutrition thus obtained as tools in studies of the genetic control of mineral nutrition.

Experimental mutagenesis as a means of inducing genetic variations in characters of mineral nutrition

There remains an unlimited, yet overlooked, possibility of utilizing mutagenesis for inducing variations in characters of mineral nutrition. The possibility is truly unlimited because the plastic processes supplying plants with nutrients along metabolic pathways are easily amenable to mutational innovations. A large number of mutants showing variations in mineral metabolism have been identified among bacteria, unicellular algae and lower fungi. With their identification and study came much of our knowledge of the prinicples underlying the genetic control of plant mineral metabolism[2,22].

Mutagenesis may be applied to higher plants as well. From general considerations it has been inferred that the successful use of this approach would require a method of identifying mutant individuals in a population consisting of not less than 10^4 plants[21,30].

The use of mutants in the genetics of plant mineral nutrition originated with the discovery of pronounced differences in chlorosis characteristic of iron deficiency in soybeans in 1922 by Weiss[36]. Since then mutations affecting the steps of the absorption–transport–utilization process of mineral elements have been widely used in genetic investigations.

It is pertinent to recall the efficient method of Feenstra c.s. which ensured the identification of mutants resistant to the toxic action of chlorate in *Arabidopsis thaliana* L.[10]. These investigators distinguished two groups of mutants with impaired utilization of nitrate. One group lost its capacty for nitrate uptake, while the other was unable to assimilate absorbed nitrate[5].

Complementary mutation groups were identified in crosses between the mutant lines[10]. The revertants, which were obtained by treating some lines with mutagens, assimilated nitrate normally. It was later shown they were mutations in modifier genes.

Our method of induction and selection of Arabidopsis by means of salinized nutrition culture (addition of 1% NaCl) has provided us with salt-tolerant lines. The frequencies of mutations resulting in salinity tolerance were of the order of 10^{-5} to 10^{-6} only. The salt tolerant mutants were identifiable because the rest of the population died under salt conditions of culture[28].

The experimental approach based on mutagenesis and selection against a provocative background has been extended to barley with a view to study altered capacities of mutants to absorb and reduce nitrate. We treated barley seeds (normal line from variety Viner) with ethylemethanesulphonate; the M_2 plants were screened out from segregating population on the basis of 1) nitrate reductase activity of the primary leaf[31] or 2) selective effects of potassium chlorate. When the screening was based on the latter, it was assumed that the reduction of chlorate to chlorite by nitrate reductase is lethal to plants[24].

Under this assumption, mutant plants unable to assimilate chlorate and, hence, nitrate, are viable[31]. Table 1 presents the data obtained in one of the experiments with barley. As a result of induction and selection, we obtained several lines incapable of growing normally on a nitrate source.

However, these lines grew and produced seeds when the ammonium form of nitrogen was added to the culture medium. Three of the mutant lines, selected in the first experimental series (Chlo18, Chlo19 and Chlo29) did not segregate in the subsequent generations. These 3 lines were crossed with normal plants and segregating nitrate reductase deficient plants were thus obtained from the F_2

Table 1. Results of one of the experiments with selection of nitrate reductase (NR)-deficient mutants from the M_2 segregating population of barley seedlings

Selection was based on	Population size in the M_2	Number of plants selected in the M_2	Number of families in the M_3		Number of the low NR lines
			Total	Low NR	
In vivo assays of NR activity	5,450	71	50	5	1
Resistance in the KClO$_3$ toxicity test	12,903	98	90	16	3

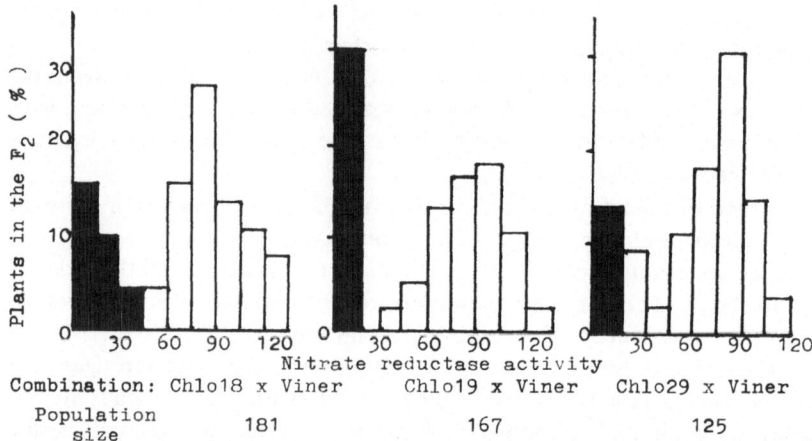

Fig. 2. Segregation for nitrate reductase activity (NRA) in the F_2 generation after hybridization of mutant lines with their normal parent. The NRA is expressed as a percentage of NRA in normal plants (Viner).

populations. The F_2 backrosses between Chlo18, Chlo19, Chlo29 and their parent line (Viner) showed segregation patterns for nitrate reductase activity close to monohybrid ones (Fig. 2). The results obtained with the nitrate reductase mutants in Viner are in good agreement with those obtained in Steptoe variety by Kleinhofs et al.[17]. Crosses involving the 3 mutant lines allowed us to establish that the mutations in Chlo18 and Chlo19 lines are allelic and that Chlo29 has a complementary mutation at a locus not linked with those of Chlo18/19.

The physiological and biochemical effects of these mutations are also different.

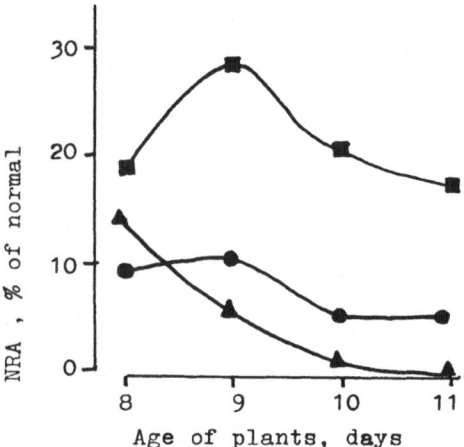

Fig. 3. The time course of nitrate reductase activity in the first leaf of mutant plants under 30°C, continuous light 25,000 lux. ● Chlo18, ▲ Chlo19, ■ Chlo29.

Table 2. The effect of temperature on the NR activity in the first leaf of 10-day-old barley NR mutant seedlings (% of normal NR activity)

Genotype of the seedlings	Temperature, °C		
	20	26	30
Chlo18	12	9	5
Chlo19	13	12	1
Chlo29	6	22	20
Viner	100	100	100

The time course of residual nitrate reductase activity in the primary leaves of Chlo mutants during early development is depicted in Fig. 3. The nitrate reductase activity is consistently higher in Chlo29 as compared with Chlo18 and Chlo19.

Increase in environmental temperature to more than 20°C lowers nitrate reductase activity in Chlo18 and Chlo19, but raises it in Chlo19 (Table 2). At 30°, starving Chlo29 plants are still capable of growth on pure nitrate nitrogen source, while Chlo18 and Chlo19 plants wilt.

The rate of nitrate uptake from the nutrient solutions in the normal variety and the mutant lines showed large differences (Table 3). Chlo18, Chlo19 and Chlo29 showed faster uptake and higher concentrations of nitrate in leaf tissues than normal. Thus is seems that nitrate uptake by barley plants depends rather on the level of nitrogen starvation than on its reduction system.

The molecular and genetic aspects of nitrogen nutrition have been widely discussed at the International Symposium on Nitrate Assimilation in Gaters-

Table 3. Nitrate uptake from nutrient solution with 10 mM KNO$_3$ (in 5 mM CaSO$_4$, pH 6.0) and nitrate accumulation in the plant tissues of 13-day-old barley seedlings. Total uptake for 5 days at 18°C, continuous light 25,000 lux

Genotype of the seedlings	NO_3^-, micromoles·plant^{-1}	
	Total uptake	Accumulated in plant tissues
Chlo18	104.4	90.9
Chlo19	126.0	97.7
Chlo29	116.4	94.8
Viner	63.6	34.5
LSD Ø.95	12.5	8.7

leben, DDR, 1982, and now the control of nitrate absorption and utilization appears to be a more attainable goal, especially in plant tissue cultures.

We believe that further studies with mutants in nitrate assimilation in different species of plants would be very promising and informative for our understanding of the general prinicples for genetic control of mineral nutrition in higher plants.

References

1. Asher C J and Ozanne P G 1977 Individual plant variability in susceptibility to potassium deficiency: Some observations on capeweed (*Arctotheca calendula* (L.) Levyns). Aust. J. Plant Physiol. 4, 499–503.
2. Carlson P S 1975 The fungal-like genetics of higher plants. Genetics 79, 353–356.
3. Clarkson D T and Hanson J B 1980 The mineral nutrition of higher plants. Ann. Rev. Plant Physiol. 31, 239–298.
4. Deckard E L, Lucken K A, Joppa L R and Hammond J J 1977 Nitrate reductase activity, nitrogen distribution, grain yield and grain protein of tall and semidwarf near-isogenic lines of *Triticum aestivum* and *T. turgidum*. Crop Sci. 17, 293–295.
5. Doddema H, Hofstra J J and Feenstra W J 1976 Uptake of nitrate by *Arabidopsis thaliana* mutants. Arabidopsis Inform. Service 13, 136–140.
6. Epstein E 1976 Genetic potentials for solving problems of soil mineral stress: adaptation of crops to salinity. *In* M J Wright (Ed.). Plant adaptation to mineral stress in problem soils. Cornell Univ. Agr. Expt. Sta, Ithaca, NY, pp 73–82.
7. Epstein E and Norlyn J D 1977 Seawater based crop production: a feasibility study. Science 197, 249–251.
8. Epstein E, Norlyn J D, Rush D W, Kingbury R W, Kelley D B, Cunningham G A and Wrona A F 1980 Saline culture of crops: a genetic approach. Science 210, 399–404.
9. Fawole I, Gabelman W H, Gerloff G C and Nordheim E V 1982 Heritability of efficiency in phosphorus utilization in beans (*Phaseolus vulgaris* L.). J. Am Soc. Hortic. Sci. 107, 94–97.
10. Feenstra W J and Oostinder-Braaksma F J 1976 Genetic control of nitrate reduction in Arabidopsis. Arabidopsis Inf. Service 12, 141–152.
11. Foy C D, Lafever N N, Schwartz J W and Fleming A L 1974 Aluminium tolerance of wheat cultures related to region of origin. Agron. J. 66, 751–758.
12. Foy C D, Jones J E and Webb H W 1980 Adaptation of cotton genotypes to acid aluminium-toxic soils. Agron. J. 72, 833–839.
13. Gabelman W H and Gerloff G C 1978 Isolation plant germplasm with altered efficiencies in mineral nutrition. Hortic. Sci. 13, 682–684.
14. Goodman P J 1977 Selection for nitrogen responses in Lolium. Ann Bot 41, 243–256.
15. Goodman P J 1979 Genetic variation in inorganic nitrogen metabolism in barley. *In* Welsh Plant Breed. Sta. Ann. Rep. 149–150.
16. Greidanus T, Peterson L E and Dana M N 1972 Essentiality of ammonium for cranberry nutrition. J. Am. Soc. Hortic. Sci. 97, 272–277.
17. Kleinhofs A, Warner R L, Muehlbauer F J and Nilan R A 1978 Induction and selection of specific gene mutations in Hordeum and Pisum. Mut. Res 51, 29–35.
18. Kleinhofs A, Kuo T and Warner R L 1979 Nitrate reductase deficient mutants on barley. Barley Genet. Newslett. 9, 55–57.
19. Klimashevsky E L 1974 Problema genotipicheskoi specifiki kornevogo pitaniya rasteniy (The

problem of genotypic specificity of plant nutrition). *In* Sort i Udobreniye, Irkutsk pp 11–54. (USSR, *In Russian*).

20 Klimashevsky E L and Chernyshova N F 1980 Actualniye voprosy geneticheskoi variabilnosti mineralnogo pitaniya culturnich rasteniy (Urgent problems of genetic variability of cultivated plants mineral nutrition). Fiziol. i Biochim. Kultur. Rast. (*In Russian*) 4, 375–388.

21 Koval S F and Tokarev B I 1980 Primeneniie complexa provokacionnych phonov dla otbora rasteniy (The selection of plants on the complex provocative backgrounds). *In* Problemy otbora i ocenki selekcionnogo materiala, Kiev Naukova Dumka (*In Russian*), 32–36.

22 Lacroute F 1975 The use of mutants in metabolic studies. *In* Methods of Cell. Biol, New York 11, 235–245.

23 Lafever H N, Campbell L G and Foy C D 1977 Differential response of wheat cultivars to aluminium. Agron. J. 69, 563–568.

24 Liljeström S and Aberg B 1966 Studies on the mechanisms of chlorate toxicity. Lantbruks-Högsk. Ann. 32, 93–112.

25 Oostinder-Braaksma F J and Feenstra W J 1973 Isolation and characterization of chlorate-resistant mutants of *Arabidopsis thaliana*. Mut. Res. 19, 175–185.

26 Oostinder-Braaksma F J and Feenstra W J 1982 Nitrate reduction on the wildtype and a nitrate reductase deficient mutant of *Arabidopsis thaliana*. Physiol. Plant. 54, 351–380.

27 Sarić M R 1981 Genetic specificity in relation to plant mineral nutrition. J. Plant Nutr. 3, 743–766.

28 Shumny V K, Tokarev B I and Dedov V M 1981 Genetico-seleccionniye aspekti mineralnogo pitaniya rastenii (Genetical and breeding aspects of plant mineral nutrition). Selsko-chozaistvennaia bioligiia (*In Russian*), 16, 185–192.

29 Tokarev B I 1977 Metodi opredeleniia nitratreductaznoy activnosti u pchenici i yachmena (Methods for determination of the level of nitrate reductase activity in wheat and barley). *In* Effectivnost chimizacii selskogo chozaistva v Sibiri (Effects of Chemical Agriculture in Siberia), (in Russian). Siberian Dept. of All-Union Acad. of Agr. Sci., Novosibirsk 58–65.

30 Tokarev B I 1982 Sostoianiie i perspectivi razvitiia genetici mineralnogo pitaniia rastenii (The state and perspectives for development in genetics of plant mineral nutrition). *In* Tezisi docladov IY sezda VOGIS (Proc. of the Meeting of All-Union Soc. of Genetics and Plant Breeders) (*In Russian*) Kishenev 166–167.

31 Tokarev B I and Shumny V K 1977 Viyavlenie mutantov jachmenya s ponijennoi nitratreductaznoi activnostyu posle obrabotki semyan ethylmetansulphonatom (Detection of barley mutants with a low level of nitrate reductase activity after seed treatement with ethylemethanesulphonate). Genetica (*In Russian*) 13, 2098–2103.

32 Tokarev B I and Shumny V K 1981 Dwa lokusa kontroliruiushie nitratreductasnuiu aktivnost u yachmenya (Two loci for nitrate reductase activity control in barley). Genetica (*In Russian*) 17, 852–857.

33 Vavilov N I 1932 Genetika na slujbe socialisticheskogo zemledeliya (Genetics on the service for socialist soil management). Socialisticheskoie rasteniyevodstvo (*In Russian*) 4, 19–24.

34 Vose P B 1974 Ocenka i ispolzovaniye otzivchivosti sortov selskochozjajstvennih rastenii na usloviia mineralnogo pitaniia (Utilization and recognition of nutritional variations in crop plants). *In* Sort i udobrenie (*In Russian*), Ircutsk 61–71.

35 Warner R L, Lin C J and Kleinhofs A 1977 Nitrate reductase deficient mutants in barley. Nature, London 269, 406–407.

36 Weiss M G 1943 Inheritance and physiology of efficiency in iron utilization in soybeans. Genetics 18, 253–268.